Current Topics in Microbiology and Immunology

Volume 390

More information about this series at http://www.springer.com/series/82

Christian Münz
Editor

Epstein Barr Virus Volume 1

One Herpes Virus: Many Diseases

Responsible Series Editor: Peter K. Vogt

Editor
Christian Münz
Institute of Experimental Immunology
University of Zürich
Zürich
Switzerland

ISSN 0070-217X ISSN 2196-9965 (electronic)
Current Topics in Microbiology and Immunology
ISBN 978-3-319-22821-1 ISBN 978-3-319-22822-8 (eBook)
DOI 10.1007/978-3-319-22822-8

Library of Congress Control Number: 2015948721

Springer Cham Heidelberg New York Dordrecht London

Printed on acid-free paper

Springer International Publishing AG Switzerland is part of Springer Science+Business Media (www.springer.com)

Preface

We celebrated the 50th anniversary of the discovery of Epstein Barr virus (EBV) in Burkitt's lymphoma last year. During these 50 years of research on EBV, this first human candidate tumor virus has been found associated with many more malignant diseases in addition to Burkitt's lymphoma, including Hodgkin's lymphoma, nasopharyngeal carcinoma, a subset of gastric carcinomas, rare T/NK cell lymphomas, and many more. However, not only malignant diseases, but also some autoimmune diseases and the lymphocytosis of infectious mononucleosis have been found to be linked to EBV. In addition, we have learned to appreciate that continuous cell-mediated immune control prevents these EBV associated diseases, but cannot inhibit persistent infection, which the virus establishes in more than 90 % of the human adult population. Thus, EBV serves both as a paradigm for viral oncogenesis in humans and life-long immune control of chronic infection at the same time. The changes in the viral host cell and the host's immune control that determine the switch between these two states, continue to fascinate us and new experimental developments allow us to address this question in much more detail. Our ability to sequence EBV genomes faster and at lower cost allows us to explore the genetic diversity of EBV and its possible disease association for the first time. The recombinant EBV system allows us to generate mutant viruses to address the functional relevance of this diversity and new in vivo models of EBV infection, tumorigenesis, and immune control provide valuable insights into the pathologic relevance of the EBV characteristics that we have mapped during the last 50 years. With these tools in hand we should be able to unravel many more secrets that this human tumor virus keeps and develop vaccines against some of the EBV associated diseases in the next 50 years.

This exciting journey is summarized in the two book volumes in front of you. It starts with personal accounts of the discovery, tumor association, and immune control by pioneers of EBV research (Anthony Epstein, George Klein, Vivianna Lutzky, and Dennis Moss). It then continues with the knowledge on EBV genetics and epigenetics that has been gained (Paul Farrell, Paul Lieberman, Wolfgang Hammerschmidt, Regina Feederle, Olaf Klinke, Anton Kuthikin, Remy Poirey, Ming-Han Tsai, and Henri-Jacques Delecluse). An overview of EBV associated diseases ranging from infectious mononucleosis and primary immune deficiencies to EBV associated tumors and autoimmune diseases completes the first volume (David Thorley-Lawson, Kristin Hogquist, Samantha Dunmire, Henri Balfour, Jeffrey Cohen, Ann Moormann, Rosemary Rochford, Paul Murray, Andrew Bell, Jane Healy, Sandeep Dave, Nancy Raab-Traub, Kassandra Munger, and Alberto Ascherio). In the second volume individual latent EBV gene products are then discussed (Lori Frappier, Bettina Kempkes, Paul Ling, Martin Allday, Quentin Bazot, Robert White, Arnd Kieser, Kai Sterz, Osman Cen, Richard Longnecker, Rebecca Skalsky, and Bryan Cullen). Viral entry and exit complete the virology chapters (Lindsey Hutt-Fletcher, Luidmila Chesnokova, Ru Jiang, Jessica McKenzie, and Ayman El-Guindy). The remainder of volume two is dedicated to the EBV specific immune response (Martin Rowe, Anna Lünemann, David Nadal, Jaap Middeldorp, Andrew Hislop, Graham Taylor, Maaike Ressing, Michiel van Gent, Anna M. Gram, Marjolein Hooykaas, Sytse Piersma, and Emmanuel Wiertz), in vivo models of EBV infection (Fred Wang, Janine Mühe, and Christian Münz), and EBV specific therapies (Stephen Gottschalk, Cliona Rooney, Corey Smith, Rajiv Khanna, Jennifer Kanakry, and Richard Ambinder). The resulting picture of 32 chapters on EBV biology will hopefully inspire many more young scientists to join research on this paradigmatic human tumor virus.

Indeed we might just have now the toolbox in hand not only to transfer discoveries in preclinical infection models to EBV, but also use EBV itself as a human model pathogen to learn more about the human immune system, viral dynamics in the human population, and the intricacies of EBV infection.

Zürich, Switzerland Christian Münz

Contents

Part I History

Why and How Epstein-Barr Virus Was Discovered 50 Years Ago 3
Anthony Epstein

Tumor Associations of EBV—Historical Perspectives 17
George Klein

EBV-Specific Immune Response: Early Research and Personal Reminiscences 23
D.J. Moss and V.P. Lutzky

Part II Virus Genetics and Epigenetics

Epstein–Barr Virus Strain Variation 45
Paul J. Farrell

Chromatin Structure of Epstein–Barr Virus Latent Episomes 71
Paul M. Lieberman

The Epigenetic Life Cycle of Epstein–Barr Virus 103
Wolfgang Hammerschmidt

Epstein–Barr Virus: From the Detection of Sequence Polymorphisms to the Recognition of Viral Types 119
Regina Feederle, Olaf Klinke, Anton Kutikhin, Remy Poirey, Ming-Han Tsai and Henri-Jacques Delecluse

Part III Viral Infection and Associated Diseases

EBV Persistence—Introducing the Virus 151
David A. Thorley-Lawson

Infectious Mononucleosis 211
Samantha K. Dunmire, Kristin A. Hogquist and Henry H. Balfour

Primary Immunodeficiencies Associated with EBV Disease 241
Jeffrey I. Cohen

Burkitt's Lymphoma 267
Rosemary Rochford and Ann M. Moormann

Contribution of the Epstein-Barr Virus to the Pathogenesis of Hodgkin Lymphoma 287
Paul Murray and Andrew Bell

The Role of EBV in the Pathogenesis of Diffuse Large B Cell Lymphoma 315
Jane A. Healy and Sandeep S. Dave

Nasopharyngeal Carcinoma: An Evolving Role for the Epstein–Barr Virus 339
Nancy Raab-Traub

EBV and Autoimmunity 365
Alberto Ascherio and Kassandra L. Munger

Index 387

Part I
History

Why and How Epstein-Barr Virus Was Discovered 50 Years Ago

Anthony Epstein

Abstract An account is given of the experiences and events which led to a search being undertaken for a causative virus in the recently described Burkitt's lymphoma and of the steps which ultimately culminated in the discovery of the new human herpesvirus which came to be known as Epstein-Barr virus (EBV).

Contents

1 Introduction 4
  1.1 Early Chance Events Essential for Both "Why" and "How" 4
  1.2 A Subsequent Key Chance Leading to "Why" 4
2 The Search for a Virus 6
  2.1 Reflections on Research Funding in the 1960s 7
  2.2 The Beginning of "How"—Persistent Early Failures 7
  2.3 An Idea Giving a Glimmer of Hope for "How" 8
  2.4 Chance Provides the Key to "How" 8
  2.5 The End of the Beginning to "How" 10
3 The Final Breakthrough to "How" 11
  3.1 "How" the Virus Was Found 11
  3.2 Naming the Virus 13
  3.3 Characterization of the Uniqueness of the Virus 13
4 Concluding Remarks 14
References 14

Based on a lecture presented at EBV@50, the International Meeting in Oxford held in the week of 28 March 2014 to celebrate the 50th Anniversary of the first publication on the virus by Epstein, Achong and Barr on 28 March 1964.

A. Epstein (✉)
Wolfson College, University of Oxford, Linton Road, Oxford OX2 6UD, UK
e-mail: anthony.epstein@wolfson.ox.ac.uk

C. Münz (ed.), *Epstein Barr Virus Volume 1*, Current Topics in Microbiology and Immunology 390, DOI 10.1007/978-3-319-22822-8_1

1 Introduction

The story I am recalling here arose from a sequence of linked chances which followed from one to the next in an extraordinary chain, with each coming at exactly the right moment for its significance to be recognized. As was memorably pointed out by Louis Pasteur 160 years ago, "Dans les champs de l'observation le hasard ne favorise que les esprits préparés" (*In the field of observation chance favours only the prepared mind*).

1.1 Early Chance Events Essential for Both "Why" and "How"

The chain started some 65 years ago when I began research at the Middlesex Hospital Medical School in London (founded 1836; since 1987, incorporated into University College London) and a quite unexpected death gave me access to one of the very earliest electron microscopes at a time when such things were exceptionally rare. Made in UK by Metropolitan-Vickers in Manchester in 1946 it was the first commercially available instrument of this kind, but is now an exhibit in the Manchester Museum of Science and Industry; sadly it was not persisted with and the subsequent market went to Holland, Germany, Japan and the USA.

It was also a lucky chance that the Middlesex Hospital Medical School had an interest in the then deeply unfashionable chicken cancer viruses. So it was that I came to work on the Rous fowl sarcoma virus, the first virus known to cause malignant tumours. It was studied then by only a handful of people worldwide; indeed, so unfashionable at that time was the idea of viruses causing cancer in general that Peyton Rous (1879–1970) only got the Nobel Prize (1966) 55 years after he made his discovery (Rous 1911) because he lived to 86 by which time views on this subject had changed radically.

With the Rous virus I was able, using the electron microscope, to demonstrate its morphology and show for the first time that it was an RNA not a DNA virus (Epstein 1958; Epstein and Holt 1958). All this made me keenly aware of viruses which cause cancer and of the possibilities of electron microscopy.

1.2 A Subsequent Key Chance Leading to "Why"

This further chance was critically significant. In the 1950s a British Colonial Service medical officer based in Uganda came, when on leave in UK, to the Middlesex Hospital, London (Fig. 1), where he had a connection with the Academic Department of Surgery and his enthralling seminars were usually about

Fig. 1 The Middlesex Hospital, London, UK (founded 1745; 2005, replaced by the new University College London Hospital). Image courtesy of University College London Hospital NHS Trust archive

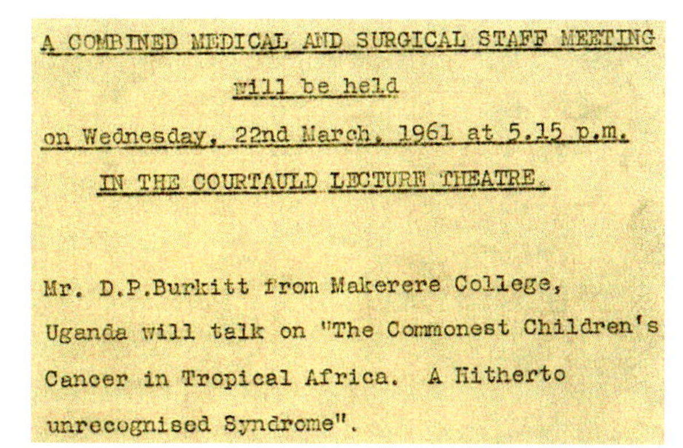

A COMBINED MEDICAL AND SURGICAL STAFF MEETING

will be held

on Wednesday, 22nd March, 1961 at 5.15 p.m.

IN THE COURTAULD LECTURE THEATRE.

Mr. D.P.Burkitt from Makerere College, Uganda will talk on "The Commonest Children's Cancer in Tropical Africa. A Hitherto unrecognised Syndrome".

Fig. 2 Photograph of the notice of a talk by Denis Burkitt in 1961 at which he gave the first account outside Africa of the lymphoma which came to carry his name. The original notice is still extant

the exotic and extreme cases encountered in a developing country. Early in 1961 he came again, but this time he gave a very different kind of talk—the speaker was in fact Denis Burkitt (1911–1993), an unknown bush surgeon as he described himself, and his lecture (Fig. 2) was the first account he had ever given outside Africa of the lymphoma which brought him worldwide fame. Quite by chance I saw the notice of Burkitt's talk, and probably out of curiosity, I went.

After the first 20 min I was greatly excited by this strange malignant tumour of children in Africa affecting bizarre sites and fatal in a few months (reviewed in Burkitt 1963). But even stranger, Burkitt went on to present unprecedented preliminary data which showed that geographical distribution depended on temperature and rainfall. This suggested to me that a biological agent must play a part in causation and with my knowledge of tumour viruses I immediately postulated a climate-dependant arthropod vector spreading a cancer-causing virus. It turned out later that it was a cofactor which was arthropod borne (Burkitt 1969), but my idea focused correctly on the need to search for a viral cause.

Even as Burkitt was talking, I decided to stop my current work in order to seek for viruses in what became known as Burkitt's lymphoma—so excited was I that after the talk I took the notice off a board (Fig. 2) and I have had it ever since. When Burkitt finished speaking, I was introduced to him, I invited him to my laboratory, and we agreed to collaborate. It was these quite unrelated chances which were responsible for "Why" the virus was discovered.

2 The Search for a Virus

So what about "How" the virus was discovered? That started with generous support from the then British Empire Cancer Campaign (founded 1923, became the Cancer Research Campaign 1970, became Cancer Research UK 2002) which funded me to visit Uganda a few weeks later. A first visit to Africa was quite daunting in the 1960s for unlike now, when even teenagers backpack widely, exotic travel was rare then. However, here too chance leant a hand because after World War II ended in Europe I had been posted to the Far East (Fig. 3) where the conflict continued with Japan, and having learned how things were done in the Indian Empire under the British Raj, it was easy to find my way around the British Ugandan Colonial Administration modelled on it.

Fig. 3 Inspection at Bareilly Cantonment in 1946 by Field Marshall Sir Claude Auchinleck, C-in-C India Command. Capt. M.A. Epstein (*right*)

The purpose of my visit to Uganda was, of course, to work out how a regular supply of lymphoma samples from Burkitt's patients in the capital Kampala could be flown overnight to my laboratory in London.

2.1 Reflections on Research Funding in the 1960s

Commenting on these events a much later Editorial aptly remarked "It is hard to imagine any current funding agency supporting a project based on the gut feeling of a young worker without any supporting data. Thank goodness that was not the case 40 years ago!"

2.2 The Beginning of "How"—Persistent Early Failures

For 2 years I applied the virus isolation techniques then in use to lymphoma samples with depressing negative results. Tumour material was inoculated into test cell cultures, embryonated hen eggs and newborn mice but without effects and direct examination in the electron microscope also proved fruitless. Failure to gain anything with this tool in relation to the lymphoma was especially disappointing in view of my early access to it. But additionally so since in 1956 I had gone, thanks to the Anna Fuller Fund of New Haven, to the Rockefeller Institute in New York (now the Rockefeller University) specifically to learn from George Palade (1912–2008; Nobel Prize 1974) at the time of his outstanding contributions to the earliest phases of biological electron microscopy and, indeed, to the very foundations of the whole of modern cell biology.

This long period without results was extremely alarming at a very insecure stage in my career. There was no employment law at that time—I had no letter of appointment, no terms and conditions, and no idea from year to year whether the Head of Department would feel inclined to reappoint me.

At this very low point I managed, unusually for a UK scientist then, to get a very small grant from the US National Cancer Institute. This $45,000 gave me some very modest independence and enabled me at the end of 1963 to recruit Dr Bert Achong (1928–1996) to help, once I had taught him, with the electron microscopy and Miss Yvonne Barr (as she then was) to assist with tissue culture of which she already had some experience. Before this I had worked for 15 years only with George Ball (Fig. 4), an absolutely reliable and completely unflappable young laboratory technician, who had provided indispensable support and continued to do so in the decades to come.

Fig. 4 George Ball in 1961, the absolutely reliable young laboratory technician who provided indispensable support before, throughout and long after the search for EBV. Image courtesy of Mr. G.R. Ball

2.3 An Idea Giving a Glimmer of Hope for "How"

In the event, an idea at this time proved more important than the grant. It occurred to me that if the tumour cells could be grown in culture away from host defences, a latent cancer virus might be activated and become apparent as I knew happened with certain chicken tumours (Bonar et al. 1960). However, doing this with a human lymphoma seemed unlikely since no type of human lymphocytic cell had ever been maintained in vitro for more than an hour or two (Woodliffe 1964). Nevertheless I tried repeatedly with the lymphoma using fragments in plasma clots, fragments floating on rafts and so on, but depressingly and quite predictably all failed.

2.4 Chance Provides the Key to "How"

Yet once again chance intervened in a big way. On Friday 5 December 1963 the overnight flight from Kampala was diverted to Manchester by fog and we were only able to retrieve our biopsy in the afternoon after the plane finally reached London. As usual the tissue was floating in transit fluid, but unusually this was cloudy. As it was getting late and the cloudiness was likely to be due to bacterial contamination, the feeling was that we could leave the laboratory for the weekend. But instead of discarding the specimen and going home I put a drop of the cloudy fluid on a slide and examined it with the light microscope as a wet preparation.

Rather than seeing the expected contaminating bacteria I was astonished to find that the cloudiness was due to large numbers of viable-looking free-floating tumour cells (Fig. 5) which had been shaken from the cut edges of the lymphoma sample during the flight.

This chance was in turn assisted by another, for I was immediately reminded that earlier that year on a visit to Yale Medical School I had learned that their Mouse Lymphoma Research Group had only succeeded in culturing mouse lymphoma cells by starting with suspensions of free-floating single cells (Fischer 1957, 1958) obtained in their case in ascitic fluid after growing the tumours in the abdominal cavities of mice.

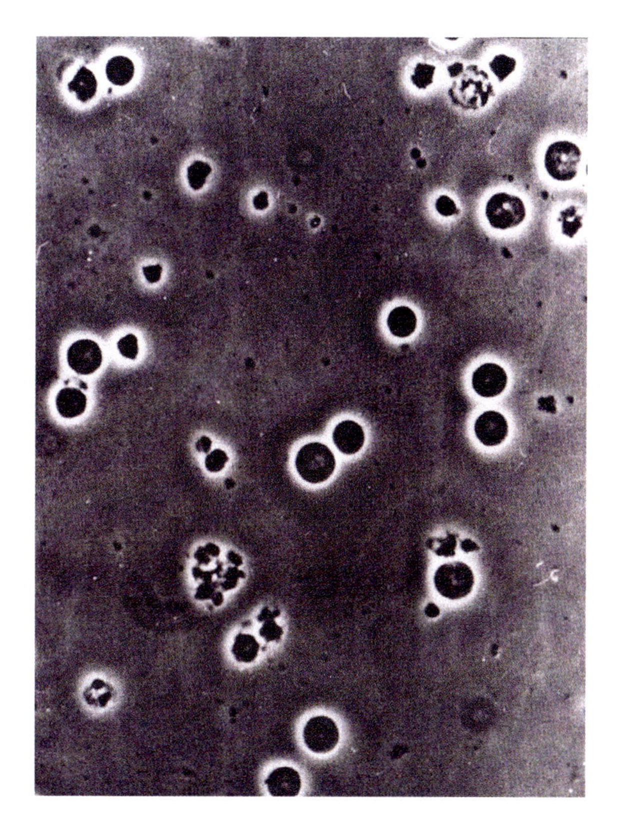

Fig. 5 Wet preparation of free-floating viable-looking lymphoma cells shaken from the cut edges of the lymphoma sample sent overnight from Uganda 4/5 December 1963. The appearance was reminiscent of a mouse ascites tumour. Phase-contrast light micrograph

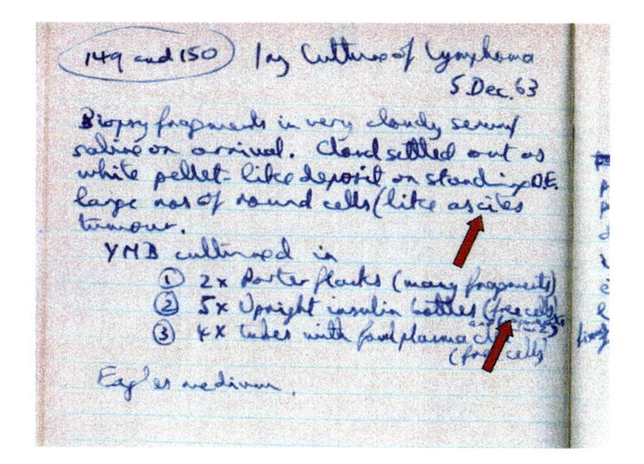

149 and 150 | 1ry Cultures of Lymphoma
5 Dec. 63
Biopsy fragments in very cloudy serum/saline on arrival. Cloud settled out as white pellet-like deposit on standing. O.E. large nos of round cells (like ascites tumour.
YMB cultured in
① 2x Porter flasks (many fragments)
② 5x Upright insulin bottles (free cells)
③ 4x tubes with fowl plasma clot (free cells)
Eagle's medium.

Fig. 6 Photograph of the author's laboratory notebook for 5 December 1963. Note the delayed lymphoma sample described as "like ascites tumour" (*arrow*) and set up for the first time in suspension culture—"free cells" (*arrow*)

Because of this, the free-floating cells in our delayed sample were described in my laboratory book (Fig. 6) as "like ascites tumour" and were set up for the first time in suspension culture (Fig. 6) as "free cells".

2.5 *The End of the Beginning to "How"*

Shortly after setting up the suspension culture a continuous lymphoma-derived cell line grew out (Fig. 7) which we labelled EB (Epstein and Barr) to distinguish its containers from the HELA, OMK, BHK and other banal cell lines we had in the laboratory, and suspension culture rapidly gave us more such lines from further Burkitt's lymphomas. It should be noted that 50 years ago there were no hoods and we worked on the open bench with rigorous aseptic technique in the updraft of a lighted Bunsen burner which carried atmospheric contaminants away. Very early on we used extremely small conical flasks which allowed more culture fluid without increasing the depth than with straight-sided containers, since depth critically affected the diffused oxygen tension around the cells resting on the bottom, and this system had the advantage that it could readily be scaled up as the cells became plentiful.

This was the first time that human lymphocytic cells had ever been grown long term in vitro, and when the account of the successful procedure was sent for publication, a leading journal's expert referees were unwilling to believe that human lymphocytic cells could be cultured at all. Yet suspension is now the standard technique to grow such cells used worldwide today for a huge number of different types of research.

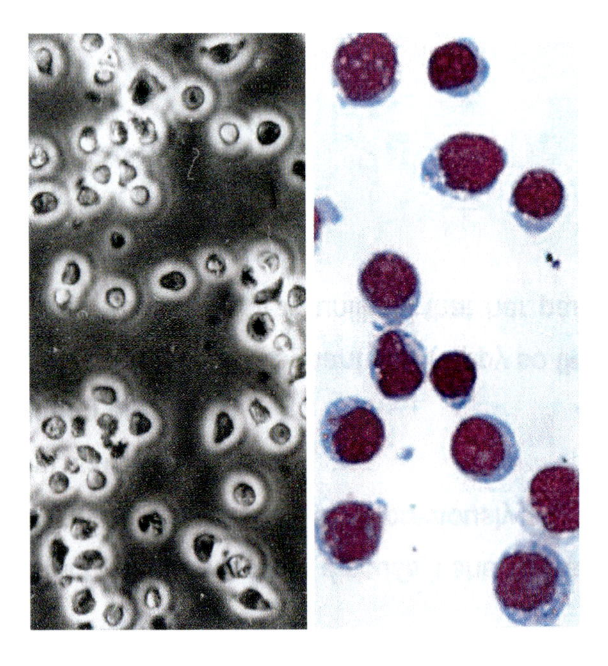

Fig. 7 Light micrographs of the first ever culture of Burkitt's lymphoma-derived lymphoblasts designated EB1. Phase contrast of live cells (*left*) and Giemsa-stained fixed cells (*right*)

3 The Final Breakthrough to "How"

All efforts to show a virus in EB cells using standard contemporary biological tests failed, so as soon as material could be spared, some cells were fixed, pelleted and embedded for electron microscopy. But it should be emphasized that this was not accepted then as a method for demonstrating viruses; dogma required that they should be shown by their biological activity or by finding the antibodies they induced. It was not credited that they could be recognized morphologically. Indeed, at this time when electron microscopy was rare and little understood, the images obtained of biological material were considered by many as artefacts of fixation and processing.

It is worth mentioning here that a notable exception to such views was provided by Oxford's Professor Sir Howard Florey (1898–1968; penicillin Nobel laureate 1945, later Lord Florey of Adelaide and Marston); not one to miss a new and important advance he had come himself to my very small laboratory in London on 21 January 1959 to see what electron microscopy was about in preparation for setting up a unit in his department.

3.1 "How" the Virus Was Found

As regards images of viruses, my time with George Palade had convinced me that they could be recognized, and classified at least into families, by their appearance as had been done for bacteria with the light microscope for 100 years.

I examined the first EB cell preparation with the electron microscope on 24 February 1964 and was exhilarated to see unequivocal virus particles in a cultured lymphoma cell in the very first grid square I searched. I was extremely agitated in case the specimen might burn up in the electron beam—I switched off, I walked round the block in the snow without a coat, and when somewhat calmer I returned to record what I had seen.

I recognized at once that I was looking at a typical member of the herpesvirus group (Fig. 8) with which I was already very familiar and noted it as such in my electron microscope laboratory book (Fig. 9) "virus, like herpes", but there was no means of knowing which herpesvirus it might be. However, it did seem quite extraordinary that a herpesvirus was producing virus particles in a cell line yet was so biologically inert that it did not destroy the whole culture as the known herpesviruses would have. Accordingly I rapidly set about reporting the discovery with my new assistants Bert Achong and Yvonne Barr (Fig. 10). The resulting paper (Epstein et al. 1964) appeared on 28 March 1964 and became a Citation Classic in 1979, and the 50th anniversary of its publication was celebrated at an International Meeting in the week of 28 March 2014.

The unusual inertness was reinforced when biological tests for herpesviruses were applied to the EB cells and all proved negative. At this point I became

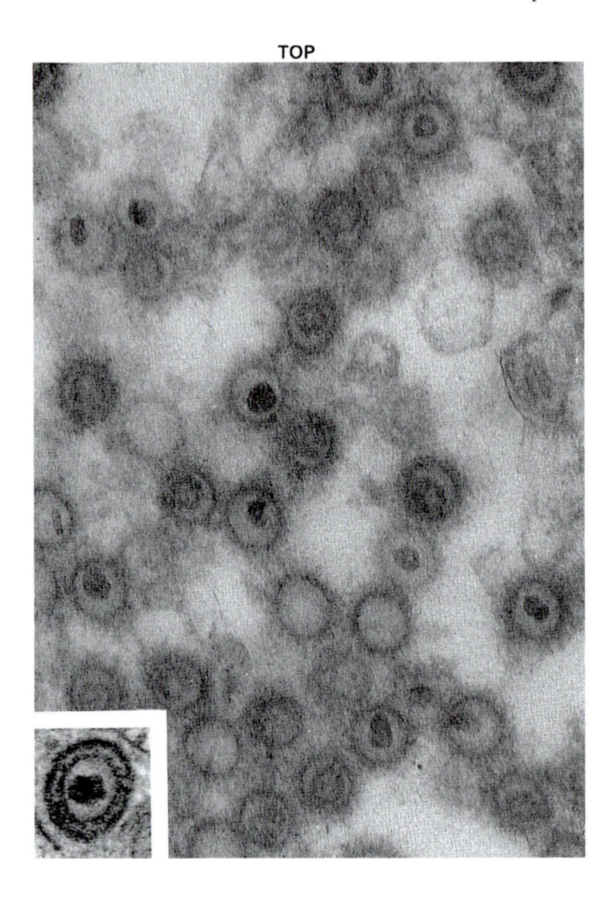

Fig. 8 The first electron micrographs of EBV. Immature virions assembling in the cytoplasm of a cultured EB1 lymphoma cell; *inset*, a mature enveloped particle. These images were recognized at once as a herpesvirus

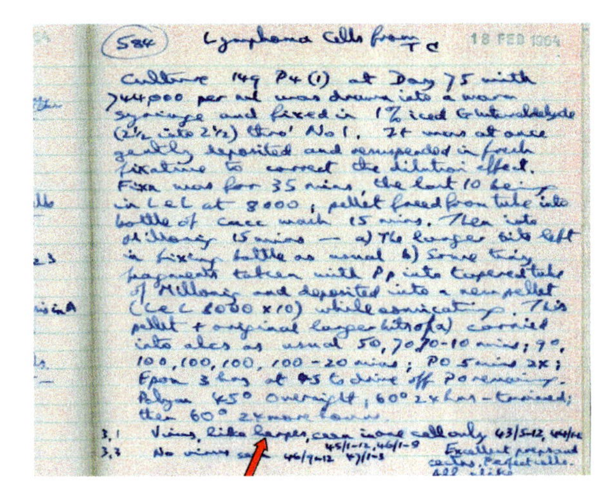

(584) Lymphoma Cells from TC 18 FEB 1964

Culture 149 P4 (1) at Day 75 with 744,000 per ml was drawn into a warm syringe and fixed in 1% iced Glutaraldehyde (2½ into 2½) thro' No 1. It was at once gently deposited and resuspended in fresh fixative to correct the dilution effect. Fixn was for 35 mins, the last 10 being in LeL at 8000; pellet freed from tube into bottle of conc wash 15 mins. Then into [illegible] 15 mins — a) The larger bits left in fixing bottle as usual b) Some tiny fragments taken with Pp into tapered tube of Millonig and deposited into a new pellet (LeL 8000 x 10) while sonicating. This pellet + original larger bits of (a) carried into alcs as usual 50, 70, 70-10 mins; 90, 100, 100, 100, 100 - 20 mins; PO 5 mins 2x; Epon 3 hrs at 45 to drive off PO remaining. Polymn 45° overnight, 60° 24 hrs - trimmed; then 60° 24 more hours.

3,1 Virus, like herpes, seen in one cell only 43/5-12, 44/1-12, 45/1-12, 46/1-8

3,3 No virus seen 46/9-12 47/1-3 Excellent preservation, Perfect cells. All alike

Fig. 9 Photograph of the author's electron microscopy notebook dealing with the EB1 cells harvested on 18 February 1964 and examined, after the usual delays for processing, on 24 February 1964. Note entry "virus, like herpes" (*arrow*)

Fig. 10 M.A. Epstein, B.G. Achong (1928–1996) and Y.M. Barr in 1964 at the time of the first publication on EBV (Epstein et al. 1964)

concerned that something unnoticed in our procedures was inactivating the virus and it was clearly urgent to have the tests repeated in some other laboratory.

I approached two leading British herpes virologists, but neither was interested in our unorthodox findings, and so it came about that I contacted my friends the husband and wife virologists Werner and Gertrude Henle (1910–1987; 1912–2006) at the Children's Hospital in Philadelphia.

EB cells were flown from my laboratory to Philadelphia, the Henles rapidly confirmed the biological inertness of the virus, and we then reported jointly that it was a new member of the herpes family (Epstein et al. 1965).

3.2 *Naming the Virus*

Following my sending the virus to the Henles, they soon subsequently referred to it as "EBV" (Henle et al. 1968) after the EB cells in which it had come to them, and this name caught on and was rapidly universally adopted.

3.3 *Characterization of the Uniqueness of the Virus*

In addition to the biological inertness of the virus, its immunological singularity was soon demonstrated in Philadelphia (Henle and Henle 1966) and in my laboratory using quite different techniques (Epstein and Achong 1967). Shortly after this its novel biochemical nature was also established (zur Hausen et al. 1970), and 14 years later the complete viral genome was sequenced (Baer et al. 1984).

In the light of subsequent knowledge of the very limited range of cells with receptors for the virus, the failure to show biological activity is readily understandable, but it was very puzzling at the time. It was indeed fortunate that work on the lymphoma cells and the search for a virus was undertaken in a laboratory where a rare electron microscope was in daily use (yet another chance) as otherwise the extreme inertness could have left it undiscovered.

Table 1 Epstein-Barr virus—research publications (from PubMed)

28 March 1964–28 March 2014	30,995
1984	525
2004	1079
2013	25/week

4 Concluding Remarks

EBV was in fact the first virus to be found solely by electron microscopy, and the story of its discovery thus acted out a little joke published over 100 years ago before viruses were known or electron microscopes dreamt of:

> The microbe is so very small
> You cannot make him out at all
> But many sanguine people hope
> To see him through the microscope
>
> (Belloc 1897)

But the huge extent of work on EBV following its finding by electron microscopy is not generally realized even by experts in the field and is therefore worth a comment. In the first 50 years since the discovery there were more than 30,000 peer-reviewed publications on EBV (Table 1). Of course in the early years the numbers were very small, but as the decades went by they increased dramatically (Table 1, cf 1984 and 2004) and finally in 2013 they were running at 25 per week. It is an arresting thought that each author of a chapter in the present book is making a contribution, however small, to the vast worldwide undertaking of accumulated EBV research.

The reason for the wide interest in EBV has been, of course, because it was the first putative and then the first definitive human cancer virus. Interestingly, in the 50 years that the virus has been known to science human tumour virology has moved from the distant margins of the biomedical agenda to the very centre and in recent years to the very top with the introduction of anti-tumour virus vaccine programmes to prevent significant human cancers.

References

Baer R, Bankier AT, Biggin MD et al (1984) DNA sequence and expression of the B95-8 Epstein-Barr virus genome. Nature 310:207–211

Belloc H (1897) The microbe in more beasts for worse children. Duckworth & Co., London

Bonar RA, Weinstein D, Sommer JR et al (1960) Virus of avian myeloblastosis. XVII. Morphology of progressive virus-myeloblast interactions in vitro. Natl Cancer Inst Monogr 4:251–290

Burkitt DP (1963) A lymphoma syndrome in tropical Africa. In: Richter GW, Epstein MA (eds) International review of experimental pathology, vol 2. Academic Press, New York, pp 69–138

Burkitt DP (1969) Etiology of Burkitt's lymphoma—an alternative hypothesis to a vectored virus. J Natl Cancer Inst 42:19–28

Epstein MA (1958) Composition of the Rous virus nucleoid. Nature 181:1808
Epstein MA, Achong BG (1967) Immunologic relationships of the herpes-like EB virus of cultured Burkitt lymphoblasts. Cancer Res 27:2489–2493
Epstein MA, Achong BG, Barr YM (1964) Virus particles in cultured lymphoblasts from Burkitt's lymphoma. Lancet 1:702–703
Epstein MA, Henle G, Achong BG, Barr YM (1965) Morphological and biological studies on a virus in cultured lymphoblasts from Burkitt's lymphoma. J Exp Med 121:761–770
Epstein MA, Holt SJ (1958) Observations on the Rous virus; integrated electron microscopical and cytochemical studies of fluorocarbon purified preparations. Br J Cancer 12:363–369
Fischer GA (1957) Tissue culture of mouse leukemic cells. Proc Am Assn Cancer Res 2:201
Fischer GA (1958) Studies of the culture of leukemic cells in vitro. Ann NY Acad Sci 76:673–680
Henle G, Henle W (1966) Studies on cell lines derived from Burkitt's lymphoma. Trans NY Acad Sci 29:71–79
Henle G, Henle W, Diehl V (1968) Relation of Burkitt's tumor-associated herpes-type virus to infectious mononucleosis. Proc Natl Acad Sci USA 59:94–101
Rous P (1911) A sarcoma of the fowl transmissible by an agent separable from the tumor cells. J Exp Med 13:397–411
Woodliffe HJ (1964) Blood and bone marrow cell culture. Eyre and Spottiswoode, London
zur Hausen H, Schulte-Holthausen H, Klein G et al (1970) EBV DNA in biopsies of Burkitt tumours and anaplastic carcinomas of the nasopharynx. Nature 228:1056–1058

Tumor Associations of EBV—Historical Perspectives

George Klein

Abstract This is a brief history of our collaborative work with Werner and Gertrude Henle, Francis Wiener, George and Yanke Manolov, and others on the association of Epstein-Barr virus (EBV) with Burkitt lymphoma and other human tumors. Special emphasis is put on the question where EBV is a true cancer virus.

Contents

1 Introduction ... 17
2 Is EBV a Cancer Virus? ... 21
3 What Is the Role of EBV in the Genesis of BL? ... 21
4 The Lessons of HART Therapy ... 21
References ... 22

1 Introduction

The inspiring articles of Dennis Burkitt and Dennis Wright in the early 60s made the scientific community aware of the African childhood lymphoma prevalent in hot and humid regions of Africa and the "starry sky" like histology. The suggestion that an insect transmitted virus may cause the disease triggered researchers in numerous laboratories to look for the hypothetical agent. The search was facilitated by the fact that the tumor readily fell apart into single cell suspensions without any trypsinization and grew readily into cell lines.

The Virus Cancer Program of the NIH was in full swing. One day—probably in 1963 or 1964—I visited John Moloney, who headed the program, to tell him the latest news about our project on virus-induced mouse tumors. Tony Epstein

G. Klein (✉)
Department of Microbiology, Tumor and Cell Biology (MTC),
Karolinska Institutet, 17177 Stockholm, Sweden
e-mail: gk-secretary@mtc.ki.se; georg.klein@ki.se

C. Münz (ed.), *Epstein Barr Virus Volume 1*, Current Topics in Microbiology and Immunology 390, DOI 10.1007/978-3-319-22822-8_2

was the other visitor. I knew Tony from his earlier visits to Torbjörn Caspersson's department in Stockholm where I worked. He showed EM images of cell lines from what was now called Burkitt lymphoma (BL) to John. In some lines, herpes-type virus particles could be seen in a small minority of the cells that were clearly degenerating and dying. John and I thought that the tumor cells may have picked a common herpesvirus as a passenger. But Tony said: It may be a wild goose but it is a goose that has to be chased. Right he was.

Soon thereafter, Werner and Gertrud Henle in Philadelphia performed immunofluorescence tests on the same lines and showed that the virus containing cells failed to react with antibodies against any known herpesvirus. This was, therefore, a new human herpesvirus. The Henles and we decided to call it EBV.

We were ready to join the adventure of looking for the footsteps of a virus in proliferating BL cells, using the experience we had from work with virus-induced mouse lymphomas. We were fortunate to establish an "air bridge" with Peter Clifford, Head of the ENT Department at the Kenyatta National Hospital in Nairobi. Getting in touch with him, we followed the percept that if you look for a collaborator to do a really hard job with you, find the busiest person and he will do it.

Peter was the only ENT surgeon between Johannesburg and Cairo with an immense working load, but passionately interested in BL. He has developed its chemotherapy in parallel with Dennis Burkitt. Unlike Burkitt, he gave only moderate doses to spare the immune system. The frequency of long-term survivors—or, as it turned out later, cures—was higher in his material than in Burkitt's more drastically treated patients.

On my request for biopsies and sera, the material started coming with clockwork regularity every Tuesday, with the only weekly SAS plane from Nairobi, frozen sera in dry ice, live tumor tissue in wet ice, in great abundance. Every Tuesday night was Burkitt night at our laboratory for about ten years. In addition to what we did with the material, we also fanned it out to other laboratories in Europe, Japan, and the USA.

Our most significant finding was the discovery of EBNA, the EBV-encoded nuclear antigen which later turned out to be a conglomerate of six different proteins. When we first detected EBNA, it was still not clear whether the virus was present in some hidden form in the proliferating cells of the tumor that have not entered the lytic viral cycle which inevitably led to cell death.

To detect EBNA, Beverly Reedman and I departed from the observation of John Pope in Australia, showing that an EBV-specific complement fixing antigen was present in a BL line that did not make any virus. We decided to look for it by anticomplement fluorescence. EBNA soon appeared in all its magnificence (Reedman and Klein 1973).

The detection of EBNA by anticomplement fluorescence was tricky, and sometimes, it did not work. Years later, under the rule of Idi Amin, a note appeared in Newsweek saying that the African radiotherapist, Charles Olweny, Head of the Uganda Cancer Center at that time, was found in the forest with his head cut-off together with two other colleagues, because they opposed the renaming of

Makerere Medical College to Idi Amin University. I tried to call Charles. He was not there. I called a week later and he was still not there. But he was going to come the week after. When I called the third time, he came on the line. Charles! I shouted. Are you all right? No George, the EBNA test is not working, he said.

Lloyd Old and Herbert Oettgen at Sloan Kettering were among the recipient laboratories in the USA. They detected an EBV-specific soluble antigen by immunoprecipitation in BL and also nasopharyngeal carcinoma (NPC) specimens. The EBV/NPC association was confirmed by the serology of the Henles. But it was not clear whether the virus was carried by the carcinoma cells or by the abundant lymphoid infiltrate. We were inclined to blame the latter until we found that the lymphoid infiltrate of NPC consisted mostly of T cells, not known as EBV harboring cells at the time. Importantly, the EBNA test clearly showed that latent virus was carried by the carcinoma cells themselves. With Harald zur Hausen, we could also confirm the presence of multiple EBV genomes in both BL and NPC cells, by DNA–DNA hybridization (zur Hausen et al. 1970).

The fanning out of the Nairobi material had many other interesting and some amusing byproducts. Blood and biopsies were sent to Philip Fialkow, human geneticist at the University of Seattle who was looking for unusual isozyme markers. One of his letters came with a big red label saying "top secret"! Not to be opened by anyone, but Professor Klein. Inside there was another envelope with a wax seal on it and the text "under no circumstances must this letter be opened by anyone but Professor Klein". The letter said: Dear George, destroy this letter immediately when you have read it and make sure that its content is not communicated to Nairobi. You have sent us the blood of a Masai chief, living on the southern slope of the Kilimanjaro. On our request you have also supplied us with blood of his three wives and nine children. So far, we have tested eight of the children. None of them can be the descendants of the Masai chief. PS We just completed the ninth test. Paternity excluded.

Knowing a little about the Masai, I made a copy of the letter and sent it to the Svensson sisters, our Swedish secretaries in Nairobi, girls with a good sense of humor. They immediately wrote to Fialkow as follows: Dear Dr. Fialkow, the Masai are a highly ethical and moral people but there customs differ from ours. After the Masai chief has entertained a close friend of his age group for dinner, he walks around with the friend among the huts of his wives. He puts down his spear in front of one of the huts. That is where he invites his friend to spend the night. However, if anyone would try to get into a wife without being invited, he would be immediately killed. PS there is no need to keep this information confidential.

Following the confirmation that BL and NPC are the two most regularly EBV-carrying malignant tumors, their dominating position remained. The low differentiated or anaplastic form of NPC was found to carry latent EBV in nearly 100 %. This was independent of geography. It was equally true for the high-incidence Southern Chinese and the rare Western cases. Such a regular association must have a profound significance.

Today, several decades after the discovery of the association, the role of the virus for the etiology of the tumor is still unknown.

In BL, the situation is somewhat different. The virus is associated with about 90 % of the high endemic African tumors, but the worldwide sporadic BLs carry EBV in only 20–30 %. The rest contains no virus and has no traces of its genome.

An important missing link in the genesis of BL was provided by cytogenetic studies. A curious, unexpected convergence occurred at our laboratory. On the second floor, Francis Wiener, just about the best mouse tumor cytogeneticist of this time, worked on mouse plasmacytomas, together with Shinsuke Ohno from Japan. On the third floor, George Manolov and his wife Yanka Manolova, human cytogeneticists from Bulgaria, looked at the chromosomes of the incoming BL biopsies from Nairobi. The two cytogenetic teams did not have much communication. Frankly, they did not have much respect for one another. The mouse cytogeneticist did not think much about the work on human material that did not permit experimentation. For the human cytogeneticist, the mouse work appeared irrelevant.

And then suddenly, the two floors converged. Almost precisely, the same chromosomal translocations were discovered in mouse plasmacytoma and human BL. Subsequently, many other laboratories became involved in this work. The outcome was that the translocations act by juxtaposing the powerful c-myc oncogene to one of the three immunoglobulin loci. To my great surprise, the work of Phil Leder and Carlo Croce on BL and of Susan Cory and Ken Marcu on MPC confirmed my speculation that an eminent molecular biologist friend, Lennart Philipsson, called the most hair-raising extrapolation from the centimorgans to the kilobases, postulating that these translocations act by juxtaposing an oncogene to a highly active normal gene in that particular cell type (Klein 1981). It was the only hypothesis I ever made that turned out to be correct. The work was crowned by Michel Cole's demonstration that the oncogene was c-myc in both BL and MPC. I do not know of any other example in cancer biology where tumors of two different species, with entirely different pathogenetic histories, arise due to basically the same oncogene activation event in the same broad cell lineage as the only common denominator.

Thus, the Ig/myc translocation, rather than EBV, is the common feature of all BLs. But what is the role of EBV?
Prior to and in parallel with the cytogenetic developments, some monumental discoveries were made in the EBV field. They include the transforming and immortalizing activity of the virus for B cells, the role of the virally encoded growth transformation-associated proteins, and particularly EBNA2 and LMP1 for sustained B-cell proliferation and the causative role of the virus for mononucleosis. Moreover, the virus turned out to be the driving force for the immunoblastomas that arise from it genetically, iatrogenically (e.g., in transplant recipients) or by coinfection (HIV immunosuppressed patients). Importantly, EBV-carrying African BL cells do not express the full growth program. They do not need it because they are driven by myc.

In addition to BL and anaplastic NPC, other consistent tumor-EBV associations were discovered later. The relationship with Hodgkin's lymphoma (HL) is particularly noteworthy since there is both epidemiological and molecular evidence that the virus may be essential for the proliferative potential of the HLs that carry it.

Similar evidence is emerging for the EBV-carrying form of diffuse large B-cell lymphoma. The virus can also be carried by T-cell lymphomas. Its association with the T-cell-derived lethal midline granuloma is particularly consistent.

2 Is EBV a Cancer Virus?

Having completed the sequencing of the powerfully tumorigenic mouse polyoma virus, a considerable feat at the time, Fred Sanger felt that a human tumor virus should come next. On the advice of Beverly Griffin, he chose EBV, considered as the first human tumor virus. Paul Farrell carried out the job which had important consequences.

But is EBV really a tumor virus?
The immunoblastomas of the immunosuppressed are clearly driven by EBV as already mentioned. But are the immunoblastomas "truly malignant"? They tend to remain diploid and do not carry characteristic tumor-associated mutations. The frequently mutated p53 pathway remains wild type, as a rule.

3 What Is the Role of EBV in the Genesis of BL?

This question has no straightforward answer. But the presence of the virus in 90 % of the high endemic tumors cannot be a coincidence. I also attach considerable significance to the argument of Bill Sugden. He pointed out that the lack of synchrony between cell division and EBV episome duplication must lead to the loss of the viral genomes unless they are needed for the sustainability of the tumor. In fact, EBV-carrying BL lines can loose the virus in vitro, but a similar loss has never been observed in vivo. This indicates that some viral function may provide the in vivo tumor cell with a selective growth advantage.

4 The Lessons of HART Therapy

Malignant lymphomas have been a frequent complication of untreated AIDS. Part of them were EBV-driven immunoblastomas, but a substantial portion were Ig/myc-carrying BL or BL-like lymphomas. This picture changed radically after the introduction of HART therapy. In parallel with increase of the CD4+ T-cell count, the EBV-driven immunoblastomas practically disappeared. This makes sense in view of the highly efficient T-cell-mediated immune response against EBV-driven immunoblasts. The Ig–myc-carrying BL and BL-like lymphomas showed no similar tendency to decrease and are by now the most frequent lymphoma complication in HART-treated HIV carriers.

Ig–myc translocation-carrying cells with constitutively activated myc genes are prone to apoptosis. They need to be rescued by the cognate antigen or by B-cell stimulatory lymphokines and cytokines. HIV proteins linger around in the lymph nodes for a long time even under HART therapy. They may contribute to the survival of the Ig/myc translocation-carrying cells that are known to arise continuously as accidents of normal B-cell differentiation.

This argument is also pertinent to the question why chronic hyperendemic malaria and HIV are the only two conditions where translocation-carrying BLs arise in a relatively high frequency. A B-cell stimulatory environment is the characteristic for both conditions. Rescue from apoptosis by B-cell stimulatory cytokines may be the common responsible mechanism.

References

Klein G (1981) Nature 294:313–318
Reedman BM, Klein G (1973) Int J Cancer 11:499–520
zur Hausen H et al (1970) Nature 228:1056–1058

EBV-Specific Immune Response: Early Research and Personal Reminiscences

D.J. Moss and V.P. Lutzky

Abstract Early research on Epstein–Barr virus (EBV) developed from serological observations that were made soon after the discovery of the virus. Indeed, the definition of the humoral response to a variety of EBV proteins dominated the early literature and was instrumental in providing the key evidence for the association of the virus with infectious mononucleosis (IM), Burkitt's lymphoma (BL), and nasopharyngeal carcinoma (NPC). Each of these disease associations involved a distinct pattern of serological reactivity to the EBV membrane antigens (MA), early antigens (EA), and the EBV nuclear antigen (EBNA). When it became generally accepted that the marked lymphocytosis, which is a hallmark of acute IM, was dominated by T cells, considerable effort was directed toward untangling the specificities that might be associated with restricting the proliferation of newly infected B cells. Early evidence was divided between support for both EBV non-specific and/or HLA non-restricted components. However, all results needed to be reassessed in light of the observation that T cells died by apoptosis within hours of separation from fresh blood from acute IM patients. The observation that EBV-infected cultures from immune (but not non-immune) individuals began to die (termed regression) about 10 days post-seeding, provided the first evidence of a specific memory response which was apparently capable of controlling the small pool of latently infected B cells which all immune individuals possess. In this early era, $CD8^+$ T cells were thought to be the effector population responsible for this phenomenon, but later studies suggested a role for $CD4^+$ cells. This historical review includes reference to key early observations in regard to both the specific humoral and cellular responses to EBV infection from the time of the discovery of the virus until 1990. As well, we have included personal recollections in regard to the events surrounding the discovery of the memory T cell response since we believe they add a human dimension to a chapter focussed on early history.

D.J. Moss (✉) · V.P. Lutzky
QIMR Berghofer Medical Research Institute, Brisbane, Australia
e-mail: denis.moss@qimrberghofer.edu.au

V.P. Lutzky
e-mail: Viviana.lutzky@qimrberghofer.edu.au

C. Münz (ed.), *Epstein Barr Virus Volume 1*, Current Topics in Microbiology and Immunology 390, DOI 10.1007/978-3-319-22822-8_3

Contents

1 Humoral Responses to EBV Infection 24
1.1 Antibody to the Virus Capsid Antigen 24
1.2 Antibody to the EBV Membrane Antigen 25
1.3 Antibody to the EBV Early Antigen 26
1.4 Antibody to EBV Nuclear Antigen 26
1.5 EBV-Specific IgA Response as a Serological NPC Marker 27
2 EBV Seroepidemiology 27
3 Cellular Response to EBV Infection 29
3.1 Early Studies to Define an EBV-Specific T Cell Response in Acute EBV Infection 29
3.2 Memory T Cell Response in Healthy Immune Individuals: Personal Reminiscences of Denis Moss from 1977 to 1980 30
3.3 Defining the First EBV CTL Epitope: Personal Reminiscences of Denis Moss from 1988 to 1990 36
3.4 Early Contributions of Other Investigators 36
4 Final Remarks 37
References 37

1 Humoral Responses to EBV Infection

1.1 Antibody to the Virus Capsid Antigen

In 1966, the Henle laboratory observed that a small proportion of cells in a series of Burkitt's lymphoma (BL) cell lines (EB-1 and EB-2) showed strong immunofluorescence when stained with sera from a variety of patient groups as well as most (but not all) healthy Americans (Henle and Henle 1966). The authors noted in this publication that while it was not possible to absolutely ascribe Epstein–Barr virus (EBV) specificity to this strong immunofluorescent staining, several observations supported this conclusion. Firstly, both virus particles and strong immunofluorescence were associated with cells undergoing degeneration. Secondly, exposure of cell lines to the anti-herpesvirus drug 5-methylamino-2′-deoxy-uridine also reduced the proportion of stainable cells. Irrefutable proof of the EBV specificity of the immunofluorescence was provided by Harald zur Hausen, who demonstrated by electron microscopy that individual fluorescent cells all contained virus particles, while no virus particles were demonstrated in non-immunofluorescent cells (zur Hausen et al. 1967). This and related studies clearly established that this staining was directed to the EBV capsid antigen (VCA) and was subsequently extensively used to establish a prior EBV infection (clinical or silent). However, the anti-VCA IgG immunofluorescence assay had diagnostic limitations since (a) the low titers seen in acute infectious mononucleosis (IM) overlapped with the high titers seen in some healthy individuals (b) only 10–20 % of sera showed a significant difference between acute and convalescent sera. Of particular

Table 1 Typical antibody responses against the EBV proteins VCA, EA and EBNA in IM, BL and NPC patients along with healthy controls

Disease	Antibody to					
	VCA			EA		EBNA
	IgM	IgA	IgG	D	R	
IM	++	0	++	+	0	0
BL	0	0	++	0	++	+
NPC	0	++	++	++	0	++
Controls	0	0	+	0	0	+

0, Not detectable as a rule; +, Antibody titers ≤1:160; ++, Antibody titers ≥1:320
Table reproduced (with permission) from Henle and Henle (1979)

diagnostic importance, however, was the observation that an IgM response to EBV VCA was detectable only during the acute phase of primary infection and this response faded within two months of clinical presentation (Edwards and McSwiggan 1974; Henle and Henle 1979; Nikoskelainen and Hanninen 1975; Nikoskelainen et al. 1974; Schmitz and Scherer 1972). A summary of the antibody patterns against VCA and other EBV antigens (see below) for different EBV-associated diseases is presented in Table 1.

1.2 Antibody to the EBV Membrane Antigen

Early studies indicated that some EBV antibody-positive human sera detected a specific membrane antigen (MA) on the surface of BL cells (Klein et al. 1966, 1967). Further investigation indicated two components associated with this reactivity—an early EBV antigen (early MA) on live BL tumor cells in vivo and a late MA antigen (late MA) in later stages of virus replication. This latter reactivity was seen in association with VCA. Indeed, Henle in delivering a special lecture at the First International Symposium on EBV in Greece (Henle and Henle 1984) recalled that in 1967 George Klein reported to a conference of the American Cancer Society that there was a strong correlation between those cell lines that exhibited MA reactivity and those, in which Henle demonstrated VCA immunofluorescence, thus inferring that MA was virus-induced. These MA reactivities were originally demonstrated using three different serological assays (1) immunofluorescence using sera from BL as well as other sources (Klein et al. 1966, 1967, 1968), (2) antibody-dependent cellular cytotoxicity (Pearson and Orr 1976), and (3) ^{125}I-labeled protein A binding (Pearson and Qualtiere 1978). The immunofluorescence test for the detection of anti-MA was later replaced by a blocking procedure to avoid interactions with membrane antigens unrelated to EBV (Klein et al. 1969). Because MA is a complex of several components (early and late MA), the utility of this assay was limited. Sera used to detect this reactivity included a wide spectrum of EBV specificities, but subsequent studies indicated that MA reactivity was closely associated with neutralization of the virus (Pearson et al. 1970; Thorley-Lawson and Geilinger 1980). The early literature refers to three different methods of measuring virus neutralization: (1) inhibition of EA antigens (Henle et al. 1970b), (2) inhibition of colony formation (Rocchi and Hewetson 1973), and (3) inhibition of transformation (Henle et al. 1967; Miller et al. 1972; Pope et al. 1968). When neutralization levels were assessed in a group of human sera by each of these assays, there was a reasonable, but not a perfect, correlation between each of these technologies (de Schryver et al. 1974).

Since it was well established that anti-MA reactivity closely correlated with virus neutralization, it is not surprising that there was an early attempt to develop a vaccine to EBV based on the high molecular weight glycoprotein component of MA (gp340) (Epstein and Morgan 1986; North et al. 1980, 1982; Thorley-Lawson and Geilinger 1980). This effort was based largely on the ability of immunized cotton-top tamarins to inhibit the formation of malignant lymphomas following challenge with EBV. These efforts to develop a vaccine were based on the hypothesis that a specific humoral response would be protective from the appearance of clinical symptoms.

1.3 Antibody to the EBV Early Antigen

A further antigen and antibody specificity was discovered following superinfection of lymphoblasts from non-producer lymphoblastoid cell lines (LCLs) with EBV derived from the P_3HR-1 cell line. This superinfection was shown to reduce cell growth but failed to stain for VCA in immunofluorescence assays using sera from healthy individuals (Henle et al. 1970b), yet stained strongly in up to 90 % of cells when sera from IM, BL, or nasopharyngeal carcinoma (NPC) patients were used. This serum reactivity (referred to as early antigen, EA) disappeared when clinical symptoms abated (Henle et al. 1971b). Later studies (Henle et al. 1971a) divided this reactivity on the basis of methanol sensitivity (R and D). Since the anti-D reactivity was strongly associated with acute IM and rarely seen in healthy individuals, this assay was historically valuable in the diagnosis of IM (Table 1) (Henle et al. 1974).

1.4 Antibody to EBV Nuclear Antigen

The EBV nuclear antigen (EBNA) was first detected using EBV immune serum in an anti-complement immunofluorescence (ACIF) assay. The high sensitivity and specificity of this three-step assay relies on the ability of complement from an EBV-negative donor serum to bind to the EBNA antibody/EBNA protein complex followed by conjugated anti-human β_1C/β_1A (Reedman and Klein 1973). Antibody to this protein is widely found in EBV immune sera except in acute IM where it appears weeks or months after onset of illness (Henle et al. 1974). Interestingly, high levels of EBNA antibody are detected in NPC patients (Table 1).

1.5 EBV-Specific IgA Response as a Serological NPC Marker

The discovery of a strong IgA response to EBV in NPC patients heralded the beginning of its serological utility as an outstanding feature of this malignancy (Wara et al. 1975). Stimulated by this early report, the Henle and Klein laboratories (Henle and Henle 1976) tested a total of 372 sera from patients with NPC, other carcinomas of the head and neck, BL, IM, and healthy controls. The results demonstrated that prior to therapy, 93 % of NPC patients had an IgA VCA response and that these titers increased from stages I or II to stages III or IV, i.e., a strong correlation with the total tumor burden. Conversely, many of the NPC patients examined 2–6 years after initial therapy had only low levels of EBV-specific IgA or none at all, and the majority of those with high titers were known to have residual or recurrent disease. In contrast to untreated NPC patients, less than 5 % of 73 patients with other carcinomas and none of 76 healthy individuals revealed a specific IgA response. These observations provided a stepping stone for the widespread use (particularly in mainland China) of the detection of IgA to VCA and EA as a means of mass population screening (Zeng et al. 1979, 1982, 1983).

2 EBV Seroepidemiology

Within a decade of the discovery of the virus, most of the important serological trends linked with different EBV-associated diseases had been defined. Initial interest in this field was intense in regard to BL since this was the first human malignancy shown to have an apparent viral etiology. Anti-VCA assays were used to determine whether the serological titers in BL endemic areas were significantly higher than those in non-endemic areas. These epidemiological surveys confirmed that infection with EBV among healthy individuals was extremely widespread, with higher titers being recorded among east African (typically by three years of age) as well as in low socioeconomic individuals in other regions of the world, compared to those in higher socioeconomic regions (Gerber and Monroe 1968; Henle and Henle 1966; Moore et al. 1966). When these studies were extended to defining anti-VCA titers in a range of other EBV-associated diseases along with controls including non-EBV-associated malignancies (Fig. 1), it became apparent that anti-VCA levels in IM, BL, and NPC patients were generally higher than

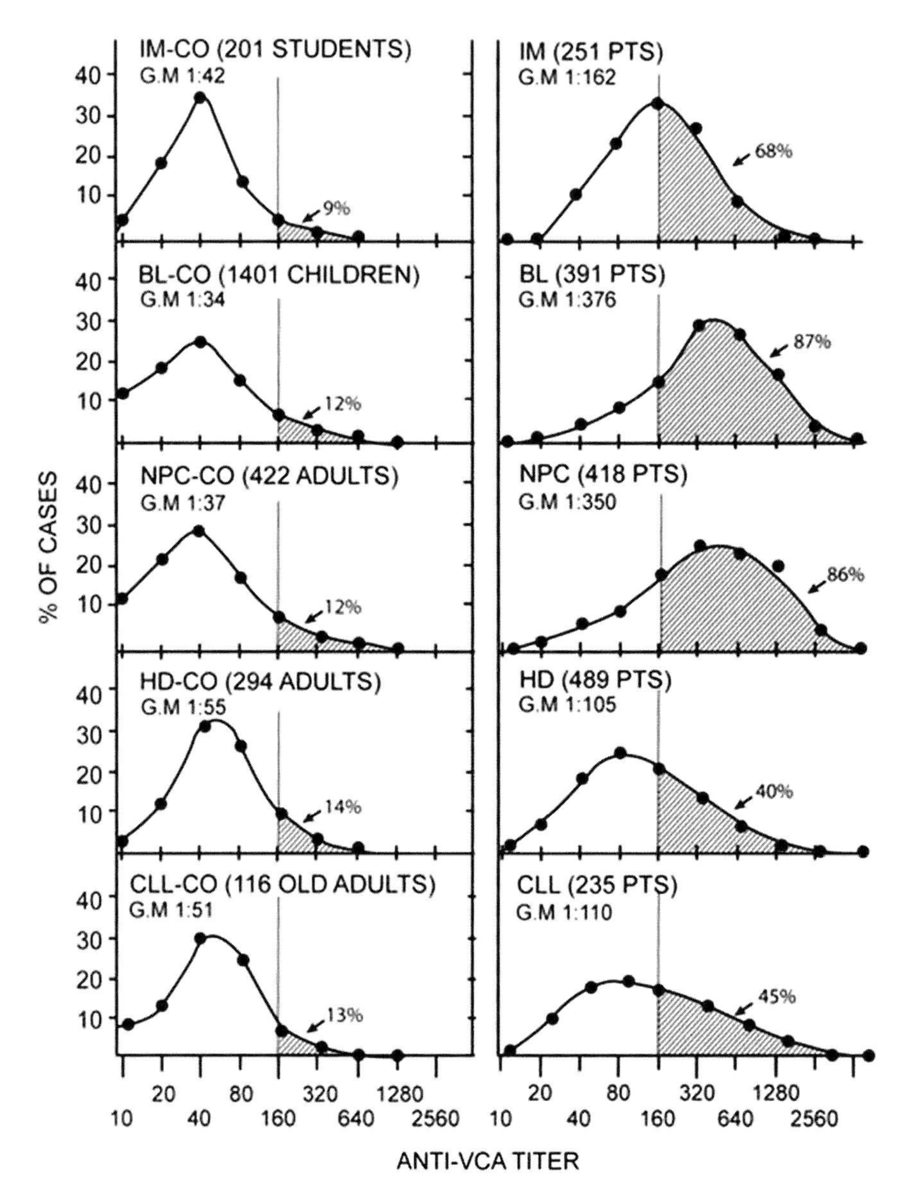

Fig. 1 Distribution (expressed as a percentage of cases) of antibody titers to VCA in patients (PTS) with various diseases and appropriate control (CO) groups. *IM* infectious mononucleosis; *BL* Burkitt's lymphoma; *NPC* nasopharyngeal carcinoma; *HD* Hodgkin's disease; *CLL* chronic lymphocytic leukemia; *GM* geometric mean. Figure adapted (with permission) from Henle and Henle (1979)

those recorded in healthy individuals living in the same area (de-The et al. 1978; de Schryver et al. 1969; Henle and Henle 1979; Niederman et al. 1968). Evidence for the association of EBV with these diseases was further strengthened when these assays were extended to include the EA complex (Henle and Henle 1979; Henle et al. 1970a, 1971b).

3 Cellular Response to EBV Infection

3.1 Early Studies to Define an EBV-Specific T Cell Response in Acute EBV Infection

One of the pathological hallmarks of acute IM is the presence of atypical mononuclear cells which permeate not only into the peripheral blood but also into many tissues (Carter 1975). It should be pointed out that this heterogenous population is not specific for acute IM but was referred to much earlier in relation to bacterial infection (Turk 1898) and graft rejection (Parker and Mowbray 1971). It was generally understood that these atypical mononuclear cells would be dominated by recently transformed B cells, and it was thus somewhat unexpected when it was established that this population possessed sheep erythrocytes receptors rather than B cell markers (Sheldon et al. 1973). This result was confirmed in a study of acute IM in which five independent B and T cell markers were assessed (surface immunoglobulin, complement receptors, B and T lymphocyte-specific markers, and sheep erythrocyte receptors). It was concluded that these atypical mononuclear cells were indeed largely of T cell origin (Pattengale et al. 1974) resulting in intense interest in the possibility that they might either include a specific component capable of limiting the proliferation of virus-infected B cells or were homeostatic, non-specific suppressor cells. Attempts were made to identify EBV-specific T cell activity by exposing cultured lymphocytes from acute IM donors to specific EBV antigens and monitoring the response in a leukocyte migration inhibition (LMI) assay (Lai et al. 1977) and leukocyte adherence inhibition (LAI) assay (Chan et al. 1977). Both of these approaches showed some indication of a specific response but were somewhat awkward to use and did not find wide-spread application. Subsequently, two distinct T cell activities were identifiable in vitro (1) a suppressor cell activity that was demonstrated using pokeweed mitogen-induced B cell activation (Henderson et al. 1977; Johnsen et al. 1979; Tosato et al. 1979), which was also later revealed to suppress the proliferative responses of T cells to mitogenic and antigenic stimuli (Reinherz et al. 1980) and (2) an apparently EBV-selective cytotoxicity following removal of conventional NK cell activity (Bakacs et al. 1978; Royston et al. 1975; Svedmyr and Jondal 1975). Thus, for example, the publication from Royston and colleagues was based on a collection of mononuclear cells from 21 acute IM patients and demonstrated a higher level of lysis toward autologous LCLs than non-EBV-infected target cells (non-EBV tumor cells). It should be pointed out, however, that the effector to target ratio used was high (100:1) and the level of lysis of the autologous cell line fairly modest (10 %) compared to the non-EBV-infected control (3 %). These conditions of assay would currently be regarded as extreme, particularly when compared with the high levels of NK cell activity included in the blood of acute IM patients. One of the criticisms that could reasonably be leveled at these and other earlier studies (Bakacs et al. 1978; Lipinski et al. 1979; Seeley et al. 1981; Tursz et al. 1977) is that they purport to detect EBV specificity but lack obvious HLA restriction (Zinkernagel and Doherty 1979). Killing of this type led to the concept of a lymphocyte-determined membrane antigen (LYDMA), which was said to be expressed on the surface of all LCLs (Klein et al. 1976; Svedmyr and Jondal 1975). The concept that cytotoxic T cells (CTLs) recognized a surface protein was quite understandable and predated the historic discovery that CTL recognized small peptides in association with class I MHC and that these small peptides were frequently derived from proteins, which by serological criteria appeared to have an intracellular location (Townsend et al. 1985).

It should be pointed out that these early studies needed to be assessed carefully in light of the observation that T cells from acute IM patients die by apoptosis (Fig. 2) within hours of isolation (Bishop et al. 1985; Moss et al. 1985) and that the early assays to detect bioactivity were almost certainly significantly affected by this phenomenon. Indeed, when T cells from IM patients were isolated after 15 h, less than 5 % of T cells were recovered which is in stark contrast to that seen from blood of healthy donors (Moss et al. 1985). Interestingly, apoptosis can be largely avoided by inclusion of IL-2 in the culture medium, both at the time of isolation of the mononuclear cells as well as in all of the steps involved in the analysis of these cells (Bishop et al. 1985).

The contemporary view regarding the T cell response has clarified much of the early uncertainty regarding its presence or otherwise during acute infection. It is now generally accepted that the $CD8^+$ lymphocytosis seen during acute infection includes a large specific T cell component and that following resolution of the disease there is a disproportionate culling of the $CD8^+$ population and restoration of the CD4/CD8 ratio back into the normal range (Annels et al. 2000; Callan et al. 1996, 2000; Roos et al. 2000).

3.2 Memory T Cell Response in Healthy Immune Individuals: Personal Reminiscences of Denis Moss from 1977 to 1980

In 1973, Beverley Reedman from the EBV laboratory at QIMR, spent a year in Stockholm at the Karolinska Institute with George and Eva Klein. This collaboration resulted in the first description of the presence of the EBNA protein (Reedman and Klein 1973), which was very important in that it allowed a ready visualization of the EBV protein present in all endemic BL cell lines and LCLs (Pope et al. 1969). Based on this momentous discovery of an EBV protein

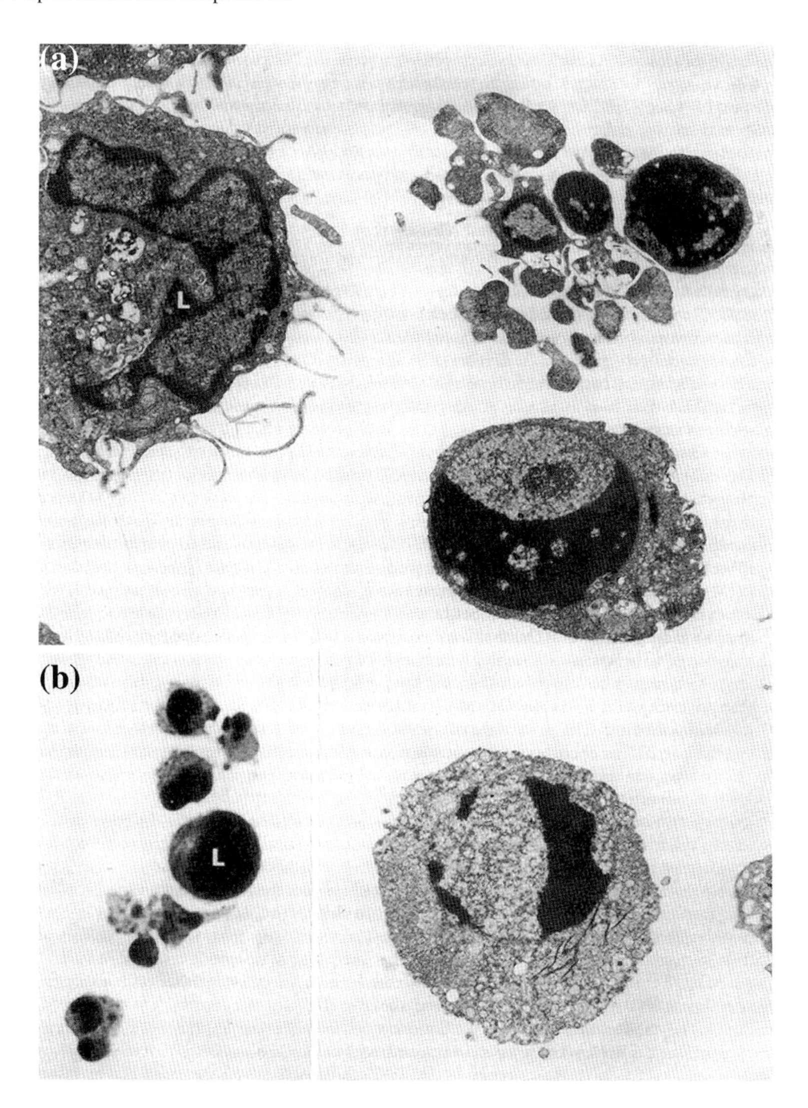

Fig. 2 **a** Electron micrograph of T cells 15 h after fractionation from a representative acute IM patient. A lymphocyte of normal appearance (*L*) is shown with three dying lymphocytes displaying characteristic features of apoptosis (loss of microvilli, cytoplasmic and nuclear condensation, *crescent* shaped margination of chromatin, nuclear fragmentation, and cellular budding). Magnification ×10,000. **b** Light micrograph of T cells 15 h after fractionation from the same patient as above. A lymphocyte of normal appearance (*L*) is surrounded by dying cells showing features associated with cell death by apoptosis (condensation and basophilic nuclear fragments). Magnification ×1300. Figure reproduced with permission from Moss et al. (1985)

invariably associated with EBV infection, our laboratory decided to embark on a project to map the kinetics with which EBNA was induced following infection of human lymphocytes. I had anticipated that this should not present significant technical problems as the EBNA staining assay was both sensitive and specific.

We demonstrated that by three days, about 10 % of lymphocytes showed strong EBNA staining, while the majority of cells were similarly stained by five days (Moss and Pope 1975). This study was exclusively carried out using lymphocytes from EBV immune adults. It came as somewhat a surprise that these strongly proliferating cultures began to show cell concentration-dependent microscopic evidence of cell death by day 10, followed by a total collapse of the lymphoblastoid clumps by day 21, together with the complete disappearance of EBNA staining (Fig. 3). In the discussion section of this publication, reference was made to an earlier observation (Moss and Pope 1972) reporting the difficulty in establishing cell lines in some instances and we hypothesized that this phenomenon of cell death might be due to "an immune response to developing EBV antigens" (Moss and Pope 1975). Indeed, we went on to show in preliminary experiments (unpublished) that lymphocyte cultures from an EBV seronegative individual continued to expand without any evidence of cell death. This concept of the effect of the immune status of the lymphocyte donor on the course of EBV transformation was further reinforced by our later observation that transformation of lymphocytes from EBV seropositive (but not seronegative) individuals was reversibly blocked when cultured on adult fibroblasts (Moss et al. 1977). Unfortunately, the mechanism involved in this inhibition was never determined since the phenomenon proved to be very sensitive to minor changes in culture conditions. By the time Alan Rickinson (Bristol) arrived in John Pope's laboratory in 1977, I was able to discuss with him a working hypothesis compatible with these published and unpublished observations. In particular, we suggested that cell death in these cultures from EBV immune individuals may well have an immunological basis. Our laboratory had not regularly seen this cell death phenomenon earlier because at that time we had only recently switched from using cord blood

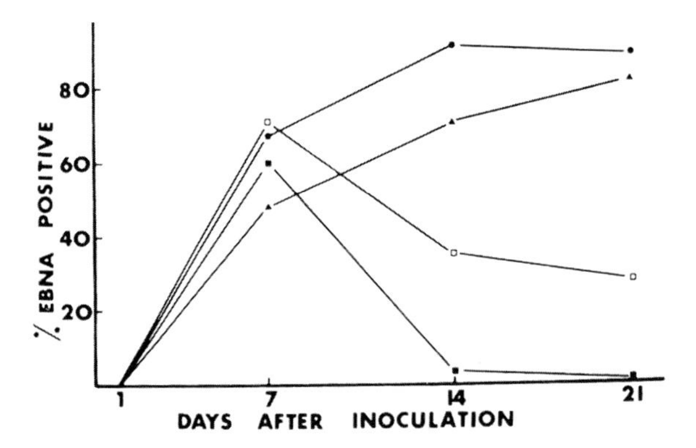

Fig. 3 The relationship between EBNA production (shown as the percentage of EBNA-positive cells/ml) and initial cell concentration following infection of human leukocytes with EBV at a multiplicity of 0.003 TD_{50}/cell. Initial cell concentrations were 10^6/ml (■—■); 5×10^5/ml (□—□); 2.5×10^5/ml (●—●); 10^5/ml (▲—▲). Figure reproduced with permission from Moss and Pope (1975)

(always EBV seronegative) to adult blood (usually EBV seropositive) in our experiments due to new hospital regulations regarding the use of cord blood for experimental purposes. Alan and I decided to work together to either prove or disprove the immunological basis of regression. We were fortunate in that I had only recently returned from a visit to the laboratory of Tony Basten (University of Sydney) who had demonstrated to me the new technique of separating human T cells rosettes using sheep red blood cells coated with the highly purified sulfhydryl reagent, 2 aminoethylisothiouronium bromide hydrobromide (AET) (Kaplan and Clark 1974). It was unfortunate that the first experiment we jointly set up included my regular EBV seronegative donor as a control and the results indicated that these cells began to die after 10 days of apparent proliferation. We were greatly relieved to learn that this individual had recently seroconverted. Indeed, this single experiment convinced us of the validity of our hypothesis and during that year we unequivocally proved that cell death (which Alan termed regression) was mediated by a population of EBV-specific T cells present in the blood of all healthy immune individuals (Fig. 4) (Moss et al. 1978; Rickinson et al. 1979). In a subsequent study, we proposed that previously infected individuals possessed a pool of virus-specific memory T cells capable of mounting a specific response when challenged in vitro (Moss et al. 1979). Given the marked T cell expansion associated with primary infection, we were surprised to subsequently demonstrate that the lymphocyte population from acute IM patients showed weak regression (Rickinson et al. 1980). Indeed, more than five years were to elapse before an explanation for this result was forthcoming (Fig. 2) (Bishop et al. 1985; Moss et al. 1985).

One of the useful features of the regression assay is that it can be used to provide a semiquantitative estimate of the overall level of EBV-specific CTL immunity in the blood of any immune individual based on a 50 % endpoint of the incidence of transformation in EBV-infected cultures (Moss et al. 1978). In our experience, this regression endpoint was surprisingly reproducible and was applied to compare the level of EBV CTL immunity in malaria endemic and non-endemic individuals in Papua New Guinea (Moss et al. 1983a), as well as in NPC patients (Moss et al. 1983b).

A series of publications sought to investigate the mechanism of regression. The results of double chamber experiments argued against the involvement of any soluble factors including antibody, whereas there was an absolute requirement for T cells in the initial cultures along with a requirement for proliferation of effector T cells within the first 14 days (Rickinson et al. 1979). Furthermore, T cells prepared by ammonium chloride lysis of T cell rosettes from regressing cultures showed growth inhibition of autologous and HLA-related target cells (Moss et al. 1979). This lysis was neither dependent on the target cells expressing serologically defined envelope antigens on their cell surface nor did it involve an artifact in the culture medium (e.g., fetal calf serum). Moreover, lysis was not intrinsically associated with B cell proliferation per se.

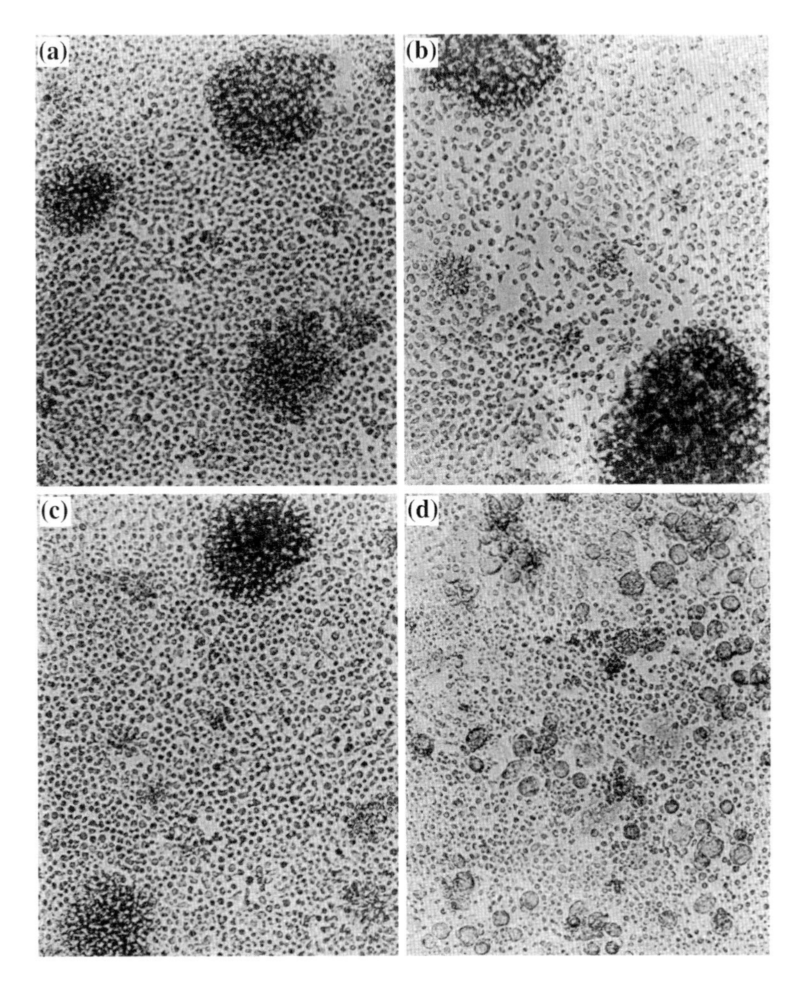

Fig. 4 Photomicrographs of EBV-infected cultures of blood mononuclear cells seeded initially at 10^6 cells/ml (×125). **a** Seronegative adult donor cells 14 days post-infection. **b** Seronegative adult donor cells 28 days post-infection. **c** Seropositive adult donor cells 14 days post-infection. **d** Seropositive adult donor cells 28 days post-infection. Figure reproduced with permission from Moss et al. (1978)

Overall, this year's collaboration with Alan was very important since it provided for the first time experimental proof for the widely held notion that an immune memory response must exist in all previously infected individuals to prevent uncontrolled proliferation of the latently infected B cells which are a found in all immune donors (Nilsson et al. 1971).

After Alan returned to Bristol, I realized that regression, although most likely mediated by CTLs, may have been due to some non-cytotoxic, suppressor mechanism. At this stage, Ihor Misko (Australian National University, Canberra) joined the Pope's laboratory and brought with him additional immunological skills including experience in HLA restriction and the use of the ^{51}chromium release assay. We thus set about formally proving that regressing cultures included a CTL cell component capable of killing autologous and HLA-related LCLs (Misko et al. 1980). Interestingly, this finding in healthy immune donors contrasted with the earlier result in acute IM patients in which no HLA-associated killing was reported (Svedmyr and Jondal 1975; Svedmyr et al. 1979).

In 1979, the Queensland Cancer Council provided me with the opportunity of working in Tony Epstein's laboratory in Bristol for 12 months, once again with Alan but also with Lesley Wallace and Martin Rowe. Notable findings during this era included detailed investigations into EBV-specific HLA restriction (Fig. 5) (Moss et al. 1981b; Rickinson et al. 1981b; Wallace et al. 1981) as well as demonstrating the use of irradiated autologous LCLs to activate a specific CTL response in vitro (Rickinson et al. 1981a). This method went on to become a cornerstone in the activation of specific CTLs in patients with various EBV-associated lymphoid malignancies (Gottschalk et al. 2005; Rooney et al. 1998).

In order to derive some notion of the possible nature of the target for T cell recognition, we went on to demonstrate that LYDMA was expressed on the cell surface soon after the appearance of the EBNA protein(s) and coincident with, but not dependent on, the initiation of cellular DNA synthesis (Moss et al. 1981a). This was a significant finding in relation to the development of an EBV vaccine since it

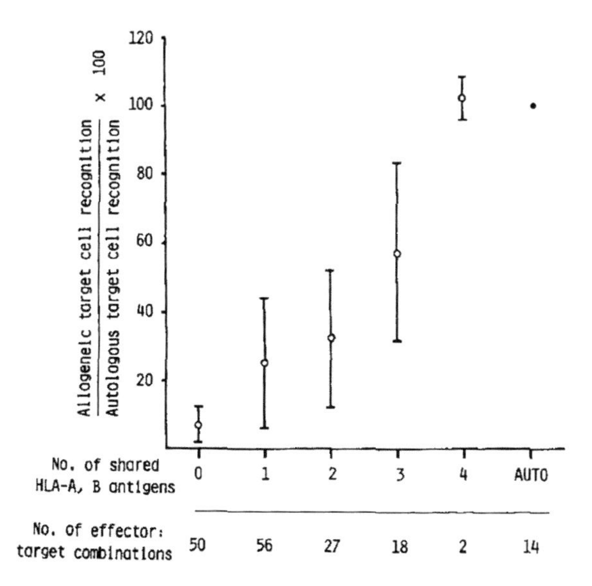

Fig. 5 Allogeneic target cell recognition, measured either in a ^{51}chromium release assay or in a growth inhibition assay and expressed each time as a percentage of that activity shown by the same effector cells against their autologous cell line, plotted against the number of HLA-A and HLA-B antigens shared between effector and target cells. Composite data from 14 donors covering 153 allogeneic combinations; mean values (±1 standard deviation) are shown for groups of effector: target cell combination sharing 0, 1, 2, 3, or 4 HLA-A and HLA-B antigens. Figure reproduced with permission from Moss et al. (1981b)

provided evidence that immunological control over EBV-infected B cells appeared to be exercised soon after infection, raising the possibility that a vaccine might provide protection before clinical symptoms appeared. I found the intellectual environment in Bristol very stimulating and scientifically productive. As well, the close proximity to America and Europe gave me the chance of meeting a host of EBV luminaries including George and Eva Klein and Maria Masucci (Karolinska Institute, Stockholm) along with Guy de Thé (Pasteur Institute, Paris) and Werner Henle (Children's Hospital of Philadelphia).

3.3 *Defining the First EBV CTL Epitope: Personal Reminiscences of Denis Moss from 1988 to 1990*

Following the discovery by the Oxford group (Townsend et al. 1985) that T cells recognized small peptides (typically nine amino acids long), it became clear that our laboratory would need to develop a similar molecular approach if our long-term goal of developing a vaccine to EBV was to be realized. Up to this point, peptide epitopes had only been defined in relation to comparatively small viruses such as influenza and it was not clear how a similar technology could be directly applied to EBV which included more than 10 times as many proteins. In hindsight, the critical observation involved the isolation of a series of $CD4^{+}$ and $CD8^{+}$ CTL clones that recognized autologous type 1 (formerly known as type A) but not type 2 (formerly known as type B) LCLs (Moss et al. 1988). These two strains were believed at the time to differ on the basis sequence divergence within the *Bam*H1 WYH region of the genome encoding EBNA2 (Dambaugh et al. 1984). Subsequently, these clones were used to screen a series of peptides showing sequence divergence between both strains and which also corresponded to predicted epitope algorithms (Delisi and Berzofsky 1985; Rothbard and Taylor 1988). This resulted in the identification of the first EBV CTL epitope (Burrows et al. 1990).

3.4 *Early Contributions of Other Investigators*

The literature includes a significant body of work in this early era from other investigators. Notable among these was the observation that T cells from adult individuals could delay the appearance of foci of EBV-transformed cells in comparison with corresponding cultures from cord blood donors (Thorley-Lawson et al. 1977). Despite the problems inherent in the interpretation of a morphological study of this kind, the data indicated that the effect was observable over a range of initial cell concentrations as well as different B:T cell ratios. Furthermore, a subsequent study using EBNA and ^{3}thymidine induction confirmed the initial observation (Thorley-Lawson 1980). Curiously, it was subsequently shown that mitomycin C-treated T cells could delay transformation suggesting that this delay was independent of T cell proliferation. It should be pointed out that the serological status of the individuals involved in these studies was not recorded. However, subsequent studies demonstrated that this effect occurred with both seropositive and seronegative donors (Shope and Kaplan 1979), inferring that this delay did not have an immunological basis and was quite distinct from the observation of regression of transformation. Indeed, a role for α interferon has been suggested raising the possibility of a contribution of NK cells to this delay in transformation (Shope and Kaplan 1979).

The sensitivity of BL cells to different mechanisms of killing was another important aspect of research in this early era. These studies involved a comparison of the killing sensitivity of so-called paired cell lines, i.e., BL and LCLs derived from the same patient. It appeared that BL cells, although sensitive to allo-specific lysis, were generally insensitive to EBV-specific T cell killing (Rooney et al. 1985a, b). Subsequent studies revealed that this phenomenon was related to the different pattern of latency present in BL cell lines compared to LCLs (Rowe et al. 1985), thus excluding the possibility that BL arose from an EBV-specific T cell defect.

Although there was widespread agreement in the early literature that regression was mediated by virus-specific CTLs that had been reactivated in vitro from memory T cells, there existed a degree of confusion as to whether all the components of the phenomenon were included within the $CD8^+$ subset or whether $CD4^+$ may also play a role (Crawford et al. 1983; Rickinson et al. 1984; Tsoukas et al. 1982). More contemporary investigations indicate that $CD4^+$ cells do indeed play a role in controlling the expansion of newly infected $CD23^+$ B cells (Nikiforow et al. 2001).

4 Final Remarks

The link between EBV with various human diseases provided the impetus for the use of EBV serology as a biomarker. This was particularly the case in IM, NPC, and BL. Subsequently, the discovery of a memory T cell response to the virus provided an additional perspective of the dynamics of the control of the latent EBV infection in all immune individuals. In conclusion, this early era provided a basis for the subsequent developments in EBV immunobiology as well as a platform for immunological intervention in EBV-associated diseases.

References

Annels NE, Callan MF, Tan L, Rickinson AB (2000) Changing patterns of dominant TCR usage with maturation of an EBV-specific cytotoxic T cell response. J Immunol 165:4831–4841

Bakacs T, Svedmyr E, Klein E (1978) EBV-related cytotoxicity of Fc receptor negative T lymphocytes separated from the blood of infectious mononucleosis patients. Cancer Lett 4:185–189

Bishop CJ, Moss DJ, Ryan JM, Burrows SR (1985) T lymphocytes in infectious mononucleosis. II. Response in vitro to interleukin-2 and establishment of T cell lines. Clin Exp Immunol 60:70–77
Burrows SR, Sculley TB, Misko IS, Schmidt C, Moss DJ (1990) An Epstein-Barr virus-specific cytotoxic T cell epitope in EBV nuclear antigen 3 (EBNA 3). J Exp Med 171:345–349
Callan MF, Steven N, Krausa P, Wilson JD, Moss PA, Gillespie GM, Bell JI, Rickinson AB, McMichael AJ (1996) Large clonal expansions of $CD8^+$ T cells in acute infectious mononucleosis. Nat Med 2:906–911
Callan MF, Fazou C, Yang H, Rostron T, Poon K, Hatton C, McMichael AJ (2000) $CD8^+$ T-cell selection, function, and death in the primary immune response in vivo. J Clin Investig 106:1251–1261
Carter RL (1975) Infectious mononucleosis: model for self-limiting lymphoproliferation. Lancet 1:846–849
Chan SH, Wallen WC, Levine PH, Periman P, Perlin E (1977) Lymphocyte responses to EBV-associated antigens in infectious mononucleosis, and Hodgkin's and non-Hodgkin's lymphoma patients, with the leukocyte adherence inhibition assay. Int J Cancer J 19:356–363
Crawford DH, Iliescu V, Edwards AJ, Beverley PC (1983) Characterisation of Epstein-Barr virus-specific memory T cells from the peripheral blood of seropositive individuals. Br J Cancer 47:681–686
Dambaugh T, Hennessy K, Chamnankit L, Kieff E (1984) U2 region of Epstein-Barr virus DNA may encode Epstein-Barr nuclear antigen 2. Proc Natl Acad Sci USA 81:7632–7636
de Schryver A, Friberg S Jr, Klein G, Henle W, Henle G, De-The G, Clifford P, Ho HC (1969) Epstein-Barr virus-associated antibody patterns in carcinoma of the post-nasal space. Clin Exp Immunol 5:443–459
de Schryver A, Klein G, Hewetson J, Rocchi G, Henle W, Henle G, Moss DJ, Pope JH (1974) Comparison of EBV neutralization tests based on abortive infection or transformation of lymphoid cells and their relation to membrane reactive antibodies (anti-MA). Int J Cancer 13:353–362
Delisi C, Berzofsky JA (1985) T-cell antigenic sites tend to be amphipathic structures. Proc Natl Acad Sci USA 82:7048–7052
de-The G, Geser A, Day NE, Tukei PM, Williams EH, Beri DP, Smith PG, Dean AG, Bronkamm GW, Feorino P et al (1978) Epidemiological evidence for causal relationship between Epstein-Barr virus and Burkitt's lymphoma from Ugandan prospective study. Nature 274:756–761
Edwards JM, McSwiggan DA (1974) Studies on the diagnostic value of an immunofluorescence test for EB virus-specific IgM. J Clin Pathol 27:647–651
Epstein MA, Morgan AJ (1986) Progress with subunits vaccines against the virus. In: Epstein MA, Achong BG (eds) The Epstein-Barr virus: recent advances. William Heinemann medical books, London, pp 271–289
Gerber P, Monroe JH (1968) Studies on leukocytes growing in continuous culture derived from normal human donors. J Natl Cancer Inst 40:855–866
Gottschalk S, Rooney CM, Heslop HE (2005) Post-transplant lymphoproliferative disorders. Annu Rev Med 56:29–44
Henderson E, Miller G, Robinson J, Heston L (1977) Efficiency of transformation of lymphocytes by Epstein-Barr virus. Virology 76:152–163
Henle G, Henle W (1966) Immunofluorescence in cells derived from Burkitt's lymphoma. J Bacteriol 91:1248–1256
Henle G, Henle W (1976) Serum IgA antibodies to Epstein-Barr virus (EBV)-related antigens a new feature of nasopharyngeal carcinoma. Bibl Haematol 43:322–325
Henle W, Henle G (1979) Seroepidemiology of the virus. In: Epstein MA, Achong BG (eds) The Epstein-Barr virus. Springer, New York, pp 61–78
Henle G, Henle W (1984) Epstein-Barr virus: past, present and future. In: Levine PH, Ablashi DV, Pearson GR, Kottaridis SD (eds) Epstein-Barr virus and associated diseases. Martinus Nijhoff Publishing, Boston/Dordrecht/Lancaster, pp 677–686

Henle W, Diehl V, Kohn G, Zur Hausen H, Henle G (1967) Herpes-type virus and chromosome marker in normal leukocytes after growth with irradiated Burkitt cells. Science 157:1064–1065

Henle W, Henle G, Ho HC, Burtin P, Cachin Y, Clifford P, de Schryver A, de-The G, Diehl V, Klein G (1970a) Antibodies to Epstein-Barr virus in nasopharyngeal carcinoma, other head and neck neoplasms, and control groups. J Nat Cancer Inst 44:225–231

Henle W, Henle G, Zajac BA, Pearson G, Waubke R, Scriba M (1970b) Differential reactivity of human serums with early antigens induced by Epstein-Barr virus. Science 169:188–190

Henle G, Henle W, Klein G (1971a) Demonstration of two distinct components in the early antigen complex of Epstein-Barr virus-infected cells. Int J Cancer 8:272–282

Henle W, Henle G, Niederman JC, Klemola E, Haltia K (1971b) Antibodies to early antigens induced by Epstein-Barr virus in infectious mononucleosis. J Infect Dis 124:58–67

Henle W, Henle GE, Horwitz CA (1974) Epstein-Barr virus specific diagnostic tests in infectious mononucleosis. Hum Pathol 5:551–565

Johnsen HE, Madsen M, Kristensen T (1979) Lymphocyte subpopulations in man: suppression of PWM-induced B-cell proliferation by infectious mononucleosis T cells. Scand J Immunol 10:251–255

Kaplan ME, Clark C (1974) An improved rosetting assay for detection of human T lymphocytes. J Immunol Methods 5:131–135

Klein G, Clifford P, Klein E, Stjernsward J (1966) Search for tumor-specific immune reactions in Burkitt lymphoma patients by the membrane immunofluorescence reaction. Proc Natl Acad Sci USA 55:1628–1635

Klein G, Clifford P, Klein E, Smith RT, Minowada J, Kourilsky FM, Burchenal JH (1967) Membrane immunofluorescence reactions of Burkitt lymphoma cells from biopsy specimens and tissue cultures. J Natl Cancer Inst 39:1027–1044

Klein G, Pearson G, Nadkarni JS, Nadkarni JJ, Klein E, Henle G, Henle W, Clifford P (1968) Relation between Epstein-Barr viral and cell membrane immunofluorescence of Burkitt tumor cells. I. Dependence of cell membrane immunofluorescence on presence of EB virus. J Exp Med 128:1011–1020

Klein G, Pearson G, Henle G, Henle W, Goldstein G, Clifford P (1969) Relation between Epstein-Barr viral and cell membrane immunofluorescence in Burkitt tumor cells. 3. Comparison of blocking of direct membrane immunofluorescence and anti-EBV reactivities of different sera. J Exp Med 129:697–705

Klein E, Becker S, Svedmyr E, Jondal M, Vanky F (1976) Tumor infiltrating lymphocytes. Ann N Y Acad Sci 276:207–216

Lai PK, Alpers MP, MacKay-Scollay EM (1977) Development of cell-mediated immunity to Epstein-Barr herpesvirus in infectious mononucleosis as shown by leukocyte migration inhibition. Infect Immun 17:28–35

Lipinski M, Fridman WH, Tursz T, Vincent C, Pious D, Fellous M (1979) Absence of allogeneic restriction in human T-cell-mediated cytotoxicity to Epstein-Barr virus-infected target cells. Demonstration of an HLA-linked control at the effector level. J Exp Med 150:1310–1322

Miller G, Niederman JC, Stitt DA (1972) Infectious mononucleosis: appearance of neutralizing antibody to Epstein-Barr virus measured by inhibition of formation of lymphoblastoid cell lines. J Infect Dis 125:403–406

Misko IS, Moss DJ, Pope JH (1980) HLA antigen-related restriction of T lymphocyte cytotoxicity to Epstein-Barr virus. Proc Natl Acad Sci USA 77:4247–4250

Moore GE, Grace JT Jr, Citron P, Gerner R, Burns A (1966) Leukocyte cultures of patients with leukemia and lymphomas. N Y State J Med 66:2757–2764

Moss DJ, Pope JH (1972) Assay of the infectivity of Epstein-Barr virus by transformation of human leucocytes in vitro. J General Virol 17:233–236

Moss DJ, Pope JH (1975) EB Virus-associated nuclear antigen production and cell proliferation in adult peripheral blood leukocytes inoculated with the QIMR-WIL strain of EB virus. Int J Cancer 15:503–511

Moss DJ, Scott W, Pope JH (1977) An immunological basis for inhibition of transformation of human lymphocytes by EB virus. Nature 268:735–736
Moss DJ, Rickinson AB, Pope JH (1978) Long-term T-cell-mediated immunity to Epstein-Barr virus in man. I. Complete regression of virus-induced transformation in cultures of seropositive donor leukocytes. Int J Cancer 22:662–668
Moss DJ, Rickinson AB, Pope JH (1979) Long-term T-cell-mediated immunity to Epstein-Barr virus in man. III. Activation of cytotoxic T cells in virus-infected leukocyte cultures. Int J Cancer 23:618–625
Moss DJ, Rickinson AB, Wallace LE, Epstein MA (1981a) Sequential appearance of Epstein-Barr virus nuclear and lymphocyte-detected membrane antigens in B cell transformation. Nature 291:664–666
Moss DJ, Wallace LE, Rickinson AB, Epstein MA (1981b) Cytotoxic T cell recognition of Epstein-Barr virus-infected B cells. I. Specificity and HLA restriction of effector cells reactivated in vitro. Eur J Immunol 11:686–693
Moss DJ, Burrows SR, Castelino DJ, Kane RG, Pope JH, Rickinson AB, Alpers MP, Heywood PF (1983a) A comparison of Epstein-Barr virus-specific T-cell immunity in malaria-endemic and -nonendemic regions of Papua New Guinea. Int J Cancer 31:727–732
Moss DJ, Chan SH, Burrows SR, Chew TS, Kane RG, Staples JA, Kunaratnam N (1983b) Epstein-Barr virus specific T-cell response in nasopharyngeal carcinoma patients. Int J Cancer 32:301–305
Moss DJ, Bishop CJ, Burrows SR, Ryan JM (1985) T lymphocytes in infectious mononucleosis. I. T cell death in vitro. Clin Exp Immunol 60:61–69
Moss DJ, Misko IS, Burrows SR, Burman K, McCarthy R, Sculley TB (1988) Cytotoxic T-cell clones discriminate between A- and B-type Epstein-Barr virus transformants. Nature 331:719–721
Niederman JC, McCollum RW, Henle G, Henle W (1968) Infectious mononucleosis. Clinical manifestations in relation to EB virus antibodies. JAMA, J Am Med Assoc 203:205–209
Nikiforow S, Bottomly K, Miller G (2001) CD4+ T-cell effectors inhibit Epstein-Barr virus-induced B-cell proliferation. J Virol 75:3740–3752
Nikoskelainen J, Hanninen P (1975) Antibody response to Epstein-Barr virus in infectious mononucleosis. Infect Immun 11:42–51
Nikoskelainen J, Leikola J, Klemola E (1974) IgM antibodies specific for Epstein-Barr virus in infectious mononucleosis without heterophil antibodies. Br Med J 4:72–75
Nilsson K, Klein G, Henle W, Henle G (1971) The establishment of lymphoblastoid lines from adult and fetal human lymphoid tissue and its dependence on EBV. Int J Cancer 8:443–450
North JR, Morgan AJ, Epstein MA (1980) Observations on the EB virus envelope and virus-determined membrane antigen (MA) polypeptides. Int J Cancer 26:231–240
North JR, Morgan AJ, Thompson JL, Epstein MA (1982) Purified Epstein-Barr virus Mr 340,000 glycoprotein induces potent virus-neutralizing antibodies when incorporated in liposomes. Proc Natl Acad Sci USA 79:7504–7508
Parker JR, Mowbray JF (1971) Peripheral blood leucocyte changes during human renal allograft rejection. Transplantation 11:201–209
Pattengale PK, Smith RW, Perlin E (1974) Atypical lymphocytes in acute infectious mononucleosis. Identification by multiple T and B lymphocyte markers. N Engl J Med 291:1145–1148
Pearson GR, Orr TW (1976) Antibody-dependent lymphocyte cytotoxicity against cells expressing Epstein-Barr virus antigens. J Natl Cancer Inst 56:485–488
Pearson GR, Qualtiere LF (1978) Papain solubilization of the Epstein-Barr virus-induced membrane antigen. J Virol 28:344–351
Pearson G, Dewey F, Klein G, Henle G, Henle W (1970) Relation between neutralization of Epstein-Barr virus and antibodies to cell-membrane antigens-induced by the virus. J Natl Cancer Inst 45:989–995
Pope JH, Horne MK, Scott W (1968) Transformation of foetal human leukocytes in vitro by filtrates of a human leukaemic cell line containing herpes-like virus. Int J Cancer 3:857–866

Pope JH, Horne MK, Wetters EJ (1969) Significance of a complement-fixing antigen associated with herpes-like virus and detected in the Raji cell line. Nature 222:186–187

Reedman BM, Klein G (1973) Cellular localization of an Epstein-Barr virus (EBV)-associated complement-fixing antigen in producer and non-producer lymphoblastoid cell lines. Int J Cancer 11:499–520

Reinherz EL, O'Brien C, Rosenthal P, Schlossman SF (1980) The cellular basis for viral-induced immunodeficiency: analysis by monoclonal antibodies. J Immunol 125:1269–1274

Rickinson AB, Moss DJ, Pope JH (1979) Long-term C-cell-mediated immunity to Epstein-Barr virus in man. II. Components necessary for regression in virus-infected leukocyte cultures. Int J Cancer 23:610–617

Rickinson AB, Moss DJ, Pope JH, Ahlberg N (1980) Long-term T-cell-mediated immunity to Epstein-Barr virus in man. IV. Development of T-cell memory in convalescent infectious mononucleosis patients. Int J Cancer 25:59–65

Rickinson AB, Moss DJ, Allen DJ, Wallace LE, Rowe M, Epstein MA (1981a) Reactivation of Epstein-Barr virus-specific cytotoxic T cells by in vitro stimulation with the autologous lymphoblastoid cell line. Int J Cancer 27:593–601

Rickinson AB, Moss DJ, Wallace LE, Rowe M, Misko IS, Epstein MA, Pope JH (1981b) Long-term T-cell-mediated immunity to Epstein-Barr virus. Cancer Res 41:4216–4221

Rickinson AB, Rowe M, Hart IJ, Yao QY, Henderson LE, Rabin H, Epstein MA (1984) T-cell-mediated regression of "spontaneous" and of Epstein-Barr virus-induced B-cell transformation in vitro: studies with cyclosporin A. Cell Immunol 87:646–658

Rocchi G, Hewetson JF (1973) A practical and quantitative microtest for determination of neutralizing antibodies against Epstein-Barr virus. J General Virol 18:385–391

Rooney CM, Rickinson AB, Moss DJ, Lenoir GM, Epstein MA (1985a). Cell-mediated immunosurveillance mechanisms and the pathogenesis of Burkitt's lymphoma. IARC Scientific Publications, pp 249–264

Rooney CM, Rowe M, Wallace LE, Rickinson AB (1985b) Epstein-Barr virus-positive Burkitt's lymphoma cells not recognized by virus-specific T-cell surveillance. Nature 317:629–631

Rooney CM, Smith CA, Ng CY, Loftin SK, Sixbey JW, Gan Y, Srivastava DK, Bowman LC, Krance RA, Brenner MK et al (1998) Infusion of cytotoxic T cells for the prevention and treatment of Epstein-Barr virus-induced lymphoma in allogeneic transplant recipients. Blood 92:1549–1555

Roos MT, van Lier RA, Hamann D, Knol GJ, Verhoofstad I, van Baarle D, Miedema F, Schellekens PT (2000) Changes in the composition of circulating CD8+ T cell subsets during acute Epstein-Barr and human immunodeficiency virus infections in humans. J Infect Dis 182:451–458

Rothbard JB, Taylor WR (1988) A sequence pattern common to T cell epitopes. The EMBO journal 7:93–100

Rowe M, Rooney CM, Rickinson AB, Lenoir GM, Rupani H, Moss DJ, Stein H, Epstein MA (1985) Distinctions between endemic and sporadic forms of Epstein-Barr virus-positive Burkitt's lymphoma. Int J Cancer 35:435–441

Royston I, Sullivan JL, Periman PO, Perlin E (1975) Cell-mediated immunity to Epstein-Barr-virus-transformed lymphoblastoid cells in acute infectious mononucleosis. N Engl J Med 293:1159–1163

Schmitz H, Scherer M (1972) IgM antibodies to Epstein-Barr virus in infectious mononucleosis. Archiv fur die gesamte Virusforschung 37:332–339

Seeley J, Svedmyr E, Weiland O, Klein G, Moller E, Eriksson E, Andersson K, van der Waal L (1981) Epstein Barr virus selective T cells in infectious mononucleosis are not restricted to HLA-A and B antigens. J Immunol 127:293–300

Sheldon PJ, Hemsted EH, Papamichail M, Holborow EJ (1973) Thymic origin of atypical lymphoid cells in infectious mononucleosis. Lancet 1:1153–1155

Shope TC, Kaplan J (1979) Inhibition of the in vitro outgrowth of Epstein-Barr virus-infected lymphocytes by TG lymphocytes. Journal of Immunology 123:2150–2155

Svedmyr E, Jondal M (1975) Cytotoxic effector cells specific for B Cell lines transformed by Epstein-Barr virus are present in patients with infectious mononucleosis. Proc Natl Acad Sci USA 72:1622–1626

Svedmyr E, Klein G, Weiland O (1979) The EBV-carrying, beta2 M/HLA deficient Burkitt lymphoma line Daudi is sensitive to EBV-specific killer T-cells of mononucleosis patients. Cancer Lett 7:15–20

Thorley-Lawson DA (1980) The suppression of Epstein-Barr virus infection in vitro occurs after infection but before transformation of the cell. J Immunol 124:745–751

Thorley-Lawson DA, Geilinger K (1980) Monoclonal antibodies against the major glycoprotein (gp350/220) of Epstein-Barr virus neutralize infectivity. Proc Natl Acad Sci USA 77:5307–5311

Thorley-Lawson DA, Chess L, Strominger JL (1977) Suppression of in vitro Epstein-Barr virus infection. A new role for adult human T lymphocytes. J Exp Med 146:495–508

Tosato G, Magrath I, Koski I, Dooley N, Blaese M (1979) Activation of suppressor T cells during Epstein-Barr-virus-induced infectious mononucleosis. N Engl J Med 301:1133–1137

Townsend AR, Gotch FM, Davey J (1985) Cytotoxic T cells recognize fragments of the influenza nucleoprotein. Cell 42:457–467

Tsoukas CD, Carson DA, Fong S, Vaughan JH (1982) Molecular interactions in human T cell-mediated cytotoxicity to EBV II. Monoclonal antibody OKT3 inhibits a post-killer-target recognition/adhesion step. J Immunol 129:1421–1425

Turk W (1898) Klinische Untersuchungen uberdas verhalten des Blutes bei akuten Infektions Krankheiten. Braumuiller, Vienna

Tursz T, Fridman WH, Senik A, Tsapis A, Fellous M (1977) Human virus-infected target cells lacking HLA antigens resist specific T-lymphocyte cytolysis. Nature 269:806–808

Wallace LE, Moss DJ, Rickinson AB, McMichael AJ, Epstein MA (1981) Cytotoxic T cell recognition of Epstein-Barr virus-infected B cells. II. Blocking studies with monoclonal antibodies to HLA determinants. Eur J Immunol 11:694–699

Wara WM, Wara DW, Phillips TL, Ammann AJ (1975) Elevated IGA in carcinoma of the nasopharynx. Cancer 35:1313–1315

Zeng Y, Liu YX, Wei JN, Zhu JS, Cai SL, Wang PZ, Zhong JM, Li RC, Pan WJ, Li EJ et al (1979) Serological mass survey of nasopharyngeal carcinoma (author's transl). Zhongguo yi xue ke xue yuan xue bao Acta Academiae Medicinae Sinicae 1:123–126

Zeng Y, Zhang LG, Li HY, Jan MG, Zhang Q, Wu YC, Wang YS, Su GR (1982) Serological mass survey for early detection of nasopharyngeal carcinoma in Wuzhou City, China. Int J Cancer 29:139–141

Zeng Y, Zhong JM, Li LY, Wang PZ, Tang H, Ma YR, Zhu JS, Pan WJ, Liu YX, Wei ZN et al (1983) Follow-up studies on Epstein-Barr virus IgA/VCA antibody-positive persons in Zangwu County, China. Intervirology 20:190–194

Zinkernagel RM, Doherty PC (1979) MHC-restricted cytotoxic T cells: studies on the biological role of polymorphic major transplantation antigens determining T-cell restriction-specificity, function, and responsiveness. Adv Immunol 27:51–177

zur Hausen H, Henle W, Hummeler K, Diehl V, Henle G (1967) Comparative study of cultured Burkitt tumor cells by immunofluorescence, autoradiography, and electron microscopy. J Virol 1:830–837

Part II
Virus Genetics and Epigenetics

Epstein–Barr Virus Strain Variation

Paul J. Farrell

Abstract What is wild-type Epstein–Barr virus and are there genetic differences in EBV strains that contribute to some of the EBV-associated diseases? Recent progress in DNA sequencing has resulted in many new Epstein–Barr virus (EBV) genome sequences becoming available. EBV isolates worldwide can be grouped into type 1 and type 2, a classification based on the EBNA2 gene sequence. Type 1 transforms human B cells into lymphoblastoid cell lines much more efficiently than type 2 EBV and molecular mechanisms that may account for this difference in cell transformation are now becoming understood. Study of geographic variation of EBV strains independent of the type 1/type 2 classification and systematic investigation of the relationship between viral strains, infection and disease are now becoming possible. So we should consider more directly whether viral sequence variation might play a role in the incidence of some EBV-associated diseases.

Contents

1 Introduction 46
2 History of EBV Sequencing 47
3 Broad Aspects of Genome Variation—EBV Types, Selection Forces and Recombination ... 49
4 Variation in EBV Genes 51
4.1 LMP1 51
4.2 LMP2A 53
4.3 EBNAL 53
4.4 EBNA3 Family 54
4.5 Other Latent Cycle Elements 54
4.6 Lytic Cycle 55

P.J. Farrell (✉)
Section of Virology, Department of Medicine, Imperial College, St Mary's Campus, Norfolk Place, London W2 1PG, UK
e-mail: p.farrell@imperial.ac.uk

C. Münz (ed.), *Epstein Barr Virus Volume 1*, Current Topics in Microbiology and Immunology 390, DOI 10.1007/978-3-319-22822-8_4

5 Functional Difference Between Type 1 and Type 2 EBNA2 ... 56
6 How to Identify Variation Potentially Relevant to EBV-Associated Diseases ... 58
References ... 59

1 Introduction

EBV infection is prevalent all over the world, but some EBV-associated diseases have unusual geographic distributions, for example nasopharyngeal carcinoma (NPC) or endemic Burkitt's lymphoma (BL) (Rickinson 2014). Genetic variation of the host, local environmental co-factors and co-infections all play a role in this, but natural variation in the EBV may also be important. There have been many studies which show substantial geographic variation in the virus sequence in normal infected populations, so the endemic strain of EBV in some parts of the world might be inherently more able to contribute to cancers. Heterogeneity of the endemic virus circulating in the general population might also result in multiple strains of EBV being present in one person and selection of a specific variant in a cancer, as proposed for an LMP1 variant in a case of NPC (Edwards et al. 2004).

It is additionally possible that specific mutations in EBV might arise in an infected person. The normal life cycle of EBV is thought to involve infected cells passing through a germinal centre in a lymph node, which is a highly mutagenic environment, and the virus replication itself might also accumulate errors (although herpesviruses are generally thought to have very stable genomes). Examples of infrequent mutations that appear to be relevant to disease could be the deletions in some BL EBV that allow BHRF1 expression from Wp to prevent apoptosis (Kelly et al. 2009) and the mutation of EBNA-3B causing diffuse large B cell lymphoma (DLBCL) in mice, with some EBNA-3B mutants also found in human DLBCL (White et al. 2012).

All these speculative ideas highlight the importance of knowing more about natural variation of EBV, particularly identifying what constitutes wild-type EBV so that mutation or variation from this can be recognised. Most EBV strains that have been characterised come from cancer cell lines or have been selected by B cell transformation to make a lymphoblastoid cell line (LCL). The recent description of M81 EBV from a Hong Kong nasopharyngeal carcinoma (NPC) patient emphasises this point; M81 BAC (bacterial artificial chromosome) EBV is more spontaneously lytic and infects epithelial cells better than the B95-8 reference strain (Tsai et al. 2013); its properties were interpreted as being relevant to development of NPC.

The renewed interest in an EBV vaccine (Cohen et al. 2011) also emphasises the importance of ensuring that this will be directed to the wild-type EBV sequence. In general, we need to know what is wild-type EBV. This chapter integrates and updates some previously published reviews on EBV sequence variation (Jenkins and Farrell 1996; Tzellos and Farrell 2012).

2 History of EBV Sequencing

Because of the relatively large genome for a virus (175 kb) and the presence of several tandem repeat arrays (which are difficult to sequence), sequencing of EBV genomes was initially quite limited. The sequences of some small fragments of B95-8 EBV were published in 1982 (Cheung and Kieff 1982; Dambaugh and Kieff 1982), but the first complete EBV sequence (accession number V01555) of the B95-8 strain was published in 1984 (Baer et al. 1984). At that time, it was the largest DNA sequence that had been determined.

The B95-8 strain was sequenced because B95-8 was the only EBV cell line available that produced virus at a sufficient level to make it practical to clone the viral restriction fragments. B95-8 cells were originally derived (Miller et al. 1972) using EBV from 883L (a spontaneous human LCL from an infectious mononucleosis patient) to transform lymphocytes from the cotton top marmoset (*Saguinus oedipus*). EBV is secreted into the medium spontaneously, giving useful amounts of infectious, transforming virus but B95-8 cells can also be induced into the lytic cycle with phorbol myristate acetate (PMA, also called TPA). To obtain the DNA for sequencing, B95-8 EBV was produced from PMA-treated B95-8 cells and the EBV DNA was cloned as restriction fragments, Eco RI fragments in a cosmid vector pHC79 or Bam HI fragments in pBR322 (Arrand et al. 1981). Sequencing was by the Sanger method (Baer et al. 1984).

It was already known from restriction site mapping that there is a 13.6-kb deletion in B95-8 relative to other known EBV strains (Raab-Traub et al. 1980) and this sequence was subsequently determined from cloned restriction fragments of EBV from the Raji BL cell line (Parker et al. 1990). The error rate in the original sequencing proved to be very low (about 1/50,000), so predictions made of the open reading frames (Baer et al. 1984) turned out to be an accurate guide to the genetic content and these have been the basis of much of the subsequent investigation of EBV. Three single nucleotide errors discovered in the genome sequence were corrected (de Jesus et al. 2003), but the sequence has not been adjusted for loss of a small part of the repeat array in oriP (Fruscalzo et al. 2001; Kanda et al. 2011) selected by cloning in plasmids. To facilitate studies on the whole viral genome, a "wild-type" EBV sequence (EBVwt, AJ507799) was assembled from the corrected B95-8 and Raji sequence, the number of major internal repeat units in the sequence was reduced to 7.5 to be more typical and annotation was updated (de Jesus et al. 2003).

A more standard annotation of wild-type EBV including three additional small open reading frames that could now be recognised from sequence comparison was released by AJ Davison and PJ Farrell in 2010 as the RefSeq HHV4 (EBV) sequence NC_007605. This is the current standard reference sequence.

Key insights into EBV sequence variation have come from publication of further complete EBV sequences from Africa, China, Japan and the USA. The EBV sequence with accession number DQ279927 from the AG876 African BL cell line (Dolan et al. 2006) is a type 2 EBV strain (see below). EBV from the African

Mutu BL cell line (KC207814) and the Japanese Akata BL cell line (KC207813) were reported in 2013 (Lin et al. 2013), the same year as two isolates K4413-Mi (KC440851) and K4123-Mi (KC440852) from spontaneous LCLs are produced in the USA (Lei et al. 2013). The African BL Raji EBV has also been fully sequenced (KF717093).

The Chinese GD1 isolate (AY961628) was from a LCL (LCL) made by immortalising cord B cells with EBV from saliva of an NPC patient in Guangzhou (Zeng et al. 2005). The GD2 sequence (HQ020558) is a direct determination of EBV sequence from a Guangzhou NPC biopsy (Liu et al. 2011). Further EBV sequences from Hong Kong NPCs, HKNPC1 from a biopsy (JQ009376) (Kwok et al. 2012) and an additional series of 8 Hong Kong NPCs from the same group (Kwok et al. 2014) have been reported. EBV from two cell lines derived from Hong Kong has also been sequenced; M81 (KF373730.1) is from a marmoset LCL prepared using EBV from an NPC patient (Tsai et al. 2013; Desgranges et al. 1976) and C666.1 (KC617875.1) is the only epithelial cell line derived from an NPC that is known to retain its EBV in cell culture (Tso et al. 2013).

The methods used to determine these EBV sequences have advanced with technological development. The AG876 sequence was determined using Sanger sequencing of cosmid-cloned fragments, supplemented by PCR amplification of selected regions. GD1 and HKNPC1 were amplified as a set of PCR fragments, which were then sequenced. Many of the more recent sequences were obtained by selecting the EBV reads from "next-generation" Illumina sequencing of the whole cell and viral DNA.

The small amount of starting material required for determination of a complete EBV sequence by the Illumina methods makes it realistic to analyse large numbers of primary human EBV strains, but the cost of this without first enriching the EBV DNA is still quite high. Although next-generation sequencing has greatly increased the production of sequences, the number of tandem repeat arrays in the EBV genome and relatively high GC content still make sequencing of EBV relatively difficult.

An interesting approach to concentrating sequencing effort on the EBV content of human samples involves first enriching the viral sequences on custom SureSelect beads (Depledge et al. 2011). This can greatly reduce the cost of the sequencing if many samples are to be tested. This approach has been used by a group at the Sanger Institute to sequence 71 novel EBV strains (Palser et al. 2015) and was also used on some of the Hong Kong samples (Kwok et al. 2014). Combining that data with the sequences already published give sufficient information to draw some general conclusions about natural EBV variation, even though many parts of the word are still not represented in the data. It is already clear that the genetic standard map of the EBV genome annotated in the RefSeq genome NC_007605 is generally consistent with all the known strains, although a few strains have deletions in various parts of the viral genome. The open reading frames are mostly conserved, providing further evidence for their validity.

3 Broad Aspects of Genome Variation—EBV Types, Selection Forces and Recombination

The first major variation to be identified in EBV was the type 1 or type 2 classification based on differences in EBNA2 (Dambaugh et al. 1984; Adldinger et al. 1985; Rickinson and Kieff 2007). The types were also known as type A and B, respectively. Most EBV genes differ by less than 5 % in sequence in different isolates, but EBNA2 clearly sorts into type 1 or type 2, with only 70 % identity at the nucleotide level and 54 % identity in the protein sequence. There is linked variation in EBNA3 genes (Rowe et al. 1989), but the degree of sequence difference is less. Type 1 is the main EBV prevalent worldwide, but in sub-Saharan Africa, type 2 EBV is equally abundant and there are many mixed infections. In Argentina, type 1 was present in 76 % of healthy carriers, type 2 in 15 %, and co-infections with both types in 7 % (Correa et al. 2004); in Mexico, 33 % had type 1, 57 % had type 2 and 10 % had mixed infection (Palma et al. 2013), but in Australia, 98 % of infections were type 1 (Lay et al. 2012).

Sequencing of type 2 EBV from the AG876 cell line enabled a comparison between type 1 and type 2 EBV genomes (Dolan et al. 2006). This revealed that the two types are co-linear and very similar, with the exception of the known divergent alleles. The type variation has a clear phenotypic consequence in cell culture; type 2 EBV is much less effective at establishing LCLs than type 1 EBV (Rickinson et al. 1987). There has been considerable progress in understanding the mechanism of this, described in detail separately below. There is strong but not complete genetic linkage of EBNA2 and EBNA3 genes in relation to the types. The significance of these types has recently been put on a much firmer basis by analysis of the 71 novel strains and the previously published 12 strains (Palser et al. 2015).

Recent comparison of the genome sequences of 20 HSV isolates demonstrated the extensive historical recombination that has occurred (Szpara et al. 2014). Preliminary results from the Sanger Institute study of 71 sequences also demonstrate this in EBV (Palser et al. 2015). At this point, it is not possible to identify preferred sites of recombination or haplotype blocks of sequence in the whole group of EBV sequences that are known. Considering smaller sets of sequences from single geographic areas, however, makes it possible to recognise regions of similarity to other strains. This extensive recombination means that it is not appropriate to use phylogenetic trees to compare the whole genomes directly. Principal component analysis can be used on the whole genomes and this has shown that the type 1/type 2 classification is quantitatively the greatest variation in EBV strains worldwide (Palser et al. 2015).

The extent to which EBV can be meaningfully classified into types based on individual gene markers depends partly on the extent of inter-typic recombination that may occur. There is clear evidence for inter-typic recombination based on the polymorphisms distributed along the genome (Midgley et al. 2000; Yao et al. 1996; Kim et al. 2006). A more direct study of 83 EBV genome sequences (Palser et al. 2015) showed that 69 had EBNA2 and EBNA3s both type 1, 12 EBVs had both EBNA2 and EBNA3s type 2, and 2 EBVs had type 2 EBNA3s but type 1 EBNA2. This describes the genetic linkage between EBNA2 and EBNA3s. A comparison of the B95-8, GD1 and AG876 sequences led to a proposal for a minimum theoretical number of recombination events that would be required to result in the current genome arrangements of those viruses (McGeoch and Gatherer 2007), but the extent of historical recombination accumulated in the panel of 83 EBV genomes studied (Palser et al. 2015) is very complex. It will be simpler to study recombination in a less diverse panel of strains, perhaps from a single geographic area.

Points of variation have been studied within type 1 EBNA2 (Schuster et al. 1996), but type 1 and type 2 EBNA2 sequences are sufficiently different to preclude their recombination within EBNA2. So the type 1 and type 2 EBNA2 characteristics will tend to survive irrespective of recombination that may occur elsewhere in the genome. Interpreting the significance of the persistence of these types will therefore depend on understanding the mechanism and phenotypic consequences of the type differences in vivo. The recent discovery that APOBEC3 cytidine deaminase RNA editing can modify EBV genomes in cell culture demonstrates another possible mechanism for generating virus heterogeneity in vivo, in addition to infection with multiple strains and virus recombination (Suspene et al. 2011).

Factors that affect in vivo selection of viral recombinants and variants might be expected to include immune surveillance and the ability to infect and persist through the complex life cycle of EBV. Immune surveillance would be expected to correlate with MHC type since functional epitopes will vary according to the presentation on MHC. Since predominant MHC types differ between racial groups and geographically, this could be a major factor in world wide variation of EBV. Many epitopes for CTL surveillance have been mapped in EBV antigens and correlated with MHC type. There is some clear evidence for epitope selection based on immune surveillance (Burrows et al. 2004; Midgley et al. 2003a, b; Nagamine et al. 2007; Tang et al. 2008; Lin et al. 2004, 2005; Duraiswamy et al. 2003). Evidence for positive selection also comes from ratios of non-synonymous to synonymous changes in the SNPs present in reading frames (Palser et al. 2015). This showed an excess of non-synonymous changes in EBNA3 genes and LMP1, as predicted from the immunological studies, but also in the immune evasion protein BNLF2a (Horst et al. 2012), which binds TAP and in the glycoproteins gp350 (BLLF1) and gL (BKRF2). So far little is known about the extent to which EBV selection in vivo may be affected by other polymorphisms that could affect, for example, ability to promote cell proliferation or viral replication. Detailed study of the non-synonymous variation in EBNA3 genes (Palser et al. 2015) surprisingly revealed that most of the amino acids contributing to the positive selection were not in known T cell epitopes. This result and the apparent clustering of genes showing positive selection in the genome suggest that there may be additional mechanisms determining the positive selection, which are not understood at present.

There are several examples of deletions of part of the EBV genome in endemic BL cell lines. The best characterised of these are the approximately similar deletions in P3HR1, Daudi, Sav, Oku and Ava BL cells (Kelly et al. 2002) that remove EBNA2, most of BHLF1 and the C-terminal part of EBNA-LP. The significance of these was interpreted as a mechanism for avoiding EBNA2 antagonism of c-MYC function (Kelly et al. 2002) in the tumour cells. Further investigation indicated that enhanced expression of BHRF1 (the anti-apoptotic, viral BCL2 homologue) as a result of the deletion could also play a role in some BL cells harbouring EBV with this type of deletion (Kelly et al. 2009).

Some other well-characterised EBV deletions are in the Raji BL cell line. Two separate deletions result in a loss of EBNA3C and some genes essential for lytic DNA replication (Hatfull et al. 1988). Complementation of the defective lytic cycle genes by expression of BALF2 was sufficient to restore lytic DNA replication in Raji cells (Decaussin et al. 1995). As mentioned above, the B95-8 strain of EBV also has a deletion relative to most EBV isolates (Raab-Traub et al. 1980; Parker et al. 1990); the B95-8 deletion removes many of the BART miRNA sequences and one of the lytic origins of replication, but this does not seem to adversely affect lytic replication or immortalisation of B cells.

Rearranged, defective EBV genomes (known as het DNA) have been characterised in detail in the P3HR1 BL cell line (Jenson et al. 1986). The rearrangements cause constitutive expression of BZLF1 (Rooney et al. 1988) and the resulting lytic cycle activation allows persistence of the defective genomes in the cell population. There have been reports of similar rearranged EBV genomes in vivo (Patton et al. 1990; Gan et al. 2002), but a recent study concluded that they are only present quite rarely in vivo (Ryan et al. 2009).

4 Variation in EBV Genes

4.1 LMP1

The locus of variation that has attracted most investigation in relation to disease is LMP1, which contains a higher degree of polymorphism than most EBV genes (Palser et al. 2015). LMP1 is a membrane protein which makes many interactions through its C-terminal region that mediate important signal transduction. These interactions regulate NF-kB and cell survival in several ways, so it is easy to envisage how sequence variation could alter those processes. A key step forward in understanding the many points of polymorphism in LMP1 came with classification of LMP1 variants into 7 main groups (Mainou and Raab-Traub 2006) and development of a rapid heteroduplex assay that allows classification of LMP1 type in a large number of samples. The sequence variants of LMP1 relative to B95-8 were named Alaskan, China 1, China 2, Med+, Med−, and NC. This more general insight into LMP1 variation taking into account the whole LMP1 sequence (Miller et al. 1994; Walling et al. 2004; Mainou and Raab-Traub 2006) has also been used by many other groups (Sung et al. 1998; Kanai et al. 2007; Nagamine et al. 2007; Lin et al. 2005; Zhao et al. 2005; Shibata et al. 2006; Saechan et al. 2006, 2010; Dardari et al. 2006; Li et al. 2009; Nguyen-Van et al. 2008; Diduk et al. 2008; Pavlish et al. 2008; Wang et al. 2007), but there was generally little evidence for a specific disease association of variants. Investigation of variant LMP1 sequences in HIV patients in Switzerland also identified polymorphisms (I124 V/I152L and F144I/D150A/L151I), which were markers of increased NF-κB activation in vitro but were not associated with EBV-associated Hodgkin's lymphoma (Zuercher et al. 2012).

It is still not clear how many EBV variants are present within one individual or even whether being infected offers immune protection against acquiring additional EBV strains. Studies using a heteroduplex assay confirmed that individuals can be infected with multiple variants (Edwards et al. 2004; Tierney et al. 2006). Multiple LMP1 variants can be found in people with infectious mononucleosis (Fafi-Kremer et al. 2005), Hodgkin's lymphoma or NPC (Rey et al. 2008), and there is also evidence from people who are immunosuppressed, for example AIDS patients, for infection with multiple EBV strains (Palefsky et al. 2002; Walling et al. 2003). Based on LMP1 analysis, variants differ in abundance between throat wash samples and peripheral blood samples in a variety of conditions (Sitki-Green et al. 2002, 2003, 2004; Tierney et al. 2006), although a recent study reported the polymorphisms in EBV from throat washings and tumour in Chinese NPC patients to be the same (Nie et al. 2013). Evidence for a specific variant of LMP1 being involved in a cancer could be provided by finding selective presence of that allele in cancer cells relative to the virus in the saliva or peripheral circulation. This is what was found in an analysis of an NPC patient (Edwards et al. 2004), but the interpretation made was in the context of evasion of immune surveillance of the LMP1 in the MHC background of the patient rather than specifically enhanced transforming activity of the LMP1.

Interest in LMP1 variation and function was stimulated by reports that a variant with a 30-bp deletion (Cao LMP1) isolated from an NPC tumour had a greater transforming activity than the reference LMP1 (Hu et al. 1991, 1993). There are many points of sequence difference between Cao LMP1 and the reference B95-8 protein, but attention was focussed on the 30-bp deletion in Cao LMP1. The 30-bp deletion (amino acids 346–355) includes part of C-terminal activating region 2 (CTAR2, amino acids 351–386) of LMP1. CTAR2 (Huen et al. 1995) mediates signalling to NF-kB and AP-1, but the 30-bp deletion does not alter the parts of LMP1 that activate NF-kB (Farrell 1998). An analysis of 249 patients in Taiwan showed that patients with the Cao CTAR2 variant had an increased risk of distant metastasis compared with the non-Cao variant (Pai et al. 2007). The Cao CTAR2 was also a negative predictor for overall survival and post-metastasis disease-specific survival in that series. Another study indicated enhanced ability in tumour-derived variants to activate ERK kinase and induce c-Fos (Vaysberg et al. 2008), although that was accounted for by G212S or S366T rather than the 30-bp deletion.

Since references given in a previous review (Jenkins and Farrell 1996), many groups have investigated the presence of the 30-bp deletion LMP1 in normal carriers (Correa et al. 2004) or a variety of EBV-associated diseases (Cheung et al. 1998; Hayashi et al. 1998; Zhang et al. 2002; Tai et al. 2004; Jen et al. 2005; Chang et al. 2006; Zhao et al. 2005; Correa et al. 2007; See et al. 2008; Lorenzetti et al. 2012; Banko et al. 2012; Giron et al. 2013). A rarer 69-bp deletion in the C-terminus has also been studied (Hadhri-Guiga et al. 2006); it was reported to have a reduced ability to activate the cell AP1 transcription factor (Larcher et al. 2003). In general, it is clear that these LMP1 variants are widely distributed with somewhat different frequencies in different parts of the world, but in most studies, there was no evidence for a specific association with disease (Senyuta et al. 2014; Gantuz et al. 2013). Other reports of LMP1 variation associated with specific populations and disease contexts, for example gastric and oral carcinoma (Chen et al. 2010, 2011; BenAyed-Guerfali et al. 2011; Higa et al. 2002), have mostly lacked sufficient numbers or control samples to interpret in the context of the relationship between variation and disease, but some studies have suggested an association with disease (Corvalan et al. 2006).

4.2 LMP2A

Low levels of LMP2A enhance cell survival, but high levels of LMP2A can also interfere with signalling from the B cell receptor by binding of lyn and fyn tyrosine kinases to the hydrophilic N-terminal part of LMP2A. Sequence polymorphism has been described in the N-terminal region but it does not affect the key phosphorylated residues (Busson et al. 1995). Although sequence polymorphism has been detected in isolates from various parts of the world and various diseases (Berger et al. 1999; Tanaka et al. 1999; Wang et al. 2010a; Han et al. 2012), there is no evidence for a disease association at present. Sequence variation is present, for example in South-east Asia and New Guineau (Lee et al. 1993), in mapped epitopes for class I restricted cytotoxic T cells, but this is not sufficient to prevent the possibility of LMP2A being a target for immunotherapy of EBV-associated cancers (Rickinson and Moss 1997; Khanna et al. 1999; Taylor et al. 2014).

4.3 EBNAL

EBNA1 proteins frequently differ in size due to variation in the length of the Gly–Ala repeat, but further differences in the unique parts of EBNA1 have been used to define P (prototype, B95-8) and V (variant) EBNA1, which differ at about 15 amino acids (Bhatia et al. 1996). These each have two subtypes defined by the amino acid at position 487 (P-ala, P-thr, V-pro and V-leu). In the initial report (Bhatia et al. 1996), P-thr was most frequent in peripheral blood lymphocytes of

African and American samples and in African tumours, but most American EBV-associated lymphomas had V-leu EBNA1. A subsequent report confirmed the subtypes but found no association with lymphoma (Habeshaw et al. 1999). The variation might affect immune recognition of EBNA1 (Wrightham et al. 1995; Chen et al. 1999; Bell et al. 2008) and has been noted in Chinese NPC samples (Ai et al. 2012; Snudden et al. 1995; Zhang et al. 2004; Wang et al. 2010a) and gastric cancer (Chen et al. 2012), but there is no substantial evidence of disease association. It has been reported that the V-val subtype of EBNA1 has increased ability to activate the enhancer functions of oriP in transfection assays (Do et al. 2008; Mai et al. 2010).

4.4 EBNA3 Family

Some of the polymorphism in the EBNA3 genes is linked to EBNA2-type variation (Rowe et al. 1989; Palser et al. 2015), but additional subvariation in EBNA3A and 3C has been noted with no relationship to disease (Gorzer et al. 2006; Wu et al. 2012).

Since EBNA-3B is not required for immortalisation by EBV, variation in EBNA-3B was originally considered mainly in the context of immune surveillance. For example (Chu et al. 1999), polymorphisms of EBNA-3B (called EBNA4 in that publication) were found to be frequent in EBV-associated Hodgkin's lymphoma, gastric carcinoma and AIDS-lymphoma but not related to patients' HLA-A11 status. Sequence variation in EBNA-3B is now being re-examined since it has been realised that EBNA-3B acts as a tumour suppressor gene (White et al. 2012). Loss of function mutants of EBNA-3B have much higher B cell transforming activity than wild type and a propensity to cause DLBCL in a mouse model (White et al. 2012). It will therefore be important to determine which of the many polymorphisms in EBNA-3B affect its function so as to increase the viral transforming activity.

4.5 Other Latent Cycle Elements

The EBER RNAs are strongly expressed in all types of EBV-associated cancer. The extent of sequence variation in EBERs is quite small and has been linked to the type 1/type 2 EBNA2 status (Arrand et al. 1989; Schuster et al. 1996). A recent report identified some new variants in Chinese samples (Wang et al. 2010b). At present, there is little information on sequence variation in the BART or BHRF1 miRNAs, although attention has been drawn to a G155849A polymorphism near the RPMS1 open reading frame in the BART region of EBV from EBV associated with NPC in China (Li et al. 2005). Variation in promoter activity affecting expression in reporter assays has also been reported for the Cp and Qp promoters in EBV isolates from Chinese NPC samples (Wang et al. 2012a), but there is no evidence for a role in disease. In contrast, it is clear that the number of copies of the major internal repeat (IR1), which contains the Wp promoter, does affect the ability of EBV to express EBNA2 and EBNA-LP efficiently and has a major effect on transformation efficiency (Tierney et al. 2011).

4.6 Lytic Cycle

BZLF1 is the transcription factor that initiates the lytic cycle reactivation in B cells and its promoter is tightly regulated since it mediates the switch from latency to the lytic cycle. There has been considerable interest in BZLF1 variants that might affect its function or expression (Lorenzetti et al. 2012; Jin et al. 2010; Imajoh et al. 2012; Yang et al. 2014). Although there are suggestions of variant sequence in Zp correlating with disease (Martini et al. 2007; Lorenzetti et al. 2014), further functional analysis would be required to substantiate this.

Natural variation in the BZLF1 protein sequence has been described in Chinese isolates but showed no relationship to disease (Ji et al. 2008; Luo et al. 2011). Variation in the BZLF1 dimerisation sequence was studied using synthetic peptides but found to have only a small effect on its activity (Hicks et al. 2001). Polymorphism has also been noted in BRLF1 (Jia et al. 2010; Yang et al. 2014), the EBV transcription factor which cooperates with BZLF1 to activate the early lytic cycle genes. Although a possible disease association was noted, this would require substantiation with more samples and controls.

Variation has also been reported in the early lytic cycle genes BHRF1 (Jing et al. 2010) and BNLF2a (Horst et al. 2012). BHRF1 is similar to BCL2 with anti-apoptosis activity and BNLF2a has a role in immune evasion, reducing cell surface HLA class I levels. Protein function was not affected by these sequence polymorphisms. In an earlier study, sequence variants of the BHRF1 protein were identified (Khanim et al. 1997), but no effect of the variation was found in the ability to protect against apoptosis induced by cisplatin.

BARF1 is expressed in the early lytic cycle in B cell reactivation but is also expressed in NPC cells. It is secreted and binds to colony stimulating factor 1 (CSF-1), inhibiting the binding of CSF-1 to its receptor (Strockbine et al. 1998). Sequence polymorphism has been identified in BARF1 in Indonesian and Chinese samples (Hutajulu et al. 2010; Wang et al. 2012c) but most likely reflects natural selection of EBV strains unconnected to carcinogenesis.

The late lytic cycle gene gp350 encodes a surface glycoprotein on the EBV particle which is the target of neutralising antibodies for EBV infection. Sequence variation of this protein has been identified but again appears to show geographic restriction rather than tumour specific polymorphism (Kawaguchi et al. 2009; Luo et al. 2012).

5 Functional Difference Between Type 1 and Type 2 EBNA2

At present, there is little evidence for a disease relationship to the EBV types. One study found that type 1 EBV was more likely to cause infectious mononucleosis than type 2 EBV (Crawford et al. 2006), but a second investigation found no significant difference (Tierney et al. 2006). Although type 2 EBV is prevalent in the same sub-Saharan region of Africa as endemic BL, the frequency of type 1 or type 2 EBV in BL from that region seems to reflect the incidence in the population rather than a specific-type association (Rickinson and Kieff 2007). Some subtype variation of EBNA2 has been noted in Chinese EBV-associated carcinomas (Wang et al. 2012b) and in European lymphomas (Schuster et al. 1996), but again there is little evidence that it is related to disease.

The greatest biological and functional difference between the two viral types is that type 1 EBV immortalises B cells in vitro much more efficiently than type 2 EBV (Rickinson et al. 1987). When LCLs are transformed with type 1 EBV, they grow more quickly and to a higher saturation cell density in comparison with type 2 transformants (Rickinson et al. 1987). This difference in in vitro transforming efficiency between type 1 and type 2 EBV has been mapped to the EBNA2 locus (Cohen et al. 1989). When a type 2 P3HR1 EBV strain was engineered to carry a type 1 EBNA2 sequence, this virus gained the type 1 immortalisation phenotype (Cohen et al. 1989). In contrast, EBNA3 type does not affect the immortalisation ability of the virus, as replacing the type 2 EBNA3 gene locus with corresponding type 1 sequences in the P3HR1 EBV genome showed no difference in primary B lymphocyte growth transformation (Sample et al. 1990; Tomkinson and Kieff 1992).

This in vitro transformation phenotype of type 1 and type 2 EBV correlates with tumour formation frequency in SCID mice that were inoculated intraperitoneally with type 1 or type 2 in vitro-transformed LCLs (Rowe et al. 1991; Cohen et al. 1992). Also, similar rates of tumour induction were observed for EBV-LCLs generated in vitro with a wild-type type 1 strain or with a type 2 P3HR1 strain carrying a type 1 EBNA2 in the SCID mice model (Cohen et al. 1992).

More recently, a transfection assay with an LCL (EREB2.5) infected with EBV containing conditional EBNA2 function was used to compare the abilities of type 1 and type 2 EBNA2 to maintain cell proliferation (Lucchesi et al. 2008). Type 1 EBNA2 maintained the normal growth of the cells, but the type 2 EBNA2 did not, providing a simple cell growth assay for this aspect of EBNA2 activity. The reduced proliferation in cells expressing type 2 EBNA2 correlated with loss of expression of some cell genes that are known to be targets of type 1 EBNA2. Microarray analysis of EBNA2 target genes identified a small number of genes that are more strongly induced by type 1 than by type 2 EBNA2, and one of these genes (CXCR7) was shown to be required for proliferation of LCLs. The EBV LMP1 gene was also more strongly induced by type 1 EBNA2 than by type 2, but this effect was transient. The results indicated that differential gene regulation by EBV type 1 and type 2 EBNA2 might be the basis for the much weaker B cell transformation activity of type 2 EBV strains compared to type 1 strains (Lucchesi et al. 2008).

To map the part of the EBNA2 protein responsible for the enhanced cell growth activity of type 1 EBNA2, the effect on EREB2.5 cell growth of chimaeras of type 1 and type 2 EBNA2 was tested in the EREB2.5 cell growth assay (Cancian et al. 2011). Although the major sequence differences between type 1 and type 2 EBNA-2 lie in N-terminal parts of the protein, the superior ability of type 1 EBNA-2 to induce proliferation of EBV-infected lymphoblasts was found to be mostly determined by the C-terminus of EBNA-2. Substitution of the C-terminus of type 1 EBNA-2 into the type 2 protein was sufficient to confer a type 1 growth phenotype in the EREB2.5 cell growth assay and type 1 expression levels of LMP-1 and CXCR7. Within this region, the RG, CR7 and TAD domains were the minimum type 1 sequences required. The results indicated that the C-terminus of EBNA-2 accounts for the greater ability of type 1 EBV to promote B cell proliferation, through mechanisms that include higher induction of genes (LMP-1 and CXCR7) required for proliferation and survival of EBV-LCLs (Cancian et al. 2011).

More detailed analysis showed that converting a single amino acid of type 2 EBNA2 from serine to the aspartate found in type 1 EBNA2 (S442D) was sufficient to give the type 1 growth maintenance of LCLs (Tzellos et al. 2014). That amino acid is in the transactivation domain (TAD) of EBNA2, and the TAD of type 1 EBNA2 was about 2-fold more active in a standard TAD reporter assay. However, this did not explain the specific superior induction of a small group of cell genes and the LMP1 gene by type 1 EBNA2. About 9 genes were directly induced more strongly by type 1 EBNA2 (Lucchesi et al. 2008), including CXCR7, which was shown to be required for LCL proliferation. Recent ChIP-seq analysis of binding sites for EBNA2 in LCL and BL cell lines has shown that many EBNA2 binding sites are quite distant from the transcription start sites of genes and regulated genes frequently have several EBNA2 binding loci within about 50 kb of the transcription start sites. By comparing the DNA sequences of EBNA2 binding sites in differentially regulated genes, a conserved motif was identified that is present in the differentially regulated promoters, including the LMP1 promoter (Tzellos et al. 2014). The motif contains the consensus binding sites for PU.1 and IRF transcription factors, and it is thought that the function of EBNA2 at these sites may be mediated by its association with these factors rather than the usual RBP-Jk DNA binding protein.

It seems surprising that such a complex functional difference between the EBNA2 types would be mediated by a single amino acid when the proteins are only 56 % identical. Perhaps there are other phenotypes mediated by other sequence differences, but these have not yet been measured in these B cell assays.

The physiological significance of the type 1/type 2 variation is not known at present. One interesting speculation is that type 2 EBV might be favoured in conditions of chronic immune activation. This might be present in the parts of Africa where type 2 EBV is most abundant due to other co-infections including malaria.

Other situations where detection of type 2 seems relatively common also involve disturbance of the immune system, for example in AIDS patients. It has also recently been reported that type 2 EBV can infect T lymphocytes and cause a transient cell proliferation with concomitant cytokine secretion (Coleman et al. 2015).

6 How to Identify Variation Potentially Relevant to EBV-Associated Diseases

This can be considered using NPC in southern China as an example. Chinese strains of EBV have characteristic polymorphisms in several regions of the genome, including some of the regions relevant to NPC, such as the BART miRNA region and LMP1. At present, there is little direct evidence that these polymorphisms contribute to the high incidence of NPC in southern China, but it has recently been shown that the BART miRNAs can play an important role in growth of NPC cells (Qiu et al. 2015). To investigate the role of the sequence polymorphisms, two types of experiment will be needed. First, it will be necessary to sequence large numbers of EBV samples from people with or without NPC from the endemic area, sampling directly from the tumour in NPC patients and from their saliva. A correlation could indicate whether there are tumour specific variants within individuals—studies of this type are currently in progress. Secondly, it would be necessary to test whether some of these specific polymorphisms alter the tumorigenic properties of the virus in an experimental system. There is a great need for a meaningful animal model for EBV-associated NPC so that mutations could be introduced into the virus and the role of polymorphisms could be analysed. At present, no system of that type exists, but it represents an important challenge for future research on NPC. The detailed analysis of differences between type 1 and type 2 EBNA2 described above, with emphasis on a single amino acid (S442D) determining transcriptional function on a specific set of genes important for cell growth, shows how challenging it may be to identify the crucial sequence differences among a large number of inconsequential polymorphisms. If the mechanism became clear from model systems and were obviously relevant, this might be sufficient to convince. A more rigorous proof through prevention of infection by immunisation seems a distant possibility at present, but it is possible that some form of immunisation would modify disease without actually blocking infection. In general to understand natural variation of EBV, there is currently a great need for more viral genome sequences, much better worldwide coverage and more whole EBV genome analysis relative to disease to give the statistical power that will be required to identify variation relevant to disease.

References

Adldinger HK, Delius H, Freese UK, Clarke J, Bornkamm GW (1985) A putative transforming gene of Jijoye virus differs from that of Epstein-Barr virus prototypes. Virology 141(2):221–234

Ai J, Xie Z, Liu C, Huang Z, Xu J (2012) Analysis of EBNA-1 and LMP-1 variants in diseases associated with EBV infection in Chinese children. Virol J 9:13. doi:10.1186/1743-422X-9-13

Arrand JR, Rymo L, Walsh JE, Bjorck E, Lindahl T, Griffin BE (1981) Molecular cloning of the complete Epstein-Barr virus genome as a set of overlapping restriction endonuclease fragments. Nucleic Acids Res 9(13):2999–3014

Arrand JR, Young LS, Tugwood JD (1989) Two families of sequences in the small RNA-encoding region of Epstein-Barr virus (EBV) correlate with EBV types A and B. J Virol 63(2):983–986

Baer R, Bankier AT, Biggin MD, Deininger PL, Farrell PJ, Gibson TJ, Hatfull G, Hudson GS, Satchwell SC, Seguin C, Tuffnell P, Barrell B (1984) DNA sequence and expression of the B95-8 Epstein-Barr virus genome. Nature 310(5974):207–211

Banko A, Lazarevic I, Cupic M, Stevanovic G, Boricic I, Jovanovic T (2012) Carboxy-terminal sequence variation of LMP1 gene in Epstein-Barr-virus-associated mononucleosis and tumors from Serbian patients. J Med Virol 84(4):632–642. doi:10.1002/jmv.23217

Bell MJ, Brennan R, Miles JJ, Moss DJ, Burrows JM, Burrows SR (2008) Widespread sequence variation in Epstein-Barr virus nuclear antigen 1 influences the antiviral T cell response. J Infect Dis 197(11):1594–1597. doi:10.1086/587848

BenAyed-Guerfali D, Ayadi W, Miladi-Abdennadher I, Khabir A, Sellami-Boudawara T, Gargouri A, Mokdad-Gargouri R (2011) Characteristics of Epstein Barr virus variants associated with gastric carcinoma in southern Tunisia. Virol J 8:500. doi:10.1186/1743-422X-8-500

Berger C, Rothenberger S, Bachmann E, McQuain C, Nadal D, Knecht H (1999) Sequence polymorphisms between latent membrane proteins LMP1 and LMP2A do not correlate in EBV-associated reactive and malignant lympho-proliferations. Int J Cancer 81(3):371–375

Bhatia K, Raj A, Guitierrez MI, Judde JG, Spangler G, Venkatesh H, Magrath IT (1996) Variation in the sequence of Epstein Barr virus nuclear antigen 1 in normal peripheral blood lymphocytes and in Burkitt's lymphomas. Oncogene 13(1):177–181

Burrows JM, Bromham L, Woolfit M, Piganeau G, Tellam J, Connolly G, Webb N, Poulsen L, Cooper L, Burrows SR, Moss DJ, Haryana SM, Ng M, Nicholls JM, Khanna R (2004) Selection pressure-driven evolution of the Epstein-Barr virus-encoded oncogene LMP1 in virus isolates from Southeast Asia. J Virol 78(13):7131–7137. doi:10.1128/JVI.78.13.7131-7137.2004

Busson P, Edwards RH, Tursz T, Raab-Traub N (1995) Sequence polymorphism in the Epstein-Barr virus latent membrane protein (LMP)-2 gene. J Gen Virol 76(Pt 1):139–145

Cancian L, Bosshard R, Lucchesi W, Karstegl CE, Farrell PJ (2011) C-terminal region of EBNA-2 determines the superior transforming ability of type 1 Epstein-Barr virus by enhanced gene regulation of LMP-1 and CXCR7. PLoS Pathog 7(7):e1002164. doi:10.1371/journal.ppat.1002164

Chang KP, Hao SP, Lin SY, Ueng SH, Pai PC, Tseng CK, Hsueh C, Hsieh MS, Yu JS, Tsang NM (2006) The 30-bp deletion of Epstein-Barr virus latent membrane protein-1 gene has no effect in nasopharyngeal carcinoma. Laryngoscope 116(4):541–546. doi:10.1097/01.mlg.0000201993.53410.40

Chen JN, Ding YG, Feng ZY, Li HG, He D, Du H, Wu B, Shao CK (2010) Association of distinctive Epstein-Barr virus variants with gastric carcinoma in Guangzhou, southern China. J Med Virol 82(4):658–667. doi:10.1002/jmv.21731

Chen JN, Jiang Y, Li HG, Ding YG, Fan XJ, Xiao L, Han J, Du H, Shao CK (2011) Epstein-Barr virus genome polymorphisms of Epstein-Barr virus-associated gastric carcinoma in gastric remnant carcinoma in Guangzhou, southern China, an endemic area of nasopharyngeal carcinoma. Virus Res 160(1–2):191–199. doi:10.1016/j.virusres.2011.06.011

Chen JN, Zhang NN, Jiang Y, Hui DY, Wen ZJ, Li HG, Ding YG, Du H, Shao CK (2012) Variations of Epstein-Barr virus nuclear antigen 1 in Epstein-Barr virus-associated gastric carcinomas from Guangzhou, southern China. PLoS ONE 7(11):e50084. doi:10.1371/journal.pone.0050084

Chen MR, Tsai CH, Wu FF, Kan SH, Yang CS, Chen JY (1999) The major immunogenic epitopes of Epstein-Barr virus (EBV) nuclear antigen 1 are encoded by sequence domains which vary among nasopharyngeal carcinoma biopsies and EBV-associated cell lines. J Gen Virol 80(Pt 2):447–455

Cheung A, Kieff E (1982) Long internal direct repeat in Epstein-Barr virus DNA. J Virol 44(1):286–294

Cheung ST, Leung SF, Lo KW, Chiu KW, Tam JS, Fok TF, Johnson PJ, Lee JC, Huang DP (1998) Specific latent membrane protein 1 gene sequences in type 1 and type 2 Epstein-Barr virus from nasopharyngeal carcinoma in Hong Kong. Int J Cancer 76(3):399–406

Chu PG, Chang KL, Chen WG, Chen YY, Shibata D, Hayashi K, Bacchi C, Bacchi M, Weiss LM (1999) Epstein-Barr virus (EBV) nuclear antigen (EBNA)-4 mutation in EBV-associated malignancies in three different populations. Am J Pathol 155(3):941–947. doi:10.1016/S0002-9440(10)65193-0

Cohen JI, Fauci AS, Varmus H, Nabel GJ (2011) Epstein-Barr virus: an important vaccine target for cancer prevention. Sci Transl Med 3(107):107fs107. doi:10.1126/scitranslmed.3002878

Cohen JI, Picchio GR, Mosier DE (1992) Epstein-Barr virus nuclear protein 2 is a critical determinant for tumor growth in SCID mice and for transformation in vitro. J Virol 66(12):7555–7559

Cohen JI, Wang F, Mannick J, Kieff E (1989) Epstein-Barr virus nuclear protein 2 is a key determinant of lymphocyte transformation. Proc Natl Acad Sci USA 86(23):9558–9562

Coleman CB, Wohlford EM, Smith NA, King CA, Ritchie JA, Baresel PC, Kimura H, Rochford R (2015) Epstein-barr virus type 2 latently infects T cells, inducing an atypical activation characterized by expression of lymphotactic cytokines. J Virol 89(4):2301–2312. doi:10.1128/JVI.03001-14

Correa RM, Fellner MD, Alonio LV, Durand K, Teyssie AR, Picconi MA (2004) Epstein-Barr virus (EBV) in healthy carriers: distribution of genotypes and 30 bp deletion in latent membrane protein-1 (LMP-1) oncogene. J Med Virol 73(4):583–588. doi:10.1002/jmv.20129

Correa RM, Fellner MD, Durand K, Redini L, Alonio V, Yampolsky C, Colobraro A, Sevlever G, Teyssie A, Benetucci J, Picconi MA (2007) Epstein Barr virus genotypes and LMP-1 variants in HIV-infected patients. J Med Virol 79(4):401–407. doi:10.1002/jmv.20782

Corvalan A, Ding S, Koriyama C, Carrascal E, Carrasquilla G, Backhouse C, Urzua L, Argandona J, Palma M, Eizuru Y, Akiba S (2006) Association of a distinctive strain of Epstein-Barr virus with gastric cancer. Int J Cancer 118(7):1736–1742. doi:10.1002/ijc.21530

Crawford DH, Macsween KF, Higgins CD, Thomas R, McAulay K, Williams H, Harrison N, Reid S, Conacher M, Douglas J, Swerdlow AJ (2006) A cohort study among university students: identification of risk factors for Epstein-Barr virus seroconversion and infectious mononucleosis. Clin Infect Dis 43(3):276–282

Dambaugh T, Hennessy K, Chamnankit L, Kieff E (1984) U2 region of Epstein-Barr virus DNA may encode Epstein-Barr nuclear antigen 2. Proc Natl Acad Sci USA 81(23):7632–7636

Dambaugh TR, Kieff E (1982) Identification and nucleotide sequences of two similar tandem direct repeats in Epstein-Barr virus DNA. J Virol 44(3):823–833

Dardari R, Khyatti M, Cordeiro P, Odda M, ElGueddari B, Hassar M, Menezes J (2006) High frequency of latent membrane protein-1 30-bp deletion variant with specific single mutations in Epstein-Barr virus-associated nasopharyngeal carcinoma in Moroccan patients. Int J Cancer 118(8):1977–1983. doi:10.1002/ijc.21595 de Jesus O, Smith PR, Spender L, Karstegl CE, Niller H, Huang D, Farrell PJ (2003) Updated Epstein-Barr virus (EBV) DNA sequence and analysis of a promoter for the BART (CST, BARF0) RNAs of EBV. J Gen Virol 84:1443–1450

Decaussin G, Leclerc V, Ooka T (1995) The lytic cycle of Epstein-Barr virus in the nonproducer Raji line can be rescued by the expression of a 135-kilodalton protein encoded by the BALF2 open reading frame. J Virol 69(11):7309–7314

Depledge DP, Palser AL, Watson SJ, Lai IY, Gray ER, Grant P, Kanda RK, Leproust E, Kellam P, Breuer J (2011) Specific capture and whole-genome sequencing of viruses from clinical samples. PLoS ONE 6(11):e27805. doi:10.1371/journal.pone.0027805

Desgranges C, Lenoir G, de-The G, Seigneurin JM, Hilgers J, Dubouch P (1976) In vitro transforming activity of EBV. I-Establishment and properties of two EBV strains (M81 and M72) produced by immortalized Callithrix jacchus lymphocytes. Biomedicine 25(9):349–352

Diduk SV, Smirnova KV, Pavlish OA, Gurtsevitch VE (2008) Functionally significant mutations in the Epstein-Barr virus LMP1 gene and their role in activation of cell signaling pathways. Biochemistry Biokhimiia 73(10):1134–1139

Do NV, Ingemar E, Phi PT, Jenny A, Chinh TT, Zeng Y, Hu L (2008) A major EBNA1 variant from Asian EBV isolates shows enhanced transcriptional activity compared to prototype B95.8. Virus Res 132(1–2):15–24. doi:10.1016/j.virusres.2007.10.020

Dolan A, Addison C, Gatherer D, Davison AJ, McGeoch DJ (2006) The genome of Epstein-Barr virus type 2 strain AG876. Virology 350(1):164–170

Duraiswamy J, Burrows JM, Bharadwaj M, Burrows SR, Cooper L, Pimtanothai N, Khanna R (2003) Ex vivo analysis of T-cell responses to Epstein-Barr virus-encoded oncogene latent membrane protein 1 reveals highly conserved epitope sequences in virus isolates from diverse geographic regions. J Virol 77(13):7401–7410

Edwards RH, Sitki-Green D, Moore DT, Raab-Traub N (2004) Potential selection of LMP1 variants in nasopharyngeal carcinoma. J Virol 78(2):868–881

Fafi-Kremer S, Morand P, Germi R, Ballout M, Brion JP, Genoulaz O, Nicod S, Stahl JP, Ruigrok RW, Seigneurin JM (2005) A prospective follow-up of Epstein-Barr virus LMP1 genotypes in saliva and blood during infectious mononucleosis. J Infect Dis 192(12):2108–2111. doi:10.1086/498215

Farrell P (1998) Signal transduction from the Epstein-Barr virus LMP-1 transforming protein. Trends Microbiol 6:175–177

Fruscalzo A, Marsili G, Busiello V, Bertolini L, Frezza D (2001) DNA sequence heterogeneity within the Epstein-Barr virus family of repeats in the latent origin of replication. Gene 265(1–2):165–173

Gan YJ, Razzouk BI, Su T, Sixbey JW (2002) A defective, rearranged Epstein-Barr virus genome in EBER-negative and EBER-positive Hodgkin's disease. Am J Pathol 160(3):781–786. doi:10.1016/S0002-9440(10)64900-0

Gantuz M, Lorenzetti MA, Altcheh J, De Matteo E, Moscatelli G, Moroni S, Chabay PA, Preciado MV (2013) LMP1 promoter sequence analysis in Epstein Barr virus pediatric infection reveals preferential circulation of B95. 8 related variants in Argentina. Infection Genet Evol (journal of molecular epidemiology and evolutionary genetics in infectious diseases) 14:275–281. doi:10.1016/j.meegid.2012.12.025

Giron LB, Ramos da Silva S, Barbosa AN, Monteiro de Barros Almeida RA, Rosario de Souza L, Elgui de Oliveira D (2013) Impact of Epstein-Barr virus load, virus genotype, and frequency of the 30 bp deletion in the viral BNLF-1 gene in patients harboring the human immunodeficiency virus. J Med Virol 85(12):2110–2118. doi:10.1002/jmv.23722

Gorzer I, Niesters HG, Cornelissen JJ, Puchhammer-Stockl E (2006) Characterization of Epstein-Barr virus type I variants based on linked polymorphism among EBNA3A, -3B, and -3C genes. Virus Res 118(1–2):105–114. doi:10.1016/j.virusres.2005.11.020

Habeshaw G, Yao QY, Bell AI, Morton D, Rickinson AB (1999) Epstein-barr virus nuclear antigen 1 sequences in endemic and sporadic Burkitt's lymphoma reflect virus strains prevalent in different geographic areas. J Virol 73(2):965–975

Hadhri-Guiga B, Khabir AM, Mokdad-Gargouri R, Ghorbel AM, Drira M, Daoud J, Frikha M, Jlidi R, Gargouri A (2006) Various 30 and 69 bp deletion variants of the Epstein-Barr virus LMP1 may arise by homologous recombination in nasopharyngeal carcinoma of Tunisian patients. Virus Res 115(1):24–30. doi:10.1016/j.virusres.2005.07.002

Han J, Chen JN, Zhang ZG, Li HG, Ding YG, Du H, Shao CK (2012) Sequence variations of latent membrane protein 2A in Epstein-Barr virus-associated gastric carcinomas from Guangzhou, southern China. PLoS ONE 7(3):e34276. doi:10.1371/journal.pone.0034276

Hatfull G, Bankier AT, Barrell BG, Farrell PJ (1988) Sequence analysis of Raji Epstein-Barr virus DNA. Virology 164(2):334–340

Hayashi K, Chen WG, Chen YY, Murakami I, Chen HL, Ohara N, Nose S, Hamaya K, Matsui S, Bacchi MM, Bacchi CE, Chang KL, Weiss LM (1998) Deletion of Epstein-Barr virus latent membrane protein 1 gene in Japanese and Brazilian gastric carcinomas, metastatic lesions, and reactive lymphocytes. Am J Pathol 152(1):191–198

Hicks MR, Balesaria S, Medina-Palazon C, Pandya MJ, Woolfson DN, Sinclair AJ (2001) Biophysical analysis of natural variants of the multimerization region of Epstein-Barr virus lytic-switch protein BZLF1. J Virol 75(11):5381–5384. doi:10.1128/JVI.75.11.5381-5384.2001

Higa M, Kinjo T, Kamiyama K, Iwamasa T, Hamada T, Iyama K (2002) Epstein-Barr virus (EBV) subtype in EBV related oral squamous cell carcinoma in Okinawa, a subtropical island in southern Japan, compared with Kitakyushu and Kumamoto in mainland Japan. J Clin Pathol 55(6):414–423

Horst D, Burrows SR, Gatherer D, van Wilgenburg B, Bell MJ, Boer IG, Ressing ME, Wiertz EJ (2012) Epstein-Barr virus isolates retain their capacity to evade T cell immunity through BNLF2a despite extensive sequence variation. J Virol 86(1):572–577. doi:10.1128/JVI.05151-11

Hu LF, Chen F, Zheng X, Ernberg I, Cao SL, Christensson B, Klein G, Winberg G (1993) Clonability and tumorigenicity of human epithelial cells expressing the EBV encoded membrane protein LMP1. Oncogene 8(6):1575–1583

Hu LF, Zabarovsky ER, Chen F, Cao SL, Ernberg I, Klein G, Winberg G (1991) Isolation and sequencing of the Epstein-Barr virus BNLF-1 gene (LMP1) from a Chinese nasopharyngeal carcinoma. J Gen Virol 72(Pt 10):2399–2409

Huen DS, Henderson SA, Croom-Carter D, Rowe M (1995) The Epstein-Barr virus latent membrane protein-1 (LMP1) mediates activation of NF-kappa B and cell surface phenotype via two effector regions in its carboxy-terminal cytoplasmic domain. Oncogene 10(3):549–560

Hutajulu SH, Hoebe EK, Verkuijlen SA, Fachiroh J, Hariwijanto B, Haryana SM, Stevens SJ, Greijer AE, Middeldorp JM (2010) Conserved mutation of Epstein-Barr virus-encoded BamHI-A Rightward Frame-1 (BARF1) gene in Indonesian nasopharyngeal carcinoma. Infect Agents Cancer 5:16. doi:10.1186/1750-9378-5-16

Imajoh M, Hashida Y, Murakami M, Maeda A, Sato T, Fujieda M, Wakiguchi H, Daibata M (2012) Characterization of Epstein-Barr virus (EBV) BZLF1 gene promoter variants and comparison of cellular gene expression profiles in Japanese patients with infectious mononucleosis, chronic active EBV infection, and EBV-associated hemophagocytic lymphohistiocytosis. J Med Virol 84(6):940–946. doi:10.1002/jmv.23299

Jen KY, Higuchi M, Cheng J, Li J, Wu LY, Li YF, Lin HL, Chen Z, Gurtsevitch V, Fujii M, Saku T (2005) Nucleotide sequences and functions of the Epstein-Barr virus latent membrane protein 1 genes isolated from salivary gland lymphoepithelial carcinomas. Virus Genes 30(2):223–235. doi:10.1007/s11262-004-5630-5

Jenkins P, Farrell P (1996) Are particular Epstein-Barr virus strains linked to disease? Semin Cancer Biol 7:209–215

Jenson HB, Rabson MS, Miller G (1986) Palindromic structure and polypeptide expression of 36 kilobase pairs of heterogeneous Epstein-Barr virus (P3HR-1) DNA. J Virol 58(2):475–486

Ji KM, Li CL, Meng G, Han AD, Wu XL (2008) New BZLF1 sequence variations in EBV-associated undifferentiated nasopharyngeal carcinoma in southern China. Arch Virol 153(10):1949–1953. doi:10.1007/s00705-008-0195-6

Jia Y, Wang Y, Chao Y, Jing Y, Sun Z, Luo B (2010) Sequence analysis of the Epstein-Barr virus (EBV) BRLF1 gene in nasopharyngeal and gastric carcinomas. Virol J 7:341. doi:10.1186/1743-422X-7-341

Jin Y, Xie Z, Lu G, Yang S, Shen K (2010) Characterization of variants in the promoter of BZLF1 gene of EBV in nonmalignant EBV-associated diseases in Chinese children. Virol J 7:92. doi:10.1186/1743-422X-7-92

Jing YZ, Wang Y, Jia YP, Luo B (2010) Polymorphisms of Epstein-Barr virus BHRF1 gene, a homologue of bcl-2. Chin J Cancer 29(12):1000–1005

Kanai K, Satoh Y, Saiki Y, Ohtani H, Sairenji T (2007) Difference of Epstein-Barr virus isolates from Japanese patients and African Burkitt's lymphoma cell lines based on the sequence of latent membrane protein 1. Virus Genes 34(1):55–61. doi:10.1007/s11262-006-0010-y

Kanda T, Shibata S, Saito S, Murata T, Isomura H, Yoshiyama H, Takada K, Tsurumi T (2011) Unexpected instability of family of repeats (FR), the critical cis-acting sequence required for EBV latent infection, in EBV-BAC systems. PLoS ONE 6(11):e27758. doi:10.1371/journal.pone.0027758

Kawaguchi A, Kanai K, Satoh Y, Touge C, Nagata K, Sairenji T, Inoue Y (2009) The evolution of Epstein-Barr virus inferred from the conservation and mutation of the virus glycoprotein gp350/220 gene. Virus Genes 38(2):215–223. doi:10.1007/s11262-008-0323-0

Kelly G, Bell A, Rickinson A (2002) Epstein-Barr virus-associated Burkitt lymphomagenesis selects for downregulation of the nuclear antigen EBNA2. Nat Med 8(10):1098–1104

Kelly GL, Long HM, Stylianou J, Thomas WA, Leese A, Bell AI, Bornkamm GW, Mautner J, Rickinson AB, Rowe M (2009) An Epstein-Barr virus anti-apoptotic protein constitutively expressed in transformed cells and implicated in burkitt lymphomagenesis: the Wp/BHRF1 link. PLoS Pathog 5(3):e1000341. doi:10.1371/journal.ppat.1000341

Khanim F, Dawson C, Meseda CA, Dawson J, Mackett M, Young LS (1997) BHRF1, a viral homologue of the Bcl-2 oncogene, is conserved at both the sequence and functional level in different Epstein-Barr virus isolates. J Gen Virol 78(Pt 11):2987–2999

Khanna R, Moss DJ, Burrows SR (1999) Vaccine strategies against Epstein-Barr virus-associated diseases: lessons from studies on cytotoxic T-cell-mediated immune regulation. Immunol Rev 170:49–64

Kim SM, Kang SH, Lee WK (2006) Identification of two types of naturally-occurring intertypic recombinants of Epstein-Barr virus. Mol Cells 21(2):302–307

Kwok H, Tong AH, Lin CH, Lok S, Farrell PJ, Kwong DL, Chiang AK (2012) Genomic sequencing and comparative analysis of Epstein-Barr virus genome isolated from primary nasopharyngeal carcinoma biopsy. PLoS ONE 7(5):e36939. doi:10.1371/journal.pone.0036939

Kwok H, Wu CW, Palser AL, Kellam P, Sham PC, Kwong DL, Chiang AK (2014) Genomic diversity of Epstein-Barr virus genomes isolated from primary nasopharyngeal carcinoma biopsies. J Virol 88:10662–10672. doi:10.1128/JVI.01665-14

Larcher C, Bernhard D, Schaadt E, Adler B, Ausserlechner MJ, Mitterer M, Huemer HP (2003) Functional analysis of the mutated Epstein-Barr virus oncoprotein LMP1(69del): implications for a new role of naturally occurring LMP1 variants. Haematologica 88(12):1324–1335

Lay ML, Lucas RM, Toi C, Ratnamohan M, Ponsonby AL, Dwyer DE (2012) Epstein-Barr virus genotypes and strains in central nervous system demyelinating disease and Epstein-Barr virus-related illnesses in Australia. Intervirology 55(5):372–379. doi:10.1159/000334693

Lee SP, Thomas WA, Murray RJ, Khanim F, Kaur S, Young LS, Rowe M, Kurilla M, Rickinson AB (1993) HLA A2.1-restricted cytotoxic T cells recognizing a range of Epstein-Barr virus isolates through a defined epitope in latent membrane protein LMP2. J Virol 67(12):7428–7435

Lei H, Li T, Hung GC, Li B, Tsai S, Lo SC (2013) Identification and characterization of EBV genomes in spontaneously immortalized human peripheral blood B lymphocytes by NGS technology. BMC Genom 14:804. doi:10.1186/1471-2164-14-804

Li A, Zhang XS, Jiang JH, Wang HH, Liu XQ, Pan ZG, Zeng YX (2005) Transcriptional expression of RPMS1 in nasopharyngeal carcinoma and its oncogenic potential. Cell Cycle 4(2):304–309

Li DJ, Bei JX, Mai SJ, Xu JF, Chen LZ, Zhang RH, Yu XJ, Hong MH, Zeng YX, Kang T (2009) The dominance of China 1 in the spectrum of Epstein-Barr virus strains from Cantonese patients with nasopharyngeal carcinoma. J Med Virol 81(7):1253–1260. doi:10.1002/jmv.21503

Lin HJ, Cherng JM, Hung MS, Sayion Y, Lin JC (2005) Functional assays of HLA A2-restricted epitope variant of latent membrane protein 1 (LMP-1) of Epstein-Barr virus in nasopharyngeal carcinoma of southern China and Taiwan. J Biomed Sci 12(6):925–936. doi:10.1007/s11373-005-9017-y

Lin JC, Cherng JM, Lin HJ, Tsang CW, Liu YX, Lee SP (2004) Amino acid changes in functional domains of latent membrane protein 1 of Epstein-Barr virus in nasopharyngeal carcinoma of southern China and Taiwan: prevalence of an HLA A2-restricted 'epitope-loss variant'. J Gen Virol 85(Pt 7):2023–2034. doi:10.1099/vir.0.19696-0

Lin Z, Wang X, Strong MJ, Concha M, Baddoo M, Xu G, Baribault C, Fewell C, Hulme W, Hedges D, Taylor CM, Flemington EK (2013) Whole-genome sequencing of the Akata and Mutu Epstein-Barr virus strains. J Virol 87(2):1172–1182. doi:10.1128/JVI.02517-12

Liu P, Fang X, Feng Z, Guo YM, Peng RJ, Liu T, Huang Z, Feng Y, Sun X, Xiong Z, Guo X, Pang SS, Wang B, Lv X, Feng FT, Li DJ, Chen LZ, Feng QS, Huang WL, Zeng MS, Bei JX, Zhang Y, Zeng YX (2011) Direct sequencing and characterization of a clinical isolate of Epstein-Barr virus from nasopharyngeal carcinoma tissue by using next-generation sequencing technology. J Virol 85(21):11291–11299. doi:10.1128/JVI.00823-11

Lorenzetti MA, Gantuz M, Altcheh J, De Matteo E, Chabay PA, Preciado MV (2012) Distinctive Epstein-Barr virus variants associated with benign and malignant pediatric pathologies: LMP1 sequence characterization and linkage with other viral gene polymorphisms. J Clin Microbiol 50(3):609–618. doi:10.1128/JCM.05778-11

Lorenzetti MA, Gantuz M, Altcheh J, De Matteo E, Chabay PA, Preciado MV (2014) Epstein-Barr virus BZLF1 gene polymorphisms: malignancy related or geographically distributed variants? Clin Microbiol Infect (the official publication of the European Society of Clinical Microbiology and Infectious Diseases) 20:O861–O869. doi:10.1111/1469-0691.12631

Lucchesi W, Brady G, Dittrich-Breiholz O, Kracht M, Russ R, Farrell PJ (2008) Differential gene regulation by Epstein-Barr virus type 1 and type 2 EBNA2. J Virol 82(15):7456–7466

Luo B, Liu M, Chao Y, Wang Y, Jing Y, Sun Z (2012) Characterization of Epstein-Barr virus gp350/220 gene variants in virus isolates from gastric carcinoma and nasopharyngeal carcinoma. Arch Virol 157(2):207–216. doi:10.1007/s00705-011-1148-z

Luo B, Tang X, Jia Y, Wang Y, Chao Y, Zhao C (2011) Sequence variation of Epstein-Barr virus (EBV) BZLF1 gene in EBV-associated gastric carcinomas and nasopharyngeal carcinomas in northern China. Microbes Infect (Institut Pasteur) 13(8–9):776–782. doi:10.1016/j.micinf.2011.04.002

Mai SJ, Xie D, Huang YF, Wang FW, Liao YJ, Deng HX, Liu WJ, Hua WF, Zeng YX (2010) The enhanced transcriptional activity of the V-val subtype of Epstein-Barr virus nuclear antigen 1 in epithelial cell lines. Oncol Rep 23(5):1417–1424

Mainou BA, Raab-Traub N (2006) LMP1 strain variants: biological and molecular properties. J Virol 80(13):6458–6468. doi:10.1128/JVI.00135-06

Martini M, Capello D, Serraino D, Navarra A, Pierconti F, Cenci T, Gaidano G, Larocca LM (2007) Characterization of variants in the promoter of EBV gene BZLF1 in normal donors, HIV-positive patients and in AIDS-related lymphomas. J Infect 54(3):298–306. doi:10.1016/j.jinf.2006.04.015

McGeoch DJ, Gatherer D (2007) Lineage structures in the genome sequences of three Epstein-Barr virus strains. Virology 359(1):1–5. doi:10.1016/j.virol.2006.10.009

Midgley RS, Bell AI, McGeoch DJ, Rickinson AB (2003a) Latent gene sequencing reveals familial relationships among Chinese Epstein-Barr virus strains and evidence for positive selection of A11 epitope changes. J Virol 77(21):11517–11530

Midgley RS, Bell AI, Yao QY, Croom-Carter D, Hislop AD, Whitney BM, Chan AT, Johnson PJ, Rickinson AB (2003b) HLA-A11-restricted epitope polymorphism among Epstein-Barr virus strains in the highly HLA-A11-positive Chinese population: incidence and immunogenicity of variant epitope sequences. J Virol 77(21):11507–11516

Midgley RS, Blake NW, Yao QY, Croom-Carter D, Cheung ST, Leung SF, Chan AT, Johnson PJ, Huang D, Rickinson AB, Lee SP (2000) Novel intertypic recombinants of Epstein-Barr virus in the Chinese population. J Virol 74(3):1544–1548

Miller G, Shope T, Lisco H, Stitt D, Lipman M (1972) Epstein-Barr virus: transformation, cytopathic changes, and viral antigens in squirrel monkey and marmoset leukocytes. Proc Natl Acad Sci USA 69(2):383–387

Miller WE, Edwards RH, Walling DM, Raab-Traub N (1994) Sequence variation in the Epstein-Barr virus latent membrane protein 1. J Gen Virol 75(Pt 10):2729–2740

Nagamine M, Kishibe K, Takahara M, Nagato T, Ishii H, Bandoh N, Ogino T, Harabuchi Y (2007) Selected amino acid change encoding Epstein-Barr virus-specific T cell epitope of the LMP2A gene in Japanese nasal NK/T cell lymphoma patients. Intervirology 50(5):319–322. doi:10.1159/000106462

Nguyen-Van D, Ernberg I, Phan-Thi Phi P, Tran-Thi C, Hu L (2008) Epstein-Barr virus genetic variation in Vietnamese patients with nasopharyngeal carcinoma: full-length analysis of LMP1. Virus Genes 37(2):273–281. doi:10.1007/s11262-008-0262-9

Nie Y, Sun Y, Wang Y, Liu C, Zhao C, Luo B (2013) Epstein-Barr virus gene polymorphism in different parts of the same nasopharyngeal carcinoma patient. Arch Virol 158(5):1031–1037. doi:10.1007/s00705-012-1578-2

Pai PC, Tseng CK, Chuang CC, Wei KC, Hao SP, Hsueh C, Chang KP, Tsang NM (2007) Polymorphism of C-terminal activation region 2 of Epstein-Barr virus latent membrane protein 1 in predicting distant failure and post-metastatic survival in patients with nasopharyngeal carcinoma. Head Neck 29(2):109–119. doi:10.1002/hed.20483

Palefsky JM, Berline J, Greenspan D, Greenspan JS (2002) Evidence for trafficking of Epstein-Barr virus strains between hairy leukoplakia and peripheral blood lymphocytes. J Gen Virol 83(Pt 2):317–321

Palma I, Sanchez AE, Jimenez-Hernandez E, Alvarez-Rodriguez F, Nava-Frias M, Valencia-Mayoral P, Salinas-Lara C, Velazquez-Guadarrama N, Portilla-Aguilar J, Pena RY, Ramos-Salazar P, Contreras A, Alfaro A, Espinosa AM, Najera N, Gutierrez G, Mejia-Arangure JM, Arellano-Galindo J (2013) Detection of Epstein-Barr virus and genotyping based on EBNA2 protein in Mexican patients with hodgkin lymphoma: a comparative study in children and adults. Clin Lymphoma Myeloma Leuk 13(3):266–272. doi:10.1016/j.clml.2012.11.010

Palser AL, Grayson NE, White RE, Corton C, Correia S, Ba abdullah MM, Watson SJ, Cotten M, Arrand JR, Murray P, Allday MJ, Rickinson AB, Young LS, Farrell PJ, Kellam P (2015) Genome diversity of Epstein-Barr virus from multiple tumour types and normal infection. J Virol 89:5222–5237

Parker BD, Bankier A, Satchwell S, Barrell B, Farrell PJ (1990) Sequence and transcription of Raji Epstein-Barr virus DNA spanning the B95-8 deletion region. Virology 179(1):339–346

Patton DF, Shirley P, Raab-Traub N, Resnick L, Sixbey JW (1990) Defective viral DNA in Epstein-Barr virus-associated oral hairy leukoplakia. J Virol 64(1):397–400

Pavlish OA, Diduk SV, Smirnova KV, Shcherbak LN, Goncharova EV, Shalginskikh NA, Arkhipov VV, Kichigina M, Stepina VN, Belousova NV, Osmanov EA, Iakovleva LS, Gurtsevich VE (2008) Mutations of the Epstein-Barr virus LMP1 gene mutations in Russian patients with lymphoid pathology and healthy individuals. Vopr Virusol 53(1):10–16

Qiu J, Smith P, Leahy L, Thorley-Lawson DA (2015) The Epstein-Barr virus encoded BART miRNAs potentiate tumor growth in vivo. PLoS Pathog 11(1):e1004561. doi:10.1371/journal.ppat.1004561

Raab-Traub N, Dambaugh T, Kieff E (1980) DNA of Epstein-Barr virus VIII: B95-8, the previous prototype, is an unusual deletion derivative. Cell 22(1 Pt 1):257–267

Rey J, Xerri L, Bouabdallah R, Keuppens M, Brousset P, Meggetto F (2008) Detection of different clonal EBV strains in Hodgkin lymphoma and nasopharyngeal carcinoma tissues from the same patient. Br J Haematol 142(1):79–81. doi:10.1111/j.1365-2141.2008.07162.x

Rickinson AB (2014) Co-infections, inflammation and oncogenesis: future directions for EBV research. Semin Cancer Biol 26C:99–115. doi:10.1016/j.semcancer.2014.04.004

Rickinson AB, Kieff E (2007) Epstein-Barr virus. In: Knipe D, Howley PM (eds) Fields virology, vol 2, 5th edn. Lippincott-Raven, Philadelphia, pp 2680–2700

Rickinson AB, Moss DJ (1997) Human cytotoxic T lymphocyte responses to Epstein-Barr virus infection. Annu Rev Immunol 15:405–431

Rickinson AB, Young LS, Rowe M (1987) Influence of the Epstein-Barr virus nuclear antigen EBNA 2 on the growth phenotype of virus-transformed B cells. J Virol 61(5):1310–1317

Rooney C, Taylor N, Countryman J, Jenson H, Kolman J, Miller G (1988) Genome rearrangements activate the Epstein-Barr virus gene whose product disrupts latency. Proc Natl Acad Sci USA 85(24):9801–9805

Rowe M, Young LS, Cadwallader K, Petti L, Kieff E, Rickinson AB (1989) Distinction between Epstein-Barr virus type A (EBNA 2A) and type B (EBNA 2B) isolates extends to the EBNA 3 family of nuclear proteins. J Virol 63(3):1031–1039

Rowe M, Young LS, Crocker J, Stokes H, Henderson S, Rickinson AB (1991) Epstein-Barr virus (EBV)-associated lymphoproliferative disease in the SCID mouse model: implications for the pathogenesis of EBV-positive lymphomas in man. J Exp Med 173(1):147–158

Ryan JL, Jones RJ, Elmore SH, Kenney SC, Miller G, Schroeder JC, Gulley ML (2009) Epstein-Barr virus WZhet DNA can induce lytic replication in epithelial cells in vitro, although WZhet is not detectable in many human tissues in vivo. Intervirology 52(1):8–16. doi:10.1159/000210833

Saechan V, Mori A, Mitarnun W, Settheetham-Ishida W, Ishida T (2006) Analysis of LMP1 variants of EBV in southern Thailand: evidence for strain-associated T-cell tropism and pathogenicity. J Clin Virol 36(2):119–125. doi:10.1016/j.jcv.2006.01.018

Saechan V, Settheetham-Ishida W, Kimura R, Tiwawech D, Mitarnun W, Ishida T (2010) Epstein-Barr virus strains defined by the latent membrane protein 1 sequence characterize Thai ethnic groups. J Gen Virol 91(Pt 8):2054–2061. doi:10.1099/vir.0.021105-0

Sample J, Young L, Martin B, Chatman T, Kieff E, Rickinson A, Kieff E (1990) Epstein-Barr virus types 1 and 2 differ in their EBNA-3A, EBNA-3B, and EBNA-3C genes. J Virol 64(9):4084–4092

Schuster V, Ott G, Seidenspinner S, Kreth HW (1996) Common Epstein-Barr virus (EBV) type-1 variant strains in both malignant and benign EBV-associated disorders. Blood 87(4):1579–1585

See HS, Yap YY, Yip WK, Seow HF (2008) Epstein-Barr virus latent membrane protein-1 (LMP-1) 30-bp deletion and Xho I-loss is associated with type III nasopharyngeal carcinoma in Malaysia. World J Surg Oncol 6:18. doi:10.1186/1477-7819-6-18

Senyuta N, Yakovleva L, Goncharova E, Scherback L, Diduk S, Smirnova K, Maksimovich D, Gurtsevitch V (2014) Epstein-barr virus latent membrane protein 1 polymorphism in nasopharyngeal carcinoma and other oral cavity tumors in Russia. J Med Virol 86(2):290–300. doi:10.1002/jmv.23729

Shibata Y, Hoshino Y, Hara S, Yagasaki H, Kojima S, Nishiyama Y, Morishima T, Kimura H (2006) Clonality analysis by sequence variation of the latent membrane protein 1 gene in patients with chronic active Epstein-Barr virus infection. J Med Virol 78(6):770–779. doi:10.1002/jmv.20622

Sitki-Green D, Covington M, Raab-Traub N (2003) Compartmentalization and transmission of multiple Epstein-Barr virus strains in asymptomatic carriers. J Virol 77(3):1840–1847

Sitki-Green D, Edwards RH, Webster-Cyriaque J, Raab-Traub N (2002) Identification of Epstein-Barr virus strain variants in hairy leukoplakia and peripheral blood by use of a heteroduplex tracking assay. J Virol 76(19):9645–9656

Sitki-Green DL, Edwards RH, Covington MM, Raab-Traub N (2004) Biology of Epstein-Barr virus during infectious mononucleosis. J Infect Dis 189(3):483–492. doi:10.1086/380800

Snudden DK, Smith PR, Lai D, Ng MH, Griffin BE (1995) Alterations in the structure of the EBV nuclear antigen, EBNA1, in epithelial cell tumours. Oncogene 10(8):1545–1552

Strockbine LD, Cohen JI, Farrah T, Lyman SD, Wagener F, DuBose RF, Armitage RJ, Spriggs MK (1998) The Epstein-Barr virus BARF1 gene encodes a novel, soluble colony-stimulating factor-1 receptor. J Virol 72(5):4015–4021

Sung NS, Edwards RH, Seillier-Moiseiwitsch F, Perkins AG, Zeng Y, Raab-Traub N (1998) Epstein-Barr virus strain variation in nasopharyngeal carcinoma from the endemic and non-endemic regions of China. Int J Cancer 76(2):207–215

Suspene R, Aynaud MM, Koch S, Pasdeloup D, Labetoulle M, Gaertner B, Vartanian JP, Meyerhans A, Wain-Hobson S (2011) Genetic editing of herpes simplex virus 1 and Epstein-Barr herpesvirus genomes by human APOBEC3 cytidine deaminases in culture and in vivo. J Virol 85(15):7594–7602. doi:10.1128/JVI.00290-11

Szpara ML, Gatherer D, Ochoa A, Greenbaum B, Dolan A, Bowden RJ, Enquist LW, Legendre M, Davison AJ (2014) Evolution and diversity in human herpes simplex virus genomes. J Virol 88(2):1209–1227. doi:10.1128/JVI.01987-13

Tai YC, Kim LH, Peh SC (2004) High frequency of EBV association and 30-bp deletion in the LMP-1 gene in CD56 lymphomas of the upper aerodigestive tract. Pathol Int 54(3):158–166. doi:10.1111/j.1440-1827.2003.01602.x

Tanaka M, Kawaguchi Y, Yokofujita J, Takagi M, Eishi Y, Hirai K (1999) Sequence variations of Epstein-Barr virus LMP2A gene in gastric carcinoma in Japan. Virus Genes 19(2):103–111

Tang YL, Lu JH, Cao L, Wu MH, Peng SP, Zhou HD, Huang C, Yang YX, Zhou YH, Chen Q, Li XL, Zhou M, Li GY (2008) Genetic variations of EBV-LMP1 from nasopharyngeal carcinoma biopsies: potential loss of T cell epitopes. Braz J Med Biol Res (Revista brasileira de pesquisas medicas e biologicas/Sociedade Brasileira de Biofisica [et al]) 41(2):110–116

Taylor GS, Jia H, Harrington K, Lee LW, Turner J, Ladell K, Price DA, Tanday M, Matthews J, Roberts C, Edwards C, McGuigan L, Hartley A, Wilson S, Hui EP, Chan AT, Rickinson AB, Steven NM (2014) A recombinant modified vaccinia ankara vaccine encoding Epstein-Barr virus (EBV) target antigens: a phase I trial in UK patients with EBV-positive cancer. Clin Cancer Res. doi:10.1158/1078-0432.CCR-14-1122-T

Tierney RJ, Edwards RH, Sitki-Green D, Croom-Carter D, Roy S, Yao QY, Raab-Traub N, Rickinson AB (2006) Multiple Epstein-Barr virus strains in patients with infectious mononucleosis: comparison of ex vivo samples with in vitro isolates by use of heteroduplex tracking assays. J Infect Dis 193(2):287–297. doi:10.1086/498913

Tierney RJ, Kao KY, Nagra JK, Rickinson AB (2011) Epstein-Barr virus BamHI W repeat number limits EBNA2/EBNA-LP coexpression in newly infected B cells and the efficiency of B-cell transformation: a rationale for the multiple W repeats in wild-type virus strains. J Virol 85(23):12362–12375. doi:10.1128/JVI.06059-11

Tomkinson B, Kieff E (1992) Second-site homologous recombination in Epstein-Barr virus: insertion of type 1 EBNA 3 genes in place of type 2 has no effect on in vitro infection. J Virol 66(2):780–789

Tsai MH, Raykova A, Klinke O, Bernhardt K, Gartner K, Leung CS, Geletneky K, Sertel S, Munz C, Feederle R, Delecluse HJ (2013) Spontaneous lytic replication and epitheliotropism define an Epstein-Barr virus strain found in carcinomas. Cell Rep 5(2):458–470. doi:10.1016/j.celrep.2013.09.012

Tso KK, Yip KY, Mak CK, Chung GT, Lee SD, Cheung ST, To KF, Lo KW (2013) Complete genomic sequence of Epstein-Barr virus in nasopharyngeal carcinoma cell line C666-1. Infect Agents Cancer 8(1):29. doi:10.1186/1750-9378-8-29

Tzellos S, Correia PB, Karstegl CE, Cancian L, Cano-Flanagan J, McClellan MJ, West MJ, Farrell PJ (2014) A single amino acid in EBNA-2 determines superior B lymphoblastoid cell line growth maintenance by Epstein-Barr virus type 1 EBNA-2. J Virol 88:8743–8753. doi:10.1128/JVI.01000-14

Tzellos S, Farrell PJ (2012) Epstein-Barr virus sequence variation—biology and disease. Pathogens 1:156–174

Vaysberg M, Hatton O, Lambert SL, Snow AL, Wong B, Krams SM, Martinez OM (2008) Tumor-derived variants of Epstein-Barr virus latent membrane protein 1 induce sustained Erk activation and c-Fos. J Biol Chem 283(52):36573–36585. doi:10.1074/jbc.M802968200

Walling DM, Brown AL, Etienne W, Keitel WA, Ling PD (2003) Multiple Epstein-Barr virus infections in healthy individuals. J Virol 77(11):6546–6550

Walling DM, Etienne W, Ray AJ, Flaitz CM, Nichols CM (2004) Persistence and transition of Epstein-Barr virus genotypes in the pathogenesis of oral hairy leukoplakia. J Infect Dis 190(2):387–395. doi:10.1086/421708

Wang FW, Wu XR, Liu WJ, Liang YJ, Huang YF, Liao YJ, Shao CK, Zong YS, Mai SJ, Xie D (2012a) The nucleotide polymorphisms within the Epstein-Barr virus C and Q promoters from nasopharyngeal carcinoma affect transcriptional activity in vitro. Eur Arch oto-Rhino-Laryngol (official journal of the European Federation of Oto-Rhino-Laryngological Societies) 269(3):931–938. doi:10.1007/s00405-011-1862-x

Wang X, Liu X, Jia Y, Chao Y, Xing X, Wang Y, Luo B (2010a) Widespread sequence variation in the Epstein-Barr virus latent membrane protein 2A gene among northern Chinese isolates. J Gen Virol 91(Pt 10):2564–2573. doi:10.1099/vir.0.021881-0

Wang X, Wang Y, Wu G, Chao Y, Sun Z, Luo B (2012b) Sequence analysis of Epstein-Barr virus EBNA-2 gene coding amino acid 148-487 in nasopharyngeal and gastric carcinomas. Virol J 9:49. doi:10.1186/1743-422X-9-49

Wang Y, Kanai K, Satoh Y, Luo B, Sairenji T (2007) Carboxyl-terminal sequence variation of latent membrane protein 1 gene in Epstein-Barr virus-associated gastric carcinomas from Eastern China and Japan. Intervirology 50(3):229–236. doi:10.1159/000100566

Wang Y, Wang XF, Sun ZF, Luo B (2012c) Unique variations of Epstein-Barr virus-encoded BARF1 gene in nasopharyngeal carcinoma biopsies. Virus Res 166(1–2):23–30. doi:10.1016/j.virusres.2012.02.022

Wang Y, Zhang X, Chao Y, Jia Y, Xing X, Luo B (2010b) New variations of Epstein-Barr virus-encoded small RNA genes in nasopharyngeal carcinomas, gastric carcinomas, and healthy donors in northern China. J Med Virol 82(5):829–836. doi:10.1002/jmv.21714

White RE, Ramer PC, Naresh KN, Meixlsperger S, Pinaud L, Rooney C, Savoldo B, Coutinho R, Bodor C, Gribben J, Ibrahim HA, Bower M, Nourse JP, Gandhi MK, Middeldorp J, Cader FZ, Murray P, Munz C, Allday MJ (2012) EBNA3B-deficient EBV promotes B cell lymphomagenesis in humanized mice and is found in human tumors. J Clin Investig 122(4):1487–1502. doi:10.1172/JCI58092

Wrightham MN, Stewart JP, Janjua NJ, Pepper SD, Sample C, Rooney CM, Arrand JR (1995) Antigenic and sequence variation in the C-terminal unique domain of the Epstein-Barr virus nuclear antigen EBNA-1. Virology 208(2):521–530. doi:10.1006/viro.1995.1183

Wu G, Wang Y, Chao Y, Jia Y, Zhao C, Luo B (2012) Characterization of Epstein-Barr virus type 1 nuclear antigen 3C sequence patterns of nasopharyngeal and gastric carcinomas in northern China. Arch Virol 157(5):845–853. doi:10.1007/s00705-012-1241-y

Yang Y, Jia Y, Wang Y, Wang X, Sun Z, Luo B (2014) Sequence analysis of EBV immediate-early gene BZLF1 and BRLF1 in lymphomas. J Med Virol 86:1788–1795. doi:10.1002/jmv.23911

Yao QY, Tierney RJ, Croom-Carter D, Cooper GM, Ellis CJ, Rowe M, Rickinson AB (1996) Isolation of intertypic recombinants of Epstein-Barr virus from T-cell-immunocompromised individuals. J Virol 70(8):4895–4903

Zeng MS, Li DJ, Liu QL, Song LB, Li MZ, Zhang RH, Yu XJ, Wang HM, Ernberg I, Zeng YX (2005) Genomic sequence analysis of Epstein-Barr virus strain GD1 from a nasopharyngeal carcinoma patient. J Virol 79(24):15323–15330

Zhang XS, Song KH, Mai HQ, Jia WH, Feng BJ, Xia JC, Zhang RH, Huang LX, Yu XJ, Feng QS, Huang P, Chen JJ, Zeng YX (2002) The 30-bp deletion variant: a polymorphism of latent membrane protein 1 prevalent in endemic and non-endemic areas of nasopharyngeal carcinomas in China. Cancer Lett 176(1):65–73

Zhang XS, Wang HH, Hu LF, Li A, Zhang RH, Mai HQ, Xia JC, Chen LZ, Zeng YX (2004) V-val subtype of Epstein-Barr virus nuclear antigen 1 preferentially exists in biopsies of nasopharyngeal carcinoma. Cancer Lett 211(1):11–18. doi:10.1016/j.canlet.2004.01.035

Zhao S, Liu WP, Wang XL, Zhang WY, Jiang W, Tang Y, Li GD (2005) Detection of the 30 base pair deletion of Epstein-Barr virus latent membrane protein 1 in extranodal nasal type NK/T-cell lymphoma and its prognostic significance. Zhonghua bing li xue za zhi Chin J Pathol 34(11):720–723

Zuercher E, Butticaz C, Wyniger J, Martinez R, Battegay M, Boffi El Amari E, Dang T, Egger JF, Fehr J, Mueller-Garamvogyi E, Parini A, Schaefer SC, Schoeni-Affolter F, Thurnheer C, Tinguely M, Telenti A, Rothenberger S, Swiss HIVCS (2012) Genetic diversity of EBV-encoded LMP1 in the Swiss HIV Cohort Study and implication for NF-Kappab activation. PLoS ONE 7(2):e32168. doi:10.1371/journal.pone.0032168

Chromatin Structure of Epstein–Barr Virus Latent Episomes

Paul M. Lieberman

Abstract EBV latent infection is characterized by a highly restricted pattern of viral gene expression. EBV can establish latent infections in multiple different tissue types with remarkable variation and plasticity in viral transcription and replication. During latency, the viral genome persists as a multi-copy episome, a non-integrated-closed circular DNA with nucleosome structure similar to cellular chromosomes. Chromatin assembly and histone modifications contribute to the regulation of viral gene expression, DNA replication, and episome persistence during latency. This review focuses on how EBV latency is regulated by chromatin and its associated processes.

Contents

1 Introduction ... 72
2 History and Discovery of the EBV Latent Episomes ... 73
3 Nuclear Entry, DNA Recognition, and Chromatinization ... 73
4 Early Viral Gene Expression ... 76
5 Chromatin Organization of the Latent Episome ... 78
6 OriP as a Chromosome Organizing Element ... 80
7 EBNA1 as a Critical Regulator of EBV Chromatin ... 82
8 Barriers to Lytic Reactivation ... 83
9 Late Gene Transcription and Replication ... 85
10 Network and Homeostatic Control of Viral Chromatin ... 86
11 Chromatin Modifiers that Disrupt Latency—Environment, Pathogenesis, and Therapeutics ... 88
12 Variation and Heterogeneity of Viral Chromatin Control ... 89
13 Conclusions ... 89
References ... 90

P.M. Lieberman (✉)
The Wistar Institute, Philadelphia, PA 19104, USA
e-mail: Lieberman@wistar.org

C. Münz (ed.), *Epstein Barr Virus Volume 1*, Current Topics in Microbiology and Immunology 390, DOI 10.1007/978-3-319-22822-8_5

Abbreviations

3C	Chromatin conformation capture
BAH	Bromo-adjacent homology
BL	Burkitt's lymphoma
CBD	Chromosome binding domain
ChIP	Chromatin immunoprecipitation
CMV	Cytomegalovirus
CTD	Carboxy-terminal domain
DDR	DNA damage response
DS	Dyad symmetry
EBV	Epstein–Barr virus
FGARAT	Phosphoribosylformylglycinamidine synthase
FR	Family of repeats
HDAC	Histone deacetylase
INR	Initiator element
KSHV	Kaposi's sarcoma-associated herpesvirus
LCL	Lymphoblastoid cell line
MCFA	Medium chain fatty acids
MNase I	Micrococcal nuclease I
NPC	Nasopharyngeal carcinoma
NK/T	Natural killer/T cell
OriP	Origin of plasmid replication
PML-NB	Promyelocytic leukemia-nuclear body
SCFA	Small chain fatty acids
SUMO	Small ubiquitin-like modifier
TPA	Phorbol ester 12-O-Tetradecanoylphorbol-13-acetate
VPA	Valproic acid

1 Introduction

Epstein–Barr Virus (EBV) can act as a cofactor in numerous and diverse human cancers, ranging from B- and NK/T-cell lymphomas to epithelial carcinomas of the stomach and nasopharynx. In all instances, EBV-associated pathogenesis correlates with the persistence of the viral genome in the majority of tumor cells. This persistence is referred to as a "latent" infection, because infectious viral particles are rarely produced. Importantly, EBV latent infection is a highly active, dynamic, and "programmed" process, where the viral DNA co-opts host-cell fate decisions, including proliferation, differentiation, and survival. During latency, the viral genome expresses a restricted set of latency-associated genes, replicates once and only once, and segregates faithfully to newly divided daughter cells. The latency gene expression programs are highly dynamic and respond to host-cell-specific factors and environmental signals.

This review considers how viral latency is controlled by chromatin. In it broadest definition, chromatin refers to the nucleoprotein complexes that assemble on DNA to form a functional chromosome. At the most basic level, chromatin is composed of the histone core proteins that form nucleosomes by wrapping two turns of ~145 bp of DNA around an octamer of H2A-H2B/H3-H4 (review in Zentner and Henikoff 2013). The histones can be extensively post-translationally modified to produce epigenetic marks that have profound effects on DNA metabolism and gene expression. These modifications are commonly referred to as the "marks" of a histone code that is interpreted by protein "readers," "writers," and "erasers" (Ruthenburg et al. 2007; Jenuwein and Allis 2001). Viruses, such as EBV, utilize and manipulate the histone code and chromatin structure to execute their own lifecycle programs. For EBV, chromatin formation is critical for the persistence of the viral genome and the control of viral gene expression during latent infection.

2 History and Discovery of the EBV Latent Episomes

EBV was identified in 1964 as a latent infection associated with Burkitt lymphoma (BL) tissue and BL-derived cell lines (Epstein et al. 1964). EBV DNA from latently infected cells could be separated from chromosomal DNA by gradient centrifugation (Nonoyama and Pagano 1972) and sedimented as closed circular episomes (Adams and Lindahl 1975). The chromatin structure of EBV was first examined MNase I nuclease digestion assays that revealed a nucleosome organization that was indistinguishable from the host bulk chromosome (Shaw et al. 1979). Later studies identified several regions of the genome that were DNase I sensitive and likely to be nucleosome-free regions (Dyson and Farrell 1985). This was confirmed by Sexton and Pagano who showed that the regions within the origin of plasmid replication (OriP) were nucleosome free, suggesting that chromatin structure was not uniform throughout the latent genome (Sexton and Pagano 1989) and therefore subject to regulation.

3 Nuclear Entry, DNA Recognition, and Chromatinization

The early events that lead to a stable latent infection have been referred to as the pre-latency phase, since latency is the predominant outcome of primary infection in resting B-lymphocytes (Kalla and Hammerschmidt 2012). Several major events occur on the path to latency, all of which are highly regulated, both by the cell and virus. The major events that lead to assembling a functional latent episome include membrane receptor engagement and signal transduction, capsid transport to the nucleus, nuclear entry of viral DNA, genome circularization, genome chromatinization, transcription of latency-associated genes, repression of lytic cycle genes, and genome copy number maintenance (Fig. 1).

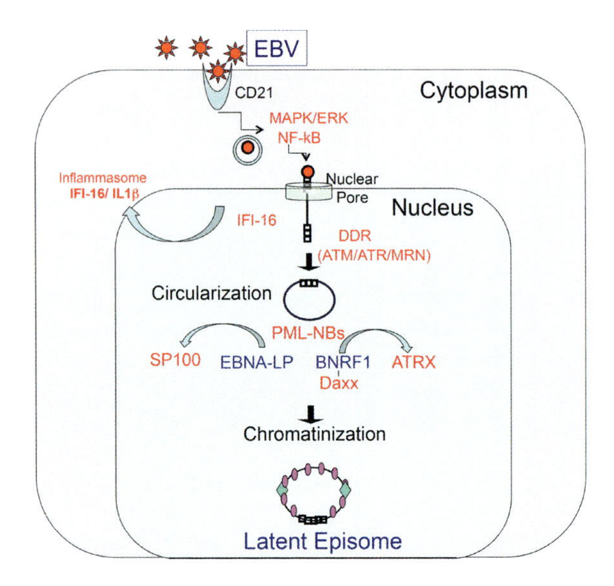

Fig. 1 Early events in the establishment of EBV episomal chromatin. EBV virions enter through CD21 receptor and activate signal transduction pathways that can influence nuclear transcription factors, including NF-kB and AP1. Linear, viral DNA lacking nucleosomes enter the nuclear pore and confront host-cell factors, including IFI-16, PML-NBs, and components of the DDR. Viral genomes circularize and assemble into chromatin (chromatinization). At least one viral tegument protein, BNRF1, and early latency gene product, EBNA-LP, can alter the PML-NBs and influence the chromatinization of the viral genome to promote latency-associated gene expression

Like other herpesviruses, EBV is thought to enter the nucleus as naked linear DNA without any associated chromatin proteins. Mass-spectrometry and proteomic studies of EBV virions fail to detect histone proteins (Johannsen et al. 2004). This is consistent with previous studies showing that newly replicated DNA and virion-associated DNA lack typical nucleosome structure (Dyson and Farrell 1985). It is also consistent with a more recent study finding that EBV productive replication occurs in histone-deficient replication compartments and in cells depleted of histone chaperones (Chiu et al. 2013). This strongly suggests that histones are not incorporated into the newly synthesized and encapsidated viral genomes. How the viral DNA gets inserted into the nucleus is also not known for EBV. For HSV-1, the DNA is thought to enter the nucleus through nuclear pores (Meyring-Wosten et al. 2014; Liashkovich et al. 2011). Nuclear pore interactions may be highly significant, since nuclear pore proteins can function in chromatin assembly and transcriptional regulation (Ptak et al. 2014).

Among the first nuclear events known to occur when viral genomes enter the nucleus is the confrontation with the host intrinsic resistances and anti-viral responses (Weitzman et al. 2010; Boutell and Everett 2013). These include the DNA damage response (DDR), the inflammatory response, and the formation of PML-nuclear bodies. The promyelocytic leukemia (PML) protein assembles into a PML-nuclear body (PML-NB) at sites where viral DNA localizes in the nucleus (reviewed in Van Damme and Van Ostade 2011; Everett and Chelbi-Alix 2007). PML-NB formation depends on SUMO modification of PML (Cuchet-Lourenco et al. 2011) and the recruitment of SUMO interacting proteins, such as Daxx, and other interacting proteins, like ATRX and SP100, that collectively form a nuclear body where viral DNA-structures may be captured (Reichelt et al. 2011). PML-NBs are typically destroyed or disabled by viral-encoded proteins, such as HSV-1 ICP0 and CMV pp75 (Saffert and Kalejta 2008). Destruction of PML-NBs is necessary for productive infection of HSV and CMV, and likely to be necessary for EBV lytic replication, as well.

At least one EBV tegument protein, BNRF1, has been implicated in modulating PML-NBs through its interaction with Daxx (Tsai et al. 2011). BNRF1 is a member of the FGARAT family of proteins, conserved among gammaherpesvirus, and includes KSHV ORF75, MHV68 ORF75a, b, c, and HVS ORF75 and ORF3. While these genes have been duplicated or triplicated in other gammaherpesvirus, EBV retains a single member of this family. The FGARAT domain has homology to de novo purine biosynthesis enzymes conserved from prokaryotes to humans. However, no enzymatic activity has yet been attributed to the viral FGARAT proteins and amino acid mutations in catalytic amino acids suggest that gammaherpesvirus FGARAT proteins have lost conventional purine biosynthesis capability, but may have acquired other, as yet unknown, activities.

Recent studies have revealed that all gammaherpesvirus FGARAT proteins interact with components of the PML-NBs. The EBV-encoded BNRF1 protein has been shown to interact directly with Daxx (Tsai et al. 2011). Daxx has been shown to collaborate with ATRX to assemble histone variant H3.3 into repressive chromatin at GC-rich repetitive DNA (Lewis et al. 2010; Goldberg et al. 2010). BNRF1 binding to Daxx displaces ATRX from binding Daxx (Tsai et al. 2011). The implication of these studies is that BNRF1 prevents ATRX association with Daxx, and thereby inhibits ATRX-dependent formation of repressive H3.3 on viral genomes. Interesting, other gammaherpesvirus FGARAT targets different components of the PML-NBs and in most cases leads to the degradation of one or more components (Full et al. 2014; Ling et al. 2008). The fact the BNRF1 does not cause the degradation of PML, Daxx, or ATRX suggests that EBV has a different strategy for interacting with PML-NBs than other herpesviruses. The modulation of Daxx–ATRX without its degradation may be related to the propensity of EBV to establish a latent, rather than a lytic infection.

A potentially related response to viral DNA is the interaction with interferon inducible protein 16 (IFI-16). IFI-16 was shown to recognize the newly infecting DNA genomes of KSHV as they enter the nucleus (Kerur et al. 2011). IFI-16 interaction with KSHV DNA results in the nuclear export of IFI-16 and the activation of the ASC-linked inflammasome (Singh et al. 2013). IFI-16 also interacts with the latent EBV episome in the nucleus to cause the chronic activation of the inflammasome (Ansari et al. 2013). IFI-16 interacts with structured DNA with a preference for four-way junction cruciform DNA and superhelical DNA (Brazda et al. 2012). It is proposed that IFI-16 preferentially recognizes EBV latent genome relative to cellular DNA (Ansari et al. 2013). IFI-16 has been linked to

DNA damage recognition and response (Aglipay et al. 2003). IFI-16 also interacts with the KSHV (Singh et al. 2013; Kerur et al. 2011), CMV (Horan et al. 2013; Gariano et al. 2012), and HSV-1 (Johnson et al. 2013; Orzalli et al. 2013; Conrady et al. 2012) genome upon entry in the nucleus, and leading to inflammasome signaling. In HSV-1, IFI-16 has been implicated in transcriptional repression of viral genes (Orzalli et al. 2013). IFI-16 signaling and cytoplasmic export is regulated by lysine acetylation (Li et al. 2012). Whether IFI-16 regulates chromatin formation on the EBV episome is not known.

Foreign DNA entering the nucleus is expected to elicit a DNA damage response (Weitzman et al. 2010). EBV infection has been shown to activate the DNA damage response (DDR) through the ATM/Chk2 kinasepathway during the pre-latency phase of infection (Nikitin et al. 2010). However, this DDR response occurs at ~7 days of post-infection and correlates best with EBV-induced host-cell hyperproliferative response. An earlier DDR to linear, non-chromatinized DNA may be masked by EBV-encoded factors, but these have not yet been characterized. Pharmacological inhibitors of ATM and Chk2 increase EBV transformation efficiency upon primary infection (Nikitin et al. 2010), indicating that DDR pathways are an inherent block to EBV latency. EBV encodes a number of other tegument proteins that are likely to influence these early events in chromatin assembly and gene regulation. Among these are BPLF1 which has deubiquitinating and deneddylating activity (Gastaldello et al. 2010, 2012). BPLF1 has been shown to interact with cellular protein complex Rad6/18 and deubiquinate cellular PCNA and to functionally inhibit ribonulceotide reductase activity and attenuate DNA replication lesion repair synthesis (Kumar et al. 2014; Whitehurst et al. 2009, 2012). Whether BPLF1 regulates the DDR response during early stages of infection, and how this may affect viral chromatinization is not yet known.

4 Early Viral Gene Expression

The earliest viral gene expression detected after de novo infection of resting human primary B-lymphocytes is the EBNA2 transcript derived from the Wp promoter (Tierney et al. 2000a, b, 2007; Alfieri et al. 1991; Shannon-Lowe et al. 2005). The B-cell transcription factor, BSAP/PAX5, was found to bind and regulate Wp transcription during primary infection (Tierney et al. 2000a, 2007). Wp-initiated transcriptscan make a short form of EBNA-LP and EBNA2 only. EBNA-LP and EBNA2 interact with chromatin regulatory factors to further promote the formation of the viral chromosome and latency gene expression program.

EBNA-LP can bind and inactivate SP100, a member of the PML-NBs that has been implicated in transcription repression (Echendu and Ling 2008; Ling et al. 2005). EBNA-LPtargets a different component of the PML-NBs than does BNRF1 and at a time when BNRF1 may be degraded. This suggests that temporal remodeling of PML-NBs by EBV is necessary for coordinated chromatin assembly and latency transcription during the pre-latency phase.

EBNA2 recruits several transcriptional co-activators and chromatin regulatory factors to the viral episome. Biochemical studies reveal that EBNA2 can interact with SNF2 family of ATP-dependent chromatin remodeling factors (Wu et al. 1996), p300/CBP family of histone acetylases (Wang et al. 2000), and RNA pol II initiation factor TFIIH (Chabot et al. 2014). EBNA2 does not bind to DNA directly, but associates with several sequence-specific transcription factors, most notably RBP-jK(also referred to as CBF1and CSL) (Ling et al. 1993). RBP-jKis the scaffold for intracellular Notch binding, which leads to activation of specific class of genes important for growth and differentiation in response to extracellular signaling through Delta (reviewed in Hayward 2004). EBNA2 can also interact with other transcription factors, including Pu.1, which is implicated in the regulation of LMP1 transcription (Sjoblom et al. 1995a, b). More recent genome-wide studies indicate that EBNA2 colocalizes with many B-cell-specific transcription factors throughout the host chromosome, including the BATF/IRF4or SPI1/IRF4 (Jiang et al. 2014; Zhao et al. 2011). Whetherthese sites are also found on the viral chromosome are not yet known.

Once EBNA2 is expressed, it triggers a promoter switch to the upstream start site referred to as Cp (Woisetschlaeger et al. 1991). EBNA2-driven Cp is competent for generating a much longer (~100 kb) multicistronic transcript that encodes EBNA3A, 3B, and 3C, as well as EBNA1. Enhancing RNA pol II elongation may be an important component of the transition to type III latency where all EBNA3 genes are generated from this long Cp-initiated transcript. EBNA2 binding at Cp enhances recruitment of the pTEF-b elongation factor for RNA pol II (Palermo et al. 2008, 2011). EBNA2 stimulates the CDK9-dependent phosphorylation of RNA pol II CTD and enhances its ability to overcome negative elongation factor blocks to transcription elongation. Recent studies in transcription and chromatin regulation have implicated the control of RNA pol II elongation as a major step in transcription control throughout the host genome, as well as in the formation of chromatin organization and higher-order structures (Gilchrist and Adelman 2012).

EBNA1 can be generated from an alternative promoter, referred to as Qp (Schaefer et al. 1995). Qp can be expressed prior to Cp utilization (Schlager et al. 1996), and Qp is thought to be auto-repressed by high levels of EBNA1 that are generated through Cp (Yoshioka et al. 2008). Transcription initiation at Qp may be driven by several different transcription factors, including Initiator element (INR) (Nonkwelo et al. 1997), interferon regulatory factors IRF1, IRF2 (Schaefer et al. 1997), and IRF7 (Zhang and Pagano 2000), STATs (Chen et al. 1999), Rb (Ruf and Sample 1999), histone demethylase LSD1 (Chau et al. 2008), chromatin-organizing factor CTCF (Salamon et al. 2009; Day et al. 2007), and heat shock factor 1 (HSF1) (Wang et al. 2011). EBNA1 also binds to the transcription initiation site of Qp where it can auto-inhibit transcription through a mechanism that involves binding to its own pre-mRNA (Yoshioka et al. 2008). Thus, Qp regulation depends on both EBNA1 levels and host-cell factors.

Some latency-associated viral genes can be activated by cellular, as well as viral transcriptional regulators. LMP1 transcription can be activated by EBNA2, but EBNA2-independent promoter activity has also been observed in different latency and tumor types. Factors that regulate LMP1 expression include Pu.1(Sjoblom et al. 1995a, b), ATF/CRE (Sjoblom et al. 1998), E-box proteins (Sjoblom-Hallen et al. 1999), NF-kB (Johansson et al. 2009), STAT3/5 (Chen et al. 2001, 2003), C/EBP (Noda et al. 2011), and CTCF (Chen et al. 2014a). Similar regulation can be observed for LMP2. LMP1 and LMP2 expression independent of EBNA2 is important since they are frequently expressed in EBV-positive epithelial tumors lacking EBNA2 (Tsao et al. 2002). In addition, many epithelial tumors have high levels of EBV miRNAs and BART transcripts (Pratt et al. 2009), which initiate from a regions that may not depend upon EBNA2 co-activation (Kim do and Lee 2012). The temporal regulation of BART and miRNA transcription is not known with respect to establishment of latency, but it is likely that these are activated as early transcripts in epithelial infections, which have been difficult to model efficiently in vitro. Among the BART and miRNA promoter regulatory elements are C/EBPbeta and cMyc (Kim do and Lee 2012; Chen et al. 2005). It is not known what factors control the different expression levels of BART and miRNAs in different cell and tumor types.

5 Chromatin Organization of the Latent Episome

The coordination or competition between chromatin assembly and transcription factor binding on the EBV genome is likely to play an important role in establishing the latent episomal chromosome. The conserved cellular factor CTCF has been strongly implicated in chromatin organization in all metazoan organisms (Ong and Corces 2014; Phillips and Corces 2009). CTCF is an 11-zinc-finger DNA binding protein that can organize both nucleosome position and higher-order DNA loop interactions including mediating promoter–enhancer interactions. CTCF sites are often co-occupied by cohesin subunits (SMC1, SMC3, and RAD21), which can mediate sister chromatid cohesion as well as facilitate DNA–loop interactions important for gene regulation (Merkenschlager and Odom 2013). In the established EBV episome in latently infected B-lymphocytes, CTCF and cohesin bind at ~19 binding sites throughout the viral genome, with some sites showing higher co-occupancy (Arvey et al. 2012, 2013) (Fig. 2).

CTCF binding sites can influence nucleosome position and histone modification patterns. For EBV, CTCF binding sites upstream of Cp and Qp have been implicated as chromatin boundaries that prevent the spreading of repressive heterochromatin into the promoter control regions (Tempera et al. 2010; Chau et al. 2006). Mutation of CTCF binding sites in EBV bacmids reintroduced into 293 cells showed an increase in heterochromatin formation at Cp and Qp after several weeks in culture. CTCF binding sites are CpG-rich, and DNA methylation can inhibit CTCF binding. At Qp, CTCF has been shown to prevent DNA methylation (Tempera et al. 2010), and most latently infected cells lack detectable DNA methylation at Qp (Takacs et al. 2010). Thus, one major function of CTCF is to prevent

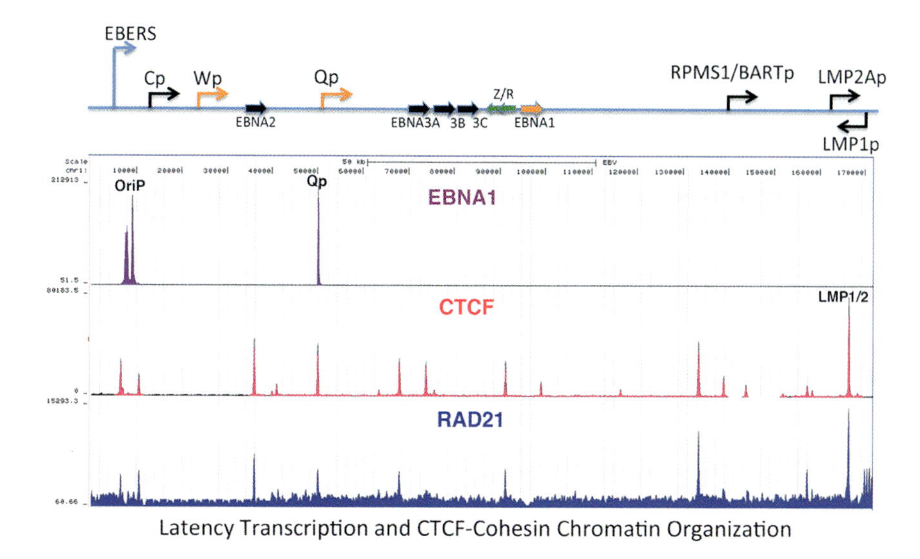

Fig. 2 EBV latent transcript promoters aligned with CTCF-cohesin chromatin organization. Schematic of the EBV genome depicting the position of transcriptionstart sites for the major latency transcripts, including the RNA pol III-dependent EBERS, and the RPMS1/BARTp promoter that expresses BART miRNAs. Lytic immediate early genesBZLF1(Z) and BRLF1 (R)are indicatedby *green arrows*. Below shows the ChIP-Seq tracks for EBNA1, CTCF, and Cohesin in EBV-infected lymphoblastoid cell lines

DNA methylation at critical promoters, such as Qp. In this capacity, CTCF functions as an insulator.

Both CTCF and cohesins were found to be highly enriched at the LMP1/LMP2 locus, binding within the first intron of LMP2A and the 3' UTR of LMP1. LMP1 and LMP2 are convergently transcribed, and CTCF may function to coordinate some of RNA pol II activity to avoid head-to-head collisions. Mutations of the CTCF binding site at the LMP1/LMP2 locus led to a loss of LMP1 and LMP2A expression, and a surprising increase in LMP2B expression initiating from downstream promoters near the terminal repeats (Chen et al. 2014a). Cohesin depletion did not have the same effect as CTCF binding site mutations, suggesting that these factors have different fundamental activities in gene regulation. Interestingly, shRNA depletion of cohesins leads to the reactivation of KSHV, but not EBV in latently infected B-lymphoma cell lines (Li et al. 2014; Chen et al. 2012). This suggests that cohesins have a different regulatory function for EBV and KSHV, or that their function is cell-type and latency-type dependent. The function of CTCF and cohesin at the LMP1/2 promoter appears to be more complex than chromatin boundary function, and DNA looping may mediate more complex 3D regulatory interactions. DNA looping has been assayed using chromosome conformation capture (3C) for the LMP1/LMP2 region, and an interaction with OriP was identified (Chen et al. 2014a; Arvey et al. 2012). DNA loop interactions were also found between OriP and Cp or Qp, depending on latency type and correlating with promoter activity (Tempera et al. 2011). These findings suggest that higher-order chromosome conformation is important for regulating viral gene expression during latency. In this capacity, CTCF functions to facilitate DNA loops and hubs between regulatory elements.

6 OriP as a Chromosome Organizing Element

The unusual chromosome structure of OriP was recognized in early studies examining EBV nucleosome patterns and DNAse sensitivity (Sexton and Pagano 1989; Dyson and Farrell 1985). More recent 3C chromatin conformation capture studies suggest that Orip may serve as a central hub in mediating multiple interactions with the viral genome (Tempera and Lieberman 2010, 2014). OriP has been shown to function as a transcriptional enhancer of Cp and LMP1 promoters (Gahn and Sugden 1995; Nilsson et al. 2001; Puglielli et al. 1996), and 3C data provide evidence for a physical interaction between OriP and these promoters when transcriptionally active. The chromatin structure around OriP has been examined by MNase I digestion and by genome-wide ChIP-Seq studies (Arvey et al. 2012, 2013) (Fig. 3). First, EBNA1 binding to FR can prevent nucleosome binding in vitro, and the FR region is either nucleosome-free or has irregular unphased nucleosomes. Purified

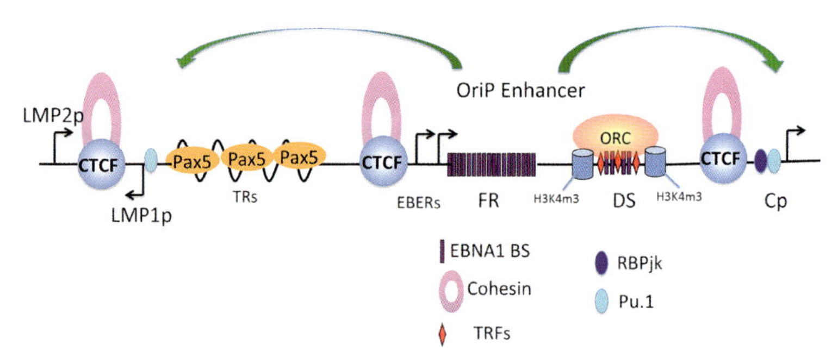

Fig. 3 Organization of the latency control region around the OriP enhancer. The EBNA1 binding sites (*purple boxes*) at the family of repeats (FR) and Dyad Symmetry (DS) that constitute OriP are depicted as the central regulator of EBV transcription and chromatin during latency. OriP functions as a transcriptional enhancer of the promoters at LMP1 and Cp (controlling EBNA multicistronic transcript). CTCF-cohesinbinding sites are also indicated as they occupy sites upstream or close to promoter elements at Cp, EBERs, and LMP1/2. The DS is shown to bind telomere repeat factors (TRFs) and ORC and have positioned nucleosomes enriched in H3K4me3. PAX5 is shown to be enriched at the terminal repeats. Binding sites for B-cell transcription factors RBP-jKand Pu.1that dock EBNA2 at Cp or LMP1 are indicated

EBNA1 can efficiently assemble onto chromatinized templates and destabilize histones bound to OriP (Avolio-Hunter and Frappier 2003; Avolio-Hunter et al. 2001).

Genome-wide ChIP-Seq studies indicate that the region surrounding OriP is enriched in euchromatic marks for H3K4me3 (Arvey et al. 2013). Some of the H3K4me3 may be due to the very high levels of RNA pol III transcription at the EBERs which are immediately adjacent to OriP. However, early genetic studies have suggested that EBER deletion has no effect on the ability of EBV to immortalize B-lymphocytes or establish latent infection (Swaminathan et al. 1991). Other factors are known to bind within, or near OriP, and these may also influence chromatin-organizing functions and related OriP activities. These include cMyc binding to the EBER promoter (Niller et al. 2003), and Oct2 and E2F factors interacting with FR elements (Borestrom et al. 2012; Almqvist et al. 2005).

The DS element of OriP can function as an efficient origin of DNA replication, but does not appear to be essential for virus genome replication during latency (Ott et al. 2011; Norio and Schildkraut 2004; Norio et al. 2000). DNA replication can initiate at several other locations in the genome, including Rep* and a region within the BamHI A locus (Kirchmaier and Sugden 1998). This is very similar to how host chromosome DNA replication initiates in loosely defined zones associated with euchromatic histone modifications. Nevertheless, the DS has complex chromatin regulation that suggests it contributes to the overall success of EBV latency in some cell types. In addition to recruiting ORC and MCMs (Ritzi et al. 2003; Chaudhuri et al. 2001; Dhar et al. 2001), essential for replication origin function, the DS binds to telomere repeat factors TRF1 and TRF2, which enhance DNA replication and episome maintenance (Deng et al. 2002; Lindner et al. 2008). The telomere repeat factor binding sites flank EBNA1 sites within OriP and contribute to nucleosome position and histone modification (Zhou et al. 2005). ORC may also contribute to chromatin regulation, as it has recently been shown to interact with specific histone modifications (e.g., H4K20me2) through the BAH domain of ORC1 (Kuo et al. 2012).

Like most cellular enhancers, OriP associates with many protein factors, and it is not clear which of these are essential for transcription activation function. However, unlike cellular enhancers, OriP is unique in its repetitive DNA structure, aberrant nucleosome structure, and ability to tether to metaphase chromosomes. Although no specific tethering sites have been identified, OriP generally associates with active regions of the nucleus (Deutsch et al. 2010). IF FISH studies reveals that viral genomes associate with chromatin regions enriched in H3K4me3 and H3K9ac. Loss of DS showed a slight increase in association of EBV genomes with H3K27me3 domains. The FR region was more critical for this association, than DS. These studies suggest that OriP can determine what type of chromatin domain EBV episomes associate with during interphase, and possibly tether to during metaphase chromosome formation and mitotic division (Fig. 3).

7 EBNA1 as a Critical Regulator of EBV Chromatin

EBNA1 has essential functions in viral transcription, DNA replication, episome maintenance, and host-cell survival (reviewed in Frappier 2012; Smith and Sugden 2013). EBNA1 binds with high affinity and selectivity to three major positions in the EBV genome: FR, DS, and Qp (Rawlins et al. 1985). EBNA1 can also bind to sequence-specific sites in the host genome (Lu et al. 2010; Dresang et al. 2009). Sequence-specific DNA binding is mediated by the C-terminal DNA binding domain, which shares structural homology to KSHV LANA and HPV E2 DNA binding domains (Bochkarev et al. 1995). EBNA1 can also tether to metaphase chromosomes through two amino-terminal chromosome binding domains termed CBD1 (aa40-54) and CBD2 (328-377) (Kanda et al. 2013). The arginine-rich (RGG-like) CBDs can interact with numerous chromatin-associated substrates including AT-rich DNA, G-quadruplex RNA, histone H1, and EBP2 (Norseen et al. 2008; Sears et al. 2004; Hung et al. 2001; Shire et al. 1999). EBNA1 CBDs can also interact with nucleosome core particles through electrostatic interactions (Kanda et al. 2013), similar to that observed for KSHV LANA (Barbera et al. 2006).

Among the many functions of EBNA1 is its ability to induce higher-order chromatin structures. Recent studies suggest that EBNA1 can condense host chromosome by mimicking the architectural protein HMG1A (Coppotelli et al. 2013). EBNA1 can also form homo-typic oligomeric interactions through a Zn-hook formed by conserved cysteine and histidine residues in the amino-terminal domain (Singh et al. 2009; Sears et al. 2004). This interaction is likely to be critical for OriP transcriptional enhancer function as mutations in these amino acid residues caused the loss of EBNA1-dependent transcription activation of Cp-driven EBNA2 during primary B-cell infection (Altmann et al. 2006). How these different chromatin interacting and chromosome organizing activities contribute to each of EBNA1's functions in viral latency remain to be delineated.

Proteomic studies reveal that EBNA1 interacts with several chromatin modulatory proteins, including NAP1, TAF1b, USP7, PRMT5, nucleophosmin (Malik-Soni and Frappier 2012, 2014), and nucleolin (Chen et al. 2014b). Nucleolin affects EBNA1 DNA binding and transcription activation function, and this is dependent on RNA-binding domain of both nucleolin and EBNA1. USP7 and GMP synthase (GMPS) can be recruited by EBNA1 to OriP as a complex that regulates histone H2B deubiquitination (Sarkari et al. 2009). USP7-dependent deubiquitination of H2B correlates with a loss of EBNA1-dependent transcription activation (Sarkari et al. 2009). Thus, USP7 and GMPS may downregulate OriP enhancer function at some stages of EBV latency.

EBNA1 is an essential chromatin regulatory protein that has multiple functions on and off the viral genome. During latency, EBNA1 binds to several cellular genes (Lu et al. 2014; Smith and Sugden 2013) and colocalizes with regions of the interphase chromatin that is newly replicated (Ito et al. 2002). During lytic replication, EBNA1 localizes to and degrades PML nuclear bodies, and this is facilitated by EBNA1 interaction partners CK2 and USP7 (Sarkari et al. 2011). Thus, there remain many unanswered questions as to how EBNA1 and associated cellular factors mediate the essential functions of OriP, including enhancer–promoter interactions and chromosome tethering during latency. Notably, most of these EBNA1 functions are chromatin-associated processes.

8 Barriers to Lytic Reactivation

To maintain latency, lytic gene expression must be repressed. Leaky lytic gene expression, especially of the immediate early gene BZLF1 (also known as EB-1, Zta, Z, ZEBRA), would enable progression into the lytic cycle. BZLF1 expression and function is regulated at multiple levels and reviewed elsewhere (reviewed in Murata and Tsurumi 2013; Miller et al. 2007; Kenney and Mertz 2014). The chromatin regulation of the BZLF1 promoter is closely linked to transcription factor binding and signal transduction pathways involved in reactivation. Transcription of the BZLF1 gene can be activated by various signaling pathways depending on the latently infected cell type. Among the more commonly studied lytic-inducing agents are phorbol esters (TPA), halogenated nucleotides, calcium ionophores, sodium butyrate, and anti-IgG activation of the B-cell receptor in latently infected B-lymphocytes (reviewed in Speck et al. 1997; Amon and Farrell 2005; Tsurumi et al. 2005). Numerous transcription factors can bind BZLF1 promoter and respond to these signaling pathways. Transcription activators that bind BZLF1 promoter include MEF2 (Liu et al. 1997b; Murata et al. 2013) (Gruffat et al. 2002), Sp1/3 (Liu et al. 1997a), CREB, ATF, AP1, C/EBP (Wu et al. 2004), XBP-1 (Bhende et al. 2007), JDP2 (Murata et al. 2011), YY1 (Montalvo et al. 1995), SMAD (Iempridee et al. 2011), and ZEB1/2 (Ellis et al. 2010; Kraus et al. 2003). Other transcription factors, such as STAT3 (Hill et al. 2013) and Ikaros (Iempridee et al. 2014), have been shown to regulate EBV reactivation indirectly (Fig. 4).

In the absence of positive activation signals, several of these factors can repress BZLF1 transcription. MEF2 recruitment of class II HDAC to EBV BZLF1 links latency and histone modification (Gruffat et al. 2002). Depletion of ZEB1/2 or mutation of ZEB binding sites in BZLF1 promoter induces BZLF1 transcription and EBV reactivation (Ellis et al. 2010; Kraus et al. 2003). ZEB1/2 represses transcription through interactions with corepressor CtBP and HDACs (Postigo and Dean 1999). CTCF, which can function as a transcriptional repressor, was also found to bind the BZLF1 promoter (Holdorf et al. 2011; Arvey et al. 2012). However, a mutation in this CTCF site did not effect BZLF1 expression levels, at least in 293 HEK cells (Murata and Tsurumi 2013).

A search for epigenetic marks that control EBV latency has been initiated on several fronts. DNA methylation plays an important role in the epigenetic regulation of EBV latency, and this is reviewed extensively elsewhere (Woellmer and Hammerschmidt 2013; Takacs et al. 2010; Miller et al. 2007; Ambinder et al. 1999).

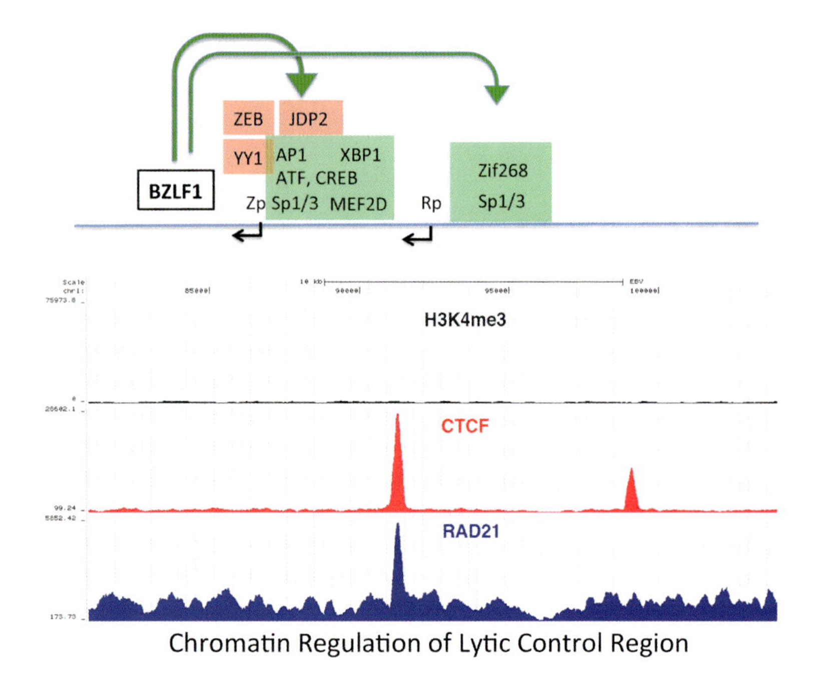

Fig. 4 Lytic control region chromatin regulatory factors. Schematic of the BZLF1 promoter regulatory region and some of the factors that have been implicated in its regulation. These include the direct DNA binding factors SP1/3, MEF2D, AP1, ATF, CREB, XBP1, and Zif268 that have been implicated in transcription activation (*green boxes*) and the factors implicated in repressing Zp (YY1, ZEB, JDP2). BZLF1-encoded protein (also referred to as Z, Zta, ZEBRA, EB1) is an important auto-activator of Zp and Rp. Rp-encoded transcription activator Rta (or R) is a potent activator of Zp, and other viral lytic genes, especially in epithelial cell latency. ChIP-Seq tracks for histone H3K4me3 show an absence of euchromatic at Zp in LCLs, with enrichment of CTCF and Rad21 immediately upstream of Zp regulatory factors

DNA methylation levels at the BZLF1 promoter were shown to be relatively low compared to most regions of the latent genome (Hernando et al. 2013). However, treatment of some latently infected cell types with demethylating agents such as 5' azacytidine can induce BZLF1 transcription and reactivation (Murata et al. 2012). It is possible that demethylating agents may also induce DNA damage and DDR which may function as more effective triggers of EBV reactivation.

Histone modifications have been implicated in the control of BZLF1 gene expression and EBV lytic reactivation. HDAC inhibitors are known to stimulate reactivation in many, but not all, latently infected cells (Luka et al. 1979; Jenkins et al. 2000). EBV chromatin histone acetylation levels are generally low during latency and increased uniformly during reactivation, including the region encompassing the BZLF1 promoter (Murata et al. 2012). ChIP assays have revealed that H3K27me3, H3K9me2/3, and H4K20me3 are enriched at the BZLF1 promoter in latently infected Raji cells (Murata et al. 2012), and H3K27me3 and H3K9me3 in

Akata cells (Ramasubramanyan et al. 2012). In Raji cells, HP1 binding and H2A ubiquitination were associated with BZLF1 in latency, while histone acetylation and H3K4me3 dominate during reactivation (Murata et al. 2012). Interestingly, while treatment of Raji cells with DZNep (inhibitor of the EZH2 H3K27me3 methylase) or TSA alone had only a modest effect on EBV reactivation, the combination of these two drugs led to a synergistic activation of BZLF1 transcription and viral reactivation. This suggests that both derepression of H3K27me3 and activation through histone acetylation (HDAC inhibition) are required for efficient activation of BZLF1 promoter in Raji cells. H3K27me3-mediated repression has also been implicated in KSHV latency, as inhibitors of EZH2, such as DZNep, are sufficient to induce reactivation as single agents (Toth et al. 2010, 2013a). However, these effects may be cell-type dependent. DZNep can also inhibit the methylation of H3K20me3 through SUV420h1, and knockdown of this gene could also lead to induction of BZLF1 transcription (Murata et al. 2012). H3K9me3 was also found to be elevated at the BZLF1 promoter, but inhibitors of G9a (BIX01294) and chaetocin (inhibitor of SUV39a) did not induce BZLF1 transcription. However, it was not clearly demonstrated that BIX01294 or chaetocin efficiently removed H3K9me3 from BZLF1 promoter. Histones at the BZLF1 promoter were also found to be enriched in histone H2A monoubiquitination on lysine 119 as well as enriched for HP1 (Murata et al. 2012). These findings suggest that multiple different forms of chromatin repression contribute to Zp repression during latency.

The DNA methylation state of the virus has also been investigated in great detail, and while this plays an integral role in the epigenetic control of gene expression and is deeply coordinated with chromatin structure and function, much of this subject is the focus of other chapters in this volume and reviewed elsewhere (Woellmer and Hammerschmidt 2013; Takacs et al. 2010; Ambinder et al. 1999).

9 Late Gene Transcription and Replication

The EBV lyticgenome evades histone assembly during lytic cycle replication (Chiu et al. 2013). During lytic replication, the nuclear morphology is reorganized, and replication compartments containing viral replication proteins, as well as cellular proteins, such as PCNA, form higher-order structures. Cellular DNA and histones are excluded from these compartments, and cellular PCNA does not colocalize with replicating viral DNA. Since PCNA is implicated in histone chaperone recruitment and histone assembly during cellular DNA replication, the segregation of PCNA away from viral replicating DNA may partly account for the escape from histone assembly. This study also found that EBV lytic cycle leads to the selective loss of cellular histone chaperones CAF1 and ASF1a at the protein and mRNA levels, providing an additional mechanism for the reduction of histone assembly on viral DNA during lytic amplification (Chiu et al. 2013).

EBV, like all herpesviruses, requires lytic cycle DNA replication to enable transcription of viral late genes. In gammaherpesviruses, late genes have an unusual TATT sequence present in the promoter regions where the TATA box is typically found (Serio et al. 1998; Wong-Ho et al. 2014). Recent studies have identified a herpes-encoded TATA-box protein (TBP)-related factor that can bind directly to the TATT element and is required for late gene transcription (Wyrwicz and Rychlewski 2007; Gruffat et al. 2012). The viral TBP is found among gamma and beta-herpesviruses, but is not found in the alpha-herpesviruses. This has several implications, including potential differences among herpesvirus family members in the regulation of late gene transcription. For EBV, late gene transcription appears to require a specialized viral-encoded transcription initiation complex that may function in the absence of normal cellular histone assembly, as suggested by the live cell imaging of histone-free replication compartments.

10 Network and Homeostatic Control of Viral Chromatin

Several signaling pathways are known to control EBV gene expression, and these are almost always linked to changes in chromatin structure. Inflammatory and DNA damage response pathways have particularly important roles in regulating EBV latency, and recent studies have revealed how these pathways intersect at the level of viral chromatin (Fig. 5).

STAT-associatedinflammatory pathways have been implicated in several different control nodes of EBV latency. STAT1/3/5 binding sites have been identified in the viral genome, and best characterized at the Qp and at sites upstream of LMP1, regulating standard ED-L1 promoter, and the alternative promoter, L1-TR, in the terminal repeats (Chen et al. 1999, 2001). LMP1 can stimulate STAT3 activation, suggesting that it provides a positive feedback loop (Chen et al. 2003; Kung et al. 2011; Buettner et al. 2006). STAT3 was identified as a cellular factor that prevents lytic reactivation when upregulated in BL and LCL cells (Daigle et al. 2010; Hill et al. 2013). Several cellular genes activated by STAT3 were shown to repress EBV lytic gene transcription, including several Zn-finger repressors (ZNF253, ZNF257, SNF589) and SETDB1. STAT3 is also shown to be important for EBV-induced B-cell immortalization, as hypomorphic mutations lead to a reduction of in vitro transformation (Koganti et al. 2014a). STAT3 was found to disrupt the DDR signaling by interfering with ATR activation of Chk1 (Koganti et al. 2014b). In these studies, activation of STAT3 is found to be an early event prior to viral gene expression, and STAT3 modifies DDR response through activation of Caspase 7 and cleavage of DDR component Claspin.

CpGactivation of TLR9 induces a G1/S cell cycle arrest that restricts EBV-induced B-cell proliferation. TLR9 signaling in BL cells was found to suppress BZLF1 transcription through histone modification (Zauner et al. 2010; Ladell et al. 2007). This response was mediated partly by IL12 and INFgamma. EBV latent genomes can also induce chronic stimulation of the IFI-16-mediated

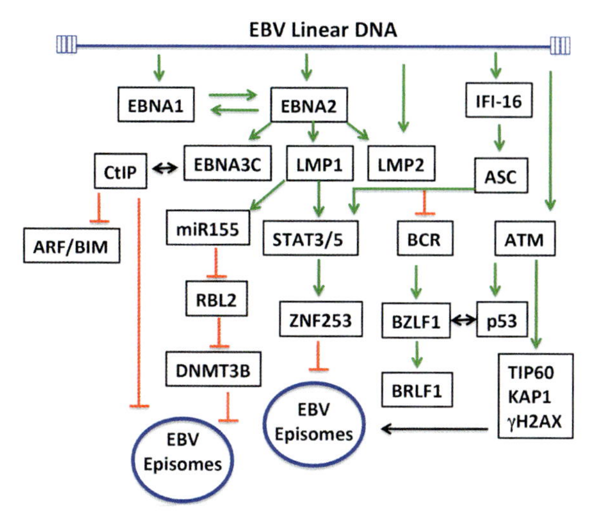

Fig. 5 Network interactions that impact the chromatin control of EBV latent/lytic switch. The interaction network regulating the decision to establish latent chromosome formation rather than lytic productive infection is depicted schematically and highlighting only a few of the many known regulatory interactions. Regulation of latency through DDR and ATM signaling, as well as inflammatory pathways mediated by STAT3 are likely to play central roles in the regulation of EBV latent to lytic switch at the level of chromatin control

inflammasome (Ansari et al. 2013), but it is not yet known whether this inflammatory signal restricts lytic reactivation or otherwise influences latent gene expression.

The DNA damage response (DDR)appears to play a major role in regulating EBV latency and reactivation. As DNA damage signaling often results in the modification of chromatin, it is likely that this pathway modifies the EBV chromosome and chromatin. Consistent with this, DNA damaging agents, especially ATM and ATR agonists, were found to be potent stimulators of EBV lytic reactivation (Feng et al. 2004; Hagemeier et al. 2012). Latency-associated EBNA3C can alter ATM/ATR signaling by interacting with downstream target Chk2 and prevent a G2/M cell cycle block (Choudhuri et al. 2007). By inhibiting ATM signaling, EBNA3C may function to block lytic cycle reactivation.

P53 plays a central role in DDR-dependent transcription. P53 has been shown to contribute to EBV reactivation induced by HDAC inhibitors in NPC cells (Chang et al. 2008). Post-translational modification of p53 at serine 46 and 392 was found to be important for this function. In contrast, p53 family member ΔNp63 is reported to inhibit EBV reactivation in epithelial cells (Kenney and Mertz 2014). Perhaps, HDAC inhibitors lead to a DNA damage response that is p53 dependent, and this has an important regulatory role in controlling EBV reactivation. P53 may support EBV reactivation through direct interactions with Sp1 protein bound to Zp, as well as increase in EGR1 (Kenney and Mertz 2014).

ATM kinase activity also plays a central role in regulating EBV reactivation. In addition to p53, ATM targets KAP1, TIP60, or H2AX may also effect EBV chromatin to toggle the latent to lytic switch (Kenney and Mertz 2014; Hagemeier et al. 2012; Li and Hayward 2011). The EBV protein kinase BGLF4 can mimic aspects of ATM kinase, as it can phosphorylate substrates such as H2AX and p27 that can prolong the pseudo S-phase necessary for viral DNA replication (Iwahori et al. 2009). BGLF4 can also phosphorylate topoisomerase II and condensins, which effect chromatin structure and chromosome organization (Chang et al. 2012; Lee et al. 2007, 2008). Like BGLF4, MHV68 ORF27 kinase activates the DDR through H2AX phosphorylation and promotes viral replication (Tarakanova et al. 2007). Interestingly, H2AX knockdown reduces the number of latently infected cells in mouse studies (Tarakanova et al. 2010), suggesting that lytic replication is required for MHV68 to achieve latent infection in vivo. BGLF4 kinase is thought to function at early stages of the lytic replication cycle and is important for remodeling chromatin and nuclear morphology, as well as arresting the cell cycle in a pseudo S-phase (Chang et al. 2012). The many targets of BGLF4 reflect host-cell factors that play important roles in maintaining viral chromatin during latency.

Homeostatic mechanisms also control viral latency by acting on chromatin (Fig. 5). LMP2a suppresses lytic reactivation by blocking signaling through surface IgG cross-linking and histone acetylation of BZLF1 (Miller et al. 1994, 1995). LMP1 induces cellular miR155, which in turn suppresses lytic reactivation by targeting RBL1 to derepress DNA methyltransferase 1 (DNMT1) (Lu et al. 2008). The increase in expression DNMT1 leads to de novo DNA methylation and stable epigenetic repression of the latent viral episome.

11 Chromatin Modifiers that Disrupt Latency—Environment, Pathogenesis, and Therapeutics

Many environmental factors alter EBV chromatin structure. Periodontal pathogen *Porphyromonas gingivalis* induces EBV lytic switch by histone modification of Zp (Imai et al. 2012). A major metabolite of *P. gingivalis* (and other bacteria) is small chain fatty acids (SCFAs), such as butyrates, that are potent inhibitors of class II HDACs. However, not all SCFAs stimulate EBV reactivation, and particular SCFAs have cell-type and viral-specific effects (Gorres et al. 2014). Interestingly, several SCFAs that are inhibitors of HDACs failed to reactivate EBV, while these were capable of activating KSHV. Similarly, some medium chain fatty acids (MCFAs), such as valproic acid (VPA) and phenylbutyrate, are capable of blocking EBV reactivation in B-cell lymphoma (BL) cell lines. Like VPA, the HDAC inhibitor sulforaphan can also inhibit EBV reactivation in NPC cells (Wu et al. 2013). This is consistent with related studies showing that lipophilic small molecules, including resveratrol (De Leo et al. 2012; Yiu et al. 2010) and moronic acid (Chang et al. 2010), could block EBV reactivation. Together, these studies suggest that a specific receptor or membrane perturbation may be playing a role in regulating EBV chromatin and reactivation.

Potential pathogen-associated lytic activators have been isolated from malaria membrane proteins (Chene et al. 2007) and the lignins in salted fish (Bouvier et al. 1995). These reactivators may contribute to the carcinogenesis of Burkitt's lymphoma and NPC, respectively. On the other hand, identification of safe and effective reactivators of EBV may be of clinical value for lytic therapy to treat EBV-positive tumors (Kenney 2006).

12 Variation and Heterogeneity of Viral Chromatin Control

EBV and KSHV share many common properties, including the formation of chromatinized episomes during latent infection. Nevertheless, there are many differences in the mechanisms of regulation for these two gammaherpesviruses during latency. Reactivation of EBV in response to HDACsand protein kinase C agonists requires new protein synthesis, but this is not the case for KSHV (Ye et al. 2007). KSHV latency is regulated by polycomb-mediated H3K27me3, while EBV latency is less dependent on this type of chromatin repression (Toth et al. 2010, 2013b). This was shown in co-infected cells treated with DZNep, inhibitor of EZH2, demonstrating that only KSHV, but not EBV, lytic cycle was reactivated. Similarly, KSHV latency can be disrupted by cohesin depletion, while cohesin depletion leads to a change in EBV latency gene expression, but not lytic reactivation (Li et al. 2014; Chen et al. 2014a).

Many factors may influence the chromatin structure and epigenetic regulation of EBV latency. Heterogeneity may exist among viral genomes in a single cell, among cells in a population, and between different cell and tissue types. Different BL-derived cell lines have different chromatin regulation, as Akata have reduced histone acetylation, while Raji have elevated H3K27me3 and H4K20me3. Also, lytic activators such as TPA, A23187, and sodium butyrate stimulate BZLF1 transcription in Raji, but do not alter the repressive marks for H3K27me3 and H4K20me3. Genome copy number and heterogeneity may play a factor in the complex behavior and response to reactivation (only a subset of genomes is responsive). Additional histone remodeling factors or acetylation of non-histone proteins may help to explain the refractory population of latent EBV (Murata et al. 2012; Ramasubramanyan et al. 2012).

13 Conclusions

The predominant structure of EBV genomes during latency is the chromatinized episome. However, it is also true that some EBV-positive tumors have integrated EBV (e.g., Namalwa), and many cells have both episomes, and integrated subgenomic fragments of EBV (Wuu et al. 1996; Lestou et al. 1996; Kripalani-Joshi and Law 1994; Srinivas et al. 1998). The more recent discoveries of EBV in epithelial cancers, such as nasopharyngeal carcinoma and gastric carcinoma, have been more problematic to conclude definitively that EBV is exclusively episomal since these tumors have been very difficult to expand in tissue culture without the loss of EBV genomic DNA. However, deep sequencing of EBV-positive tumors fails to find evidence of chromosomally integrated viral genomes (Khoury et al. 2013). Furthermore, the frequent loss of EBV genomes from cultured NPC cells suggests that these genomes are episomal. EBV episome maintenance in epithelial tumors may involve factors associated with the tumor microenvironment that is not recapitulated in tissue culture. Chang and Moore have proposed that tumor viruses transform cells only when their normal life-cycle pathways are disrupted (Moore and Chang 2010). It is interesting to consider this with EBV, which may establish aberrant latent infections in epithelial cells or rearrange in lymphoid cells to drive host-cell carcinogenesis.

Much remains to be discovered with respect to the chromatin structure (s) of latent EBV genomes in normal and pathogenic infection. While EBV may have chromatin structure "indistinguishable" from host-cell chromosome, it is becoming apparent that host chromosomes have very complex, dynamic, and heterogeneous chromatin structures that are only beginning to be understood at the level of histone modification code and higher-order chromatin structures. Detailed characterizations of viral-specific events will be necessary to determine whether EBV has chromatin structures distinct from the host chromosome. As EBV has unique chromosome dynamics, especially the establishment and maintenance of a chromatinized episome, it seems likely that it may shed new light and insight into the broader field of chromatin biology. It is also likely that EBV's unique chromatin structure and regulation will be an opportunity for therapeutic intervention.

References

Adams A, Lindahl T (1975) Epstein-Barr virus genomes with properties of circular DNA molecules in carrier cells. Proc Nat Acad Sci USA 72(4):1477–1481

Aglipay JA, Lee SW, Okada S, Fujiuchi N, Ohtsuka T, Kwak JC, Wang Y, Johnstone RW, Deng C, Qin J, Ouchi T (2003) A member of the Pyrin family, IFI16, is a novel BRCA1-associated protein involved in the p53-mediated apoptosis pathway. Oncogene 22(55):8931–8938. doi:10.1038/sj.onc.1207057

Alfieri C, Birkenbach M, Kieff E (1991) Early events in Epstein-Barr virus infection of human B lymphocytes. Virology 181(2):595–608

Almqvist J, Zou J, Linderson Y, Borestrom C, Altiok E, Zetterberg H, Rymo L, Pettersson S, Ernberg I (2005) Functional interaction of Oct transcription factors with the family of repeats in Epstein-Barr virus oriP. J Gen Virol 86(Pt 5):1261–1267. doi:10.1099/vir.0.80620-0

Altmann M, Pich D, Ruiss R, Wang J, Sugden B, Hammerschmidt W (2006) Transcriptional activation by EBV nuclear antigen 1 is essential for the expression of EBV's transforming genes. Proc Nat Acad Sci USA 103(38):14188–14193. doi:10.1073/pnas.0605985103

Ambinder RF, Robertson KD, Tao Q (1999) DNA methylation and the Epstein-Barr virus. Semin Cancer Biol 9(5):369–375. doi:10.1006/scbi.1999.0137

Amon W, Farrell PJ (2005) Reactivation of Epstein-Barr virus from latency. Rev Med Virol 15(3):149–156. doi:10.1002/rmv.456

Ansari MA, Singh VV, Dutta S, Veettil MV, Dutta D, Chikoti L, Lu J, Everly D, Chandran B (2013) Constitutive interferon-inducible protein 16-inflammasome activation during Epstein-Barr virus latency I, II, and III in B and epithelial cells. J Virol 87(15):8606–8623. doi:10.1128/JVI.00805-13

Arvey A, Tempera I, Tsai K, Chen HS, Tikhmyanova N, Klichinsky M, Leslie C, Lieberman PM (2012) An atlas of the Epstein-Barr virus transcriptome and epigenome reveals host-virus regulatory interactions. Cell Host Microbe 12(2):233–245. doi:10.1016/j.chom.2012.06.008

Arvey A, Tempera I, Lieberman PM (2013) Interpreting the Epstein-Barr Virus (EBV) epigenome using high-throughput data. Viruses 5(4):1042–1054. doi:10.3390/v5041042

Avolio-Hunter TM, Frappier L (2003) EBNA1 efficiently assembles on chromatin containing the Epstein-Barr virus latent origin of replication. Virology 315(2):398–408

Avolio-Hunter TM, Lewis PN, Frappier L (2001) Epstein-Barr nuclear antigen 1 binds and destabilizes nucleosomes at the viral origin of latent DNA replication. Nucleic Acids Res 29(17):3520–3528

Barbera AJ, Chodaparambil JV, Kelley-Clarke B, Joukov V, Walter JC, Luger K, Kaye KM (2006) The nucleosomal surface as a docking station for Kaposi's sarcoma herpesvirus LANA. Science 311(5762):856–861. doi:10.1126/science.1120541

Bhende PM, Dickerson SJ, Sun X, Feng WH, Kenney SC (2007) X-box-binding protein 1 activates lytic Epstein-Barr virus gene expression in combination with protein kinase D. J Virol 81(14):7363–7370. doi:10.1128/JVI.00154-07

Bochkarev A, Barwell JA, Pfuetzner RA, Furey W Jr, Edwards AM, Frappier L (1995) Crystal structure of the DNA-binding domain of the Epstein-Barr virus origin-binding protein EBNA 1. Cell 83(1):39–46

Borestrom C, Forsman A, Ruetschi U, Rymo L (2012) E2F1, ARID3A/bright and Oct-2 factors bind to the Epstein-Barr virus C promoter, EBNA1 and oriP, participating in long-distance promoter-enhancer interactions. J Gen Virol 93(Pt 5):1065–1075. doi:10.1099/vir.0.038752-0

Boutell C, Everett RD (2013) Regulation of alphaherpesvirus infections by the ICP0 family of proteins. J Gen Virol 94(Pt 3):465–481. doi:10.1099/vir.0.048900-0

Bouvier G, Hergenhahn M, Polack A, Bornkamm GW, de The G, Bartsch H (1995) Characterization of macromolecular lignins as Epstein-Barr virus inducer in foodstuff associated with nasopharyngeal carcinoma risk. Carcinogenesis 16(8):1879–1885

Brazda V, Coufal J, Liao JC, Arrowsmith CH (2012) Preferential binding of IFI16 protein to cruciform structure and superhelical DNA. Biochem Biophys Res Commun 422(4):716–720. doi:10.1016/j.bbrc.2012.05.065

Buettner M, Heussinger N, Niedobitek G (2006) Expression of Epstein-Barr virus (EBV)-encoded latent membrane proteins and STAT3 activation in nasopharyngeal carcinoma. Virchows Archiv (an international journal of pathology) 449(5):513–519. doi:10.1007/s00428-006-0294-2

Chabot PR, Raiola L, Lussier-Price M, Morse T, Arseneault G, Archambault J, Omichinski JG (2014) Structural and functional characterization of a complex between the acidic transactivation domain of EBNA2 and the Tfb1/p62 subunit of TFIIH. PLoS Pathog 10(3):e1004042. doi:10.1371/journal.ppat.1004042

Chang SS, Lo YC, Chua HH, Chiu HY, Tsai SC, Chen JY, Lo KW, Tsai CH (2008) Critical role of p53 in histone deacetylase inhibitor-induced Epstein-Barr virus Zta expression. J Virol 82(15):7745–7751. doi:10.1128/JVI.02717-07

Chang FR, Hsieh YC, Chang YF, Lee KH, Wu YC, Chang LK (2010) Inhibition of the Epstein-Barr virus lytic cycle by moronic acid. Antiviral Res 85(3):490–495. doi:10.1016/j.antiviral.2009.12.002

Chang YH, Lee CP, Su MT, Wang JT, Chen JY, Lin SF, Tsai CH, Hsieh MJ, Takada K, Chen MR (2012) Epstein-Barr virus BGLF4 kinase retards cellular S-phase progression and induces chromosomal abnormality. PLoS ONE 7(6):e39217. doi:10.1371/journal.pone.0039217

Chau CM, Zhang XY, McMahon SB, Lieberman PM (2006) Regulation of Epstein-Barr virus latency type by the chromatin boundary factor CTCF. J Virol 80(12):5723–5732. doi:10.1128/JVI.00025-06

Chau CM, Deng Z, Kang H, Lieberman PM (2008) Cell cycle association of the retinoblastoma protein Rb and the histone demethylase LSD1 with the Epstein-Barr virus latency promoter Cp. J Virol 82(7):3428–3437. doi:10.1128/JVI.01412-07

Chaudhuri B, Xu H, Todorov I, Dutta A, Yates JL (2001) Human DNA replication initiation factors, ORC and MCM, associate with oriP of Epstein-Barr virus. Proc Nat Acad Sci USA 98(18):10085–10089. doi:10.1073/pnas.181347998

Chen H, Lee JM, Wang Y, Huang DP, Ambinder RF, Hayward SD (1999) The Epstein-Barr virus latency BamHI-Q promoter is positively regulated by STATs and Zta interference with JAK/STAT activation leads to loss of BamHI-Q promoter activity. Proc Nat Acad Sci USA 96(16):9339–9344

Chen H, Lee JM, Zong Y, Borowitz M, Ng MH, Ambinder RF, Hayward SD (2001) Linkage between STAT regulation and Epstein-Barr virus gene expression in tumors. J Virol 75(6):2929–2937. doi:10.1128/JVI.75.6.2929-2937.2001

Chen H, Hutt-Fletcher L, Cao L, Hayward SD (2003) A positive autoregulatory loop of LMP1 expression and STAT activation in epithelial cells latently infected with Epstein-Barr virus. J Virol 77(7):4139–4148

Chen H, Huang J, Wu FY, Liao G, Hutt-Fletcher L, Hayward SD (2005) Regulation of expression of the Epstein-Barr virus BamHI-A rightward transcripts. J Virol 79(3):1724–1733. doi: 10.1128/JVI.79.3.1724-1733.2005

Chen HS, Wikramasinghe P, Showe L, Lieberman PM (2012) Cohesins repress Kaposi's sarcoma-associated herpesvirus immediate early gene transcription during latency. J Virol 86(17):9454–9464. doi:10.1128/JVI.00787-12

Chen HS, Martin KA, Lu F, Lupey LN, Mueller JM, Lieberman PM, Tempera I (2014a) Epigenetic deregulation of the LMP1/LMP2 locus of Epstein-Barr virus by mutation of a single CTCF-cohesin binding site. J Virol 88(3):1703–1713. doi:10.1128/JVI.02209-13

Chen YL, Liu CD, Cheng CP, Zhao B, Hsu HJ, Shen CL, Chiu SJ, Kieff E, Peng CW (2014b) Nucleolin is important for Epstein-Barr virus nuclear antigen 1-mediated episome binding, maintenance, and transcription. Proc Nat Acad Sci USA 111(1):243–248. doi:10.1073/pnas.1321800111

Chene A, Donati D, Guerreiro-Cacais AO, Levitsky V, Chen Q, Falk KI, Orem J, Kironde F, Wahlgren M, Bejarano MT (2007) A molecular link between malaria and Epstein-Barr virus reactivation. PLoS Pathog 3(6):e80. doi:10.1371/journal.ppat.0030080

Chiu YF, Sugden AU, Sugden B (2013) Epstein-Barr viral productive amplification reprograms nuclear architecture, DNA replication, and histone deposition. Cell Host Microbe 14(6):607–618. doi:10.1016/j.chom.2013.11.009

Choudhuri T, Verma SC, Lan K, Murakami M, Robertson ES (2007) The ATM/ATR signaling effector Chk2 is targeted by Epstein-Barr virus nuclear antigen 3C to release the G2/M cell cycle block. J Virol 81(12):6718–6730. doi:10.1128/JVI.00053-07

Conrady CD, Zheng M, Fitzgerald KA, Liu C, Carr DJ (2012) Resistance to HSV-1 infection in the epithelium resides with the novel innate sensor, IFI-16. Mucosal Immunol 5(2):173–183. doi:10.1038/mi.2011.63

Coppotelli G, Mughal N, Callegari S, Sompallae R, Caja L, Luijsterburg MS, Dantuma NP, Moustakas A, Masucci MG (2013) The Epstein-Barr virus nuclear antigen-1 reprograms transcription by mimicry of high mobility group A proteins. Nucleic Acids Res 41(5):2950–2962. doi:10.1093/nar/gkt032

Cuchet-Lourenco D, Boutell C, Lukashchuk V, Grant K, Sykes A, Murray J, Orr A, Everett RD (2011) SUMO pathway dependent recruitment of cellular repressors to herpes simplex virus type 1 genomes. PLoS Pathog 7(7):e1002123. doi:10.1371/journal.ppat.1002123

Daigle D, Megyola C, El-Guindy A, Gradoville L, Tuck D, Miller G, Bhaduri-McIntosh S (2010) Upregulation of STAT3 marks Burkitt lymphoma cells refractory to Epstein-Barr virus lytic cycle induction by HDAC inhibitors. J Virol 84(2):993–1004. doi:10.1128/JVI.01745-09

Day L, Chau CM, Nebozhyn M, Rennekamp AJ, Showe M, Lieberman PM (2007) Chromatin profiling of Epstein-Barr virus latency control region. J Virol 81(12):6389–6401. doi:10.112 8/JVI.02172-06

De Leo A, Arena G, Lacanna E, Oliviero G, Colavita F, Mattia E (2012) Resveratrol inhibits Epstein Barr Virus lytic cycle in Burkitt's lymphoma cells by affecting multiple molecular targets. Antiviral Res 96(2):196–202. doi:10.1016/j.antiviral.2012.09.003

Deng Z, Lezina L, Chen CJ, Shtivelband S, So W, Lieberman PM (2002) Telomeric proteins regulate episomal maintenance of Epstein-Barr virus origin of plasmid replication. Mol Cell 9(3):493–503

Deutsch MJ, Ott E, Papior P, Schepers A (2010) The latent origin of replication of Epstein-Barr virus directs viral genomes to active regions of the nucleus. J Virol 84(5):2533–2546. doi:10 .1128/JVI.01909-09

Dhar SK, Yoshida K, Machida Y, Khaira P, Chaudhuri B, Wohlschlegel JA, Leffak M, Yates J, Dutta A (2001) Replication from oriP of Epstein-Barr virus requires human ORC and is inhibited by geminin. Cell 106(3):287–296

Dresang LR, Vereide DT, Sugden B (2009) Identifying sites bound by Epstein-Barr virus nuclear antigen 1 (EBNA1) in the human genome: defining a position-weighted matrix to predict sites bound by EBNA1 in viral genomes. J Virol 83(7):2930–2940. doi:10.1128/ JVI.01974-08

Dyson PJ, Farrell PJ (1985) Chromatin structure of Epstein-Barr virus. J Gen Virol 66(Pt 9):1931–1940

Echendu CW, Ling PD (2008) Regulation of Sp100A subnuclear localization and transcriptional function by EBNA-LP and interferon. J interferon Cytokine Res (the official journal of the International Society for Interferon and Cytokine Research) 28(11):667–678. doi:10.1089/ jir.2008.0023

Ellis AL, Wang Z, Yu X, Mertz JE (2010) Either ZEB1 or ZEB2/SIP1 can play a central role in regulating the Epstein-Barr virus latent-lytic switch in a cell-type-specific manner. J Virol 84(12):6139–6152. doi:10.1128/JVI.02706-09

Epstein MA, Achong BG, Barr YM (1964) Virus Particles in cultured lymphoblasts from Burkitt's lymphoma. Lancet 1(7335):702–703

Everett RD, Chelbi-Alix MK (2007) PML and PML nuclear bodies: implications in antiviral defence. Biochimie 89(6–7):819–830. doi:10.1016/j.biochi.2007.01.004

Feng WH, Hong G, Delecluse HJ, Kenney SC (2004) Lytic induction therapy for Epstein-Barr virus-positive B-cell lymphomas. J Virol 78(4):1893–1902

Frappier L (2012) EBNA1 and host factors in Epstein-Barr virus latent DNA replication. Curr Opin Virol 2(6):733–739. doi:10.1016/j.coviro.2012.09.005

Full F, Jungnickl D, Reuter N, Bogner E, Brulois K, Scholz B, Sturzl M, Myoung J, Jung JU, Stamminger T, Ensser A (2014) Kaposi's sarcoma associated herpesvirus tegument protein ORF75 is essential for viral lytic replication and plays a critical role in the antagonization of ND10-instituted intrinsic immunity. PLoS Pathog 10(1):e1003863. doi:10.1371/ journal.ppat.1003863

Gahn TA, Sugden B (1995) An EBNA-1-dependent enhancer acts from a distance of 10 kilobase pairs to increase expression of the Epstein-Barr virus LMP gene. J Virol 69(4):2633–2636

Gariano GR, Dell'Oste V, Bronzini M, Gatti D, Luganini A, De Andrea M, Gribaudo G, Gariglio M, Landolfo S (2012) The intracellular DNA sensor IFI16 gene acts as restriction factor for human cytomegalovirus replication. PLoS Pathog 8(1):e1002498. doi:10.1371/ journal.ppat.1002498

Gastaldello S, Hildebrand S, Faridani O, Callegari S, Palmkvist M, Di Guglielmo C, Masucci MG (2010) A deneddylase encoded by Epstein-Barr virus promotes viral DNA replication by regulating the activity of cullin-RING ligases. Nat Cell Biol 12(4):351–361. doi:10.1038/ncb2035

Gastaldello S, Callegari S, Coppotelli G, Hildebrand S, Song M, Masucci MG (2012) Herpes virus deneddylases interrupt the cullin-RING ligase neddylation cycle by inhibiting the binding of CAND1. J Mol Cell Biol 4(4):242–251. doi:10.1093/jmcb/mjs012

Gilchrist DA, Adelman K (2012) Coupling polymerase pausing and chromatin landscapes for precise regulation of transcription. Biochim Biophys Acta 1819(7):700–706. doi:10.1016/j.bbagrm.2012.02.015

Goldberg AD, Banaszynski LA, Noh KM, Lewis PW, Elsaesser SJ, Stadler S, Dewell S, Law M, Guo X, Li X, Wen D, Chapgier A, DeKelver RC, Miller JC, Lee YL, Boydston EA, Holmes MC, Gregory PD, Greally JM, Rafii S, Yang C, Scambler PJ, Garrick D, Gibbons RJ, Higgs DR, Cristea IM, Urnov FD, Zheng D, Allis CD (2010) Distinct factors control histone variant H3.3 localization at specific genomic regions. Cell 140(5):678–691. doi:10.1016/j.cell.2010.01.003

Gorres KL, Daigle D, Mohanram S, Miller G (2014) Activation and repression of Epstein-Barr virus and Kaposi's sarcoma-associated herpesvirus lytic cycles by short- and medium-chain fatty acids. J Virol 88(14):8028–8044. doi:10.1128/JVI.00722-14

Gruffat H, Manet E, Sergeant A (2002) MEF2-mediated recruitment of class II HDAC at the EBV immediate early gene BZLF1 links latency and chromatin remodeling. EMBO Rep 3(2):141–146. doi:10.1093/embo-reports/kvf031

Gruffat H, Kadjouf F, Mariame B, Manet E (2012) The Epstein-Barr virus BcRF1 gene product is a TBP-like protein with an essential role in late gene expression. J Virol 86(11):6023–6032. doi:10.1128/JVI.00159-12

Hagemeier SR, Barlow EA, Meng Q, Kenney SC (2012) The cellular ataxia telangiectasia-mutated kinase promotes epstein-barr virus lytic reactivation in response to multiple different types of lytic reactivation-inducing stimuli. J Virol 86(24):13360–13370. doi:10.1128/JVI.01850-12

Hayward SD (2004) Viral interactions with the Notch pathway. Semin Cancer Biol 14(5):387–396. doi:10.1016/j.semcancer.2004.04.018

Hernando H, Shannon-Lowe C, Islam AB, Al-Shahrour F, Rodriguez-Ubreva J, Rodriguez-Cortez VC, Javierre BM, Mangas C, Fernandez AF, Parra M, Delecluse HJ, Esteller M, Lopez-Granados E, Fraga MF, Lopez-Bigas N, Ballestar E (2013) The B cell transcription program mediates hypomethylation and overexpression of key genes in Epstein-Barr virus-associated proliferative conversion. Genome Biol 14(1):R3. doi:10.1186/gb-2013-14-1-r3

Hill ER, Koganti S, Zhi J, Megyola C, Freeman AF, Palendira U, Tangye SG, Farrell PJ, Bhaduri-McIntosh S (2013) Signal transducer and activator of transcription 3 limits Epstein-Barr virus lytic activation in B lymphocytes. J Virol 87(21):11438–11446. doi:10.1128/JVI.01762-13

Holdorf MM, Cooper SB, Yamamoto KR, Miranda JJ (2011) Occupancy of chromatin organizers in the Epstein-Barr virus genome. Virology 415(1):1–5. doi:10.1016/j.virol.2011.04.004

Horan KA, Hansen K, Jakobsen MR, Holm CK, Soby S, Unterholzner L, Thompson M, West JA, Iversen MB, Rasmussen SB, Ellermann-Eriksen S, Kurt-Jones E, Landolfo S, Damania B, Melchjorsen J, Bowie AG, Fitzgerald KA, Paludan SR (2013) Proteasomal degradation of herpes simplex virus capsids in macrophages releases DNA to the cytosol for recognition by DNA sensors. J Immunol 190(5):2311–2319. doi:10.4049/jimmunol.1202749

Hung SC, Kang MS, Kieff E (2001) Maintenance of Epstein-Barr virus (EBV) oriP-based episomes requires EBV-encoded nuclear antigen-1 chromosome-binding domains, which can be replaced by high-mobility group-I or histone H1. Proc Natl Acad Sci USA 98(4):1865–1870. doi:10.1073/pnas.031584698

Iempridee T, Das S, Xu I, Mertz JE (2011) Transforming growth factor beta-induced reactivation of Epstein-Barr virus involves multiple Smad-binding elements cooperatively activating expression of the latent-lytic switch BZLF1 gene. J Virol 85(15):7836–7848. doi:10.1128/JVI.01197-10

Iempridee T, Reusch JA, Riching A, Johannsen EC, Dovat S, Kenney SC, Mertz JE (2014) Epstein-Barr virus utilizes Ikaros in regulating its latent-lytic switch in B cells. J Virol 88(9):4811–4827. doi:10.1128/JVI.03706-13
Imai K, Inoue H, Tamura M, Cueno ME, Inoue H, Takeichi O, Kusama K, Saito I, Ochiai K (2012) The periodontal pathogen Porphyromonas gingivalis induces the Epstein-Barr virus lytic switch transactivator ZEBRA by histone modification. Biochimie 94(3):839–846. doi:10.1016/j.biochi.2011.12.001
Ito S, Gotoh E, Ozawa S, Yanagi K (2002) Epstein-Barr virus nuclear antigen-1 is highly colocalized with interphase chromatin and its newly replicated regions in particular. J Gen Virol 83(Pt 10):2377–2383
Iwahori S, Murata T, Kudoh A, Sato Y, Nakayama S, Isomura H, Kanda T, Tsurumi T (2009) Phosphorylation of p27Kip1 by Epstein-Barr virus protein kinase induces its degradation through SCFSkp2 ubiquitin ligase actions during viral lytic replication. J Biol Chem 284(28):18923–18931. doi:10.1074/jbc.M109.015123
Jenkins PJ, Binne UK, Farrell PJ (2000) Histone acetylation and reactivation of Epstein-Barr virus from latency. J Virol 74(2):710–720
Jenuwein T, Allis CD (2001) Translating the histone code. Science 293(5532):1074–1080. doi:10.1126/science.1063127
Jiang S, Willox B, Zhou H, Holthaus AM, Wang A, Shi TT, Maruo S, Kharchenko PV, Johannsen EC, Kieff E, Zhao B (2014) Epstein-Barr virus nuclear antigen 3C binds to BATF/IRF4 or SPI1/IRF4 composite sites and recruits Sin3A to repress CDKN2A. Proc Natl Acad Sci USA 111(1):421–426. doi:10.1073/pnas.1321704111
Johannsen E, Luftig M, Chase MR, Weicksel S, Cahir-McFarland E, Illanes D, Sarracino D, Kieff E (2004) Proteins of purified Epstein-Barr virus. Proc Natl Acad Sci USA 101(46):16286–16291. doi:10.1073/pnas.0407320101
Johansson P, Jansson A, Ruetschi U, Rymo L (2009) Nuclear factor-kappaB binds to the Epstein-Barr Virus LMP1 promoter and upregulates its expression. J Virol 83(3):1393–1401. doi:10.1128/JVI.01637-08
Johnson KE, Chikoti L, Chandran B (2013) Herpes simplex virus 1 infection induces activation and subsequent inhibition of the IFI16 and NLRP3 inflammasomes. J Virol 87(9):5005–5018. doi:10.1128/JVI.00082-13
Kalla M, Hammerschmidt W (2012) Human B cells on their route to latent infection–early but transient expression of lytic genes of Epstein-Barr virus. Eur J Cell Biol 91(1):65–69. doi:10.1016/j.ejcb.2011.01.014
Kanda T, Horikoshi N, Murata T, Kawashima D, Sugimoto A, Narita Y, Kurumizaka H, Tsurumi T (2013) Interaction between basic residues of Epstein-Barr virus EBNA1 protein and cellular chromatin mediates viral plasmid maintenance. J Biol Chem 288(33):24189–24199. doi:10.1074/jbc.M113.491167
Kenney S (2006) Theodore E. Woodward Award: development of novel, EBV-targeted therapies for EBV-positive tumors. Transactions of the American Clinical and Climatological Association 117:55–73; discussion 73-54
Kenney SC, Mertz JE (2014) Regulation of the latent-lytic switch in Epstein-Barr virus. Semin Cancer Biol 26:60–68. doi:10.1016/j.semcancer.2014.01.002
Kerur N, Veettil MV, Sharma-Walia N, Bottero V, Sadagopan S, Otageri P, Chandran B (2011) IFI16 acts as a nuclear pathogen sensor to induce the inflammasome in response to Kaposi Sarcoma-associated herpesvirus infection. Cell Host Microbe 9(5):363–375. doi:10.1016/j.chom.2011.04.008
Khoury JD, Tannir NM, Williams MD, Chen Y, Yao H, Zhang J, Thompson EJ, Network T, Meric-Bernstam F, Medeiros LJ, Weinstein JN, Su X (2013) Landscape of DNA virus associations across human malignant cancers: analysis of 3,775 cases using RNA-Seq. J Virol 87(16):8916–8926. doi:10.1128/JVI.00340-13
Kim do N, Lee SK (2012) Biogenesis of Epstein-Barr virus microRNAs. Mol Cell Biochem 365(1–2):203–210. doi:10.1007/s11010-012-1261-7

Kirchmaier AL, Sugden B (1998) Rep*: a viral element that can partially replace the origin of plasmid DNA synthesis of Epstein-Barr virus. J Virol 72(6):4657–4666

Koganti S, de la Paz A, Freeman AF, Bhaduri-McIntosh S (2014a) B lymphocytes from patients with a hypomorphic mutation in STAT3 resist Epstein-Barr virus-driven cell proliferation. J Virol 88(1):516–524. doi:10.1128/JVI.02601-13

Koganti S, Hui-Yuen J, McAllister S, Gardner B, Grasser F, Palendira U, Tangye SG, Freeman AF, Bhaduri-McIntosh S (2014b) STAT3 interrupts ATR-Chk1 signaling to allow oncovirus-mediated cell proliferation. Proc Natl Acad Sci USA 111(13):4946–4951. doi:10.1073/pnas.1400683111

Kraus RJ, Perrigoue JG, Mertz JE (2003) ZEB negatively regulates the lytic-switch BZLF1 gene promoter of Epstein-Barr virus. J Virol 77(1):199–207

Kripalani-Joshi S, Law HY (1994) Identification of integrated Epstein-Barr virus in nasopharyngeal carcinoma using pulse field gel electrophoresis. Int J Cancer (Journal international du cancer) 56(2):187–192

Kumar R, Whitehurst CB, Pagano JS (2014) The Rad6/18 ubiquitin complex interacts with the Epstein-Barr virus deubiquitinating enzyme, BPLF1, and contributes to virus infectivity. J Virol 88(11):6411–6422. doi:10.1128/JVI.00536-14

Kung CP, Meckes DG Jr, Raab-Traub N (2011) Epstein-Barr virus LMP1 activates EGFR, STAT3, and ERK through effects on PKCdelta. J Virol 85(9):4399–4408. doi:10.1128/JVI.01703-10

Kuo AJ, Song J, Cheung P, Ishibe-Murakami S, Yamazoe S, Chen JK, Patel DJ, Gozani O (2012) The BAH domain of ORC1 links H4K20me2 to DNA replication licensing and Meier-Gorlin syndrome. Nature 484(7392):115–119. doi:10.1038/nature10956

Ladell K, Dorner M, Zauner L, Berger C, Zucol F, Bernasconi M, Niggli FK, Speck RF, Nadal D (2007) Immune activation suppresses initiation of lytic Epstein-Barr virus infection. Cell Microbiol 9(8):2055–2069. doi:10.1111/j.1462-5822.2007.00937.x

Lee CP, Chen JY, Wang JT, Kimura K, Takemoto A, Lu CC, Chen MR (2007) Epstein-Barr virus BGLF4 kinase induces premature chromosome condensation through activation of condensin and topoisomerase II. J Virol 81(10):5166–5180. doi:10.1128/JVI.00120-07

Lee CP, Huang YH, Lin SF, Chang Y, Chang YH, Takada K, Chen MR (2008) Epstein-Barr virus BGLF4 kinase induces disassembly of the nuclear lamina to facilitate virion production. J Virol 82(23):11913–11926. doi:10.1128/JVI.01100-08

Lestou VS, Strehl S, Lion T, Gadner H, Ambros PF (1996) High-resolution FISH of the entire integrated Epstein-Barr virus genome on extended human DNA. Cytogenet Cell Genet 74(3):211–217

Lewis PW, Elsaesser SJ, Noh KM, Stadler SC, Allis CD (2010) Daxx is an H3.3-specific histone chaperone and cooperates with ATRX in replication-independent chromatin assembly at telomeres. Proc Natl Acad Sci USA 107(32):14075–14080. doi:10.1073/pnas.1008850107

Li R, Hayward SD (2011) The Ying-Yang of the virus-host interaction: control of the DNA damage response. Future Microbiol 6(4):379–383. doi:10.2217/fmb.11.16

Li T, Diner BA, Chen J, Cristea IM (2012) Acetylation modulates cellular distribution and DNA sensing ability of interferon-inducible protein IFI16. Proc Natl Acad Sci USA 109(26):10558–10563. doi:10.1073/pnas.1203447109

Li DJ, Verma D, Mosbruger T, Swaminathan S (2014) CTCF and Rad21 act as host cell restriction factors for Kaposi's sarcoma-associated herpesvirus (KSHV) lytic replication by modulating viral gene transcription. PLoS Pathog 10(1):e1003880. doi:10.1371/journal.ppat.1003880

Liashkovich I, Hafezi W, Kuhn JM, Oberleithner H, Shahin V (2011) Nuclear delivery mechanism of herpes simplex virus type 1 genome. J Mol Recogn JMR 24(3):414–421. doi:10.1002/jmr.1120

Lindner SE, Zeller K, Schepers A, Sugden B (2008) The affinity of EBNA1 for its origin of DNA synthesis is a determinant of the origin's replicative efficiency. J Virol 82(12):5693–5702. doi:10.1128/JVI.00332-08

Ling PD, Rawlins DR, Hayward SD (1993) The Epstein-Barr virus immortalizing protein EBNA-2 is targeted to DNA by a cellular enhancer-binding protein. Proc Natl Acad Sci USA 90(20):9237–9241
Ling PD, Peng RS, Nakajima A, Yu JH, Tan J, Moses SM, Yang WH, Zhao B, Kieff E, Bloch KD, Bloch DB (2005) Mediation of Epstein-Barr virus EBNA-LP transcriptional coactivation by Sp100. EMBO J 24(20):3565–3575. doi:10.1038/sj.emboj.7600820
Ling PD, Tan J, Sewatanon J, Peng R (2008) Murine gammaherpesvirus 68 open reading frame 75c tegument protein induces the degradation of PML and is essential for production of infectious virus. J Virol 82(16):8000–8012. doi:10.1128/JVI.02752-07
Liu S, Borras AM, Liu P, Suske G, Speck SH (1997a) Binding of the ubiquitous cellular transcription factors Sp1 and Sp3 to the ZI domains in the Epstein-Barr virus lytic switch BZLF1 gene promoter. Virology 228(1):11–18. doi:10.1006/viro.1996.8371
Liu S, Liu P, Borras A, Chatila T, Speck SH (1997b) Cyclosporin A-sensitive induction of the Epstein-Barr virus lytic switch is mediated via a novel pathway involving a MEF2 family member. EMBO J 16(1):143–153. doi:10.1093/emboj/16.1.143
Lu F, Weidmer A, Liu CG, Volinia S, Croce CM, Lieberman PM (2008) Epstein-Barr virus-induced miR-155 attenuates NF-kappaB signaling and stabilizes latent virus persistence. J Virol 82(21):10436–10443. doi:10.1128/JVI.00752-08
Lu F, Wikramasinghe P, Norseen J, Tsai K, Wang P, Showe L, Davuluri RV, Lieberman PM (2010) Genome-wide analysis of host-chromosome binding sites for Epstein-Barr Virus Nuclear Antigen 1 (EBNA1). Virol J 7:262. doi:10.1186/1743-422X-7-262
Lu F, Tempera I, Lee HT, Dewispelaere K, Lieberman PM (2014) EBNA1 binding and epigenetic regulation of gastrokine tumor suppressor genes in gastric carcinoma cells. Virol J 11:12. doi:10.1186/1743-422X-11-12
Luka J, Kallin B, Klein G (1979) Induction of the Epstein-Barr virus (EBV) cycle in latently infected cells by n-butyrate. Virology 94(1):228–231
Malik-Soni N, Frappier L (2012) Proteomic profiling of EBNA1-host protein interactions in latent and lytic Epstein-Barr virus infections. J Virol 86(12):6999–7002. doi:10.1128/JVI.00194-12
Malik-Soni N, Frappier L (2014) Nucleophosmin contributes to the transcriptional activation function of the Epstein-Barr virus EBNA1 protein. J Virol 88(4):2323–2326. doi:10.1128/JVI.02521-13
Merkenschlager M, Odom DT (2013) CTCF and cohesin: linking gene regulatory elements with their targets. Cell 152(6):1285–1297. doi:10.1016/j.cell.2013.02.029
Meyring-Wosten A, Hafezi W, Kuhn J, Liashkovich I, Shahin V (2014) Nano-visualization of viral DNA breaching the nucleocytoplasmic barrier. J Controlled Release (official journal of the Controlled Release Society) 173:96–101. doi:10.1016/j.jconrel.2013.10.036
Miller CL, Lee JH, Kieff E, Longnecker R (1994) An integral membrane protein (LMP2) blocks reactivation of Epstein-Barr virus from latency following surface immunoglobulin crosslinking. Proc Natl Acad Sci USA 91(2):772–776
Miller CL, Burkhardt AL, Lee JH, Stealey B, Longnecker R, Bolen JB, Kieff E (1995) Integral membrane protein 2 of Epstein-Barr virus regulates reactivation from latency through dominant negative effects on protein-tyrosine kinases. Immunity 2(2):155–166
Miller G, El-Guindy A, Countryman J, Ye J, Gradoville L (2007) Lytic cycle switches of oncogenic human gammaherpesviruses. Adv Cancer Res 97:81–109. doi:10.1016/S0065-230X(06)97004-3
Montalvo EA, Cottam M, Hill S, Wang YJ (1995) YY1 binds to and regulates cis-acting negative elements in the Epstein-Barr virus BZLF1 promoter. J Virol 69(7):4158–4165
Moore PS, Chang Y (2010) Why do viruses cause cancer? Highlights of the first century of human tumour virology. Nat Rev Cancer 10(12):878–889. doi:10.1038/nrc2961
Murata T, Tsurumi T (2013) Epigenetic modification of the Epstein-Barr virus BZLF1 promoter regulates viral reactivation from latency. Frontiers Genet 4:53. doi:10.3389/fgene.2013.00053

Murata T, Noda C, Saito S, Kawashima D, Sugimoto A, Isomura H, Kanda T, Yokoyama KK, Tsurumi T (2011) Involvement of Jun dimerization protein 2 (JDP2) in the maintenance of Epstein-Barr virus latency. J Biol Chem 286(25):22007–22016. doi:10.1074/jbc.M110.199836

Murata T, Kondo Y, Sugimoto A, Kawashima D, Saito S, Isomura H, Kanda T, Tsurumi T (2012) Epigenetic histone modification of Epstein-Barr virus BZLF1 promoter during latency and reactivation in Raji cells. J Virol 86(9):4752–4761. doi:10.1128/JVI.06768-11

Murata T, Narita Y, Sugimoto A, Kawashima D, Kanda T, Tsurumi T (2013) Contribution of myocyte enhancer factor 2 family transcription factors to BZLF1 expression in Epstein-Barr virus reactivation from latency. J Virol 87(18):10148–10162. doi:10.1128/JVI.01002-13

Nikitin PA, Yan CM, Forte E, Bocedi A, Tourigny JP, White RE, Allday MJ, Patel A, Dave SS, Kim W, Hu K, Guo J, Tainter D, Rusyn E, Luftig MA (2010) An ATM/Chk2-mediated DNA damage-responsive signaling pathway suppresses Epstein-Barr virus transformation of primary human B cells. Cell Host Microbe 8(6):510–522. doi:10.1016/j.chom.2010.11.004

Niller HH, Salamon D, Ilg K, Koroknai A, Banati F, Bauml G, Rucker O, Schwarzmann F, Wolf H, Minarovits J (2003) The in vivo binding site for oncoprotein c-Myc in the promoter for Epstein-Barr virus (EBV) encoding RNA (EBER) 1 suggests a specific role for EBV in lymphomagenesis. Med Sci Monit (international medical journal of experimental and clinical research) 9(1):HY1–HY9

Nilsson T, Zetterberg H, Wang YC, Rymo L (2001) Promoter-proximal regulatory elements involved in oriP-EBNA1-independent and -dependent activation of the Epstein-Barr virus C promoter in B-lymphoid cell lines. J Virol 75(13):5796–5811. doi:10.1128/JVI.75.13.5796-5811.2001

Noda C, Murata T, Kanda T, Yoshiyama H, Sugimoto A, Kawashima D, Saito S, Isomura H, Tsurumi T (2011) Identification and characterization of CCAAT enhancer-binding protein (C/EBP) as a transcriptional activator for Epstein-Barr virus oncogene latent membrane protein 1. J Biol Chem 286(49):42524–42533. doi:10.1074/jbc.M111.271734

Nonkwelo C, Ruf IK, Sample J (1997) The Epstein-Barr virus EBNA-1 promoter Qp requires an initiator-like element. J Virol 71(1):354–361

Nonoyama M, Pagano JS (1972) Separation of Epstein-Barr virus DNA from large chromosomal DNA in non-virus-producing cells. Nature (New biology) 238(84):169–171

Norio P, Schildkraut CL (2004) Plasticity of DNA replication initiation in Epstein-Barr virus episomes. PLoS Biol 2(6):e152. doi:10.1371/journal.pbio.0020152

Norio P, Schildkraut CL, Yates JL (2000) Initiation of DNA replication within oriP is dispensable for stable replication of the latent Epstein-Barr virus chromosome after infection of established cell lines. J Virol 74(18):8563–8574

Norseen J, Thomae A, Sridharan V, Aiyar A, Schepers A, Lieberman PM (2008) RNA-dependent recruitment of the origin recognition complex. EMBO J 27(22):3024–3035. doi:10.1038/emboj.2008.221

Ong CT, Corces VG (2014) CTCF: an architectural protein bridging genome topology and function. Nat Rev Genet 15(4):234–246. doi:10.1038/nrg3663

Orzalli MH, Conwell SE, Berrios C, DeCaprio JA, Knipe DM (2013) Nuclear interferon-inducible protein 16 promotes silencing of herpesviral and transfected DNA. Proc Natl Acad Sci USA 110(47):E4492–E4501. doi:10.1073/pnas.1316194110

Ott E, Norio P, Ritzi M, Schildkraut C, Schepers A (2011) The dyad symmetry element of Epstein-Barr virus is a dominant but dispensable replication origin. PLoS ONE 6(5):e18609. doi:10.1371/journal.pone.0018609

Palermo RD, Webb HM, Gunnell A, West MJ (2008) Regulation of transcription by the Epstein-Barr virus nuclear antigen EBNA 2. Biochem Soc Trans 36(Pt 4):625–628. doi:10.1042/BST0360625

Palermo RD, Webb HM, West MJ (2011) RNA polymerase II stalling promotes nucleosome occlusion and pTEFb recruitment to drive immortalization by Epstein-Barr virus. PLoS Pathog 7(10):e1002334. doi:10.1371/journal.ppat.1002334

Phillips JE, Corces VG (2009) CTCF: master weaver of the genome. Cell 137(7):1194–1211. doi:10.1016/j.cell.2009.06.001
Postigo AA, Dean DC (1999) ZEB represses transcription through interaction with the corepressor CtBP. Proc Natl Acad Sci USA 96(12):6683–6688
Pratt ZL, Kuzembayeva M, Sengupta S, Sugden B (2009) The microRNAs of Epstein-Barr Virus are expressed at dramatically differing levels among cell lines. Virology 386(2):387–397. doi:10.1016/j.virol.2009.01.006
Ptak C, Aitchison JD, Wozniak RW (2014) The multifunctional nuclear pore complex: a platform for controlling gene expression. Curr Opin Cell Biol 28:46–53. doi:10.1016/j.ceb.2014.02.001
Puglielli MT, Woisetschlaeger M, Speck SH (1996) oriP is essential for EBNA gene promoter activity in Epstein-Barr virus-immortalized lymphoblastoid cell lines. J Virol 70(9):5758–5768
Ramasubramanyan S, Osborn K, Flower K, Sinclair AJ (2012) Dynamic chromatin environment of key lytic cycle regulatory regions of the Epstein-Barr virus genome. J Virol 86(3):1809–1819. doi:10.1128/JVI.06334-11
Rawlins DR, Milman G, Hayward SD, Hayward GS (1985) Sequence-specific DNA binding of the Epstein-Barr virus nuclear antigen (EBNA-1) to clustered sites in the plasmid maintenance region. Cell 42(3):859–868
Reichelt M, Wang L, Sommer M, Perrino J, Nour AM, Sen N, Baiker A, Zerboni L, Arvin AM (2011) Entrapment of viral capsids in nuclear PML cages is an intrinsic antiviral host defense against varicella-zoster virus. PLoS Pathog 7(2):e1001266. doi:10.1371/journal.ppat.1001266
Ritzi M, Tillack K, Gerhardt J, Ott E, Humme S, Kremmer E, Hammerschmidt W, Schepers A (2003) Complex protein-DNA dynamics at the latent origin of DNA replication of Epstein-Barr virus. J Cell Sci 116(Pt 19):3971–3984. doi:10.1242/jcs.00708
Ruf IK, Sample J (1999) Repression of Epstein-Barr virus EBNA-1 gene transcription by pRb during restricted latency. J Virol 73(10):7943–7951
Ruthenburg AJ, Li H, Patel DJ, Allis CD (2007) Multivalent engagement of chromatin modifications by linked binding modules. Nat Rev Mol Cell Biol 8(12):983–994. doi:10.1038/nrm2298
Saffert RT, Kalejta RF (2008) Promyelocytic leukemia-nuclear body proteins: herpesvirus enemies, accomplices, or both? Future Virol 3(3):265–277. doi:10.2217/17460794.3.3.265
Salamon D, Banati F, Koroknai A, Ravasz M, Szenthe K, Bathori Z, Bakos A, Niller HH, Wolf H, Minarovits J (2009) Binding of CCCTC-binding factor in vivo to the region located between Rep* and the C promoter of Epstein-Barr virus is unaffected by CpG methylation and does not correlate with Cp activity. J Gen Virol 90(Pt 5):1183–1189. doi:10.1099/vir.0.007344-0
Sarkari F, Sanchez-Alcaraz T, Wang S, Holowaty MN, Sheng Y, Frappier L (2009) EBNA1-mediated recruitment of a histone H2B deubiquitylating complex to the Epstein-Barr virus latent origin of DNA replication. PLoS Pathog 5(10):e1000624. doi:10.1371/journal.ppat.1000624
Sarkari F, Wang X, Nguyen T, Frappier L (2011) The herpesvirus associated ubiquitin specific protease, USP7, is a negative regulator of PML proteins and PML nuclear bodies. PLoS ONE 6(1):e16598. doi:10.1371/journal.pone.0016598
Schaefer BC, Strominger JL, Speck SH (1995) Redefining the Epstein-Barr virus-encoded nuclear antigen EBNA-1 gene promoter and transcription initiation site in group I Burkitt lymphoma cell lines. Proc Natl Acad Sci USA 92(23):10565–10569
Schaefer BC, Paulson E, Strominger JL, Speck SH (1997) Constitutive activation of Epstein-Barr virus (EBV) nuclear antigen 1 gene transcription by IRF1 and IRF2 during restricted EBV latency. Mol Cell Biol 17(2):873–886
Schlager S, Speck SH, Woisetschlager M (1996) Transcription of the Epstein-Barr virus nuclear antigen 1 (EBNA1) gene occurs before induction of the BCR2 (Cp) EBNA gene promoter during the initial stages of infection in B cells. J Virol 70(6):3561–3570

Sears J, Ujihara M, Wong S, Ott C, Middeldorp J, Aiyar A (2004) The amino terminus of Epstein-Barr Virus (EBV) nuclear antigen 1 contains AT hooks that facilitate the replication and partitioning of latent EBV genomes by tethering them to cellular chromosomes. J Virol 78(21):11487–11505. doi:10.1128/JVI.78.21.11487-11505.2004

Serio TR, Cahill N, Prout ME, Miller G (1998) A functionally distinct TATA box required for late progression through the Epstein-Barr virus life cycle. J Virol 72(10):8338–8343

Sexton CJ, Pagano JS (1989) Analysis of the Epstein-Barr virus origin of plasmid replication (oriP) reveals an area of nucleosome sparing that spans the 3' dyad. J Virol 63(12):5505–5508

Shannon-Lowe C, Baldwin G, Feederle R, Bell A, Rickinson A, Delecluse HJ (2005) Epstein-Barr virus-induced B-cell transformation: quantitating events from virus binding to cell outgrowth. J Gen Virol 86(Pt 11):3009–3019. doi:10.1099/vir.0.81153-0

Shaw JE, Levinger LF, Carter CW Jr (1979) Nucleosomal structure of Epstein-Barr virus DNA in transformed cell lines. J Virol 29(2):657–665

Shire K, Ceccarelli DF, Avolio-Hunter TM, Frappier L (1999) EBP2, a human protein that interacts with sequences of the Epstein-Barr virus nuclear antigen 1 important for plasmid maintenance. J Virol 73(4):2587–2595

Singh G, Aras S, Zea AH, Koochekpour S, Aiyar A (2009) Optimal transactivation by Epstein-Barr nuclear antigen 1 requires the UR1 and ATH1 domains. J Virol 83(9):4227–4235. doi:10.1128/JVI.02578-08

Singh VV, Kerur N, Bottero V, Dutta S, Chakraborty S, Ansari MA, Paudel N, Chikoti L, Chandran B (2013) Kaposi's sarcoma-associated herpesvirus latency in endothelial and B cells activates gamma interferon-inducible protein 16-mediated inflammasomes. J Virol 87(8):4417–4431. doi:10.1128/JVI.03282-12

Sjoblom A, Jansson A, Yang W, Lain S, Nilsson T, Rymo L (1995a) PU box-binding transcription factors and a POU domain protein cooperate in the Epstein-Barr virus (EBV) nuclear antigen 2-induced transactivation of the EBV latent membrane protein 1 promoter. J Gen Virol 76(Pt 11):2679–2692

Sjoblom A, Nerstedt A, Jansson A, Rymo L (1995b) Domains of the Epstein-Barr virus nuclear antigen 2 (EBNA2) involved in the transactivation of the latent membrane protein 1 and the EBNA Cp promoters. J Gen Virol 76(Pt 11):2669–2678

Sjoblom A, Yang W, Palmqvist L, Jansson A, Rymo L (1998) An ATF/CRE element mediates both EBNA2-dependent and EBNA2-independent activation of the Epstein-Barr virus LMP1 gene promoter. J Virol 72(2):1365–1376

Sjoblom-Hallen A, Yang W, Jansson A, Rymo L (1999) Silencing of the Epstein-Barr virus latent membrane protein 1 gene by the Max-Mad1-mSin3A modulator of chromatin structure. J Virol 73(4):2983–2993

Smith DW, Sugden B (2013) Potential Cellular Functions of Epstein-Barr Nuclear Antigen 1 (EBNA1) of Epstein-Barr Virus. Viruses 5(1):226–240. doi:10.3390/v5010226

Speck SH, Chatila T, Flemington E (1997) Reactivation of Epstein-Barr virus: regulation and function of the BZLF1 gene. Trends Microbiol 5(10):399–405. doi:10.1016/S0966-842X(97)01129-3

Srinivas SK, Sample JT, Sixbey JW (1998) Spontaneous loss of viral episomes accompanying Epstein-Barr virus reactivation in a Burkitt's lymphoma cell line. J Infect Dis 177(6):1705–1709

Swaminathan S, Tomkinson B, Kieff E (1991) Recombinant Epstein-Barr virus with small RNA (EBER) genes deleted transforms lymphocytes and replicates in vitro. Proc Natl Acad Sci USA 88(4):1546–1550

Takacs M, Banati F, Koroknai A, Segesdi J, Salamon D, Wolf H, Niller HH, Minarovits J (2010) Epigenetic regulation of latent Epstein-Barr virus promoters. Biochim Biophys Acta 1799(3–4):228–235. doi:10.1016/j.bbagrm.2009.10.005

Tarakanova VL, Leung-Pineda V, Hwang S, Yang CW, Matatall K, Basson M, Sun R, Piwnica-Worms H, Sleckman BP, Virgin HWt (2007) Gamma-herpesvirus kinase actively initiates

a DNA damage response by inducing phosphorylation of H2AX to foster viral replication. Cell Host Microbe 1(4):275–286. doi:10.1016/j.chom.2007.05.008
Tarakanova VL, Stanitsa E, Leonardo SM, Bigley TM, Gauld SB (2010) Conserved gammaherpesvirus kinase and histone variant H2AX facilitate gammaherpesvirus latency in vivo. Virology 405(1):50–61. doi:10.1016/j.virol.2010.05.027
Tempera I, Lieberman PM (2010) Chromatin organization of gammaherpesvirus latent genomes. Biochim Biophys Acta 1799(3–4):236–245. doi:10.1016/j.bbagrm.2009.10.004
Tempera I, Lieberman PM (2014) Epigenetic regulation of EBV persistence and oncogenesis. Semin Cancer Biol 26:22–29. doi:10.1016/j.semcancer.2014.01.003
Tempera I, Wiedmer A, Dheekollu J, Lieberman PM (2010) CTCF prevents the epigenetic drift of EBV latency promoter Qp. PLoS Pathog 6(8):e1001048. doi:10.1371/journal.ppat.1001048
Tempera I, Klichinsky M, Lieberman PM (2011) EBV latency types adopt alternative chromatin conformations. PLoS Pathog 7(7):e1002180. doi:10.1371/journal.ppat.1002180
Tierney R, Kirby H, Nagra J, Rickinson A, Bell A (2000a) The Epstein-Barr virus promoter initiating B-cell transformation is activated by RFX proteins and the B-cell-specific activator protein BSAP/Pax5. J Virol 74(22):10458–10467
Tierney RJ, Kirby HE, Nagra JK, Desmond J, Bell AI, Rickinson AB (2000b) Methylation of transcription factor binding sites in the Epstein-Barr virus latent cycle promoter Wp coincides with promoter down-regulation during virus-induced B-cell transformation. J Virol 74(22):10468–10479
Tierney R, Nagra J, Hutchings I, Shannon-Lowe C, Altmann M, Hammerschmidt W, Rickinson A, Bell A (2007) Epstein-Barr virus exploits BSAP/Pax5 to achieve the B-cell specificity of its growth-transforming program. J Virol 81(18):10092–10100. doi:10.1128/JVI.00358-07
Toth Z, Maglinte DT, Lee SH, Lee HR, Wong LY, Brulois KF, Lee S, Buckley JD, Laird PW, Marquez VE, Jung JU (2010) Epigenetic analysis of KSHV latent and lytic genomes. PLoS Pathog 6(7):e1001013. doi:10.1371/journal.ppat.1001013
Toth Z, Brulois K, Jung JU (2013a) The chromatin landscape of Kaposi's sarcoma-associated herpesvirus. Viruses 5(5):1346–1373. doi:10.3390/v5051346
Toth Z, Brulois K, Lee HR, Izumiya Y, Tepper C, Kung HJ, Jung JU (2013b) Biphasic euchromatin-to-heterochromatin transition on the KSHV genome following de novo infection. PLoS Pathog 9(12):e1003813. doi:10.1371/journal.ppat.1003813
Tsai K, Thikmyanova N, Wojcechowskyj JA, Delecluse HJ, Lieberman PM (2011) EBV tegument protein BNRF1 disrupts DAXX-ATRX to activate viral early gene transcription. PLoS Pathog 7(11):e1002376. doi:10.1371/journal.ppat.1002376
Tsao SW, Tramoutanis G, Dawson CW, Lo AK, Huang DP (2002) The significance of LMP1 expression in nasopharyngeal carcinoma. Semin Cancer Biol 12(6):473–487
Tsurumi T, Fujita M, Kudoh A (2005) Latent and lytic Epstein-Barr virus replication strategies. Rev Med Virol 15(1):3–15. doi:10.1002/rmv.441
Van Damme E, Van Ostade X (2011) Crosstalk between viruses and PML nuclear bodies: a network-based approach. Front Biosci (Landmark Ed) 16:2910–2920
Wang L, Grossman SR, Kieff E (2000) Epstein-Barr virus nuclear protein 2 interacts with p300, CBP, and PCAF histone acetyltransferases in activation of the LMP1 promoter. Proc Natl Acad Sci USA 97(1):430–435
Wang FW, Wu XR, Liu WJ, Liao YJ, Lin S, Zong YS, Zeng MS, Zeng YX, Mai SJ, Xie D (2011) Heat shock factor 1 upregulates transcription of Epstein-Barr Virus nuclear antigen 1 by binding to a heat shock element within the BamHI-Q promoter. Virology 421(2):184–191. doi:10.1016/j.virol.2011.10.001
Weitzman MD, Lilley CE, Chaurushiya MS (2010) Genomes in conflict: maintaining genome integrity during virus infection. Annu Rev Microbiol 64:61–81. doi:10.1146/annurev.micro.112408.134016
Whitehurst CB, Ning S, Bentz GL, Dufour F, Gershburg E, Shackelford J, Langelier Y, Pagano JS (2009) The Epstein-Barr virus (EBV) deubiquitinating enzyme BPLF1 reduces EBV ribonucleotide reductase activity. J Virol 83(9):4345–4353. doi:10.1128/JVI.02195-08

Whitehurst CB, Vaziri C, Shackelford J, Pagano JS (2012) Epstein-Barr virus BPLF1 deubiquitinates PCNA and attenuates polymerase eta recruitment to DNA damage sites. J Virol 86(15):8097–8106. doi:10.1128/JVI.00588-12
Woellmer A, Hammerschmidt W (2013) Epstein-Barr virus and host cell methylation: regulation of latency, replication and virus reactivation. Curr Opin Virol 3(3):260–265. doi:10.1016/j.coviro.2013.03.005
Woisetschlaeger M, Jin XW, Yandava CN, Furmanski LA, Strominger JL, Speck SH (1991) Role for the Epstein-Barr virus nuclear antigen 2 in viral promoter switching during initial stages of infection. Proc Natl Acad Sci USA 88(9):3942–3946
Wong-Ho E, Wu TT, Davis ZH, Zhang B, Huang J, Gong H, Deng H, Liu F, Glaunsinger B, Sun R (2014) Unconventional sequence requirement for viral late gene core promoters of murine gammaherpesvirus 68. J Virol 88(6):3411–3422. doi:10.1128/JVI.01374-13
Wu DY, Kalpana GV, Goff SP, Schubach WH (1996) Epstein-Barr virus nuclear protein 2 (EBNA2) binds to a component of the human SNF-SWI complex, hSNF5/Ini1. J Virol 70(9):6020–6028
Wu FY, Wang SE, Chen H, Wang L, Hayward SD, Hayward GS (2004) CCAAT/enhancer binding protein alpha binds to the Epstein-Barr virus (EBV) ZTA protein through oligomeric interactions and contributes to cooperative transcriptional activation of the ZTA promoter through direct binding to the ZII and ZIIIB motifs during induction of the EBV lytic cycle. J Virol 78(9):4847–4865
Wu CC, Chuang HY, Lin CY, Chen YJ, Tsai WH, Fang CY, Huang SY, Chuang FY, Lin SF, Chang Y, Chen JY (2013) Inhibition of Epstein-Barr virus reactivation in nasopharyngeal carcinoma cells by dietary sulforaphane. Mol Carcinog 52(12):946–958. doi:10.1002/mc.21926
Wuu KD, Chen YJ, Wuu SW (1996) Frequency and distribution of chromosomal integration sites of the Epstein-Barr virus genome. J Formos Med Assoc= Taiwan yi zhi 95(12):911–916
Wyrwicz LS, Rychlewski L (2007) Identification of Herpes TATT-binding protein. Antiviral Res 75(2):167–172. doi:10.1016/j.antiviral.2007.03.002
Ye J, Gradoville L, Daigle D, Miller G (2007) De novo protein synthesis is required for lytic cycle reactivation of Epstein-Barr virus, but not Kaposi's sarcoma-associated herpesvirus, in response to histone deacetylase inhibitors and protein kinase C agonists. J Virol 81(17):9279–9291. doi:10.1128/JVI.00982-07
Yiu CY, Chen SY, Chang LK, Chiu YF, Lin TP (2010) Inhibitory effects of resveratrol on the Epstein-Barr virus lytic cycle. Molecules 15(10):7115–7124. doi:10.3390/molecules15107115
Yoshioka M, Crum MM, Sample JT (2008) Autorepression of Epstein-Barr virus nuclear antigen 1 expression by inhibition of pre-mRNA processing. J Virol 82(4):1679–1687. doi:10.1128/JVI.02142-07
Zauner L, Melroe GT, Sigrist JA, Rechsteiner MP, Dorner M, Arnold M, Berger C, Bernasconi M, Schaefer BW, Speck RF, Nadal D (2010) TLR9 triggering in Burkitt's lymphoma cell lines suppresses the EBV BZLF1 transcription via histone modification. Oncogene 29(32):4588–4598. doi:10.1038/onc.2010.203
Zentner GE, Henikoff S (2013) Regulation of nucleosome dynamics by histone modifications. Nat Struct Mol Biol 20(3):259–266. doi:10.1038/nsmb.2470
Zhang L, Pagano JS (2000) Interferon regulatory factor 7 is induced by Epstein-Barr virus latent membrane protein 1. J Virol 74(3):1061–1068
Zhao B, Zou J, Wang H, Johannsen E, Peng CW, Quackenbush J, Mar JC, Morton CC, Freedman ML, Blacklow SC, Aster JC, Bernstein BE, Kieff E (2011) Epstein-Barr virus exploits intrinsic B-lymphocyte transcription programs to achieve immortal cell growth. Proc Natl Acad Sci USA 108(36):14902–14907. doi:10.1073/pnas.1108892108
Zhou J, Chau CM, Deng Z, Shiekhattar R, Spindler MP, Schepers A, Lieberman PM (2005) Cell cycle regulation of chromatin at an origin of DNA replication. EMBO J 24(7):1406–1417. doi:10.1038/sj.emboj.7600609

The Epigenetic Life Cycle of Epstein–Barr Virus

Wolfgang Hammerschmidt

Abstract Ever since the discovery of Epstein–Barr virus (EBV) more than 50 years ago, this virus has been studied for its capacity to readily establish a latent infection, which is the prominent hallmark of this member of the herpesvirus family. EBV has become an important model for many aspects of herpesviral latency, but the molecular steps and mechanisms that lead to and promote viral latency have only emerged recently. It now appears that the virus exploits diverse facets of epigenetic gene regulation in the cellular host to establish a latent infection. Most viral genes are transcriptionally repressed, and viral chromatin is densely compacted during EBV's latent phase, but latent infection is not a dead end. In order to escape from this phase, epigenetic silencing must be reverted efficiently and quickly. It appears that EBV has perfected a clever strategy to overcome transcriptional repression of its many lytic genes to initiate virus de novo synthesis within a few hours after induction of its lytic cycle. This review tries to summarize the known molecular mechanisms, the current models, concepts, and ideas underlying this viral strategy. This review also attempts to identify and address gaps in our current understanding of EBV's epigenetic mechanisms within the infected cellular host.

Contents

1 Introduction 104
2 The Pre-latent Phase 105
3 The Latent Phase 108
4 Lytic Induction and Virus Synthesis 109
5 Open Questions 112
References 114

W. Hammerschmidt (✉)
Research Unit Gene Vectors, Helmholtz Zentrum München, German Research Center for Environmental Health, and German Centre for Infection Research (DZIF), Partner site Munich, Marchioninistr. 25, 81377 Munich, Germany
e-mail: hammerschmidt@helmholtz-muenchen.de

C. Münz (ed.), *Epstein Barr Virus Volume 1*, Current Topics in Microbiology and Immunology 390, DOI 10.1007/978-3-319-22822-8_6

1 Introduction

The infection of resting primary human B lymphocytes is a versatile and tractable model to study EBV. As a consequence of infection with EBV, lymphoblastoid cells lines (LCLs) emerge that can be easily obtained from any donor and cultivated in bulk or as single cell clones in vitro. It is said that EBV immortalizes or growth transforms B cells, which become latently infected and proliferate for months. Infection of human B lymphocytes has been most informative to study the early steps of viral infection that lead to latent infection. EBV not only establishes a latent infection in LCLs but also in all EBV-infected tumors. From biopsies, certain tumor cell lines can be obtained and cultivated as permanent cell lines in vitro. For example, a large number of Burkitt's lymphoma cell lines are available that maintain their EBV-positive state in cell culture and remain latently infected. Together with LCLs, they are a rich source to study EBV latency and its epigenetic regulation.

Established Burkitt's lymphoma cell lines have been the preferred model to study the induction of EBV's lytic phase. Incubation with TPA and butyrate or cross-linking of the surface immunoglobulin of certain Burkitt's lymphoma cell lines readily induces the lytic phase of EBV and leads to virus de novo synthesis and egress of infectious virus. These measures induce the expression of a viral transcription factor termed BZLF1 (also called Z, Zta, ZEBRA, or EB1), which has been identified as the key molecular switch gene and inducer of EBV's lytic phase (Takada et al. 1986; Countryman and Miller 1985). Expression of BZLF1 can revert the latent phase and induce the expression of all viral lytic genes, which are epigenetically silenced during EBV's latent phase.

EBV not only infects human B cells, but also other cell types. Two human carcinomas, endemic nasopharyngeal carcinomas and a small fraction of gastric carcinomas, are characterized by latently EBV-infected tumor cells (see also the chapter by Nancy Raab-Traub in this book). EBV has also been found occasionally in normal, non-transformed epithelial cells in the oropharynx (Walling et al. 2001) indicating EBV does infect epithelial cells in vivo. Unfortunately, ex vivo infection of primary human epithelial cells with EBV is very inefficient. The cells do not synthesize virus progeny upon infection nor is the virus maintained long term in cell culture.

Latently, EBV-infected cells are characterized by the expression of certain viral genes. Different sets of viral genes, termed EBNAs (Epstein–Barr nuclear antigens) and LMPs (latent membrane proteins) together with noncoding transcripts such as viral microRNAs and long noncoding RNAs, the EBERs, are expressed in latently infected cells. Depending on the type and origin of the cell, different latency programs classify the extent of latent gene expression. For example, six EBNA and three LMP genes are expressed along with noncoding RNAs in LCLs. In Burkitt's lymphoma cells, only a small subset of the latent genes is commonly expressed. In all instances, the viral genes support the extrachromosomal maintenance of several copies of the EBV genome, which is about 160 kbps in size. In addition, the latent gene products induce and support proliferation of the latently infected cells and contribute to their transformed phenotype.

All herpesviruses deliver their genomic DNA in an epigenetically naïve form. Besides viral DNA, the herpesviral capsids do not appear to contain histones or other DNA-associated proteins of cellular or viral origin (Johannsen et al. 2004). The viral DNA is also free of 5′-methyl cytosine residues (Gibson and Roizman 1971; Kintner and Sugden 1981), which typically occur in a CpG dinucleotide context in mammalian DNA and are associated with epigenetic repression of transcription. EBV DNA has a high overall content of CpG dinucleotides of 5.4 %. In fact, DNA methylation plays an important role in the latent phase of EBV infection during which chromatin architecture and the epigenetic landscape also determine the transcriptional state of genomic EBV DNA.

Recent molecular data indicate that EBV is unusual in that it cannot replicate itself upon infection. This statement appears surprising given the fact that all other human herpesviruses readily support their own de novo synthesis in permissive cells, which release progeny within a short period after infection. Rather, EBV establishes a latent infection during which the viral genomic DNA acquires all features of cellular chromatin. They include extensive CpG DNA methylation, global epigenetic modifications, and histone marks, which are mostly repressive leading to transcriptional silencing of all but the virus' latent genes.

EBV's lifestyle seems to consist of three phases, which are illustrated in Fig. 1. Upon infection, the virus attaches to its receptors at the cell surface, internalizes and delivers its epigenetically naïve linear DNA genome to the nucleus of the cell. The circularization of the viral DNA is one of the very first events in order to protect the DNA ends from degradation and minimize induction of the DNA damage response of the infected cell (Hurley and Thorley-Lawson 1988). Within a few hours post-infection, a number of viral lytic genes are found expressed together with latent genes. About ten days later, EBV's latent gene expression prevails and the expression of lytic genes becomes nearly undetectable. The latent phase is established. Upon appropriate stimuli, i.e., the activation of the surface immunoglobulins on memory B cells, a signaling cascade induces the expression of the viral transcription factor BZLF1, which is the prologue to lytic reactivation and leads to virus synthesis, eventually.

2 The Pre-latent Phase

Upon infection of primary human B lymphocytes, the virus does not induce its de novo synthesis (Kalla et al. 2010) but initiates a "pre-latent phase" during which a subset of lytic genes together with viral latent genes are expressed (reviewed in

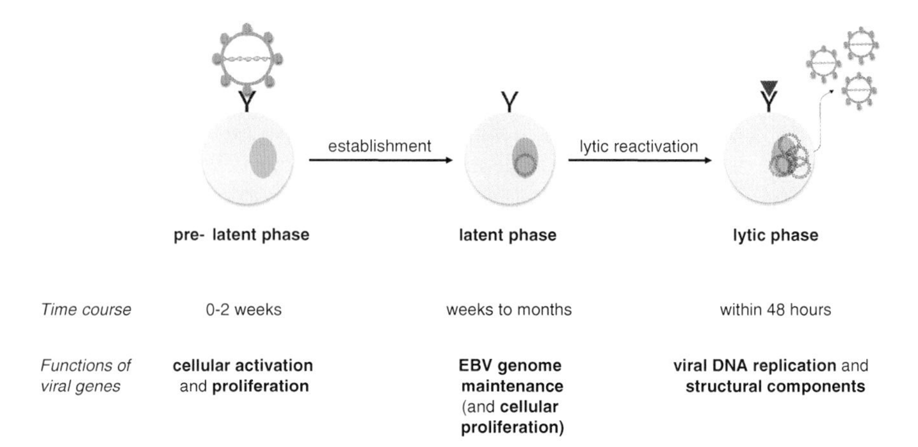

Fig. 1 EBV's lifestyle encompasses three different phases. The scheme shows a single EBV virion, which infects a B lymphocyte. The virus delivers its linear, double-stranded DNA to the nucleus of the cell where the EBV DNA is present in several copies (not shown) of an EBV minichromosome (*red circle*). During latency, this state is maintained even in proliferating cells. Upon antigen contact via the surface immunoglobulin, EBV's molecular switch gene BZLF1 is activated and induces the lytic, productive phase. The cell releases virus progeny now. The timing and length of infection during the three different phases of EBV infection are shown together with the functions of viral genes expressed in the different phases. In latently infected B cells in vitro, the virus induces their continued proliferation, which depends on the expression of certain viral latent genes. (As discussed in the review, this model is primarily based on experiments with primary human B lymphocytes and HEK293 cells, an established cell line with epithelia-like characteristics. The proposed model is debatable. In the past, the reported expression of individual lytic viral genes in epithelial cells has been interpreted to mean that they can support EBV's full-fledged viral phase immediately upon infection. However, recently published data and the current state of knowledge as discussed in this review support the three-phase model shown here.)

Woellmer and Hammerschmidt 2013; Kalla and Hammerschmidt 2012). EBV also infects primary human epithelial cells and expresses lytic as well latent viral genes upon infection (Shannon-Lowe et al. 2006 and references therein). It is likely that epithelial cells also fail to support the ready synthesis of viral progeny (Walling et al. 2001). The reason for it can be found in EBV's intricate and peculiar epigenetic lifestyle (see below).

In newly infected cells, viral transcripts are detectable prior to transcription of the incoming EBV DNA. Cells that support the synthesis of herpesviruses release exosomes, virus-like particles, and virions, which contain viral transcripts as cargo (Kalamvoki et al. 2014; Jochum et al. 2012b; Bresnahan and Shenk 2000; Bechtel et al. 2005; Cliffe et al. 2009; Greijer et al. 2000; Sciortino et al. 2001). The delivery of viral transcripts to the infected cell allows the immediate expression of viral factors that can elicit viral immune escape mechanisms and modulate cellular gene expression (Kalamvoki et al. 2014). As a consequence, EBV transcripts delivered upon infection activate the resting B lymphocytes and support their latent infection (Jochum et al. 2012a, b).

De novo transcription of the incoming EBV DNA probably starts shortly after infection of B lymphocytes. To my knowledge, no systematic study provides a complete picture of newly transcribed viral genes in these cells. It is known that as early as six hours post-infection, transcripts that originate from the so-called W promoter (Wp) encode the latent EBNA-LP and EBNA2 genes, which are also the first "latent" EBV proteins detectable in newly infected B lymphocytes (Schlager et al. 1996; Woisetschlaeger et al. 1990; Alfieri et al. 1991). It was surprising to learn that lytic viral genes are also transcribed initially and fulfill important functions in the newly infected B lymphocytes. Among these, lytic viral genes are EBV-encoded members of the anti-apoptotic BCL-2 family (Altmann and Hammerschmidt 2005), the molecular switch gene BZLF1 (Wen et al. 2007; Kalla et al. 2010), viral immunoevasins, and lytic genes with transactivating and regulatory functions such as BMRF1 and BRLF1 (Jochum et al. 2012a, b; Kalla et al. 2010). It is important to note certain viral genes essential for lytic amplification of viral DNA and genes encoding structural viral proteins are not expressed in the pre-latent phase consistent with the notion that viral progeny is not produced initially (Kalla et al. 2010).

Viral transcription early after infection appears to be chaotic. It seems that the epigenetically naïve EBV genomic DNA, which lacks histones and nucleosomes and is free of methylated CpG dinucleotides, serves as a permissive template for the cellular transcription machinery. Transcription of viral genes is not universal but seems to be restricted to viral promoters, which recruit cellular transcription factors and are transcribed by RNA polymerase II. Among them are viral promoters, which drive the expression of latent genes, e.g., EBNA1, EBNA2, and EBNA-LP (Shannon-Lowe et al. 2005; Tierney et al. 2007; Schaefer et al. 1995; Schlager et al. 1996 and references therein), but also lytic, so-called immediate early genes such as BZLF1 and BRLF1 (Wen et al. 2007; Kalla et al. 2010), and lytic early genes such as BHRF1, BALF1, BMRF1, BCRF1, and BNLF2a (Jochum et al. 2012a, b). Currently, the complete catalogue of viral genes transcribed within the first days after infection of primary B lymphocytes is not known. This promiscuous state of transcription seems to come to an end as EBV DNA acquires cellular nucleosomes and becomes methylated later during infection.

During the pre-latent phase, the resting small B lymphocytes enlarge into B cell blasts and eventually begin to proliferate. About four days post-infection, nucleosome occupancy of viral DNA becomes detectable at a select viral locus, the dyad symmetry element of the plasmid origin of DNA replication, where nucleosome positioning is accurate and highly ordered (Zhou et al. 2005). Nucleosomal occupancy increases gradually at this locus within the next days and reaches high levels genome-wide about two to three weeks post-infection (Schmeinck 2011).

In contrast, methylation of viral DNA is a slow process in infected primary B lymphocytes. After two weeks post-infection, CpG methylation becomes first detectable and gradually increases for the next two to three months to considerable levels (Kalla et al. 2010). Few regions in EBV DNA are spared from CpG methylation, which is extensive in established LCLs (Fernandez et al. 2009; Minarovits 2006;

Paulson and Speck 1999; Tierney et al. 2000; Woellmer et al. 2012) as well as in latently infected peripheral memory B lymphocytes in vivo (Woellmer et al. 2012).

EBV DNA acquires an epigenetic signature during the pre-latent phase. The early expression of a plethora of viral genes might result from the epigenetically "naked" EBV genome. During the pre-latent phase, EBV DNA nucleosomes are positioned and repressive chromatin marks are presumably introduced, which are the characteristic for EBV's chromatin during the latent phase (see below). The acquisition of cellular chromatin and a repressive epigenetic pattern driven by the host cell leads to the eventual silencing of all lytic genes but also certain promoters of latent genes. This process of epigenetic shutoff is completed about ten to 14 days post-infection (Fig. 1).

3 The Latent Phase

The programs of latent viral gene expression differ according to the type and state of differentiation of the latently EBV-infected cell. The expression of lytic genes is strictly controlled and repressed in this phase presumably to prevent immune recognition in vivo and block virus synthesis, both of which would jeopardize the survival of the latently infected cell. Recent data indicate that epigenetic modifications of viral DNA define viral latency. In other words, latency is encoded in the state of EBV's chromatin.

Four types of global epigenetic modification are known that repress EBV's lytic viral genes during latency: histone modifications, nucleosomal density and compaction of DNA, DNA methylation, and chromatin architecture.

Transcriptionally silenced lytic viral genes and their promoters are associated with H3K27me3 histone marks, which are repressive modifications introduced by the histone methyltransferase EZH2, a component of the Polycomb repressive complex 2, PRC2 (Woellmer et al. 2012). This histone modification seems to be important because trimethylation of H3K27 is erased upon induction of EBV's lytic phase (see below). Less frequent is the repressive mark H3K9me3 (Ramasubramanyan et al. 2012), which is not removed upon viral reactivation suggesting that this modification is less important for the maintenance of repressed viral chromatin during latency. In contrast to cells latently infected with Kaposi's sarcoma-associated herpesvirus (KSHV), H3K4me3 histone marks and RNA polymerase II are absent in epigenetically repressed EBV's chromatin during latency. In this respect, chromatin in latently KSHV-infected cells is bivalent at most early lytic promoters (Gunther and Grundhoff 2010; Toth et al. 2010; Lieberman 2013). It thus appears that EBV uses cellular functions of epigenetic repression and prefers static, repressive chromatin marks at transcriptionally repressed lytic genes during latency.

Repressed promoters of lytic genes are enriched in nucleosomes, and their high local concentrations argue for compact and condensed viral chromatin during latency (Woellmer et al. 2012). In fact, high nucleosome occupancy correlates with promoter silencing in many different cells latently infected with EBV (Arvey et al. 2012). Nucleosomal density is particularly high at early lytic promoters of EBV, which are controlled and transactivated by BZLF1 upon induction of EBV's lytic phase (see below) suggesting that nucleosomes provide an obstacle to transactivating factors in general and BZLF1 in particular. It is currently elusive but certain positioned nucleosomes and compacted chromatin might directly contribute to the epigenetic stability of repressed viral genes ensuring EBV's latent phase.

DNA methylation is an important epigenetic parameter and is thought to contribute to overall gene silencing. The state of CpG methylation in EBV DNA in latently infected cells is generally extremely high at repressed early lytic promoters and genes consistent with this notion (Woellmer et al. 2012; Fernandez et al. 2009). In vivo, EBV DNA in memory B cells seems to adopt a similar pattern (Woellmer et al. 2012; Paulson and Speck 1999). CpG methylation of EBV DNA is not global but bimodal because promoters of latent genes, which are active, are spared. Among others, the Fp/Qp and Cp promoters that drive EBNA1 in certain Burkitt's lymphoma cell lines, the promoters of the LMP1/LMP2A genes, and the two EBER genes, are free of methylated CpG dinucleotides similar to certain DNA sequences that co-localize with nucleosome-free regions and are suspect to mediate higher order chromatin architecture of EBV DNA (Kalla et al. 2010; Woellmer et al. 2012 and unpublished data). It thus appears that CpGs, which escape methylation, are located at functionally active sites in latently EBV-infected DNA, while the overwhelming majority of CpGs are almost entirely methylated and functionally silent. Lack of CpG methylation commonly coincides with histones bearing active or permissive histone marks indicating that key regulatory features of EBV's chromatin follow textbook rules. Promoter usage seems to be controlled by DNA methylation during latency.

The state of EBV's genomic architecture and organization in latently infected cells is just beginning to emerge. The cellular CCCTC-binding factor (CTCF) and members of the cohesion complex often co-localize on DNA and promote the formation of chromatin loops to demarkate and organize chromosomal domains. These factors seem to be involved in the architecture of genomic EBV DNA in latently infected cells (Arvey et al. 2012; Holdorf et al. 2011; Tempera et al. 2010, 2011) (see also the chapter by Paul Lieberman in this book). It is currently unclear whether binding and looping of EBV DNA lead to gene activation or repression by connecting or separating functionally distinct regulatory elements in latently infected cells. How EBV's genome architecture changes upon induction of the lytic phase of EBV's lifestyle has not been addressed, yet.

4 Lytic Induction and Virus Synthesis

During de novo virus synthesis, about 70 lytic genes of EBV are expressed that support viral asynchronous DNA amplification independent of the cellular DNA replication and encode viral structural components to allow virus morphogenesis and release of virus progeny. Two genes, BZLF1 and BRLF1, which encode viral transcription factors, orchestrate the transition from viral latency to productive, lytic infection. The former acts as a master switch regulator, which can induce the lytic phase of EBV's life cycle in latently infected cells (Sinclair 2003), and the latter is equally indispensable for the expression of all viral lytic genes (Feederle et al. 2000). Cellular differentiation of latently EBV-infected B cell or epithelial cells likely drives the expression of both BZLF1 and BRLF1 genes (Reusch et al. 2015).

EBV genuinely repressed chromatin poses the problem of efficient reactivation to support de novo virus synthesis in latently infected cells. Besides compacted chromatin with dense nucleosomes and repressive histone modifications, CpG methylation is the key features of repressed lytic genes of EBV, but, counterintuitively, DNA methylation is even a prerequisite for the escape from EBV's latent phase.

The induction of EBV's lytic phase in latently infected cells is initiated by the expression of the viral BZLF1 gene encoding the transcription factor BZLF1, which is a basic leucine zipper (bZIP) transcription factor. It is modular in structure and contains a basic domain that mediates sequence-specific binding to DNA, a coiled-coil domain that confers to homodimerization, and a transcriptional activation domain (Sinclair 2003).

BZLF1 binds sequence specifically to one class of DNA motifs, termed ZREs, but prefers a second class that contains methylated 5′-cytosine residues (5mC), termed meZREs (Bhende et al. 2004; Bergbauer et al. 2010; Kalla et al. 2010; Karlsson et al. 2008; Petosa et al. 2006; Dickerson et al. 2009). Binding to both motifs can activate viral promoters that drive a number of lytic genes. meZREs are the exclusive BZLF1-binding motifs in certain early viral promoters, which, for example, encode factors essential for the lytic amplification of EBV DNA (Bergbauer et al. 2010) such as the viral helicase, DNA polymerase, and DNA polymerase accessory protein among others (Fixman et al. 1992). The BZLF1-mediated expression of these viral genes depends on meZREs. Paradoxically, their CpG methylation is instrumental for the expression of certain essential lytic genes and indispensable for virus synthesis (Kalla et al. 2012), changing the conventional view of DNA methylation solely as a repressive epigenetic feature.

BZLF1 binds with high affinity to meZREs in a methylation-dependent fashion (Bergbauer et al. 2010). This peculiar feature also explains why the virus fails to induce synthesis of progeny virus during the initial, pre-latent phase of infection during which BZLF1 is expressed (Fig. 1). Key lytic viral promoters that depend on BZLF1 and contain meZRE sites cannot be expressed during this phase because they lack DNA methylation, which abrogates binding of BZLF1 to its meZRE target sequences and prevents progeny virus synthesis in newly infected cells. Thus, EBV invented a marvelous system to regulate its different phases with the help of DNA methylation.

In latently infected cells, binding of BZLF1 protein to methylated meZRE sites induces an epigenetic reprogramming at the repressed and silenced target promoters. Nucleosomes are removed locally, and chromatin adopts an open, accessible configuration. Epigenetic repression by Polycomb proteins is erased, activation marks like H3K4me3 are introduced, and, eventually, RNA polymerase II is bound and transcription of the target genes is increased massively. Interestingly, transcription of lytic genes occurs on a fully methylated template without the need of active DNA demethylation at promoters or gene bodies. Thus, DNA methylation per se does not prevent the binding or block the initiation and elongation activity of RNA polymerase II providing a new paradigm for gene regulation (Woellmer et al. 2012).

One of the interesting aspects of BZLF1 binding to its target promoters is the local changes that occur in nucleosome occupancy. BZLF1 binding to chromatinized DNA can evict nucleosomes in vivo indicating that the BZLF1 transcription factor directly or indirectly can induce a change in the chromatin structure (Woellmer et al. 2012). Therefore, BZLF1 could act as a pioneer factor (Woellmer and Hammerschmidt 2013). Pioneer factors can bind to compacted chromatin and trigger its local opening in order to prepare promoters or other cis-acting elements for transcriptional activation with or without the help of chromatin remodeling factors.

During their existence, herpesviruses have regularly acquired cellular genes by random recombination with host cell DNA, and certain adopted genes have evolved to serve specific viral functions. BZLF1 shares amino acid sequence homology with the AP-1 family of transcription factors. Its binding domain is very similar to that of c-FOS protein (Taylor et al. 1991; Farrell et al. 1989). It could be that BZLF1 has adopted its genuine binding properties to methylated DNA from its cellular counterpart c-FOS or other members of the AP-1 family. In fact, the prototypical c-JUN/c-FOS heterodimer is able to bind to methylated DNA sequence motifs in cellular chromatin and activate gene transcription from methylated templates as BZLF1 does (Gustems et al. 2014). DNA binding to methylated AP-1 sequence motifs is infrequent in cellular chromatin as compared to the consensus normal AP-1 target sites that lack CpG dinucleotide pairs, but it appears that EBV has optimized an existing cellular principle of gene regulation, which it can use to establish latent infections in all cells it infects.

In the lytic phase, the expression of early lytic genes precedes the induction of late lytic promoters. As with other herpesviruses, the expression of a number of late lytic genes of EBV requires newly replicated viral DNA (Amon et al. 2004), which is massively amplified via the virus-encoded replication machinery in the lytic phase. The concept of late gene regulation of herpesviruses has been elusive so far, but the viral components that drive the expression of late EBV genes have been identified, recently (Aubry et al. 2014). The virally encoded, TBP-like protein, BcRF1 is part of a set of six viral gene products, which is essential and sufficient to assemble the transcriptional pre-initiation complex including Pol II at late promoters (Gruffat et al. 2012). It appears plausible to argue that transcription of late viral genes occurs on newly replicated viral DNA, which is known to be free of methylated CpGs (Kalla et al. 2010 and references therein). Recognition of late lytic EBV promoters by this dedicated multicomponent transcription complex depends on a TATA-like viral sequence motif and replicated viral DNA, which is also free of nucleosomes (Chiu et al. 2013). Again, viral epigenetic mechanisms are crucial and timely regulated even during late viral transcription supporting escape from latency and virus morphogenesis.

5 Open Questions

How does viral DNA acquire cellular histones and/or nucleosomes?
When EBV infects a cell, it delivers the epigenetically naïve and histone-free viral DNA into the nucleus of the cell. Do cellular histone loading factors such as HIRA, DAXX, or NAP1L1 help to chromatinize the epigenetically naïve viral DNA during EBV's pre-latent phase? When are classical cellular histones synthesized, which are normally deposited in a cell cycle-dependent fashion? Virion-delivered BNRF1, a tegument protein, suppresses DAXX–ATRX-mediated H3.3 loading on viral DNA supporting latent gene expression and preventing repressive chromatin formation onto the EBV genome during primary infection of B lymphocytes (Tsai et al. 2014). Which cellular or viral factors drive de novo assembly of chromatin on epigenetically naïve EBV DNA is currently elusive.

How does latent EBV chromatin acquire repressive histone marks?
The majority of lytic viral genes and their promoters are governed by PRC2-induced trimethylation of H3K27, which is a prevalent repressive histone mark together with H3K9me3. It is unclear what directs PRC2 to these regions and whether mechanisms exist that promote the silencing of lytic genes sparing the promoters of those genes, which are active during EBV's latent phase. It is equally unclear whether additional histone methyltransferases are active during EBV's distinct phases of infection.

What causes the slow but extensive CpG methylation of viral genomic DNA?
Enzymes catalyzing the addition of a methyl group to cytosine residues are termed DNA methyltransferases (DNMTs). In newly infected B lymphocytes, the maintenance methyltransferase DNMT1 is expressed already at day one post-infection. One week post-infection expression of DNMT1 increased further together with DNMT3a and DNMT3b, which encode the two cellular de novo methyltransferases (Schmeinck 2011). EBV's latent membrane protein 1 (LMP1) induces the expression of all methyltransferases in nasopharyngeal carcinoma cell lines via the JNK pathway (Tsai et al. 2006) suggesting a role for LMP1 in upregulating DNMT expression also in newly infected B lymphocytes. The nuclear protein of 95 kDa (NP95), a cellular protein associated with maintenance and de novo DNA methylation, recruits DNMT1 to hemi-methylated CpGs, but also interacts with de novo DNMTs, histone methyltransferases, and trimethylated H3K9 connecting the DNA methylation pathway to the establishment of repressive histone marks (Meilinger et al. 2009; Rottach et al. 2010). NP95 is also upregulated in EBV-infected B lymphocytes seven days post-infection (Schmeinck 2011) indicating that NP95 could be involved in introducing methylated CpG dinucleotides in EBV DNA. The slow increase in DNA methylation of EBV's DNA could probably be

the reason for the lack of virion synthesis until twelve days post-infection (Kalla et al. 2010).

Is the promoter of BZLF1 epigenetically regulated?
BZLF1 can induce the lytic phase in latently EBV-infected cells, but its expression is tightly repressed in these cells. Several cellular repressors of transcription have been identified (Kenney and Mertz 2014 and references therein) that block the promoter of BZLF1 and prevent its reactivation. In contrast to the majority of EBV's lytic genes and their promoters, CpG dinucleotides are exceptionally rare and exempt from DNA methylation in the promoter of BZLF1 (Woellmer et al. 2012). It appears that epigenetic mechanisms, which increase or alleviate the repressed state of the BZLF1 promoter, have not been completely analyzed (Murata and Tsurumi 2014).

What causes decompaction of EBV DNA and transcriptional reactivation of higher order viral chromatin?
Reactivation of epigenetically repressed viral chromatin requires chromatin opening and removal of inactivating, repressive histone marks. How this profound change in the status of viral chromatin is achieved is not known. It is also not known how the BZLF1 factor gains access to compacted chromatin and whether it mediates, directly or indirectly, chromatin remodeling. Whether BZLF1 or other viral factors play a role in changing the architecture of viral chromatin upon lytic induction is equally uncertain.

How are CpG methylation and histones lost prior to encapsidation of virion DNA?
EBV encodes its own replication machinery of viral enzymes and cofactors that mediate unlicensed amplification of viral DNA during the virus' lytic phase. DNA amplification occurs in the nucleus at distinct sites and is accompanied by nuclear reorganization. Histones are not part of the sites of lytic DNA amplification, and PCNA is equally excluded. Moreover, cellular histone chaperone levels decrease during EBV's lytic phase indicating that EBV follows a strategy that is very different from cellular DNA replication (Chiu et al. 2013). PCNA mediates the recruitment of cytosine methyltransferases to replication forks during cellular DNA replication, but in the DNA replication machinery, PCNA is replaced by EBV's own DNA clamp BMRF1. It is speculative but BMRF1 might fail to recruit the cellular hemi-methyltransferase DNMT1 that would maintain the status of CpG methylation in newly replicated viral DNA (Chiu et al. 2013; Kalla et al. 2012).

Acknowledgments This review is based on the works of many colleagues. I sincerely apologize to all scientists whose important contributions could not be cited here due to space limitations. I would like to thank Bill Sugden for reading the manuscript and his valuable suggestions. This review and work in my laboratory is supported by Institutional Intramural Grants, grants from the Deutsche Forschungsgemeinschaft SFBTRR36/TPA04, SFB1064/TPA13, SFB1054/TPB05, grants from German Centre for Infection Research (DZIF), and National Institutes of Health Grant CA70723.

References

Alfieri C, Birkenbach M, Kieff E (1991) Early events in Epstein-Barr virus infection of human B lymphocytes. Virology 181:595–608

Altmann M, Hammerschmidt W (2005) Epstein-Barr virus provides a new paradigm: a requirement for the immediate inhibition of apoptosis. PLoS Biol 3:e404

Amon W, Binne UK, Bryant H, Jenkins PJ, Karstegl CE, Farrell PJ (2004) Lytic cycle gene regulation of Epstein-Barr virus. J Virol 78:13460–13469

Arvey A, Tempera I, Tsai K, Chen HS, Tikhmyanova N, Klichinsky M, Leslie C, Lieberman PM (2012) An atlas of the Epstein-Barr virus transcriptome and epigenome reveals host-virus regulatory interactions. Cell Host Microbe 12:233–245

Aubry V, Mure F, Mariame B, Deschamps T, Wyrwicz LS, Manet E, Gruffat H (2014) Epstein-Barr virus late gene transcription depends on the assembly of a virus-specific preinitiation complex. J Virol 88:12825–12838

Bechtel J, Grundhoff A, Ganem D (2005) RNAs in the virion of Kaposi's sarcoma-associated herpesvirus. J Virol 79:10138–10146

Bergbauer M, Kalla M, Schmeinck A, Gobel C, Rothbauer U, Eck S, Benet-Pages A, Strom TM, Hammerschmidt W (2010) CpG-methylation regulates a class of Epstein-Barr virus promoters. PLoS Pathog 6:e1001114

Bhende PM, Seaman WT, Delecluse HJ, Kenney SC (2004) The EBV lytic switch protein, Z, preferentially binds to and activates the methylated viral genome. Nat Genet 36:1099–1104

Bresnahan WA, Shenk T (2000) A subset of viral transcripts packaged within human cytomegalovirus particles. Science 288:2373–2376

Chiu YF, Sugden AU, Sugden B (2013) Epstein-Barr viral productive amplification reprograms nuclear architecture, DNA replication, and histone deposition. Cell Host Microbe 14:607–618

Cliffe AR, Nash AA, Dutia BM (2009) Selective uptake of small RNA molecules in the virion of murine gammaherpesvirus 68. J Virol 83:2321–2326

Countryman J, Miller G (1985) Activation of expression of latent Epstein-Barr herpesvirus after gene transfer with a small cloned subfragment of heterogeneous viral DNA. Proc Natl Acad Sci U S A 82:4085–4089

Dickerson SJ, Xing Y, Robinson AR, Seaman WT, Gruffat H, Kenney SC (2009) Methylation-dependent binding of the Epstein-Barr virus BZLF1 protein to viral promoters. PLoS Pathog 5:e1000356

Farrell PJ, Rowe DT, Rooney CM, Kouzarides T (1989) Epstein-Barr virus BZLF1 trans-activator specifically binds to a consensus AP-1 site and is related to c-fos. EMBO J 8:127–132

Feederle R, Kost M, Baumann M, Janz A, Drouet E, Hammerschmidt W, Delecluse HJ (2000) The Epstein-Barr virus lytic program is controlled by the co-operative functions of two transactivators. EMBO J 19:3080–3089

Fernandez AF, Rosales C, Lopez-Nieva P, Grana O, Ballestar E, Ropero S, Espada J, Melo SA, Lujambio A, Fraga MF, Pino I, Javierre B, Carmona FJ, Acquadro F, Steenbergen RD, Snijders PJ, Meijer CJ, Pineau P, Dejean A, Lloveras B, Capella G, Quer J, Buti M, Esteban JI, Allende H, Rodriguez-Frias F, Castellsague X, Minarovits J, Ponce J, Capello D, Gaidano G, Cigudosa JC, Gomez-Lopez G, Pisano DG, Valencia A, Piris MA, Bosch FX, Cahir-McFarland E, Kieff E, Esteller M (2009) The dynamic DNA methylomes of double-stranded DNA viruses associated with human cancer. Genome Res 19:438–451

Fixman ED, Hayward GS, Hayward SD (1992) Trans-acting requirements for replication of Epstein-Barr virus ori-Lyt. J Virol 66:5030–5039

Gibson W, Roizman B (1971) Compartmentalization of spermine and spermidine in the herpes simplex virion. Proc Natl Acad Sci U S A 68:2818–2821

Greijer AE, Dekkers CA, Middeldorp JM (2000) Human cytomegalovirus virions differentially incorporate viral and host cell RNA during the assembly process. J Virol 74:9078–9082

Gruffat H, Kadjouf F, Mariame B, Manet E (2012) The Epstein-Barr virus BcRF1 gene product is a TBP-like protein with an essential role in late gene expression. J Virol 86:6023–6032
Gunther T, Grundhoff A (2010) The epigenetic landscape of latent Kaposi sarcoma-associated herpesvirus genomes. PLoS Pathog 6:e1000935
Gustems M, Woellmer A, Rothbauer U, Eck SH, Wieland T, Lutter D, Hammerschmidt W (2014) c-Jun/c-Fos heterodimers regulate cellular genes via a newly identified class of methylated DNA sequence motifs. Nucleic Acids Res 42:3059–3072
Holdorf MM, Cooper SB, Yamamoto KR, Miranda JJ (2011) Occupancy of chromatin organizers in the Epstein-Barr virus genome. Virology 415:1–5
Hurley EA, Thorley-Lawson DA (1988) B cell activation and the establishment of Epstein-Barr virus latency. J Exp Med 168:2059–2075
Jochum S, Moosmann A, Lang S, Hammerschmidt W, Zeidler R (2012a) The EBV immunoevasins vIL-10 and BNLF2a protect newly infected B cells from immune recognition and elimination. PLoS Pathog 8:e1002704
Jochum S, Ruiss R, Moosmann A, Hammerschmidt W, Zeidler R (2012b) RNAs in Epstein-Barr virions control early steps of infection. Proc Natl Acad Sci USA 109:E1396–E1404
Johannsen E, Luftig M, Chase MR, Weicksel S, Cahir-McFarland E, Illanes D, Sarracino D, Kieff E (2004) Proteins of purified Epstein-Barr virus. Proc Natl Acad Sci USA 101:16286–16291
Kalamvoki M, Du T, Roizman B (2014) Cells infected with herpes simplex virus 1 export to uninfected cells exosomes containing STING, viral mRNAs, and microRNAs. Proc Natl Acad Sci USA 111:E4991–E4996
Kalla M, Gobel C, Hammerschmidt W (2012) The lytic phase of Epstein-Barr virus requires a viral genome with 5-methylcytosine residues in CpG sites. J Virol 86:447–458
Kalla M, Hammerschmidt W (2012) Human B cells on their route to latent infection–early but transient expression of lytic genes of Epstein-Barr virus. Eur J Cell Biol 91:65–69
Kalla M, Schmeinck A, Bergbauer M, Pich D, Hammerschmidt W (2010) AP-1 homolog BZLF1 of Epstein-Barr virus has two essential functions dependent on the epigenetic state of the viral genome. Proc Natl Acad Sci USA 107:850–855
Karlsson QH, Schelcher C, Verrall E, Petosa C, Sinclair AJ (2008) Methylated DNA recognition during the reversal of epigenetic silencing is regulated by cysteine and serine residues in the Epstein-Barr virus lytic switch protein. PLoS Pathog 4:e1000005
Kenney SC, Mertz JE (2014) Regulation of the latent-lytic switch in Epstein-Barr virus. Semin Cancer Biol 26:60–68
Kintner C, Sugden B (1981) Conservation and progressive methylation of Epstein-Barr viral DNA sequences in transformed cells. J Virol 38:305–316
Lieberman PM (2013) Keeping it quiet: chromatin control of gammaherpesvirus latency. Nat Rev Microbiol 11:863–875
Meilinger D, Fellinger K, Bultmann S, Rothbauer U, Bonapace IM, Klinkert WE, Spada F, Leonhardt H (2009) Np95 interacts with de novo DNA methyltransferases, Dnmt3a and Dnmt3b, and mediates epigenetic silencing of the viral CMV promoter in embryonic stem cells. EMBO Rep 10:1259–1264
Minarovits J (2006) Epigenotypes of latent herpesvirus genomes. Curr Top Microbiol Immunol 310:61–80
Murata T, Tsurumi T (2014) Switching of EBV cycles between latent and lytic states. Rev Med Virol 24:142–153
Paulson EJ, Speck SH (1999) Differential methylation of Epstein-Barr virus latency promoters facilitates viral persistence in healthy seropositive individuals. J Virol 73:9959–9968
Petosa C, Morand P, Baudin F, Moulin M, Artero JB, Muller CW (2006) Structural basis of lytic cycle activation by the Epstein-Barr virus ZEBRA protein. Mol Cell 21:565–572
Ramasubramanyan S, Osborn K, Flower K, Sinclair AJ (2012) Dynamic chromatin environment of key lytic cycle regulatory regions of the Epstein-Barr virus genome. J Virol 86:1809–1819

Reusch JA, Nawandar DM, Wright KL, Kenney SC, Mertz JE (2015) Cellular Differentiation Regulator BLIMP1 Induces Epstein-Barr Virus Lytic Reactivation in Epithelial and B Cells by Activating Transcription from both the R and Z Promoters. J Virol 89:1731–1743

Rottach A, Frauer C, Pichler G, Bonapace IM, Spada F, Leonhardt H (2010) The multi-domain protein Np95 connects DNA methylation and histone modification. Nucleic Acids Res 38:1796–1804

Schaefer BC, Strominger JL, Speck SH (1995) Redefining the Epstein-Barr virus-encoded nuclear antigen EBNA-1 gene promoter and transcription initiation site in group I Burkitt lymphoma cell lines. Proc Natl Acad Sci USA 92:10565–10569

Schlager S, Speck SH, Woisetschlager M (1996) Transcription of the Epstein-Barr virus nuclear antigen 1 (EBNA1) gene occurs before induction of the BCR2 (Cp) EBNA gene promoter during the initial stages of infection in B cells. J Virol 70:3561–3570

Schmeinck A (2011) Acquisition and loss of chromatin modifications during an Epstein-Barr Virus infection. Munich: Ludwig Maximilians-University, 1 p. Dissertation

Sciortino MT, Suzuki M, Taddeo B, Roizman B (2001) RNAs extracted from herpes simplex virus 1 virions: apparent selectivity of viral but not cellular RNAs packaged in virions. J Virol 75:8105–8116

Shannon-Lowe C, Baldwin G, Feederle R, Bell A, Rickinson A, Delecluse HJ (2005) Epstein-Barr virus-induced B-cell transformation: quantitating events from virus binding to cell outgrowth. J Gen Virol 86:3009–3019

Shannon-Lowe CD, Neuhierl B, Baldwin G, Rickinson AB, Delecluse HJ (2006) Resting B cells as a transfer vehicle for Epstein-Barr virus infection of epithelial cells. Proc Natl Acad Sci USA 103:7065–7070

Sinclair AJ (2003) bZIP proteins of human gammaherpesviruses. J Gen Virol 84:1941–1949

Takada K, Shimizu N, Sakuma S, Ono Y (1986) Transactivation of the latent Epstein-Barr virus (EBV) genome after transfection of the EBV DNA fragment. J Virol 57:1016–1022

Taylor N, Flemington E, Kolman JL, Baumann RP, Speck SH, Miller G (1991) ZEBRA and a Fos-GCN4 chimeric protein differ in their DNA-binding specificities for sites in the Epstein-Barr virus BZLF1 promoter. J Virol 65:4033–4041

Tempera I, Klichinsky M, Lieberman PM (2011) EBV latency types adopt alternative chromatin conformations. PLoS Pathog 7:e1002180

Tempera I, Wiedmer A, Dheekollu J, Lieberman PM (2010) CTCF prevents the epigenetic drift of EBV latency promoter Qp. PLoS Pathog 6:e1001048

Tierney R, Nagra J, Hutchings I, Shannon-Lowe C, Altmann M, Hammerschmidt W, Rickinson A, Bell A (2007) Epstein-Barr virus exploits BSAP/Pax5 to achieve the B-cell specificity of its growth-transforming program. J Virol 81:10092–10100

Tierney RJ, Kirby HE, Nagra JK, Desmond J, Bell AI, Rickinson AB (2000) Methylation of transcription factor binding sites in the Epstein-Barr virus latent cycle promoter Wp coincides with promoter down-regulation during virus-induced B-cell transformation. J Virol 74:10468–10479

Toth Z, Maglinte DT, Lee SH, Lee HR, Wong LY, Brulois KF, Lee S, Buckley JD, Laird PW, Marquez VE, Jung JU (2010) Epigenetic analysis of KSHV latent and lytic genomes. PLoS Pathog 6:e1001013

Tsai CL, Li HP, Lu YJ, Hsueh C, Liang Y, Chen CL, Tsao SW, Tse KP, Yu JS, Chang YS (2006) Activation of DNA methyltransferase 1 by EBV LMP1 Involves c-Jun NH(2)-terminal kinase signaling. Cancer Res 66:11668–11676

Tsai K, Chan L, Gibeault R, Conn K, Dheekollu J, Domsic J, Marmorstein R, Schang LM, Lieberman PM (2014) Viral reprogramming of the Daxx histone H3.3 chaperone during early Epstein-Barr virus infection. J Virol 88:14350–14363

Walling DM, Flaitz CM, Nichols CM, Hudnall SD, Adler-Storthz K (2001) Persistent productive Epstein-Barr virus replication in normal epithelial cells in vivo. J Infect Dis 184:1499–1507

Wen W, Iwakiri D, Yamamoto K, Maruo S, Kanda T, Takada K (2007) Epstein-Barr virus BZLF1 gene, a switch from latency to lytic infection, is expressed as an immediate-early gene after primary infection of B lymphocytes. J Virol 81:1037–1042

Woellmer A, Arteaga-Salas JM, Hammerschmidt W (2012) BZLF1 governs CpG-methylated chromatin of Epstein-Barr virus reversing epigenetic repression. PLoS Pathog 8:e1002902

Woellmer A, Hammerschmidt W (2013) Epstein-Barr virus and host cell methylation: regulation of latency, replication and virus reactivation. Curr Opin Virol 3:260–265

Woisetschlaeger M, Yandava CN, Furmanski LA, Strominger JL, Speck SH (1990) Promoter switching in Epstein-Barr virus during the initial stages of infection of B lymphocytes. Proc Natl Acad Sci USA 87:1725–1729

Zhou J, Chau CM, Deng Z, Shiekhattar R, Spindler MP, Schepers A, Lieberman PM (2005) Cell cycle regulation of chromatin at an origin of DNA replication. EMBO J 24:1406–1417

Epstein–Barr Virus: From the Detection of Sequence Polymorphisms to the Recognition of Viral Types

Regina Feederle, Olaf Klinke, Anton Kutikhin, Remy Poirey, Ming-Han Tsai and Henri-Jacques Delecluse

Abstract The Epstein–Barr virus is etiologically linked with the development of benign and malignant diseases, characterized by their diversity and a heterogeneous geographic distribution across the world. The virus possesses a 170-kb-large genome that encodes for multiple proteins and non-coding RNAs. Early on there have been numerous attempts to link particular diseases with particular EBV strains, or at least with viral genetic polymorphisms. This has given rise to a wealth of information whose value has been difficult to evaluate for at least four reasons. First, most studies have looked only at one particular gene and missed the global picture. Second, they usually have not studied sufficient numbers of diseased and control cases to reach robust statistical significance. Third, the functional significance of most polymorphisms has remained unclear, although there are exceptions such as the 30-bp deletion in LMP1. Fourth, different biological properties of the virus do not necessarily equate with a different pathogenicity. This was best illustrated by the type 1 and type 2 viruses that markedly differ in terms of their transformation abilities, yet do not seem to cause different diseases. Reciprocally, environmental and genetic factors in the host are likely to influence the outcome of infections with the same virus type. However, with recent developments in recombinant virus technology and in the availability of high throughput sequencing, the tide is now turning. The availability of 23 complete or nearly complete genomes has led to the recognition of viral subtypes, some of which possess nearly identical genotypes. Furthermore, there is growing evidence that some genetic polymorphisms among EBV strains markedly influence the biological and clinical behavior of the virus. Some virus strains are endowed with biological

R. Feederle · O. Klinke · A. Kutikhin · R. Poirey · M.-H. Tsai · H.-J. Delecluse (✉)
Unit F100, Inserm unit U1074, DKFZ, German Cancer Research Centre (DKFZ),
69120 Heidelberg, Germany
e-mail: h.delecluse@dkfz.de

R. Feederle
Helmholtz Zentrum München, German Research Center for Environmental Health,
Institute of Molecular Immunology, Marchioninistrasse 25, 81377 Munich, Germany

C. Münz (ed.), *Epstein Barr Virus Volume 1*, Current Topics in Microbiology and Immunology 390, DOI 10.1007/978-3-319-22822-8_7

properties that explain crucial clinical features of patients with EBV-associated diseases. Although we now have a better overview of the genetic diversity within EBV genomes, it has also become clear that defining phenotypic traits evinced by cells infected by different viruses usually result from the combination of multiple polymorphisms that will be difficult to identify in their entirety. However, the steadily increasing number of sequenced EBV genomes and cloned EBV BACS from diseased and healthy patients will facilitate the identification of the key polymorphisms that condition the biological and clinical behavior of the viruses. This will allow the development of preventative and therapeutic approaches against highly pathogenic viral strains.

Contents

1 Introduction 121
2 Next-Generation Sequencing Gives a Detailed Picture of Sequence Heterogeneity at the Genome Level 122
3 The Degree of Protein Heterogeneity Varies Among Viral Proteins 126
4 The EBV Genome Evinces a Large Number of Genetic Polymorphisms of Unknown Biological and Medical Significance 131
5 EBV Type 1 and Type 2, Two Strains with Different Biological Properties but Apparently not Associated with Particular Diseases 133
6 Polymorphisms in EBNA3B Modulate the Transformation Abilities of the Virus In Vitro and In Vivo 135
7 Polymorphisms Within Latent Genes are Submitted to Immune Pressure and Influence the Geographic Distribution of Virus Isolates 141
8 The NPC-Associated Virus M81 Is Endowed with Unique Properties that Distinguishes It from Other Strains In Vitro and In Vivo 142
9 Conclusions 143
References 144

Abbreviations

BAC	Bacterial artificial chromosomes
BL	Burkitt's lymphoma
EBV	Epstein–Barr Virus
EBNA	Epstein–Barr virus nuclear antigen
LMP	Latent membrane protein
miRNAs	MicroRNAs
NGS	Next-generation sequencing
NPC	Nasopharyngeal carcinoma
RFLP	Restriction fragment length polymorphism
PTLD	Posttransplant lymphoproliferative disorder

1 Introduction

The Epstein–Barr virus (EBV) possesses a double-stranded 170-kb-large DNA genome that encodes for more than 70 genes and non-coding RNAs including 25 pre-microRNAs (Rickinson 2007; Cullen 2011). EBV is etiologically linked with a large number of benign and malignant diseases, characterized by the histological diversity (Delecluse et al. 2007). Infectious mononucleosis is typically encountered in industrialized countries and results from a primary infection delayed to late teenage (Rickinson and Kieff 2007). EBV-associated tumors include lymphomas of several types, as well as carcinomas of the nasopharynx (NPC), stomach (GC), parotid, and thymus (Delecluse et al. 2007). Rare cases of sarcomas have also been reported in children with AIDS (McClain et al. 1995). These broad categories of tumors are very heterogeneous in terms of geographic distribution. While NPC occurs at high and intermediate incidence in Southeast Asia, Northern Africa, and Alaska, EBV-positive GC is encountered all around the world, although the incidence varies somehow between countries (Rickinson and Kieff 2007). EBV-positive Burkitt's lymphoma (BL) is endemic in equatorial Africa and America, and EBV-positive nodal T-cell lymphomas are frequent in Japan, and T or NK nasal-type lymphomas are mainly encountered in Asia and South America (Rickinson and Kieff 2007; Young and Rickinson 2004). Independently of their country of origin, individuals with congenital or acquired immunodeficiency are at high risk of EBV-associated lymphoproliferations (Rickinson and Kieff 2007; Young and Rickinson 2004). These posttransplant lymphoproliferative disorders (PTLD) or HIV-associated EBV-positive lymphomas or lymphoproliferations are usually B-cell lymphomas, but may also be T-cell lymphomas and Hodgkin's lymphomas (Rickinson and Kieff 2007; Young and Rickinson 2004).

The diversity in histological subtypes of tumors and their geographic heterogeneity has early been suspected to reflect infection with different virus subtypes but the large size of the virus has precluded sequencing of multiple virus isolates (Rickinson and Kieff 2007). However, it was recognized early on that sequence polymorphisms are very common all across the genome (reviewed in Chang et al. 2009). This led to a very large number of studies that investigated multiple genes from viruses present in tumors or in healthy population controls (Chang et al. 2009). However, these studies usually investigated a small number of virus isolates, lacked a sufficient number of control cases from healthy donors, and were often limited to a single gene (Chang et al. 2009). The inhomogeneity of the investigated population can also be a source of flawed results. This is particularly true for studies that were conducted in countries with a large population in which diseases might particularly affect one region such as Hong Kong and the Cantonese Province in China in which the incidence of NPC is highest worldwide. In such a case, it is very important to make sure that both the diseased and the control populations were born in the region studied since it was proposed that particular EBV polymorphisms may be restricted to some populations rather than to geographical areas (Lung et al. 1994). Crucially, the functional consequences of the viral polymorphisms for the infected hosts remained relatively unexplored until recently. In their 2009 review of the topic (Chang et al. 2009), Chang and colleagues noted that ‘definite conclusions regarding the link between EBV genotypes, disease and geography are not possible.’ We also refer to the comprehensive review by Jan Gratama and Ingemar Ernberg that covers the earlier developments in this complex field (Gratama and Ernberg 1995). The aim of the present review was to record the progress that new-generation sequencing and the increasing availability of tractable genetic systems made possible in this field. We also only briefly address the functional consequences of protein polymorphisms as these issues are more extensively addressed by Paul Farrell in another chapter of this book.

The necessary background on the biology of the virus can be found in other chapters from this book or for example in Kieff et al. (2007), Rickinson and Kieff (2007), or in Longnecker et al. (2014), and we limit ourselves to a brief summary of the virus’ defining features. EBV infects B lymphocytes with a uniquely high efficiency. Shortly after infection, B lymphocytes resume cell growth and generate so-called lymphoblastoid cell lines in vitro. This process requires expression of the set of latent proteins that belong to the Epstein–Barr nuclear antigen (EBNA 1, 2, 3A, 3B, 3C, EBNA-LP) and latent membrane protein (LMP1 and LMP2) families. EBV gene expression can also lead to new viral progeny. Viral replication requires the sequential expression of immediate-early, early, and late viral lytic genes. The protein encoded by the BZLF1 gene is endowed with transactivating properties and can initiate virus replication in permissive cells.

2 Next-Generation Sequencing Gives a Detailed Picture of Sequence Heterogeneity at the Genome Level

The increasing availability of next-generation sequencing technology has allowed sequencing of 23 EBV genomes so far, although the extent of the sequence available varies (see Table 1 and references therein). Some sequences are complete, others do not include the repeats and yet others cover only partly the viral genome. 16 sequences were obtained from viruses present in malignant tumors or from patients with these diseases (13 with NPC, 3 with BL). From these, 14 arose in Asian patients (13 Chinese and 1 Japanese) and the remaining two in African patients. The other 7 viruses were isolated from individuals with IM or from endogenous LCLs that were generated with the blood of healthy patients. Three of these viruses were isolated from individuals who reside in the USA; the others were sequenced from cell lines used in the 1000 Genomes Project and were established with blood from healthy African or European individuals. Thus, the currently available information is very much skewed toward EBV isolates associated with tumors, in particular NPC.

However, this information has already allowed a first global assessment of genetic diversity between EBV isolates isolated in Asia, Africa, in the USA, and in Europe. We have evaluated genetic distance between these strains based on an alignment of the comparable portions of the genomes (roughly 159 kb). No matter

Table 1 Currently available EBV genome sequences and common investigated polymorphisms

Strain, reference	EBV subtype 1/2 (A/B)	Type F/f	Type C/D (I/i)	wt/mut-W1/I1	Type XhoI+ /Xho−	BZLF1 protein variant	BZLF1 promoter variant	Sample origin	Sample geographical localization
AG876 (Dolan et al. 2006)	2 (B)	F	D (i)	wt-W1/I1	+	BZLF1-A2	Zp-V3	BL	Africa (Ghana)
Mutu (Lin et al. 2013)	1 (A)	F	D (i)	wt-W1/I1	+	BZLF1-C	Zp-P	BL	Africa (Kenya)
NA19114 (Santpere et al. 2014)	1 (A)	F	D (i)	wt-W1/I1	+	BZLF1-C	Zp-P	PBL	Africa (Nigeria)
NA19315 (Santpere et al. 2014)	1 (A)	F	D (i)	wt-W1/I1	+	BZLF1-C	Zp-P	PBL	Africa (Kenya)
NA19384 (Santpere et al. 2014)	1 (A)	F	D (i)	wt-W1/I1	+	BZLF1-C	Zp-P	PBL	Africa (Kenya)
GD1 (Zeng et al. 2005)	1 (A)	F	C (I)	wt-W1/I1	−	BZLF1-A1	Zp-V3	Saliva NPC	Asia (China)
GD2 (Liu et al. 2011)	1 (A)	f	C (I)	mut-W1/I1	−	BZLF1-A1	Zp-V3	NPC	Asia (China)
C666-1 (Kwok et al. 2014)	1 (A)	f	C (I)	mut-W1/I1	−	BZLF1-A1	Zp-V3	NPC	Asia (China)
HKNPC1 (Kwok et al. 2012)	1 (A)	f	C (I)	mut-W1/I1	−	BZLF1-A1	Zp-V3	NPC	Asia (China)
HKNPC2 (Kwok et al. 2014)	1 (A)	F	C (I)	mut-W1/I1	−	BZLF1-A1	Zp-V3	NPC	Asia (China)
HKNPC3 (Kwok et al. 2014)	1 (A)	f	C (I)	mut-W1/I1	−	BZLF1-A1	Zp-V3	NPC	Asia (China)
HKNPC4 (Kwok et al. 2014)	1 (A)	f	C (I)	mut-W1/I1	−	BZLF1-A1	Zp-V3	NPC	Asia (China)
HKNPC5 (Kwok et al. 2014)	1 (A)	f	C (I)	mut-W1/I1	−	BZLF1-A1	Zp-V3	NPC	Asia (China)

(continued)

Table 1 (continued)

Strain, reference	EBV subtype 1/2 (A/B)	Type F/f	Type C/D (I/i)	wt/mut-W1/I1	Type XhoI+ /Xho−	BZLF1 protein variant	BZLF1 promoter variant	Sample origin	Sample geographical localization
HKNPC6 (Kwok et al. 2014)	1 (A)	F	C (I)	mut-W1/I1	−	BZLF1-C	Zp-P	NPC	Asia (China)
HKNPC7 (Kwok et al. 2014)	1 (A)	F	C (I)	mut-W1/I1	−	BZLF1-C	Zp-P	NPC	Asia (China)
HKNPC8 (Kwok et al. 2014)	1 (A)	f	C (I)	mut-W1/I1	−	BZLF1-A1	Zp-V3	NPC	Asia (China)
HKNPC9 (Kwok et al. 2014)	1 (A)	f	C (I)	mut-W1/I1	−	BZLF1-A1	Zp-V3	NPC	Asia (China)
M81 (Tsai et al. 2013)	1 (A)	f	C (I)	mut-W1/I1	−	BZLF1-A1	Zp-V3	NPC	Asia (China)
Akata (Lin et al. 2013)	1 (A)	F	C (I)	wt-W1/I1	−	BZLF1-A1	Zp-V3	BL	Asia (Japan)
B95.8 (Baer et al. 1984, de Jesus et al. 2003)	1 (A)	F	Δ	Δ	+	BZLF1-C	Zp-P	IM	North America (USA)
K4123-Mi (Lei et al. 2013)	1 (A)	F	D (i)	wt-W1/I1	+	BZLF1-C*	Zp-P	PBL	North America (USA)
K4413-Mi (Lei et al. 2013)	1 (A)	F	D (i)	wt-W1/I1	+	BZLF1-C*	Zp-P	PBL	North America (USA)

BL Burkitt's lymphoma, *PBL* peripheral blood B lymphocytes, *NPC* nasopharyngeal carcinoma, *IM* infectious mononucleosis
Subtypes 1 and 2 (A/B) are based on the sequence of the EBNA2 protein. Type F viruses lack a BamHI site in the BamHI-F region present in the type f viruses. Type C viruses (also known as I) lacks the BamHI site at BamHI W1*/I1* boundary region (in the non-coding region between BART14 and BILF1) whereas type D viruses (also know as i) possess it. The Wt-W1/I1 polymorphism has a T at position 148,972 in the reference strain NC_007605.1 (in the non-coding region between BART14 and BILF1) whereas mut-W1/I1 has a C at the same position. Type XhoI+ viruses have an XhoI restriction site in the exon1 of LMP1 in contrast to type XhoI-viruses. The BZLF1 types are defined by clusters of amino acid variations in this protein and its promoters

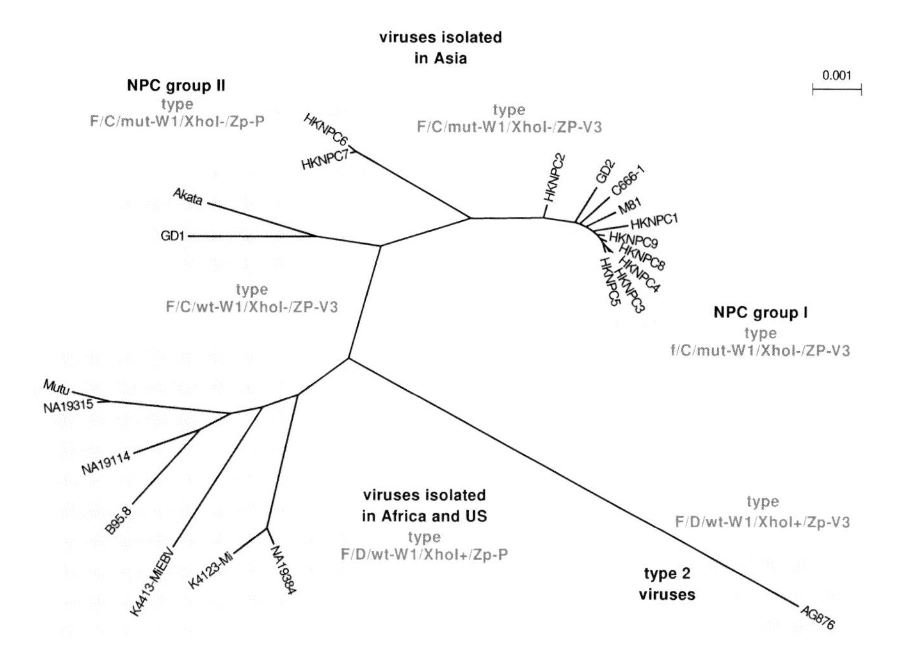

Fig. 1 Sequence heterogeneity among EBV isolates. A divergence tree of the 22 strains whose genome sequence has been published. Distances are based on a multiple alignment of 159 kb from these genomes

what model of DNA evolution was used (e.g., Felsenstein or Jukes-Cantor) and no matter how a tree was built from these distances (we used the maximum likelihood method), the overall picture remained the same. Figure 1 depicts the genetic distances as a tree but this depiction should not be mistaken for a phylogenetic tree, as many of the assumptions commonly made for phylogenies may not be valid for the Epstein–Barr virus evolution. Furthermore, this approach does not distinguish between divergence that results from polymorphisms concentrated in a small number of genes such as EBNA2 and the EBNA3 family in type 1 and type 2 viruses (see below) and divergence caused by polymorphisms that are more homogenously distributed along the genome. Nevertheless, the clustering that is apparent in Fig. 1 is globally concordant with previous analyses that looked at only a subset of these sequenced genomes (Tsai et al. 2013; Kwok et al. 2014). Indeed, we previously reported that viruses isolated from individuals who lived in different geographic area show a maximum divergence rate of approximately 1 % at the DNA level, and 1.3 % at the protein level (Tsai et al. 2013). These data are concordant with the 0.65–0.73 % variation in EBV genomes from NPC tissues relative to B95-8 observed by other authors, in particular because this analysis did not take into account repeated sequences that we found to be also polymorphic (Kwok et al. 2014).

Inspection of this tree immediately suggests the existence of three large groups of viruses, one containing viruses isolated from tumors that arose in Asian patients, one from viruses isolated in healthy or diseased individuals in Africa and in the USA and one that contains the type 2 virus AG876. Among the virus isolated in Asia, one very homogenous subgroup that comprises viruses that infect NPCs and mainly arose in the Hong Kong area is easily recognized and will be referred to as NPC I. Another 2 viruses, although clearly related to NPC I and also isolated in infected patients from Hong Kong with NPC, displayed some important differences from IR2 to IR3 genomic region and will be referred to as NPC group II. Interestingly, HKNPC2 is likely to have arisen from a recombination between a virus from NPC group I and a virus from NPC group II. Intertypic recombinants between type 1 and type 2 viruses have previously been reported in the Chinese population (Midgley et al. 2000).

Crucially, type 2 viruses and one virus from the NPC I group display particular biological properties with likely important clinical consequences as described in paragraphs 5 and 8.

However, the currently available series of viral genomes will soon be massively expanded to include 71 novel EBV genomes isolated in malignant tumors of more diverse lineages and that developed in patients from other geographic regions. Furthermore, this series will also comprise viruses isolated from EBV-positive healthy carriers from multiple regions of the world including Australia, Kenya, and the United Kingdom (P. Kellam, P. Farrell, R. White, personal communication). Therefore, our picture of viral diversity will soon be drastically enriched and will probably render this review rapidly obsolete. Nevertheless, it is very likely that an even much larger number of complete virus sequences from all continents will be needed to capture the full extent of virus diversity.

3 The Degree of Protein Heterogeneity Varies Among Viral Proteins

The EBV genome contains more than 25 types of repeated sequences, and many of which are located within coding EBV genes and are present in varying numbers across different isolates. This, combined to amino acid polymorphisms that affect the molecular weight and the presence of posttranslational modifications, gives rise to proteins of different sizes. This phenomenon strongly affects the EBNA proteins whose size can vary within a variable range and has been used to define an ‘EBNotype’ that defines different viral isolates (Gratama and Ernberg 1995). EBNA1 also displays many polymorphisms across EBV isolates but the functional significance of these mutations remains unknown (Bhatia et al. 1996; Gutierrez et al. 1997; Habeshaw et al. 1999).

In contrast, the role of polymorphisms within the EBNA2 gene has been characterized in detail and led to the definition of the type 1 and type 2 viruses that will be discussed in the sequel. Similarly, important polymorphisms have been identified within the EBNA3B gene (see paragraph 6).

The LMP1 gene carries multiple repeats that are present in different numbers in the different strains. This includes the 33-bp repeats, as well as a 15-bp insertion in some cases (Fig. 2) (Sandvej et al. 1997; Cheung et al. 1996). Finally,

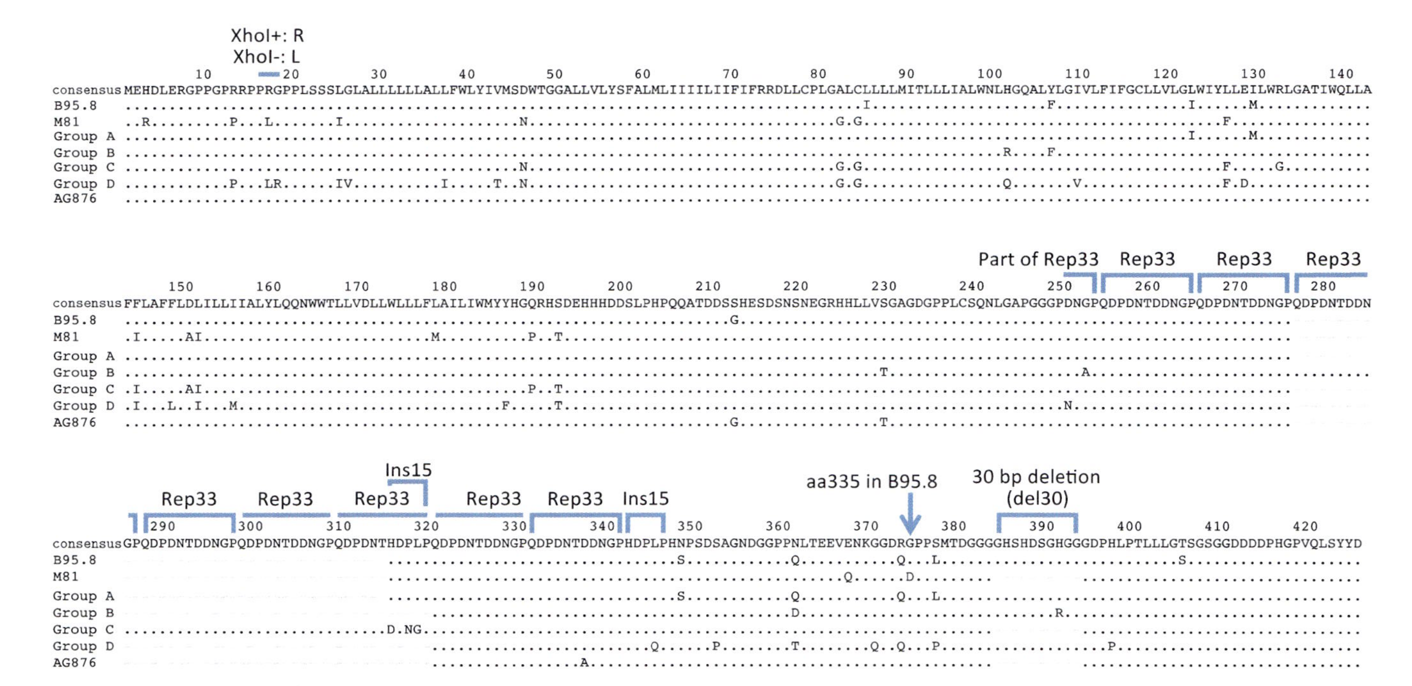

Fig. 2 Amino acid sequence of the LMP1 gene from different virus types showing polymorphisms, repeats, insertions, and deletions. ins15: insertion of 15bp/5aa, Rep33: 33bp/11aa repeat. *Dots* indicate base identity

many strains carry a 30-bp deletion at the C-terminal end of the protein. There are also multiple single amino acid polymorphisms within the carboxy terminus of the LMP1 protein. These polymorphisms have given rise to multiple variants that carry some mutations and/or the 30-bp deletion and that have been named according to the country where the virus was identified (China 1, 2, 3, Alaskan, North Carolina, Mediterranean, etc.) (Chang et al. 2009; Sung et al. 1998). A more recent paper found that these different LMP1 alleles do not vary in their ability to transform Rat-1 fibroblasts, increase mobility of Madin-Darby canine kidney cells, or induce homotypic adhesion of BJAB cells (Mainou and Raab-Traub 2006). At the molecular level, while these LMP1 variants activated the PI3K/AKT signaling pathways at similar levels, some LMP1 variants such as Alaskan induced NF-kB signaling with an approximately 2 times higher efficiency than B95-8. This explains reports from the early nineties that already found that LMP1 variants differed in their ability to activate the NF-kB pathway (reviewed in Jenkins and Farrell 1996). Altogether, although it is clear that multiple LMP1 alleles exist, there is currently no overwhelming evidence that particular LMP1 alleles are associated with particular diseases or geographical locations (Sandvej et al. 1997; Edwards et al. 1999; Mainou and Raab-Traub 2006).

BZLF1 is also a highly polymorphic EBV protein (Fig. 3a, b). The BZLF1 promoter has also been investigated in detail in an attempt to understand why some EBV isolates supported lytic replication more efficiently than others and gave rise to the current recognition of 5 variants (Zp-P, Zp-PV, Zp-V1, Zp-V3, and Zp-V4) that are themselves subdivided into multiple subvariants (Chang et al. 2009; Gutierrez et al. 2002; Tong et al. 2003; Martini et al. 2007; Lorenzetti et al. 2009; Jin et al. 2010; Yu et al. 2012; Lorenzetti et al. 2012). BZLF1 proteins have been subdivided into groups A1, A2, B1, B2, B3, B4, B5, B6, C, D, E, F, some of which such as A1, A2, and C also contain variants (Chang et al. 2009; Luo et al. 2011; Lorenzetti et al. 2014). There is growing evidence that some BZLF1 variants are endowed with different properties in recombinant viruses (Tsai et al. 2013, see paragraph 8).

The data obtained by NGS confirm and extend these data (Tsai et al. 2013; Kwok et al. 2014). Importantly, the sequence heterogeneity at the protein level is unequally distributed across the genome (Tsai et al. 2013; Kwok et al. 2014). Comparison between 22 EBV isolates reveals that most proteins diverge by less than 1 % (Fig. 4). However, other proteins, including most latent proteins, show higher divergence (Fig. 4). In case of LMP1, it reaches 8 %, making it one of the most polymorphic EBV proteins (Lin et al. 2013). Polymorphic proteins also include tegument proteins such as the large tegument proteins BOLF1 and BPLF1, or surface glycoproteins such as gp350 or gp110 that are encoded by BLLF1 and BALF4, respectively (Kwok et al. 2014). Some of these polymorphisms were found to influence the viral tropism (Tsai et al. 2013). In contrast, the early lytic proteins that perform DNA replication and include BMRF1, BMRF2, and BRLF1 are much more conserved (Tsai et al. 2013; Kwok et al. 2014). This implies that mutations within these genes are deleterious for viral propagation and are eliminated. The EBV miRNAs and BCRF1, and the viral homolog of IL10 are also well conserved (Kwok et al. 2014).

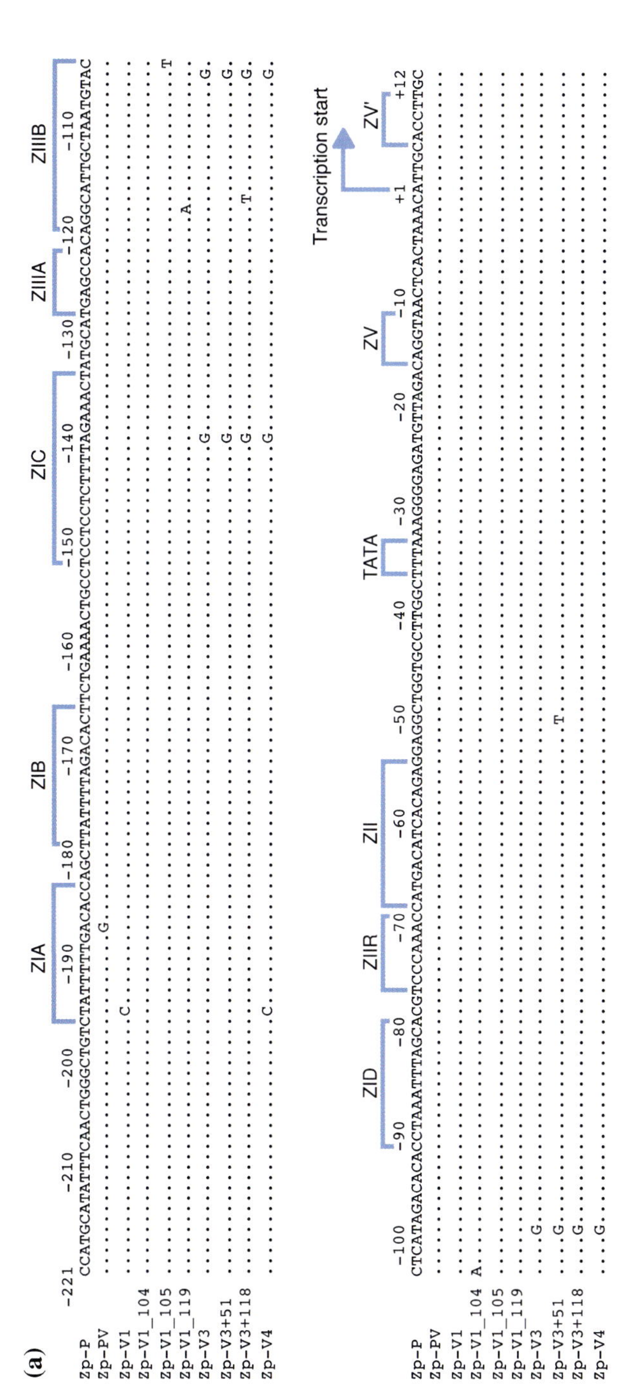

Fig. 3 **a** The promoter region of the BZLF1 gene, from position −221 to +12 relative to the transcription start and its different types. The sequence of the prototype B95.8 is compared to the published variants. *Dots* indicate base identity. **b** An alignment of the BZLF1 amino acid sequence. Representatives from A1 to F types are shown. The B95-8 sequence that corresponds to the C type is taken as a reference. *Dots* indicate amino acid identity

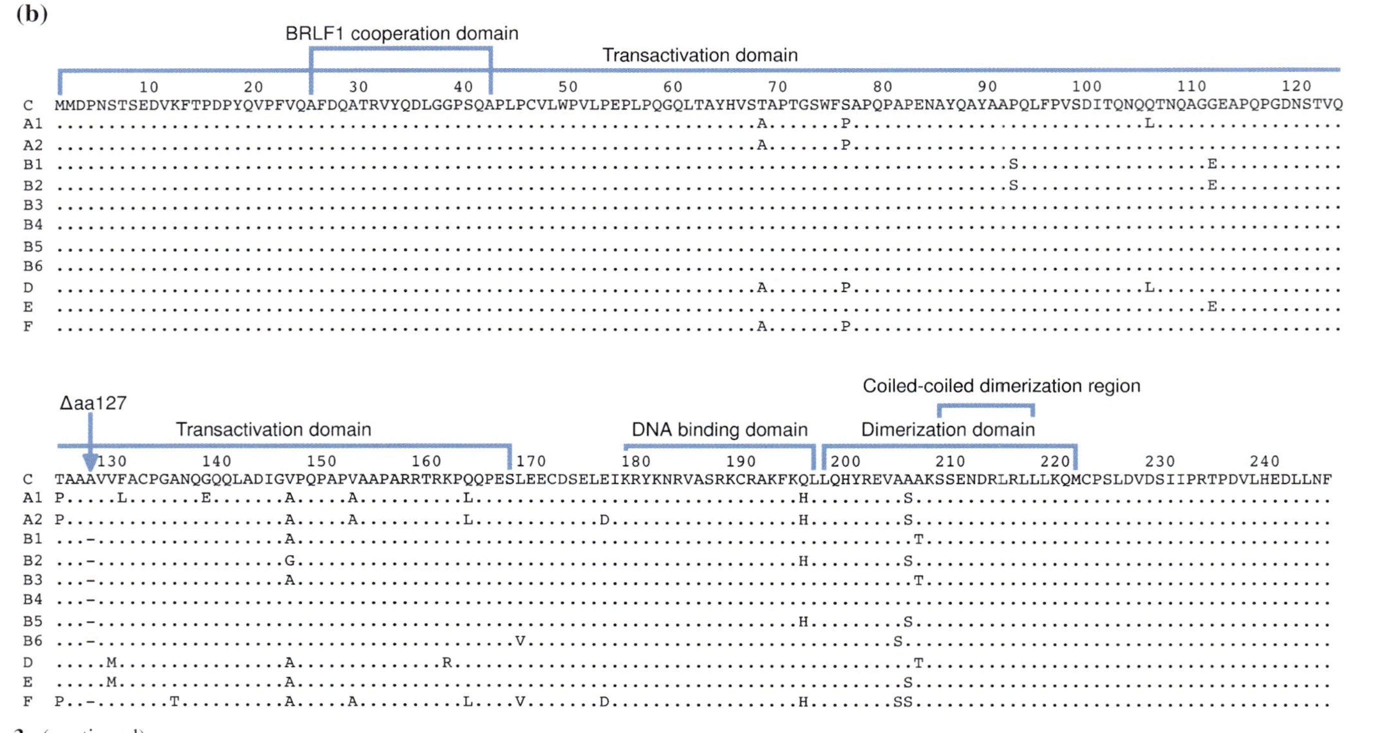

Fig. 3 (continued)

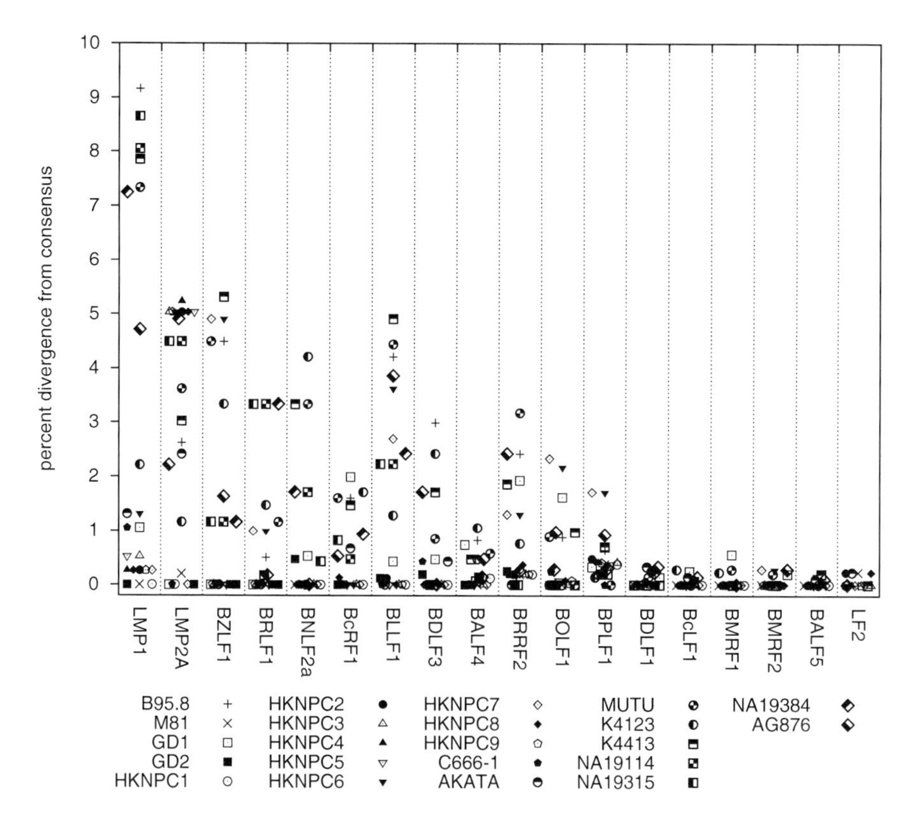

Fig. 4 The percentage of divergence at the protein level across the completely sequenced EBV strains. Proteins involved in virus replication and transformation are shown. LF2 is a putative protein

4 The EBV Genome Evinces a Large Number of Genetic Polymorphisms of Unknown Biological and Medical Significance

The genetic diversity among EBV isolates has initially been recognized by restriction analysis of viral DNA from EBV-infected cells and by sequencing of DNA obtained from any EBV-positive tissue or cell line (Chang et al. 2009). The sequence heterogeneity within isolates gives rise to various restriction sites and thus to RFLPs, some of which such as the BamHI site within the F fragment (Type F/f), the BamHI site I fragment (type C/D), or the XhoI site within the LMP1 gene (type XhoI+ versus XhoI−) have been extensively studied (Fig. 2) (Chang et al. 2009). A very high number of epidemiological studies have tried to cluster the existence of polymorphisms with EBV-associated diseases, among which the XhoI restriction site plays a prominent role (Chang et al. 2009). In all the studies that found a difference in the frequency of some LMP1 alleles in tumors relative to normal controls, this difference was mild to moderate. One recurrent problem with these studies is the paucity of information on the distribution of LMP1 alleles within the normal population (Chang et al. 2009). This is particularly true for countries with a very large population in which a cohort of patients from a large university hospital might include individuals from different regions in which the molecular characteristics of the local EBV strains might vary. A middle-sized study from Europe did not find any differences in the distribution of LMP1 variants in the control versus diseased population (Sandvej et al. 1997). Altogether, the extent to which some LMP1 variants might be more transforming than others remains to be appreciated (see paragraph 3).

Reciprocally, there are many more polymorphisms within the EBV genome that are not necessarily identifiable by RFLPs and whose importance might have not yet been recognized.

Despite the limitations inherent to sampling that have already been mentioned, we assessed the value of these markers to identify the viral strains whose sequence is known. Figure 1 and Table 1 show their distribution within the diversity tree. Interestingly, these markers clustered well with groups of viruses identified by the global genome analysis. The f variant was expressed only by the very homogenous NPC group I. The C and XhoI− markers were exclusively present in viruses isolated from Chinese or Japanese individuals; all other isolates carried the D and the XhoI+ markers. The mut-W1/I1 marker was found in all group I and group II NPC viruses but not in GD1 (Chen et al. 2012). All other isolates carried wt-W1/I1 (Chen et al. 2012). Other markers such as those present in the BZLF1 ORF or its promoter did not match perfectly the groups defined by the divergence tree. Nevertheless, it is clear that within this group of viruses, BZLF1-A1, Zp-V3, and BR1-A are always present together in a given virus, as are BZLF1-C and Zp-P. Furthermore, the first group of markers is present in all viruses found in Asian countries with the exception of the group II NPCs. What can be deducted from this analysis? First it appears that particular combinations of markers correlate very well with the groups defined by the divergence tree. This means that they might be used in the future as a set of markers to identify particular viral strains when they are used in combination. However, in agreement with previous interpretations, it remains unclear whether they identify clustering with particular groups of individuals or with particular diseases (Chang et al. 2009). As more complete virus sequences become available, their validity as predictive markers will be reevaluated.

Is there any evidence that these markers can identify EBV strains associated with particular diseases? It is certainly not the case for the gastric carcinoma in which viruses from cases that arose in different parts of the world carried different markers (reviewed in Chen et al. 2012). The viruses that infect NPC can be of type f or F. Altogether, there is currently not sufficient evidence to support the concept that these polymorphism can identify tumor types.

5 EBV Type 1 and Type 2, Two Strains with Different Biological Properties but Apparently not Associated with Particular Diseases

It has long been recognized that two viral types, type 1 and type 2 or type A and type B exist (Dambaugh et al. 1984; Adldinger et al. 1985). Both types were initially identified on the basis of extensive sequence polymorphisms within the EBNA2 gene, but it was also soon recognized that this divergence extends to the EBNA3 gene family (Rowe et al. 1989; Sculley et al. 1989; Sample et al. 1990). The divergence is strongest in EBNA2 (only 54 % homology between the 2 types), justifying the distinction of EBNA2 into type A and type B (Dambaugh et al. 1984). EBNA2A and B display many single nucleotide polymorphisms but also larger deletions/insertions (Aitken et al. 1994) (Fig. 5). Type 1 viruses transform primary B lymphocytes more readily than type 2 viruses in vitro and a recombinant type 2 virus that carries a type 1 EBNA2A acquires the transforming abilities of type 1 viruses (Rickinson et al. 1987; Cohen et al. 1989). Although the epidemiological data remain sparse, the geographic distribution of both virus types seems to vary extensively across the world. Type 1 viruses seem to largely dominate in Asia and in Europe, but both type 1 and type 2 viruses are found at more comparable rates in Africa, New Guinea, Australia, the USA, and more specifically Alaska, as well as in transplant recipients and in HIV-positive individuals (Aitken et al. 1994; Young et al. 1987; Sculley et al. 1988; Sixbey et al. 1989; Kunimoto et al. 1992; Shu et al. 1992; Kyaw et al. 1992; Apolloni and Sculley 1994). In the latter group, EBV type 2 is transmitted by sexual route in individuals with multiple partners, some of which of non-European origin. Both virus types have been found in BL NPC and HL (Young et al. 1987; Zimber et al. 1986; Sculley et al. 1988; Abdel-Hamid et al. 1992; Boyle et al. 1993; van Baarle et al. 2000). In contrast, type 1 viruses predominate in PTLD observed in US transplant recipients that express EBNA2 and EBNA3 (Frank et al. 1995). As precise information about the distribution of type A and type B viruses in the normal population of different countries is lacking, it remains unclear whether type 1 and type 2 viruses differ in the spectrum of diseases they cause or in the frequency with which they cause them. The current view is that both viruses can cause disease with the same efficiency but we would like to argue that the studied population is much too small to be able to identify subtle differences. Future large-scale studies that address viral diversity should be able to readdress these issues.

The division into type 1 and type 2 viruses does not sum up the diversity within the EBNA2 gene. Sequencing of the EBNA2 locus in a large panel of viruses isolated in diseased or healthy patients from multiple countries have revealed the existence of different strains described as 1.1a, 1.1b, 1.1c, 1.1d, 1.1e, 1.2, 1.3a, 1.3b, 1.3c, 1.3d, 1.3e (Schuster et al. 1996; Midgley et al. 2003). Strains not only 1.1a and 1.3b but also 1.3a and 1.2 are more common in Europe, whereas the strain 1.1b, also known as Wu, and 1.3e, also named Li, are frequent in Asia (Midgley et al. 2003). Interestingly, only the 1.1b allele was found in Japan

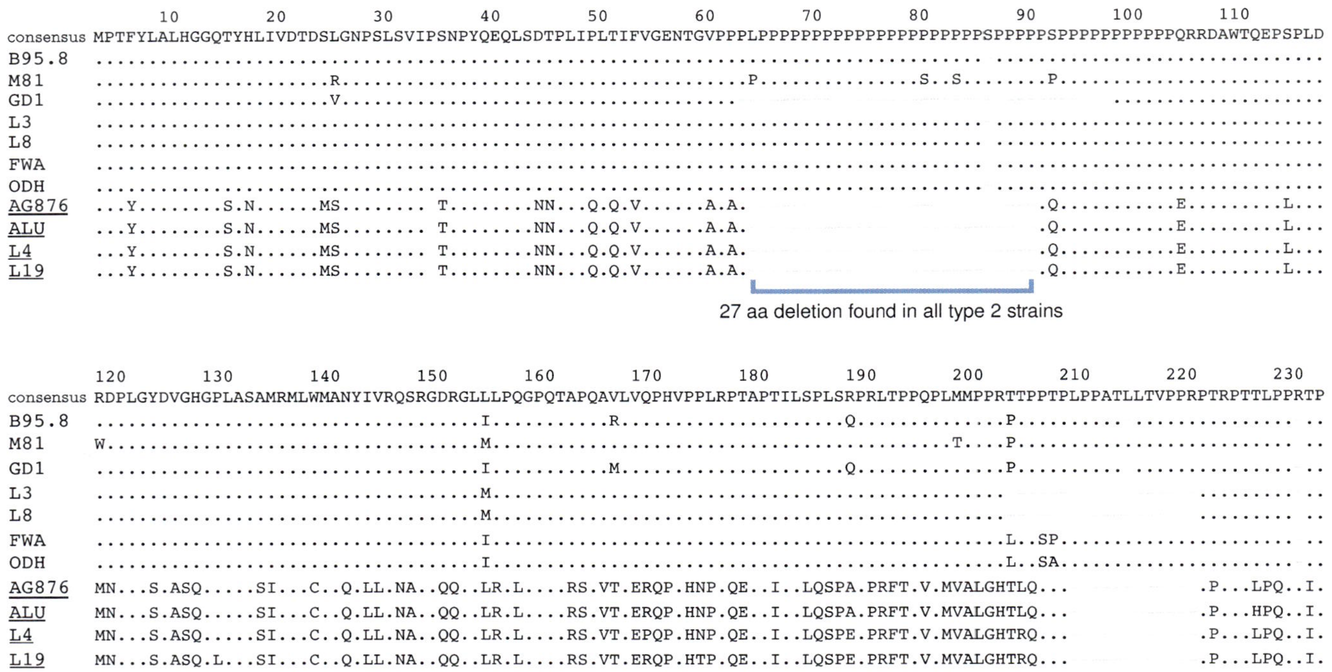

```
                10        20        30        40        50        60        70        80        90       100       110
consensus MPTFYLALHGGQTYHLIVDTDSLGNPSLSVIPSNPYQEQLSDTPLIPLTIFVGENTGVPPPLPPPPPPPPPPPPPPPPPPPPPSPPPPSPPPPPPPPPPQRRDAWTQEPSPLD
B95.8     .................................................................................... ................................
M81       ......................R.......................................P...............S..S........P..........................
GD1       ......................V......................................                                   .....................
L3        .................................................................................... ................................
L8        .................................................................................... ................................
FWA       .................................................................................... ................................
ODH       .................................................................................... ................................
AG876     ...Y........S.N......MS........ T........NN...Q.Q.V......A.A.                           .Q..........E.........L...
ALU       ...Y........S.N......MS........ T........NN...Q.Q.V......A.A.                           .Q..........E.........L...
L4        ...Y........S.N......MS........ T........NN...Q.Q.V......A.A.                           .Q..........E.........L...
L19       ...Y........S.N......MS........ T........NN...Q.Q.V......A.A.                           .Q..........E.........L...
                                                                       27 aa deletion found in all type 2 strains

          120       130       140       150       160       170       180       190       200       210       220       230
consensus RDPLGYDVGHGPLASAMRMLWMANYIVRQSRGDRGLLLPQGPQTAPQAVLVQPHVPPLRPTAPTILSPLSRPRLTPPQPLMMPPRTTPPTPLPPATLLTVPPRPTRPTTLPPRTP
B95.8     ...................................I..........R.....................Q..............P......... ............... ..
M81       W..................................M.....................................................T....P.......................... ..
GD1       ...................................I...........M....................Q..............P......... ............... ..
L3        ...................................M..................................................                  ......... ..
L8        ...................................M..................................................                  ......... ..
FWA       ...................................I...............................................L...SP....................... ..
ODH       ...................................I...............................................L...SA....................... ..
AG876     MN...S.ASQ.....SI...C..Q.LL.NA..QQ..LR.L....RS.VT.ERQP.HNP.QE..I..LQSPA.PRFT.V.MVALGHTLQ...           .P...LPQ..I.
ALU       MN...S.ASQ.....SI...C..Q.LL.NA..QQ..LR.L....RS.VT.ERQP.HNP.QE..I..LQSPA.PRFT.V.MVALGHTLQ...           .P...HPQ..I.
L4        MN...S.ASQ.....SI...C..Q.LL.NA..QQ..LR.L....RS.VT.EPQP.HNP.QE..I..LQSPE.PRFT.V.MVALGHTRQ...           .P...LPQ..I.
L19       MN...S.ASQ.L...SI...C..Q.LL.NA..QQ..LR.L....RS.VT.ERQP.HTP.QE..I..LQSPE.PRFT.V.MVALGHTRQ...           .P...LPQ..I.
```

Fig. 5 Some of the differences between type 1 and type 2 viruses. It shows the first 227 amino acids of the EBNA2 protein from different virus isolates that belong to type 1 and type 2 EBV families. The consensus sequence is also shown. The names of type 2 strains are *underlined*. *Dots* indicate amino acid identity. *Light gray* slashes signify amino acid deletion

(Sawada et al. 2011). Such a heterogeneity is hardly visible in type 2 viruses. The type 1 EBNA3 gene family was also submitted to diversification as shown by the recognition of alleles that co-segregate with EBNA2 in the infected Chinese population. Thus, Wu and Li EBNA3A, -B, and -C alleles exist and are tightly linked to the EBNA2 alleles that bear the same name. In contrast to the EBNA2 1.1b that is common between isolates from China and Japan, the EBNA3 alleles found in Japanese healthy and diseased individuals slightly differ. These sub-alleles are called Wu′ and Li′, Li″ and differ from their Chinese relatives by one or two point mutations (Sawada et al. 2011) (Fig. 6a). These polymorphisms, however, did not cluster with viruses identified from patients with T/NK lymphoproliferations.

6 Polymorphisms in EBNA3B Modulate the Transformation Abilities of the Virus In Vitro and In Vivo

Another layer of complexity has been added by the recognition that EBNA3B can display additional polymorphisms that further influence the transforming potential of the virus (White et al. 2012). Indeed, B cells transformed with EBNA3B-negative viruses have a growth advantage in vitro and in vivo in humanized mice relative to their wild-type counterparts, and this property renders them more tumorigenic in these animals (White et al. 2012). This is concordant with the previous recognition that EBNA3B is a negative regulator of EBV-induced proliferation. These data are particularly important because a virus that carries a 245-bp deletion of EBNA3B that presumably inactivates the protein was previously found in a posttransplant lymphoproliferative disorder (PTLD) that was resistant to T-cell therapy (Gottschalk et al. 2001). Another PTLD cell line also carried viruses with a truncation of EBNA3B that resulted from a missense mutation, as did an ABC-type large-cell lymphoma that arose in an HIV-infected individual (White et al. 2012) (Fig. 6b). Thus, viruses with EBNA3B deletions seem to possess a stronger transforming potential in immunosuppressed individuals (White et al. 2012). This fits with the observation that these tumors frequently express the EBNA3 genes (Rickinson 2007). Additional EBNA3B mutations have been identified in cell lines or primary EBV-positive tumors such as Hodgkin's lymphomas or Burkitt's lymphomas whose distribution differs from those observed in non-diseased controls (White et al. 2012) (Fig. 6b). However, it remains unclear at this stage whether they have biological and medical consequences, in particular because recombinant viruses that carry these mutations have not been tested yet and because these tumors are usually EBNA3-negative. Particular EBNA3 alleles could nevertheless play a role at an early stage of these diseases.

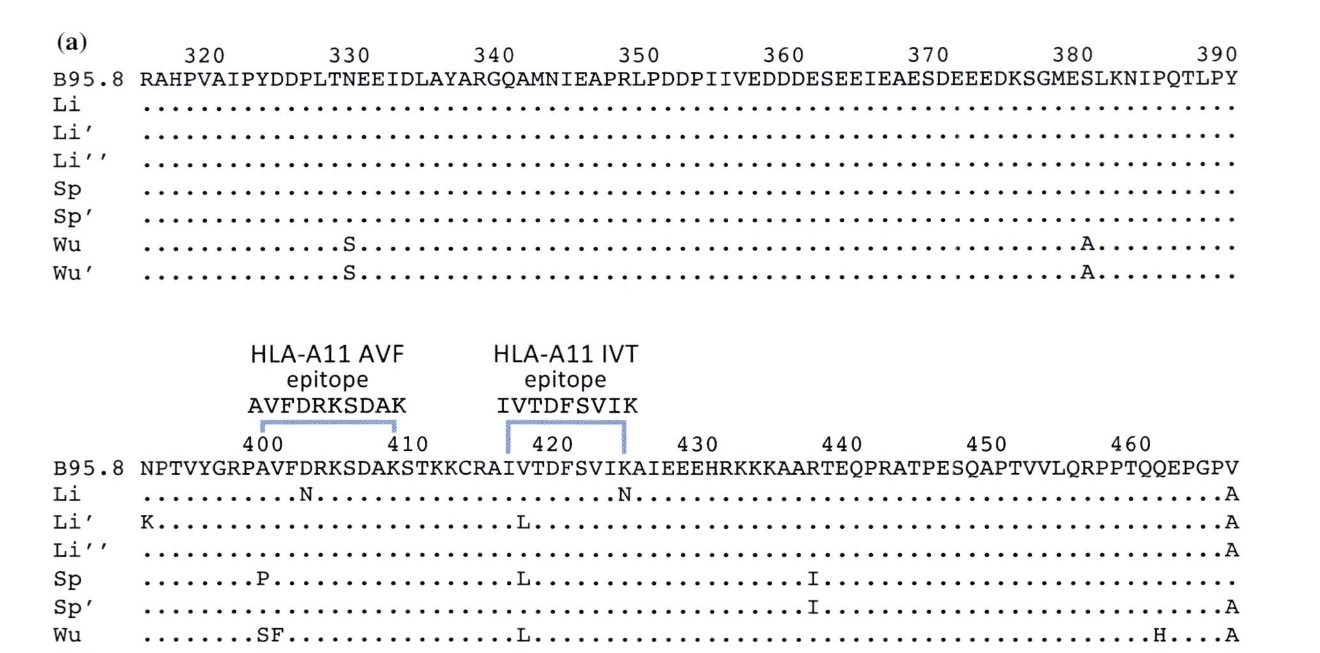

Fig. 6 **a** An alignment of the sequence of the 315-466 codons from EBNA3B with B95-8 taken as the reference. Multiple polymorphisms that define the Li, Li′, Li′′, Sp, Sp′, Wu and Wu′ alleles are shown, as are the position of two HLA A11-restricted epitopes. *Dots* indicate amino acid identity. We have also indicated the position of multiple epitopes encoded in this region. **b** EBNA3B mutations and deletions identified in the genomes of viruses isolated from B-cell tumors or from spontaneous LCLs isolated from healthy individuals (sLCL series). The EBNA3B B95-8 sequence is taken as a reference. The position of some virus epitopes is given; *dots* indicate amino acid identity; *light gray dashes* indicate amino acid deletion. **c** Some EBNA3B epitopes identified in B95-8 as well as their HLA restriction type. The figure also shows polymorphisms within these epitopes encoded by virus strains that have been completely sequenced. We then focus on the HLA-A11 restricted epitopes AVF and IVT that were found in different geographic areas: C/HK (Canton/Hong Kong), Eur/USA (European/USA), NGC (New Guinean coast), PNG (Papua New Guinean), SEA (Southeast Asia). *Dots* indicate amino acid identity and *light gray dashes* indicate amino acid deletion

(b)

```
                                                                     HLA-B27.05             HLA-A24.02  HLA-B27.02   HLA-B58
                                                                      epitope                 epitope     epitope    epitope
             6  9 14  30            44 59  84       93 104 151    157 203 213 219   225 245       253 281   287
B95.8        LSRA DAG QGNVTQVGSEPISPEI QSG GGDDPLDVHTR PTQ HRCQAIRKK LVT GTR TYSAGIVQI RRARSLSAERY VSFIEFVGW
ABC-P8       .... ... ............... ... ....T...... ... ......... ... ... ......... ........... .........
BL31         .... ... ............... ... ........... ... ......... ... ... ......... ........... .........
ABC-PDL10    .... ... ............... ... ..........H ... ......... ... ... ......... ........... .........
DLBCL-3      ..K. ... ............... ... ........... ... ......... ... ... ......... ........... .........
GC-DL44      .... ... ............... ... ........... ... ......... ... ... ........L ........... .........
HL11         .... ... ............... ... ........... ... ......... ... ... ......... ........... .........
GC-H25       .... ... R.............. ... ........... ... ......... ... ... ......... ........... .........
HL02         .... ... ............... ... ........... ... ......... ... ... ......... ........... .........
HL08         .... ... ............... ... ........... ... ......... ... ... ......... ........... .........
ABC-4        .I.. ... ....K..Y....... ... ........... ... ......... ... ... ......... ........... .........
ABC-DL67     .I.. ... ....K........DN .A. D.......S.. ... ......... .L. .P. ......... ........... .........
ABC-H9       .... ... ............... ... ........... ... ......... .L. .P. ......... ........... .........
ABC-1B       .... .V. ....K...Y...... ... ........... ... ......... ... ... ......... ........... .........
HL01         .... .V. ....K...Y...... ... ........... ... ......... ... ... ......... ........... .........
HL04         .... ... ............... ... ........... .M. ......... ... ... ......... ........... .........
HL09         .... ... ............... ... ........... ... ......... ... ... ......... ........... .........
ABC-H7       .... ... ............... ... ........... ... ......... ... ... ......... ........... .........
TLR595       .... ... ............... ... ........... ... ......... ... ... ......... ........... .........
TLR1         .... ... ............... ... ........... ... ......... ... ... ......... ........... .........

             HLA-A11 AVF   HLA-A11 IVT               HLA-B35.01                                HLA-B44
               epitope       epitope                   epitope                                 epitope
             401                      425 442    462  490           503 533  555  560 576 659      666 684   690
B95.8        AVFDRKSDAKSTKKCRAIVTDFSVIKA QPRATP QQEP AVLLHEESMQGVQVHG QPRA PASTEPVH IQP VEITPYKPTW ATMLLRQWA
ABC-P8       ........................... ...... .... ................ .... ........ ... .......... .........
BL31         ........................... ...... .... ................ .... ........ ... .......... .........
ABC-PDL10    ........................... ...... .... ................ .... ........ ... .......... .........
DLBCL-3      ........................... ...... .... ................ .... ........ ... .......... .........
GC-DL44      ........................... ...... .... ................ .... ........ ... .......... .........
HL11         ........................... ...... .... ...............S .... ........ ... .......... .........
GC-H25       ........................... ...... .... ................ .... ........ ... .......... .........
HL02         ........................... .L.... .... ................ .... ........ ... .......... .........
HL08         ........................... ....L. .... ................ .... ........ ... .......... .........
ABC-4        ........................... ...... .... ................ .SQ. L....... ... .....DE... P.K..C...
ABC-DL67     ..........................T ...... .... ................ .... L.....L. .H. ......E... .....C...
ABC-H9       ........................... ...... ..K. ................ .... L....... ... ......E... .....C.C.
ABC-1B       ........................... ...... .... ................ .S.. ........ ... .......... .........
HL01         ........................... ...... .... ................ .... ........ ... .......... .........
HL04         ........................... ...... .... ................ .... ........ ... .......... .........
HL09         ........................... ...... .... ................ .... ........ ... .......... .........
ABC-H7
TLR595
TLR1
```

Fig. 6 (continued)

```
                                          HLA-A11 AVF     HLA-A11 IVT
                                            epitope         epitope
                                          AVFDRKSDAK       IVTDFSVIK
            28   98   197  206  231  385  401                    424  439  444  460  549            567  613  638
B965.8      DQG  FVD  DNV  ATL  SDQ  NIP  AVFDRKSDAKSTKKCRAIVTDFSVIK  RTEQPRAT  PTQ  GIESDEPALTEPVHDQLLPAP  KQS  LRM
sLCL-IS1.15 ...  .M.  ...  ...  ...  ...  ..........................  ........  ...  .....................  ...  ...
sLCL-IS1.13 ...  ...  ...  ...  ...  ...  ..........................  .K....D.  ...  .....................  ...  ...
sLCL-IS1.19 ...  ...  ...  ...  ...  ...  ..........................  ........  .M.  .....................  ...  ...
sLCL-IS1.10 ...  ...  ...  ...  ...  .V.  ..........................  ........  ...  .....................  ...  ...
sLCL-IS1.12 ...  ...  ...  ...  .E.  ...  ..........................  ...L....  ...  .....................  ...  ...
sLCL-IS1.20 ...  ...  ...  ...  .E.  ...  ...N......................  ........  ...  .....................  ...  ...
sLCL10      ...  ...  ...  ...  ...  ...  ..........................  ........  ...  .....................  ...  ...
sLCL03      ...  ...  ...  .M.  ...  ...  ..........................  ........  ...  .....................  ...  ...
sLCL-IS1.04 ...  ...  ...  ...  ...  ...  ..........................  ........  ...  ....................S  ...  ...
sLCL13      ...  ...  ...  ...  ...  ...  ..........................  ........  ...  .....................  ...  .Q.
sLCL09      ...  ...  ...  ...  ...  ...  ..........................  ........  ...  .....................  ...  .Q.
sLCL05      ...  ...  ...  ...  ...  ...  ..........................  ........  ...  .....................  .P.  .Q.
sLCL-IS1.07 .Y.  ...  ...  ...  ...  ...  ..........................  ........  ...  .....................  ...  .Q.
sLCL24      ...  ...  ...  ...  ...  ...  ..........................  ........  ...  .....................  ...  .Q.
sLCL19      ...  ...  ...  ...  ...  ...  ..........................  .......M  ...  .....................  ...  ...
sLCL12      ...  ...  ...  ...  ...  ...  ..........................  .......M  ...  .............R......S  ...  ...
sLCL15      ...  ...  ...  ...  ...  ...  ..........................  ........  ...  .......D.............  ...  ...
sLCL18      ...  ...  ...  ...  ...  ...  ..........................  ........  ...  ......H..............  ...  ...
sLCL-C6     ...  ...  ...  ...  ...  ...  ..........................  I.......  ...  D....................  ...  ...
sLCL-NPC3   ...  ...  .I.  ...  ...  ...  ..........................  ........  ...  .....................  ...  ...
```

Fig. 6 (continued)

(c)

HLA-B27.05 epitope HRCQAIRKK

HLA-A24.02 epitope TYSAGIVQI

HLA-B27.02 epitope RRARSLSAERY

HLA-B58 epitope VSFIEFVGW

```
            150       160       170       180       190       200       210       220       230       240       250       260       270       280
B95.8     RHRCQAIRKKPLPIVKQRRWKLLSSCRSWRMGYRTHNLKVNSFESGGDNVHPVLVTATLGCDEGTRHATTYSAGIVQIPRISDQNQKIETAFLMARRARSLSAERYTLFFDLVSSGNTLYAIWIGLGTKNRVSFIEFVGW
M81       ............................................................................................................................................
GD1       ............................................................................................................................................
HKNPC1    ............................................................................................................................................
HKNPC6    ....................................................L....M...E..L............L..M...........................................................
K4123-Mi  ................................................................M...........................................................................
C666-1    ...................................................L........................................................................................
AG876     ........Q.....D.........PY.T.......QT.N.....T...K...L........E..L...I........L..M...............................................R...A.V.....
```

HLA-A11 AVF epitope AVFDRKSDAK

HLA-A11 IVT epitope IVTDFSVIK

HLA-B35.01 epitope AVLLHEESM

```
             390       400       410       420       430       440       450       460       470       480       490       500       510       520
B95.8     LLPYNPTVYGRPAVFDRKSDAKSTKKCRAIVTDFSVIKAIEEEHRKKKAARTEQPRATPESQAPTVVLQRPPTQQEPGPVGPLSVQARLEPWQPLPGPQVTAVLLHEESMQGVQVHGSMLDLLEKDDEVMEQRVMATLLP
M81       ...............N.....................N...............................................................T......................................
GD1       ..............................L...........................................H..........................T......................................
HKNPC1    ...............N.....................N...............................................................T......................................
HKNPC6    ............SF................L...........................................H....A.....................T......................................
K4123-Mi  .....................................N.........................................A............................................................
C666-1    ...............N.....................N.........................................A.....................T......................................
AG876     ...HS.......S..Y..P.T............L.I..V..D......T........K.D.P......R.....KVT..A.S.....Q.......SW.HE.R.I..GPPT..D.A............QH...Q.......
```

HLA-B44 epitope VEITPYKPTW

```
            600       610       620       630       640       650       660       670          680       690       700       710       720       730
B95.8     PVPVVQPSQTPDDPTKQSRPPETAAPRQWPMPLRPIPMRPLRMQPIPFNHPVGPTPHQTPQVEITPYKPTWA         QIGHIPYQPTPTGPATMLLRQWAPATMQTPPRAPTPMSPPEVPPVPRQRPR
M81       ........................................................................         ...........................KI......................
GD1       ..............M................................................L........         .M...............................................S.
HKNPC1    ........................................................................         ...........................KI......................
HKNPC6    ..............M...................................................DE....         ..............P.K..C........A......................
K4123-Mi  ........................................................................         .................H.................................
C666-1    ........................................................................         ...........................KI......................
AG876     .E..TH.....GG..T..LA...E.............LH.......S..PA.R.....P....P.F.QS..VKPPQQYQP.M.......R...HS...RP....T...P. .......P..QG..TAM..QGAPTPMPPP
```

Fig. 6 (continued)

```
                 HLA-A11 AVF     HLA-A11 IVT
                   epitope         epitope
                 400       410       420
B95.8            RPAVFDRKSDAKSTKKCRAIVTDFSVIKAI
SEA1.Eur/USA1    ....................L.........
SEA2             ..SF................L.........
PNG2             ..SF.......................R..
SEA3             ..P.................L.........
SEA6             ..PL.......................N..
SEA4.Eur/USA2    ...........................N..
SEA5             ..SL.......................N..
SEA8.PNG1        ..SL.......................T..
NGC              ...........................T..
SEA7             ...........................R..
C/HK1(AVF/A2)    ...A..........................
C/HK2(AVF/N4)    .....N........................
AG876            ..S..Y..P.T............L.I..V.
```

Fig. 6 (continued)

7 Polymorphisms Within Latent Genes are Submitted to Immune Pressure and Influence the Geographic Distribution of Virus Isolates

Although it has now been well established that EBV strains around the world are polymorphic, it remains unclear why these strains are unequally distributed. One possibility is that the virus is submitted to immune selection, i.e., mutations in some viral epitopes would alter the immune response in individuals with a given HLA type. This assertion stems from the observation that the amino acids 416-424 IVTDFSVIK nonamer epitope, and to a lesser extent 399-408 AVFDRKSDAK, both from B95-8 EBNA3B, are efficiently presented by HLA-A11 (de Campos-Lima et al. 1993) but that mutations within IVTDFSVIK, particularly at position 2 or 9, nearly completely abrogate the immune response (de Campos-Lima et al. 1993, 1994; Midgley et al. 2003) (Fig. 6c). As a consequence, infected HLA A11-positive B cells cannot present the EBNA3B allele present in type 2 viruses or in viruses isolated from Papua New Guinea or China that all have mutations at this position and/or mutations in position 1 or 2 of the AVFDRKSDAK epitope (de Campos-Lima et al. 1993). Crucially, in individuals that express HLA-A11, the cytotoxic response against the virus is mainly directed against this EBNA3B epitope (Gavioli et al. 1993). This implies that individuals with HLA-A11 can efficiently control viruses such as B95-8 that express IVTDFSVIK or AVFDRKSDAK, and presumably limit their propagation within the population. This fits with the observation that viruses whose genomes encode the IVTDFSVIK epitope are rarely found in populations that express HLA-A11 at high frequency, such as those who live in South China or in the coastal areas of Papua New Guinea (de Campos-Lima et al. 1993, 1994; Midgley et al. 2003) (Fig. 6b). Thus, a high frequency of HLA-A11 is not compatible with a high frequency of B95-8-type viruses. However, this model does not necessarily imply that Caucasian or African viruses must all be of B95-8 type. Indeed, Caucasian populations in Australia or from the Papua New Guinea highlands, only a minority of which is A11-positive, have been reported to be infected by viruses with mutations in AVFDRKSDAK and or IVTDFSVIK. This observation in our opinion does not argue against immune pressure. Indeed, the immune selection model only predicts that a high frequency of infection with B95-8 type EBNA3B in a population with a high frequency of HLA-A11 is incompatible. However, other authors interpret the equal distribution of the same EBNA3B alleles, more precisely the ones that are not presented by HLA A11, within PNG individuals who live in coastal areas and are frequently A11-positive within those who live in the highlands and are rarely A11 as an indication that immune pressure does not exist (Burrows et al. 1996). This group also extended their analysis to other EBV epitopes in EBNA3A, EBNA3B and EBNA3C as well as in LMP2A (Khanna et al. 1997). Furthermore, there was no direct correlation between the frequency of a particular HLA allele and the distribution of alleles of these viral epitopes. These data were interpreted by these authors as evidence that amino acid changes within CTL epitope regions are not influenced by the host immune system.

They did not find any evidence for a higher frequency of non-synonymous versus synonymous mutations and therefore of selection for particular viral epitopes. We feel that this argument would be more convincing if it was proven that some of the epitopes they interrogated were as immunodominant as IVTDFSVIK. If, on the contrary, the immune response is directed against multiple epitopes presented by different HLA haplotypes, it would be very difficult to argue for or against antigenic drift caused by immune pressure simply by looking at singular epitope HLA allele combinations. Moreover, Alan Rickinson and his group performed similar experiments with a panel of 31 virus isolates from China. They found that only the two A11-restricted epitopes in EBNA3B were under positive selection pressure (Midgley et al. 2003). It would be interesting to study large communities of HLA A11-positive individuals who have been living for multiple generations in areas where B95-8-type viruses are common. If these individuals remain only rarely infected by B95-8, it would definitely settle the case.

8 The NPC-Associated Virus M81 Is Endowed with Unique Properties that Distinguishes It from Other Strains In Vitro and In Vivo

We mentioned earlier that viruses in Southeast Asia, and in particular in the area around Hong Kong, have been extensively studied for the search of single nucleotide polymorphisms that might explain the high incidence of NPC in these regions and how these attempts proved altogether rather inconclusive. Nevertheless, the case for particular EBV strains involved in NPC development is compelling because of the presence of the virus in epithelial cells, which suggest a tropism of the virus for this cell type. However, EBV is generally only weakly epitheliotropic in vitro. Furthermore, individuals who display strong serological titers against viral proteins involved in replication are at a much higher risk of disease, suggesting that they are infected with strongly replicating strains.

We tested this hypothesis by studying the properties of a virus isolated from a Chinese patient (M81) in vitro and in vivo in humanized mice (Tsai et al. 2013). M81 is genetically very close to NPC group I viruses and therefore likely to be representative of this group of viruses. We found that M81 induces lytic replication in infected B cells at much higher levels than other EBV isolates in both experimental settings (Fig. 7). Indeed, infected B cells express all replication proteins, including BZLF1, gp110, and gp350 at much higher levels than B cells infected with Akata or B95-8 (Tsai et al. 2013). Furthermore, these cells produce infectious viruses, although most of these are found bound to neighbor B cells that express CD21 and CD35 at high level. M81 viruses also infect epithelial cells more efficiently than other EBV types (Tsai et al. 2013). Figure 1 shows the long genetic distance between M81 and B95-8, with mutations present in the large majority of viral genes. Construction of recombinant viruses from highly replicating viruses that carry genes from weakly replicating ones allows identification

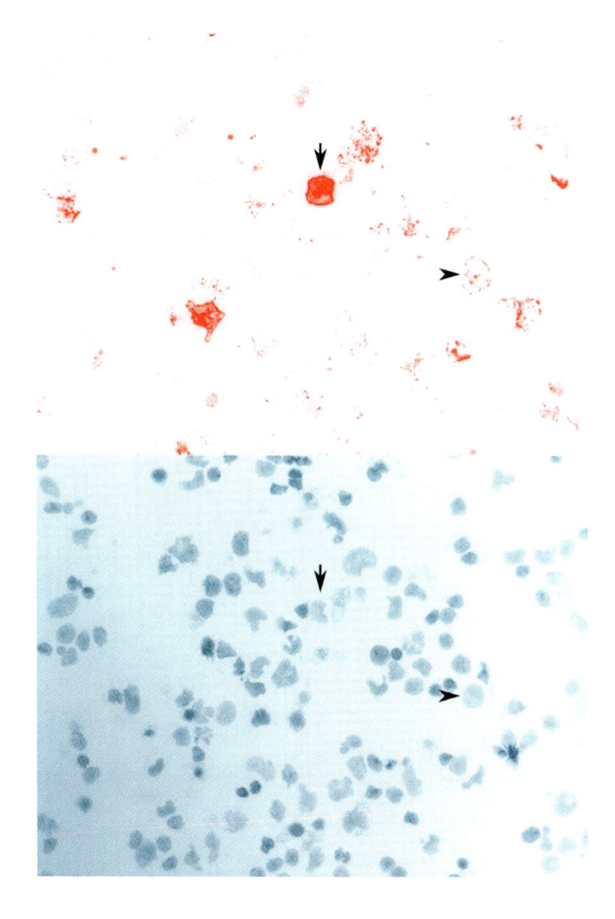

Fig. 7 B cells infected with M81 show high degree of lytic replication. *Upper picture* gp350 staining of an LCL infected with the M81 strain. The *arrow* shows a gp350-positive replicating B cell, whereas the *arrowhead* points to viruses bound to neighbor B cells. *Lower picture* DAPI staining of the same cells

of the viral proteins that are responsible for the different behavior of strains. This approach showed that BZLF1 and its promoter play a role in the acquisition of a highly replicating ability but other genes, probably many, also contribute to the phenotype (Tsai et al. 2013). The challenge of the coming years will be to identify the genes that collaborate to influence the viral behavior. Although M81's properties explain many clinical features observed in patients with NPC, it remains to be formally demonstrated that NPC development is due to a particular EBV strain. This will presumably require an animal model that allows infection of epithelial tissues in a humanized animal model.

9 Conclusions

The combination of genetic analysis and NGS strategies has allowed identification of polymorphisms in EBNA2 and EBNA3B that modulate the immortalizing potential of the virus, but also of genetic variants that radically change defining properties of the virus such as cell tropism or ability to replicate in B cells. Some of these properties are very likely to be involved in the development of EBV-associated diseases. This would represent a shift in paradigm in which the virus type is put at the center of the attention, although this does not in anyway exclude that environmental factors potentiate the effects of infection by particular viral types. It will be vital in the next years to further increase the number of viral sequences obtained from both healthy and diseased patients and to clone representative examples of these isolates on BACs to allow a functional approach of these polymorphisms. This should facilitate the definition of genotypic markers that allow unequivocal identification of the main viral strains and would eventually allow the design of a world map that gives the distribution of the different isolates. We now know that the distribution of EBV strains is heterogeneous across the different countries and or populations. It remains a tremendous challenge to understand in detail the forces that drive this heterogeneous distribution. The immunological pressure hypothesis remains in our opinion the most attractive one at this point. Although contested, we feel the EBNA3B/A11 case remains valid and could explain the relative paucity of B95-8-type isolates in China. There might be other, so far unknown, dominant epitopes that would explain the distribution of other viral strains.

This body of knowledge would have important consequences in terms of prevention strategies as vaccination with particular viral type or proteins thereof might only protect against a particular disease in particular populations. Similarly, the recognition of viral strains with enhanced transforming potential could have an important role in the monitoring of immunosuppressed patients who are at risk of PTLD.

References

Abdel-Hamid M, Chen JJ, Constantine N, Massoud M, Raab-Traub N (1992) EBV strain variation: geographical distribution and relation to disease state. Virology 190:168–175

Adldinger HK, Delius H, Freese UK, Clarke J, Bornkamm GW (1985) A putative transforming gene of Jijoye virus differs from that of Epstein-Barr virus prototypes. Virology 141:221–234

Aitken C, Sengupta SK, Aedes C, Moss DJ, Sculley TB (1994) Heterogeneity within the Epstein-Barr virus nuclear antigen 2 gene in different strains of Epstein-Barr virus. J Gen Virol 75(Pt 1):95–100

Apolloni A, Sculley TB (1994) Detection of A-type and B-type Epstein-Barr virus in throat washings and lymphocytes. Virology 202:978–981

Baer R, Bankier AT, Biggin MD, Deininger PL, Farrell PJ, Gibson TJ, Hatfull G, Hudson GS, Satchwell SC, Séguin C et al (1984) DNA sequence and expression of the B95-8 Epstein-Barr virus genome. Nature 310:207–211

Bhatia K, Raj A, Guitierrez MI, Judde JG, Spangler G, Venkatesh H, Magrath IT (1996) Variation in the sequence of Epstein Barr virus nuclear antigen 1 in normal peripheral blood lymphocytes and in Burkitt's lymphomas. Oncogene 13:177–181

Boyle MJ, Vasak E, Tschuchnigg M, Turner JJ, Sculley T, Penny R, Cooper DA, Tindall B, Sewell WA (1993) Subtypes of Epstein-Barr virus (EBV) in Hodgkin's disease: association between B-type EBV and immunocompromise. Blood 81:468–474

Burrows JM, Burrows SR, Poulsen LM, Sculley TB, Moss DJ, Khanna R (1996) Unusually high frequency of Epstein-Barr virus genetic variants in Papua New Guinea that can escape cytotoxic T-cell recognition: implications for virus evolution. J Virol 70:2490–2496
Chang CM, Yu KJ, Mbulaiteye SM, Hildesheim A, Bhatia K (2009) The extent of genetic diversity of Epstein-Barr virus and its geographic and disease patterns: a need for reappraisal. Virus Res 143:209–221
Chen JN, He D, Tang F, Shao CK (2012) Epstein-Barr virus-associated gastric carcinoma: a newly defined entity. J Clin Gastroenterol 46:262–271
Cheung ST, Lo KW, Leung SF, Chan WY, Choi PH, Johnson PJ, Lee JC, Huang DP (1996) Prevalence of LMP1 deletion variant of Epstein-Barr virus in nasopharyngeal carcinoma and gastric tumors in Hong Kong. Int J Cancer 66:711–712
Cohen JI, Wang F, Mannick J, Kieff E (1989) Epstein-Barr virus nuclear protein 2 is a key determinant of lymphocyte transformation. Proc Natl Acad Sci USA 86:9558–9562
Cullen BR (2011) Herpesvirus microRNAs: phenotypes and functions. Curr Opin Virol 1:211–215
Dambaugh T, Hennessy K, Chamnankit L, Kieff E (1984) U2 region of Epstein-Barr virus DNA may encode Epstein-Barr nuclear antigen 2. Proc Natl Acad Sci USA 81:7632–7636
de Campos-Lima PO, Gavioli R, Zhang QJ, Wallace LE, Dolcetti R, Rowe M, Rickinson AB, Masucci MG (1993) HLA-A11 epitope loss isolates of Epstein-Barr virus from a highly A11+ population. Science 260:98–100
de Campos-Lima PO, Levitsky V, Brooks J, Lee SP, Hu LF, Rickinson AB, Masucci MG (1994) T cell responses and virus evolution: loss of HLA A11-restricted CTL epitopes in Epstein-Barr virus isolates from highly A11-positive populations by selective mutation of anchor residues. J Exp Med 179:1297–1305
de Jesus O, Smith PR, Spender LC, Elgueta Karstegl C, Niller HH, Huang D, Farrell PJ (2003) Updated Epstein-Barr virus (EBV) DNA sequence and analysis of a promoter for the BART (CST, BARF0) RNAs of EBV. J Gen Virol 84:1443–1450
Delecluse H-J, Feederle R, O'Sullivan B, Taniere P (2007) Epstein Barr virus-associated tumours: an update for the attention of the working pathologist. J Clin Pathol 60:1358–1364
Dolan A, Addison C, Gatherer D, Davison AJ, McGeoch DJ (2006) The genome of Epstein-Barr virus type 2 strain AG876. Virology 350:164–170
Edwards RH, Seillier-Moiseiwitsch F, Raab-Traub N (1999) Signature amino acid changes in latent membrane protein 1 distinguish Epstein-Barr virus strains. Virology 261:79–95
Frank D, Cesarman E, Liu YF, Michler RE, Knowles DM (1995) Posttransplantation lymphoproliferative disorders frequently contain type A and not type B Epstein-Barr virus. Blood 85:1396–1403
Gavioli R, Kurilla MG, de Campos-Lima PO, Wallace LE, Dolcetti R, Murray RJ, Rickinson AB, Masucci MG (1993) Multiple HLA A11-restricted cytotoxic T-lymphocyte epitopes of different immunogenicities in the Epstein-Barr virus-encoded nuclear antigen 4. J Virol 67:1572–1578
Gottschalk S, Ng CY, Perez M, Smith CA, Sample C, Brenner MK, Heslop HE, Rooney CM (2001) An Epstein-Barr virus deletion mutant associated with fatal lymphoproliferative disease unresponsive to therapy with virus-specific CTLs. Blood 97:835–843
Gratama JW, Ernberg I (1995) Molecular epidemiology of Epstein-Barr virus infection. Adv Cancer Res 67:197–255
Gutierrez MI, Raj A, Spangler G, Sharma A, Hussain A, Judde JG, Tsao SW, Yuen PW, Joab I, Magrath IT et al (1997) Sequence variations in EBNA-1 may dictate restriction of tissue distribution of Epstein-Barr virus in normal and tumour cells. J Gen Virol 78(Pt 7):1663–1670
Gutierrez MI, Ibrahim MM, Dale JK, Greiner TC, Straus SE, Bhatia K (2002) Discrete alterations in the BZLF1 promoter in tumor and non-tumor-associated Epstein-Barr virus. J Natl Cancer Inst 94:1757–1763
Habeshaw G, Yao QY, Bell AI, Morton D, Rickinson AB (1999) Epstein-barr virus nuclear antigen 1 sequences in endemic and sporadic Burkitt's lymphoma reflect virus strains prevalent in different geographic areas. J Virol 73:965–975

Jenkins PJ, Farrell PJ (1996) Are particular Epstein-Barr virus strains linked to disease? Semin Cancer Biol 7:209–215

Jin Y, Xie Z, Lu G, Yang S, Shen K (2010) Characterization of variants in the promoter of BZLF1 gene of EBV in nonmalignant EBV-associated diseases in Chinese children. Virol J 7:92

Khanna R, Slade RW, Poulsen L, Moss DJ, Burrows SR, Nicholls J, Burrows JM (1997) Evolutionary dynamics of genetic variation in Epstein-Barr virus isolates of diverse geographical origins: evidence for immune pressure-independent genetic drift. J Virol 71:8340–8346

Kunimoto M, Tamura S, Tabata T, Yoshie O (1992) One-step typing of Epstein-Barr virus by polymerase chain reaction: predominance of type 1 virus in Japan. J Gen Virol 73(Pt 2):455–461

Kwok H, Tong AH, Lin CH, Lok S, Farrell PJ, Kwong DL, Chiang AK (2012) Genomic sequencing and comparative analysis of Epstein-Barr virus genome isolated from primary nasopharyngeal carcinoma biopsy. PLoS ONE 7:e36939

Kwok H, Wu CW, Palser AL, Kellam P, Sham PC, Kwong DL, Chiang AK (2014) Genomic diversity of Epstein-Barr virus genomes isolated from primary nasopharyngeal carcinoma biopsy samples. J Virol 88:10662–10672

Kyaw MT, Hurren L, Evans L, Moss DJ, Cooper DA, Benson E, Esmore D, Sculley TB, Mar Alba M (1992) Expression of B-type Epstein-Barr virus in HIV-infected patients and cardiac transplant recipients. AIDS Res Hum Retroviruses 8:1869–1874

Lei H, Li T, Hung GC, Li B, Tsai S, Lo SC (2013) Identification and characterization of EBV genomes in spontaneously immortalized human peripheral blood B lymphocytes by NGS technology. BMC Genom 14:804

Lin Z, Wang X, Strong MJ, Concha M, Baddoo M, Xu G, Baribault C, Fewell C, Hulme W, Hedges D et al (2013) Whole-genome sequencing of the Akata and Mutu Epstein-Barr virus strains. J Virol 87:1172–1182

Liu P, Fang X, Feng Z, Guo YM, Peng RJ, Liu T, Huang Z, Feng Y, Sun X, Xiong Z et al (2011) Direct sequencing and characterization of a clinical isolate of Epstein-Barr virus from nasopharyngeal carcinoma tissue by using next-generation sequencing technology. J Virol 85:11291–11299

Longnecker RM, Kieff E, Cohen JI (2014) Epstein-Barr virus. In: Knipe DM, Howley PM (eds) Fields of virology, vol 2, 6th edn. Lippincott Williams and Wilkins, Philadelphia, pp 1898–1958

Lorenzetti MA, Gutierrez MI, Altcheh J, Moscatelli G, Moroni S, Chabay PA, Preciado MV (2009) Epstein-Barr virus BZLF1 gene promoter variants in pediatric patients with acute infectious mononucleosis: its comparison with pediatric lymphomas. J Med Virol 81:1912–1917

Lorenzetti MAG, Gantuz M, Altcheh J, De Matteo E, Chabay PA, Preciado MV (2012) Distinctive Epstein-Barr virus variants associated with benign and malignant pediatric pathologies: LMP1 sequence characterization and linkage with other viral gene polymorphisms. J Clin Microbiol 50:609–618

Lorenzetti MA, Gantuz M, Altcheh J, De Matteo E, Chabay PA, Preciado MV (2014) Epstein-Barr virus BZLF1 gene polymorphisms: malignancy related or geographically distributed variants? Clin Microbiol Infect 20:0861–0869

Lung ML, Chang GC, Miller TR, Wara WM, Phillips TL (1994) Genotypic analysis of Epstein-Barr virus isolates associated with nasopharyngeal carcinoma in Chinese immigrants to the United States. Int J Cancer 59:743–746

Luo B, Tang X, Jia Y, Wang Y, Chao Y, Zhao C (2011) Sequence variation of Epstein-Barr virus (EBV) BZLF1 gene in EBV-associated gastric carcinomas and nasopharyngeal carcinomas in northern China. Microbes Infect 13:776–782

Mainou BA, Raab-Traub N (2006) LMP1 strain variants: biological and molecular properties. J Virol 80:6458–6468

Martini M, Capello D, Serraino D, Navarra A, Pierconti F, Cenci T, Gaidano G, Larocca LM (2007) Characterization of variants in the promoter of EBV gene BZLF1 in normal donors, HIV-positive patients and in AIDS-related lymphomas. J Infect 54:298–306

McClain KL, Leach CT, Jenson HB, Joshi VV, Pollock BH, Parmley RT, DiCarlo FJ, Chadwick EG, Murphy SB (1995) Association of Epstein-Barr virus with leiomyosarcomas in children with AIDS. N Engl J Med 332:12–18

Midgley RS, Blake NW, Yao QY, Croom-Carter D, Cheung ST, Leung SF, Chan AT, Johnson PJ, Huang D, Rickinson AB et al (2000) Novel intertypic recombinants of Epstein-Barr virus in the chinese population. J Virol 74:1544–1548

Midgley RS, Bell AI, McGeoch DJ, Rickinson AB (2003a) Latent gene sequencing reveals familial relationships among Chinese Epstein-Barr virus strains and evidence for positive selection of A11 epitope changes. J Virol 77:11517–11530

Midgley RS, Bell AI, Yao QY, Croom-Carter D, Hislop AD, Whitney BM, Chan AT, Johnson PJ, Rickinson AB (2003b) HLA-A11-restricted epitope polymorphism among Epstein-Barr virus strains in the highly HLA-A11-positive Chinese population: incidence and immunogenicity of variant epitope sequences. J Virol 77:11507–11516

Kieff E, Rickinson, AB (2007) Epstein-Barr virus and its replication. In: Knipe DM, Howley PM (eds) Fields virology, 5th ed. Lippincott Williams and Wilkins, Philadelphia, pp 2603–2654

Rickinson AB, Kieff E (2007) Epstein-Barr virus. In: Knipe DM, Howley PM, Griffin DE, Lamb RA, Martin MA, Roizman B, Straus SE (eds) Fields virology, vol 2, 5th edn. Lippincott Williams and Wilkins, Philadelphia, pp 2655–2700

Rickinson AB, Young LS, Rowe M (1987) Influence of the Epstein-Barr virus nuclear antigen EBNA 2 on the growth phenotype of virus-transformed B cells. J Virol 61:1310–1317

Rowe M, Young LS, Cadwallader K, Petti L, Kieff E, Rickinson AB (1989) Distinction between Epstein-Barr virus type A (EBNA 2A) and type B (EBNA 2B) isolates extends to the EBNA 3 family of nuclear proteins. J Virol 63:1031–1039

Sample J, Young L, Martin B, Chatman T, Kieff E, Rickinson A, Kieff E (1990) Epstein-Barr virus types 1 and 2 differ in their EBNA-3A, EBNA-3B, and EBNA-3C genes. J Virol 64:4084–4092

Sandvej K, Gratama JW, Munch M, Zhou XG, Bolhuis RL, Andresen BS, Gregersen N, Hamilton-Dutoit S (1997) Sequence analysis of the Epstein-Barr virus (EBV) latent membrane protein-1 gene and promoter region: identification of four variants among wild-type EBV isolates. Blood 90:323–330

Santpere G, Darre F, Blanco S, Alcami A, Villoslada P, Mar Alba M, Navarro A (2014) Genome-wide analysis of wild-type Epstein-Barr virus genomes derived from healthy individuals of the 1,000 Genomes Project. Genome Biol Evol 6:846–860

Sawada A, Croom-Carter D, Kondo O, Yasui M, Koyama-Sato M, Inoue M, Kawa K, Rickinson AB, Tierney RJ (2011) Epstein-Barr virus latent gene sequences as geographical markers of viral origin: unique EBNA3 gene signatures identify Japanese viruses as distinct members of the Asian virus family. J Gen Virol 92:1032–1043

Schuster V, Ott G, Seidenspinner S, Kreth HW (1996) Common Epstein-Barr virus (EBV) type-1 variant strains in both malignant and benign EBV-associated disorders. Blood 87:1579–1585

Sculley TB, Sculley DG, Pope JH, Bornkamm GW, Lenoir GM, Rickinson AB (1988a) Epstein-Barr virus nuclear antigens 1 and 2 in Burkitt lymphoma cell lines containing either 'A'- or 'B'-type virus. Intervirology 29:77–85

Sculley TB, Cross SM, Borrow P, Cooper DA (1988b) Prevalence of antibodies to Epstein-Barr virus nuclear antigen 2B in persons infected with the human immunodeficiency virus. J Infect Dis 158:186–192

Sculley TB, Apolloni A, Stumm R, Moss DJ, Mueller-Lantczh N, Misko IS, Cooper DA (1989) Expression of Epstein-Barr virus nuclear antigens 3, 4, and 6 are altered in cell lines containing B-type virus. Virology 171:401–408

Shu CH, Chang YS, Liang CL, Liu ST, Lin CZ, Chang P (1992) Distribution of type A and type B EBV in normal individuals and patients with head and neck carcinomas in Taiwan. J Virol Methods 38:123–130

Sixbey JW, Shirley P, Chesney PJ, Buntin DM, Resnick L (1989) Detection of a second widespread strain of Epstein-Barr virus. Lancet 2:761–765

Sung NSE, Edwards RH, Seillier-Moiseiwitsch F, Perkins AG, Zeng Y, Raab-Traub N (1998) Epstein-Barr virus strain variation in nasopharyngeal carcinoma from the endemic and non-endemic regions of China. Int J Cancer 76:207–215

Tong JH, Lo KW, Au FW, Huang DP (2003) To KF: Re: Discrete alterations in the BZLF1 promoter in tumor and non-tumor-associated Epstein-Barr virus. J Natl Cancer Inst 95:1008–1009 (author reply 1009)

Tsai MH, Raykova A, Klinke O, Bernhardt K, Gärtner K, Leung CS, Geletneky K, Sertel S, Münz C, Feederle R, Delecluse HJ (2013) Spontaneous lytic replication and epitheliotropism define an Epstein-Barr virus strain found in carcinomas. Cell Rep 5:458–470

van Baarle D, Hovenkamp E, Dukers NH, Renwick N, Kersten MJ, Goudsmit J, Coutinho RA, Miedema F, van Oers MH (2000) High prevalence of Epstein-Barr virus type 2 among homosexual men is caused by sexual transmission. J Infect Dis 181:2045–2049

White RE, Ramer PC, Naresh KN, Meixlsperger S, Pinaud L, Rooney C, Savoldo B, Coutinho R, Bodor C, Gribben J et al (2012) EBNA3B-deficient EBV promotes B cell lymphomagenesis in humanized mice and is found in human tumors. J Clin Invest 122:1487–1502

Young LS, Rickinson AB (2004) Epstein-Barr virus: 40 years on. Nat Rev Cancer 4:757–768

Young LS, Yao QY, Rooney CM, Sculley TB, Moss DJ, Rupani H, Laux G, Bornkamm GW, Rickinson AB (1987) New type B isolates of Epstein-Barr virus from Burkitt's lymphoma and from normal individuals in endemic areas. J Gen Virol 68(Pt 11):2853–2862

Yu XM, McCarthy PJ, Wang Z, Gorlen DA, Mertz JE (2012) Shutoff of BZLF1 gene expression is necessary for immortalization of primary B cells by Epstein-Barr virus. J Virol 86:8086–8096

Zeng MS, Li DJ, Liu QL, Song LB, Li MZ, Zhang RH, Yu XJ, Wang HM, Ernberg I, Zeng YX (2005) Genomic sequence analysis of Epstein-Barr virus strain GD1 from a nasopharyngeal carcinoma patient. J Virol 79:15323–15330

Zimber U, Adldinger HK, Lenoir GM, Vuillaume M, Knebel-Doeberitz MV, Laux G, Desgranges C, Wittmann P, Freese UK, Schneider U et al (1986) Geographical prevalence of two types of Epstein-Barr virus. Virology 154:56–66

Part III
Viral Infection and Associated Diseases

EBV Persistence—Introducing the Virus

David A. Thorley-Lawson

Abstract Persistent infection by EBV is explained by the germinal center model (GCM) which provides a satisfying and currently the only explanation for EBVs disparate biology. Since the GCM touches on every aspect of the virus, this chapter will serve as an introduction to the subsequent chapters. EBV is B lymphotropic, and its biology closely follows that of normal mature B lymphocytes. The virus persists quiescently in resting memory B cells for the lifetime of the host in a non-pathogenic state that is also invisible to the immune response. To access this compartment, the virus infects naïve B cells in the lymphoepithelium of the tonsils and activates these cells using the growth transcription program. These cells migrate to the GC where they switch to a more limited transcription program, the default program, which helps rescue them into the memory compartment where the virus persists. For egress, the infected memory cells return to the lymphoepithelium where they occasionally differentiate into plasma cells activating viral replication. The released virus can either infect more naïve B cells or be amplified in the epithelium for shedding. This cycle of infection and the quiescent state in memory B cells allow for lifetime persistence at a very low level that is remarkably stable over time. Mathematically, this is a stable fixed point where the mechanisms regulating persistence drive the state back to equilibrium when perturbed. This is the GCM of EBV persistence. Other possible sites and mechanisms of persistence will also be discussed.

D.A. Thorley-Lawson (✉)
School of Medicine, Tufts University, Boston, MA 02111, USA
e-mail: david.thorley-lawson@tufts.edu

C. Münz (ed.), *Epstein Barr Virus Volume 1*, Current Topics in Microbiology and Immunology 390, DOI 10.1007/978-3-319-22822-8_8

Contents

1 Introduction 153
2 The Germinal Center Model (GCM) of EBV Persistence—A Historical Perspective 155
3 EBV Infection in the Healthy Host—A Summary of the GCM 156
3.1 Crossing the Epithelial Barrier and the Activation/Infection of Naïve B Cells 159
3.2 Migration to the Follicle and the Germinal Center (GC) Reaction 164
3.3 EBV Persistence in the Peripheral Memory B Cell Compartment 171
3.4 Viral Replication—Plasma Cell Differentiation, Stress, and the Role of Epithelial Cells 174
4 The Cyclic Pathogen Refinement of GCM 177
5 The Model of Persistence—A Summary 180
6 Disease Pathogenesis—Insights from the GCM 180
6.1 Infectious Mononucleosis—Acute Infection (AIM) 181
6.2 Autoimmune Disease 183
6.3 Cancer 184
7 Other Sites of EBV Persistence 189
7.1 The Epithelium 189
7.2 The Tonsil Intraepithelial (Marginal Zone) B Cell—A Second Route to Persistence? 189
7.3 GC-Independent Maturation of Infected Naïve Blasts 191
7.4 Two Pathways to Persistence? 192
7.5 Direct Infection of Memory Cells 193
8 Conclusions 194
9 To Be Continued 195
10 Final Thought—EBV Is Not As Safe As You Might Think! 195
References 196

Abbreviations

AID Activation-induced cytidine deaminase
AIM Acute infectious mononucleosis
APOBEC Apolipoprotein B mRNA editing enzyme, catalytic polypeptide-like
BAFF B cell activating factor
BCR B cell receptor
BL Burkitt's lymphoma
BLC B lymphocyte chemoattractant CXCL13
CD40L CD40 ligand
cIg Cytoplasmically expressed immunoglobulin
CPM Cyclic pathogen model
CtBP C-terminal-binding protein
CTL Cytotoxic T cell
DZ Dark zone
eBL Endemic Burkitt's lymphoma
EBV Epstein-Barr virus
EBNA Epstein-Barr virus nuclear antigen
GC Germinal center
GCM Germinal center model

HD	Hodgkin's disease
HEV	High endothelial venules
HIV	Human immunodeficiency virus
IE	Immediate early
Ig	Immunoglobulin
IL	Immunoblastic lymphoma
LMP	Latent membrane protein
LZ	Light zone
RBPJk	Recombining binding protein
RTPCR	Real-time polymerase chain reaction
SDF1	Stromal cell-derived factor 1 CXCL12
sIg	Surface-expressed immunoglobulin
sBL	Sporadic Burkitt's lymphoma
Th	CD4+ T helper cell

1 Introduction

Persistent latent infection for the lifetime of the host is a defining feature of herpesviruses. Each herpesvirus has a target tissue(s) in which it persists and each has evolved a strategy for getting there and back out again. Once at the site of persistent latent infection, the strategies coalesce in the sense that the goal is to persist latently within a very small number of cells and to minimize or eliminate viral gene expression, at least at the protein level. This in turn allows the virus to evade immune regulation and persist with minimal impact on the host where it will stay for the rest of its life. Acute infection and viral reactivation to allow spread to new hosts similarly seem to have evolved for minimal impact on the host. Acute infection should occur in childhood and is often silent. It is not a coincidence that some of the human herpesviruses are so benign and non-pathogenic that they went unnoticed until the age of AIDS where chronic immunosuppression revealed their presence.

Usually, in the struggle between virus and host, one or the other wins—if it is the host, the virus is eliminated, for example influenza. Flu goes through an acute viremic stage and then is cleared within a week or two (Fig. 1a). If the virus wins, then the host dies, for example HIV. HIV also has an acute viremic stage but resolves into a low-level infection. However, this is unstable and the virus eventually returns to kill the host. EBV also has an acute viremic stage that resolves into a low-level infection, but unlike HIV the virus then simply persists stably at this very low level (something like 1 infected cell per 5 ml of blood) for the lifetime of the host (Hadinoto et al. 2009; Khan et al. 1996; Thorley-Lawson and Allday 2008). Mathematically, this is referred to as a stable fixed point. Dynamically, it is a situation that requires the mechanisms regulating the state (persistent infection) to drive it back to the fixed point whenever it is perturbed (Fig. 1b). Biologically, i.e., in the presence of perturbations, a stable fixed point is the only way to achieve stable long-term behaviors.

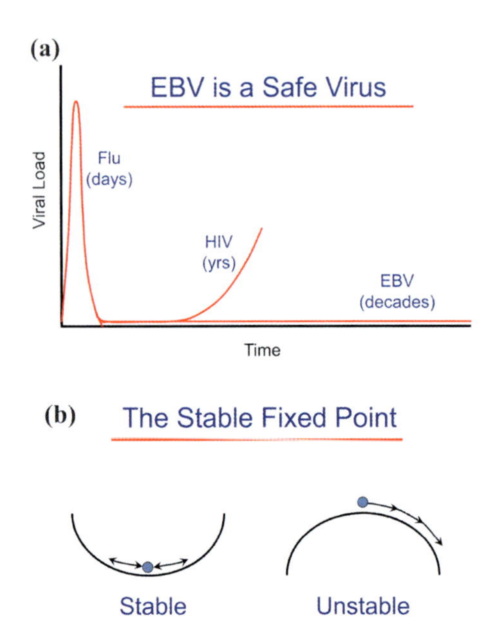

Fig. 1 EBV establishes a stable, benign, low-level, lifetime persistent infection. **a** EBV is a safe virus. EBV establishes a persistent, benign infection in virtually every human being for their entire life. This is in comparison with a virus like flu whose infection resolves in a few days or HIV which undergoes an acute infection that resolves into a long-term low-level persistent infection that eventually returns to kill the host. EBV also undergoes acute infection but then enters into a low-level persistent infection which remains stable for the life of the host. **b** The stable fixed point. The type of equilibrium EBV achieves is referred to mathematically as a stable fixed point. This means that the forces regulating the system act to return it to the same place after perturbation, e.g., a marble in the bottom of a bowl, whereas in an unstable fixed point, small perturbations irrevocably destroy the fixed point, e.g., a marble on top of the bowl. In real-life biology, where there are always perturbations, the only way to achieve long-term stability is through a stable fixed point

EBV is a paradigm for studying the mechanism by which persistent infection is maintained in vivo. It is an unlikely candidate for this status. We lack an in vitro lytic system that would allow viral genetics to be studied—the production of a single viral mutant is a laborious and technically challenging task (Delecluse and Hammerschmidt 2000). Certainly, no system exists for screening large numbers of viral variants and selecting mutants of choice. For a detailed discussion on the production of EBV recombinants, see the chapter authored by Henri-Jacques Delecluse. Similarly, we lack a malleable animal model to perform these studies. The animal models available are limited to primates which are expensive, difficult to work with, and lacking in sophisticated reagents (Wang 2013) and mouse models. For a detailed discussion of primate models, see the chapter authored by Fred Wang, and for mouse models, see the chapter authored by Christian Munz. Mouse models fall into two classes: reconstitution of genetically immunocompromised mice with human cells (Chatterjee et al. 2014) and studies on the murine gammaherpesvirus MHV68 (Barton et al. 2011). Of the two systems, the latter has proved highly tractable for studying and analyzing, at the molecular, genetic, and immunological level, the basis and details of persistent infection by a gammaherpesvirus. Of the human herpesviruses, EBV is the most amenable to study in vivo because it infects readily accessible tissue, namely B lymphocytes of the lymphoid tissue (tonsils) and peripheral blood. With the advent of sophisticated and sensitive flow cytometric techniques to characterize lymphoid populations and PCR to detect very rare infected cells and their gene expression, EBV became accessible for in vivo study.

2 The Germinal Center Model (GCM) of EBV Persistence—A Historical Perspective

Epstein-Barr virus was discovered in Burkitt's lymphoma in 1964. It is a reflection of the complex and subtle biology of the virus that 50 years later, we are only just beginning to understand the role of the virus in the development of this tumor (Speck 2002; Thorley-Lawson and Allday 2008; Vereide and Sugden 2009). By 1999, a large body of work had been accumulated pertaining to EBV's molecular and cellular biology, immunology, virology, epidemiology, clinical manifestations, and disease associations. However, this work existed as a series of independent pieces of information that did not hang together in a consistent way to explain viral biology and persistence [for a discussion of the specific issues, see Thorley-Lawson (2005)].

For example, it has long been known that, unlike most other human herpesviruses, EBV is able to establish latent persistent infection in tissue culture (Henle et al. 1967; Pope et al. 1968). The sine qua non of EBV infection in vitro is that the virus always persists latently in proliferating B lymphoblasts whose growth is driven by viral latent proteins. This process is often referred to as "immortalization." However, an apparent contradiction arose when it was discovered that the virus did not persist in this state in vivo but in a diametrically opposite type of cell, namely quiescent, resting memory B cells where viral protein expression has been extinguished (Babcock et al. 1998; Hochberg et al. 2004; Miyashita et al. 1997).

The GCM arose to resolve this contradiction (Thorley-Lawson and Babcock 1999) and in doing so provided a way to understand the complex biology of EBV. It has stood for 15 years and many tests of its reliability and predictive power (Thorley-Lawson et al. 2013). To date, it remains the only model that consistently provides a conceptual framework for understanding the complex and subtle behaviors of the virus (Thorley-Lawson and Allday 2008; Thorley-Lawson and Gross 2004). It is built on the simple idea that the virus uses the normal pathways of B cell biology in the lymphoid tissue of Waldeyer's ring (tonsils and adenoids) (Fig. 2) to establish infection, persist, and replicate. Today, the questions that arise are not as to the validity of the general model but the extent to which the virus goes along for the ride or actively manipulates the process and whether there are additional mechanisms/sites of viral persistence.

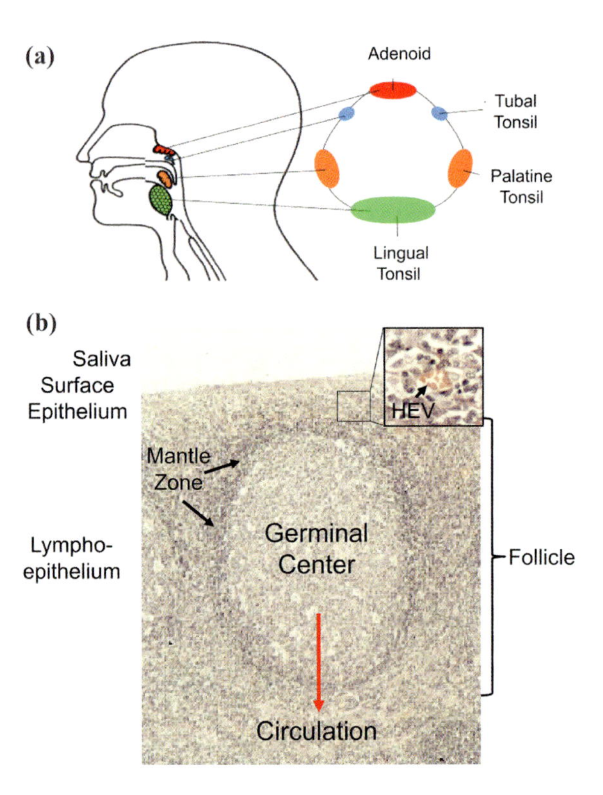

Fig. 2 The lymphoepithelium of the tonsil where EBV performs its biology. **a** Waldeyer's ring consists of the adenoids and tonsils which form a ring of lymphoid tissue at the back of the throat. **b** The structure of the lymphoepithelium underlying the saliva. *Inset* is an expanded view of the marginal zone/epithelium. B cells exit the circulation and enter the lymphoid tissue through the HEV and migrate to the mantle zone of the follicle. Here, they reside for a period of time and then either leave or, if they see antigen, enter the follicle to undergo a GC reaction which produces memory cells that can then enter the peripheral circulation. This is the B cell system that EBV exploits. For more details, see Figs. 3 and 4 and text (Figure provided by Marta Perry)

3 EBV Infection in the Healthy Host—A Summary of the GCM

A sea change in thinking about EBV was the recognition that under normal conditions, it should not be thought of as an oncogenic virus. This despite its discovery in and association with tumors and its ability to latently infect B cells in culture and continuously drive their proliferation. The essence of its biological behavior is that it initiates, establishes, and maintains persistent infection by subtly using various aspects of normal B cell biology and has evolved to minimally perturb the normal behavior of the infected B cells. A summary of normal mature B cell biology

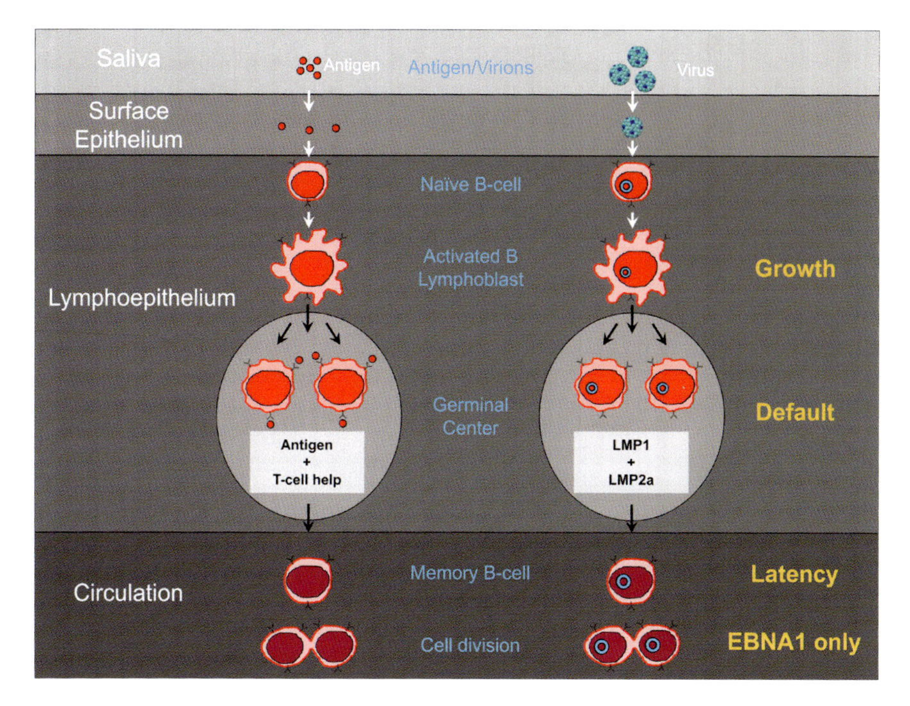

Fig. 3 EBV biology mirrors B cell biology. To the *left* is diagrammed a typical mucosal humoral immune response. Antigen in saliva crosses the epithelial barrier of the tonsil to be sampled by naïve B cells in the underlying lymphoid tissue. When naïve B cells recognize cognate antigen, they become activated blasts and migrate to the follicle to undergo a GC reaction. If they receive signals from antigen and antigen-specific Th cells, they can leave to become resting memory B cells that occasionally undergo division as part of memory B cell homeostasis. To the *right* is diagrammed how EBV uses the same pathways. EBV is spread through saliva, crosses the epithelial barrier, and infects naïve B cells. These become B cell blasts that enter the GC. Here, the viral latent proteins LMP1 and LMP2 have the capacity to provide surrogate antigen and Th survival signals that allow the latently infected B cells to leave the GC as resting memory cells that also divide through homeostasis. To the *right* are listed in orange the transcription programs used at each stage. The *blue circles* represent the viral DNA which is a circular episome

and the parallels with EBV is given in Fig. 3, and a full description of the GCM is presented in Fig. 4. A summary of the steps from Fig. 4 is as follows:

1. Oral antigens enter in saliva, are sampled by the epithelium of Waldeyer's ring , and then presented to naïve B cells in the underlying lymphoid tissue (Fig. 3). When the naïve B cells see cognate antigen, they become activated into a proliferating blast. Similarly, EBV is spread through saliva contact and crosses the epithelial barrier of Waldeyer's ring to interact with naive B cells. Upon infection of the naïve B cell, it drives the infected cell to become a proliferating blast using the growth transcription program (a summary of the viral transcription programs is provided in Table 1).

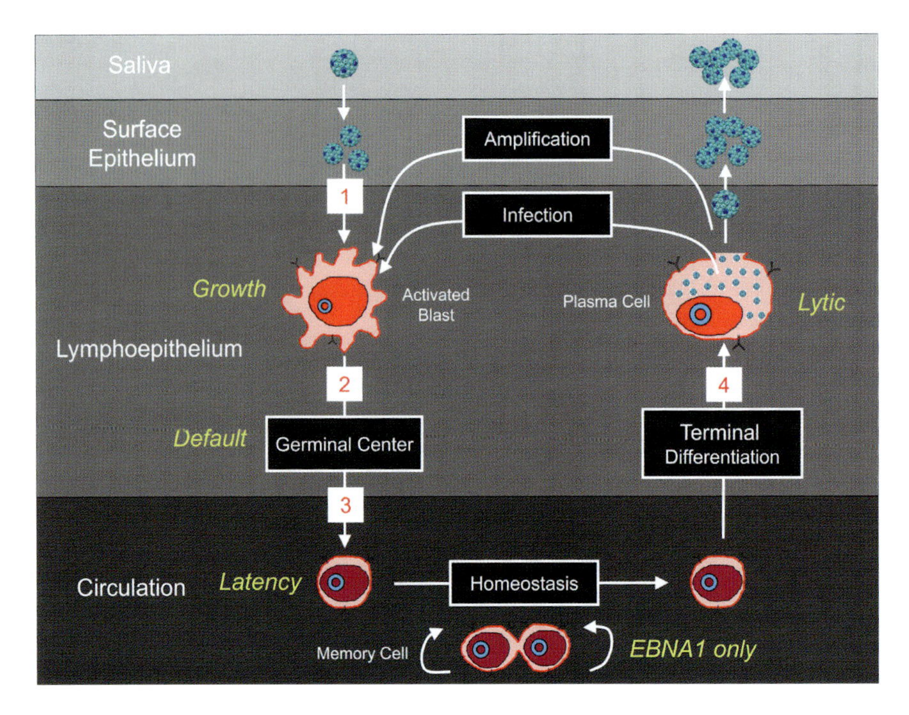

Fig. 4 The germinal center model (GCM) of EBV persistence. The stages 1–4 follow those in the text from Sect. 3. "EBV Infection in the healthy host—a summary of the GCM." For details, see the text

Table 1 The latency transcription programs of EBV in vivo

Program	Alternate	Site in vivo	Lymphoma	Expressed Proteins[a]				
Growth	Latency 3	Tonsil Naïve B	mmunoblastic	EBNA1(Cp)	EBNA2	EBNA 3A-C	LMP1	LMP2
Default	Latency 2	Tonsil GC B cell	Hodgkin's	EBNA1(Qp)[b]			LMP1	LMP2
EBNA1 only	Latency 1	Periphery dividing memory B	Burkitt's	EBNA1(Qp)[b]				
Latency	Latency 0	Periphery resting memory B						

[a]The non- coding RNAs which includes EBERs and the BART miRNAs are expressed in all normal and tumor cells irrespective of transcription program. The exception is a subset of BART miRNAs that are absent from the GC and memory compartments.
[b]EBNA1 is expressed from the Cp promoter in the growth program but a different promoter (Qp) in the default and EBNA1 only programs.

2. Antigen-activated naïve blasts migrate into the follicle to initiate a GC reaction where survival of the B cell requires signals from cognate antigen and antigen-specific helper T cells. Similarly, EBV-infected naïve blasts migrate into the follicle where they switch their transcription program to the default program which provides surrogate antigen and T cell help signals.

3. A successful, antigen-specific, GC B cell leaves the GC to enter the memory compartment as a resting, long-lived, memory B cell which is sustained through occasional homeostatic driven division. Similarly, the latently infected GC B cells leave the follicle as resting memory B cells which are quiescent with respect to viral latent protein expression (the latency transcription program). These cells occasionally divide in the periphery. Proliferation is not driven by the virus but by the normal memory homeostatic mechanisms. At this time, the virus expresses the genome tethering protein EBNA1 which allows the viral genome to replicate with the cells (EBNA1 only program).
4. Antigen-specific, memory B cells in lymphoid tissue can be signaled by cognate antigen to terminally differentiate into plasma cells and produce antibody. Similarly, if an infected, resting, memory B cell latently infected with EBV returns to Waldeyer's ring and receives signals that initiate terminal differentiation, it will trigger the release of infectious virus. The released virus can initiate a new round of naïve B cell infection or infect the epithelium. This results in transient plaques of lytic epithelial infection that greatly amplifies the amount of infectious virus that ultimately is shed into saliva for infectious spread to new hosts.

In this model, EBV gene expression is tightly regulated in a tissue-specific fashion. Dysregulation can lead to lymphomas which arise from each of the three proliferative stages of EBV infection predicted by the model. It is the context and location combined with the stage-specific viral transcription program that defines the lymphoma (Fig. 5). These are immunoblastic lymphoma (IL) from cells expressing the growth program (new infection), Hodgkin's disease (HD) from cells expressing the default program (GC cells), and Burkitt's lymphoma (BL) from cells expressing EBNA1 only (late GC cell).

The following sections will discuss evidence and relevant information for each of these 4 steps in more detail.

3.1 *Crossing the Epithelial Barrier and the Activation/Infection of Naïve B Cells*

3.1.1 Crossing the Epithelial Barrier

It is generally believed that EBV is spread through salivary contact (Hoagland 1955) and that the virus enters through the epithelium that lines the nasopharynx. The lymphoid system that surrounds the nasopharyngeal region includes the adenoids and tonsils and is called Waldeyer's ring (Fig. 2a). Together with the overlying epithelium, it forms a continuous structure referred to as the lymphoepithelium (Fig. 2b) (Perry and Whyte 1998). The epithelium is invaginated to form crypts below which resides the lymphoid tissue (Perry 1994; Tang et al. 1995). Deep in the crypts, the epithelium can be only a single epithelial cell in thickness. Environmental antigens are sampled directly through the epithelium (Perry and Whyte 1998; Brandtzaeg et al. 1999a, b). The involuted nature of the crypts allows

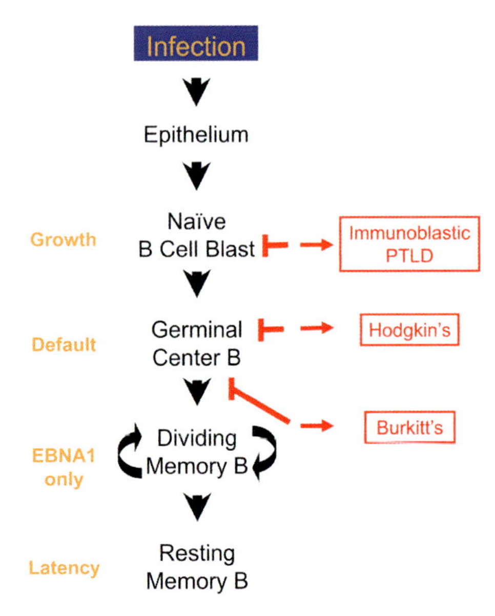

Fig. 5 The origin of EBV-positive lymphomas. EBV lymphomas arise from different stages of the infection process. The figure shows diagrammatically the flow of virus from infectious virions to latently infected resting memory B cell as detailed in Figs. 3 and 4 and the text. To the *right* are shown the 3 EBV-associated lymphomas and their proposed origin and to the *left* are listed the viral transcription programs expressed in the tumors and at the equivalent stage of infection. IL is proposed to arise from a latently infected blast that is unable to differentiate and so continues to proliferate. HD is derived from a GC B cell, and BL is a GC cell that has left the follicle. Note that a tumor is proposed to arise from each of the three stages of EBV biology that involve proliferation

for a massive surface area for detecting antigens as they come in with food and, when exposed to EBV-bearing saliva, provides a large target for EBV infection. It is likely that the virus, in saliva, enters the crypts and crosses the thin layer of epithelial cells to infect naïve B cells that reside below. The sponge-like nature and deep invaginations of the crypts ensure that all of the lymphocytes in the underlying lymphoid tissue are effectively close to the surface where EBV crosses the epithelium. How the virus crosses the epithelial barrier is unclear although there is evidence that the virus may cross passively via transcytosis (Tugizov et al. 2013). It has been speculated that the virus actually infects the epithelial cells, replicates, and then is released to infect B cells in the underlying areas, but there is no direct evidence for this and epithelial cells appear to be resistant to infection from the apical (i.e., mucosal) side (Tugizov et al. 2003).

3.1.2 The Activation/Infection of Naïve B Cells

As far as we know whenever EBV encounters and infects a resting B cells, it always latently infects it and uses the growth program to drive that cell to become a proliferating lymphoblast (Thorley-Lawson and Mann 1985). Phenotypically, the newly infected B cells look remarkably like antigen-activated B cell blasts (Thorley-Lawson et al. 1982, 1985; Nilsson 1979); however, in this case, the B cell is activated not through interaction with antigen and T cell help but through the activity of the latent proteins encoded by the growth program (Kempkes et al. 1995). The population that expresses the growth program in the tonsils of healthy carriers of the virus is activated, naive B cells (Joseph et al. 2000a; Babcock et al. 2000). These cells express CD19 (B cell lineage marker) CD23 and CD80 (B cell activation markers) and IgD (a marker of naïve B cells) and all of the latent proteins associated with the growth program. They lack CD10 (GC cell marker) and CD27 (memory B cell marker). Therefore, the target of the incoming virus is the resting naive B cell. This is the first example we will encounter of a latent gene transcription program used in lymphoma, being found in a normal infected B cell counterpart in vivo. In this case, the growth program, which is used in immunoblastic lymphoma (IL), is found in newly infected naïve B cell blasts (Table 1, Fig. 5).

Naive B cells continuously recirculate throughout the body. They extravasate from the peripheral circulation into secondary lymphoid tissue such as the tonsils through specialized structures called high endothelial vesicles (HEVs) which reside in the lymphoepithelium (inset in Figs. 2b and 6). The naïve B cells migrate through the epithelium to the mantle zone (Fig. 2b) of the follicles which resides just below the epithelium. They remain there for a few days and then reenter the circulation (Brandtzaeg et al. 1999a) unless they encounter cognate antigen in which case they migrate into the follicle.

The migration of naïve B cells from HEV in the epithelium to the mantle zone is critical for them to become exposed to the virus. This is because microdissection studies reveal that virus production and infection of new naive B cells occur in the intraepithelial layer not the mantle zone (Roughan et al. 2010). Thus, naive B cells are becoming infected, as they traverse the epithelium, by EBV that has either crossed the epithelial barrier during primary infection or been produced by the lymphoepithelium during persistent infection (Fig. 6). It follows that by the time the infected B cell arrives at the follicle, it will already be a blast so will not migrate to the mantle zone.

3.1.3 The Growth Program

Because the target for EBV infection is a resting cell, the virus must initiate latent gene transcription in a quiescent environment. It infects cells through the interaction of the viral glycoproteins gp350/220 with CD21 (Nemerow et al. 1985; Fingeroth et al. 1984) and gp42/gH/gL with MHC class II on the B cell (Li et al. 1997). For a detailed discussion of viral entry, see the chapter authored by Lindsey Hutt-Fletcher. CD21 is a receptor for C3d (a component of complement) and forms part of a multimeric signal transduction complex with CD19, CD81 (TAPA-1), and Leu-13 (Matsumoto et al. 1993). The high density of gp350/220 on the virion (Thorley-Lawson and Poodry 1982) ensures that the binding of viral particles will cause extensive cross-linking of the CD21 signaling complex which provides the signal to begin moving the resting B cells from G0 into the G1 phase of the cell

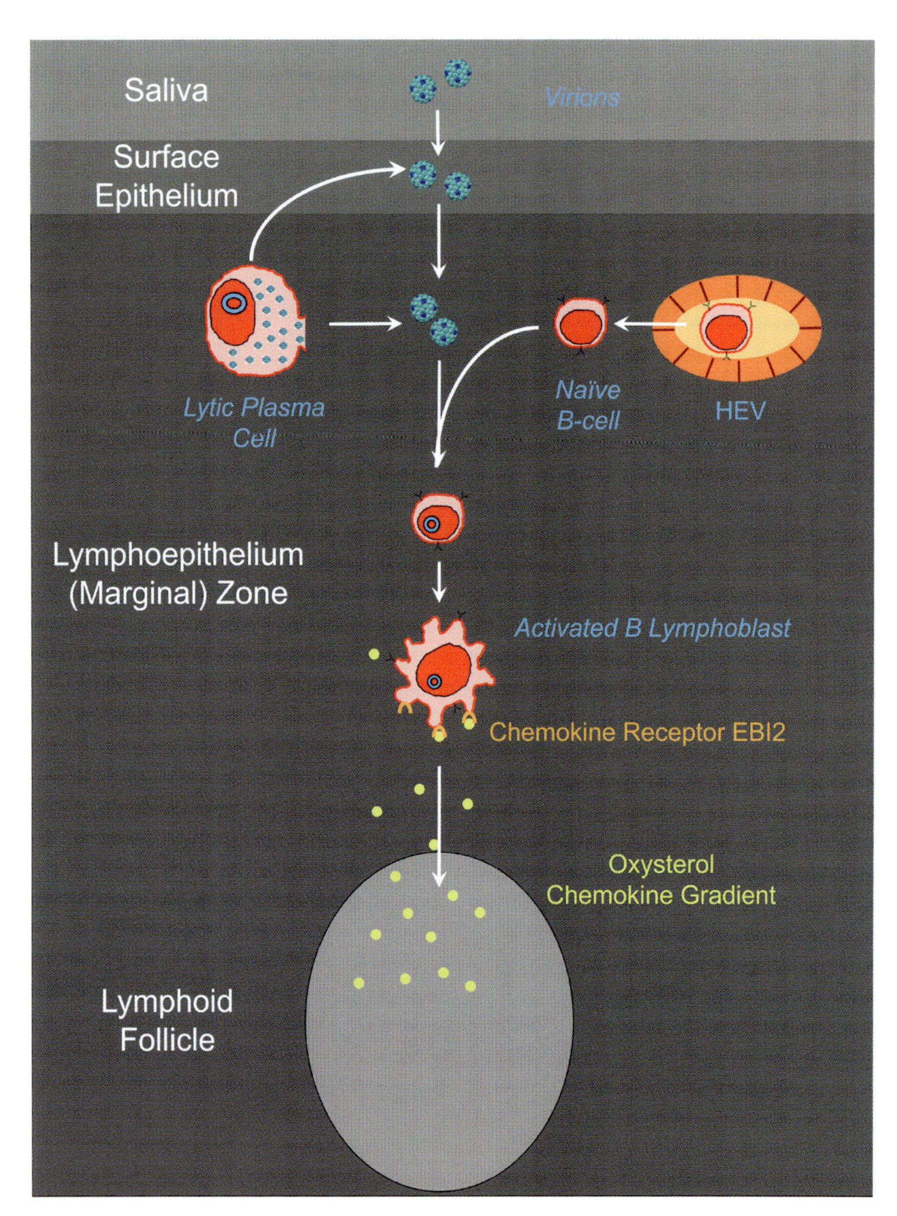

Fig. 6 The first steps of EBV infection. Naïve B cells emerge from the HEV and migrate toward the mantle zone of the follicle. On the way, they encounter EBV that either has crossed the epithelial barrier or is derived from lytically infected plasma cells. The newly infected lymphoblast upregulates the chemokine receptor EBI2 and follows a gradient of oxysterol chemokine into the follicle

cycle (Sinclair and Farrell 1995). During this time, the earliest expressed latent protein (EBNA2) is detected (Allday et al. 1989; Rooney et al. 1989). This protein is expressed from a promoter (Wp) that is present in multiple copies in the viral genome and may be designed to function in the transcriptionally sparse environment of a resting B cell (Woisetschlaeger et al. 1990). EBNA2 drives the cells through the first G1 (Sinclair et al. 1994). EBNA2 is a transcription factor that activates the promoters necessary to produce all nine of the latent proteins expressed in the growth program [reviewed in Kieff and Rickinson (2007)]. For a detailed discussion of EBNA2, see the chapter authored by Bettina Kempkes. At this point, transcription of the EBNA2 gene switches from Wp to Cp (Woisetschlaeger et al. 1990), a promoter that works optimally in B lymphoblasts and allows expression of all the EBNA proteins. The result is that infected normal B cells become activated lymphoblasts and begin to proliferate in response to the actions of viral latent proteins. Although they should not be thought of as classically transformed cells, such as are obtained with other DNA tumor viruses (e.g., SV40, papillomavirus, and adenovirus) (Allday et al. 1995), EBV-driven cells are not completely normal either as evidenced by deregulation of their cell cycle control that can result in immortal growth in culture (O'Nions and Allday 2003; Wade and Allday 2000) [reviewed in Allday (2013), O'Nions and Allday (2004)]. Thus, they rather should be thought of as undergoing a hyperplastic or preneoplastic proliferation that can develop into full-blown neoplasia if allowed to proceed unchecked and accumulate additional oncogenic mutations. However, at this point, it is necessary to mention an important caveat to these studies. Almost all have been conducted with the B95-8 strain of EBV that is often referred to as the "wild-type" "strain." In fact, this is not a wild-type strain, but a highly defective laboratory strain that is carried in marmoset cells and was selected for its ability to transform those cells in culture and make them oncogenic in marmosets, not a natural host for the virus. This virus has multiple genomic deletions (Raab-Traub et al. 1980) among which are those that deregulate expression of the major glycoproteins (Edson and Thorley-Lawson 1981) and delete virtually all of the miRNAs (Skalsky et al. 2012). The latter, in particular, are of concern for interpreting studies on how EBV makes B cells grow and how and to what extent the virus deregulates cell cycle controls.

The nine latent proteins of the growth program include six nuclear proteins (EBNAs—Epstein-Barr virus nuclear antigens—1, 2, 3a, 3b, 3c, and LP) and three membrane proteins (LMPs—latent membrane proteins) [reviewed in Kieff and Rickinson (2007) but see this textbook for the most recent information]. Several of the latent proteins have potent growth-promoting activity and can act as oncogenes. These include EBNA2 (Kempkes et al. 1995), EBNA3a (Hickabottom et al. 2002), EBNA3c (Parker et al. 1996), and LMP1 (Wang et al. 1985).

In addition to the nine latent proteins, EBV-infected lymphoblasts express two small non-polyadenylated RNAs, termed EBER1 and EBER2 (Arrand and Rymo 1982), and ~40 microRNAs. Neither EBERs nor the miRNAs are essential for EBV infection in vitro, suggesting that their functions are most important in vivo (Kuzembayeva et al. 2014). For a detailed discussion of EBV-encoded non-translated RNAs, see the chapter authored by Bryan Cullen.

The latent genes are transcribed from the viral genome which exists as a covalently closed episomal circle (Adams and Lindahl 1975). For a detailed description of genomic structure, see the chapter authored by Paul Farrell. The linear genome from the virion forms this circle when the newly infected cell begins proliferating (Hurley and Thorley-Lawson 1988). Interestingly, only a single episome forms upon initial infection, but this then begins to amplify over time as the infected cells proliferate till a steady-state distribution of episomes is found in cells that have proliferated extensively (Hurley and Thorley-Lawson 1988; Roughan et al. 2010). The forces that produce this distribution are not well understood (Nanbo et al. 2007), but it serves as a useful marker to distinguish cells that have been recently infected from ones that have proliferated extensively. Thus, the status of the viral genome in a tissue provides a considerable amount of useful information. Linear genomes indicate viral replication, whereas episomal genomes, in the absence of linear genomes, are indicative of latently infected cells (Decker et al. 2001) and the episomal copy number is a measure of proliferation history (Roughan et al. 2010).

The viral growth program has evolved to drive the activation and proliferation of new latently infected human B cells. It achieves this, not through some rare random event, such as the integration of the viral genome and disruption of cellular genes employed by retroviruses, but by a highly intricate transcriptional program that is uniquely designed to control the growth of human B cells. This ensures that EBV will efficiently and predictably establish latency and initiate cell growth whenever it encounters a resting naive B cell in the lymphoepithelium of the nasopharynx. This program puts the host, in which the virus intends to persist, at risk for developing neoplastic disease (see Sect. 6.3.1), but it is essential, so the virus can drive the newly infected cell into a state, the proliferating blast, from where it can differentiate into a resting memory B cell. Once there, the virus can shut down, become non-pathogenic, and persist for the life of the host. How does an antigen-activated B blast and, by analogy, the EBV-infected B blast become a resting memory B cell?

3.2 Migration to the Follicle and the Germinal Center (GC) Reaction

To understand how latently infected, naive B lymphoblasts expressing the growth program can become resting memory B cells, with no viral gene expression, it is first necessary to describe how a normal naive B cell blast becomes a memory cell.

3.2.1 Entering the Follicle

Naïve B cells, activated by antigen, migrate toward the GC following a gradient of the oxysterol lipid 7a,25-dihydroxycholesterol. This lipid is produced by follicular lymphoid stromal cells and is recognized by the chemokine receptor EBI2, also known as G protein-coupled receptor 183, on the activated B cell (Gatto and Brink 2013). When EBV activates the newly infected naïve B cell with the growth program, one of the phenotypic changes it causes involves induction of EBI2 (Birkenbach et al. 1993), thus insuring that the virus-infected blasts will migrate toward the follicle (Fig. 6).

3.2.2 The Germinal Center Reaction

Once an antigen-specific B cell enters the follicle as an activated blast, it undergoes a period of rapid expansion for about 3 days, with a cell division time ~8–12 h to form the GC which consists of antigen-specific B cells (Figs. 2 and 3) [reviewed in Liu and Arpin (1997), MacLennan (1994), Victora and Nussenzweig (2012)]. These cells loose surface IgD and acquire GC-specific markers including CD10, CD77, and CD38 and they express AID and bcl-6. AID is an enzyme of the APOBEC family that is highly expressed in GCs. It is the enzyme necessary to initiate somatic hypermutation (SHM) and class switch recombination (CSR) (Muramatsu et al. 2007), functions of the GC. bcl-6 on the other hand is the master transcription factor of the GC (Basso and Dalla-Favera 2010). Its expression is restricted to GC cells (Cattoretti et al. 1995), it is required for GC production (Ye et al. 1997), and its downregulation is essential for B cells to leave the GC (Calame et al. 2003). When proliferating, the cells reside in the dark zone (DZ) of the germinal center and are referred to as centroblasts. Here, the cells undergo CSR to express a single isotype, which can be IgM, IgG, IgA, or IgE and they also undergo SHM. After several divisions, the cells rest and migrate to the light zone (LZ) of the GC. These cells are referred to as centrocytes, and they compete for help delivered by antigen-specific T helper (Th) cells (Schwickert et al. 2011). The Th cell delivers its rescue signal to the B cell through the interaction of CD40 ligand on Th cells with CD40 on B cells (Banchereau et al. 1994). Signaling through CD40 also turns off expression of bcl-6 and turns on bcl-2 which allows the cell to leave the GC and differentiate (Calame et al. 2003).

Cells in the GC go through multiple rounds of proliferation, migration, and selection so that ultimately those expressing the highest affinity B cell receptor (BCR) are selected—a process referred to as affinity maturation. Migration between the light and dark zones is controlled through the expression of specific chemokine receptors CXCR4 and CXCR5 and their cognate ligands (SDF1 and BLC, respectively) (Allen et al. 2004). The cells that survive ultimately have two fates depending on the length and type of exposure to Th cells and specific lymphokines (Banchereau et al. 1994). They can either terminally differentiate into antibody-secreting plasma cells or enter the long-lived memory compartment as resting isotype-switched memory B cells. As the name implies, these cells carry immunological memory and are responsible for a heightened secondary response upon reexposure to the specific antigen.

Unswitched, IgM+/IgD+, memory cells also exist, but they do not arise through the GC (Weill et al. 2009; Weller et al. 2004). These are generally referred to as marginal zone memory B cells because they were originally described in the marginal zone of the spleen (Spencer et al. 1985, 1998) and in the circulation (Weller et al. 2004). A phenotypically related population has also been described in the epithelium of the tonsil (Dono et al. 2003; Spencer et al. 1998); however, they appear to be functionally distinct (Weill et al. 2009).

What is clear then is that a series of events must occur if an EBV-infected naive B lymphoblast, expressing the growth program, is to become a memory cell. First, the cells should enter the GC where the latent genes that drive proliferation are turned off, and then the cells must receive the requisite survival signals and finally leave as resting memory B cells.

3.2.3 EBV-Infected Cells Reside and Participate in the GC

Newly infected B cells are driven by the growth program to undergo an initial phase of rapid expansion with a division time of ~8 h for ~3 days—closely mimicking the dynamics of the early phase of GC development (Nikitin et al. 2010; Thorley-Lawson and Strominger 1978). In vitro, such cells then switch to long-term indefinite proliferation as lymphoblasts with a division time of ~24 h. However, in vivo, the cells do not continue to proliferate driven by the growth program; instead, they become GC cells and switch to a more limited form of viral gene expression—the default program.

Cells in the GC latently infected with EBV are by all measures true GC B cells. They express the classic GC surface phenotype CD10+, CD77+, CD38+, the functional markers AID and bcl-6 (Roughan and Thorley-Lawson 2009), and the correct set of chemokine receptors being $CXCR4^{+}$ $CXCR5^{+}$ and $CCR7^{-}$. The latter ensure that the cells will be retained in and migrate throughout the germinal center. They are positive for the proliferation marker Ki67 and undergo multiple rounds of cell division (≥ 20) (Roughan et al. 2010). Despite this, microdissection studies revealed that there are only on average 3–4 latently infected cells per GC (for reference, there are about 10^5 total B cells in a typical GC). Consequently, the vast majority of latently infected cells produced from the GC must die; otherwise, the memory compartment would be overwhelmed. This death could represent some version of affinity maturation/selection (if the emerging memory cells are truly antigen-selected) or simply destruction by CTL. However, functional CTLs do not appear to enter GCs (Quigley et al. 2007), so the cells would have to be continuously leaving and then killed.

Taken together, these data imply that latently infected B cells in the GC are truly undergoing a GC reaction, that the virus is having a minimal impact on the process and the cells may even be undergoing some form of affinity maturation and selection. Confirmation of this mechanism has come from studies with another B lymphotropic gammaherpesvirus: MHV68 in the mouse. The ability to genetically manipulate both host and virus in this system has allowed for a direct and convincing demonstration that latently infected B cells traverse the GC in order to enter memory (Barton et al. 2011; Collins and Speck 2014).

3.2.4 EBV-Infected Cells in the GC Express the Default Not the Growth Program

Microdissection and flow cytometric analysis have provided compelling and unequivocal evidence that the EBV-infected cells in the GC express the default program not the growth program (Babcock et al. 2000; Roughan and Thorley-Lawson 2009). The demonstration that infected GC cells express the default program means that this latency transcription program is consistent with the retention of GC phenotype and functionality in vivo. This is crucial because it identifies the critical intermediate between the lymphoblastoid growth program and the resting memory B cells. It is known that direct infection and the growth program ablate GC functionality and phenotype, i.e., they are not consistent with GC function (Babcock et al. 2000; Siemer et al. 2008). Thus, for a newly infected naïve blast to differentiate into memory, it must switch to the default program in the GC. This is the second example we will encounter of a latent gene transcription program used in lymphoma, being found in a normal infected B cell counterpart in vivo. In this case, the default program, which is used in Hodgkin's disease (HD), is found in latently infected GC cells (Table 1, Fig. 5). The default program involves only three of the nine latent proteins, EBNA1, LMP1, and LMP2a (Kieff and Rickinson 2007; Thorley-Lawson 2001). Here, the Q promoter (Qp) is employed so that EBNA1 may be expressed without the other EBNAs (Tsai et al. 1995; Schaefer et al. 1995; Nonkwelo et al. 1996). EBNA1 is essential because it is required for retaining the viral genome by tethering it to cellular DNA and allowing it to be replicated (Yates et al. 1985). For a detailed discussion of EBNA1, see the chapter authored by Lori Frapier.

3.2.5 Turning Off the Growth Program

When EBNA2 is turned off in the presence of an activated c-myc, which is expressed in GC cells (Dominguez-Sola et al. 2012; Martinez-Valdez et al. 1996), the cells downregulate surface markers' characteristic of B blasts, such as CD23, and acquire GC-specific markers, such as CD10 (Polack et al. 1996). Therefore, the infected lymphoblast appears free to acquire a GC phenotype once the differentiation block, imposed by EBNA2, is removed. One of the direct targets of EBNA2 is c-myc, a known regulator of cell growth and apoptosis (Kaiser et al. 1999). We can assume therefore that upon arrival in the follicle, the EBV lymphoblast receives a signal that turns EBNA2 and the growth program off while allowing c-myc expression to continue. How this is achieved remains unknown, but there is in vitro evidence to suggest that it may depend in part upon signals originating in the GC from cytokines such as IL-10, IL-21, and Type 1 IFN in combination with CD40 ligand (CD40L) (Kis et al. 2006, 2010; Salamon et al. 2012).

The actual mechanism by which cells switch from the growth to the default program probably depends on a negative feedback loop involving EBNA2 and the EBNA3s. These are believed to act as functional homologues of the intracellular components in the Notch signaling pathway (Kempkes et al. 1995; Speck 2002). For a detailed discussion of this hypothesis, see Thorley-Lawson and Allday (2008), and for a review of the Notch system, see Artavanis-Tsakonas et al. (1995). Upon infection of B cells, the first viral protein expressed is EBNA2 which interacts with the enhancer elements of cellular and viral latent genes to block differentiation and drive cellular proliferation. At the same time, EBNA2 activates the major EBV latent promoter Cp which leads directly to expression of all the EBNAs including EBNA3a and 3c. For a detailed discussion of the EBNA3 proteins, see the chapter authored by Martin Allday. Based on their known functions, EBNA3a and 3c could displace EBNA2 from Cp (Zimber-Strobl and Strobl 2001) and recruit repressor proteins that would lead to the stable epigenetic silencing of Cp and the suppression of EBNA2 production (Hickabottom et al. 2002; Knight et al. 2003; Radkov et al. 1999; Touitou et al. 2001). For a detailed discussion of EBV-associated chromatin and epigenetics, see the chapters authored by Paul Lieberman and Wolfgang Hammerschmidt. Cessation of EBNA2 production would cause growth arrest and allow the cells to assume a GC phenotype and express the default program. In this model, growth driven by EBV is a self-regulating feedback loop involving EBNA2 and the EBNA3s where the balance is tilted in favor of growth arrest by signaling from T cell-associated cytokines and CD40L (Kis et al. 2006, 2010; Salamon et al. 2012). It follows that the in vitro phenomenon of immortalization may be a biological artifact where the balance has been shifted in favor of EBNA2 by the absence of T cell-derived signals and the powerful selection pressure of in vitro growth.

3.2.6 EBV Can Provide the Rescue Signals—LMP1 and LMP2

Once the growth program is turned off, we have good evidence that the expression of LMP1 and LMP2 in the default program is capable of providing the two signals, T cell help and BCR, necessary to rescue the GC cell into memory.

LMP1 is a membrane protein that acts as a ligand-independent, constitutively activated receptor (Gires et al. 1997). For a detailed discussion of LMP1, see the chapter authored by Arnd Kieser. It does this by engaging signaling molecules (Izumi and Kieff 1997; Mosialos et al. 1995) which normally transmit signals from CD40 when it engages its ligand on Th cells [reviewed in Lam and Sugden (2003)]. Thus, in principle, LMP1 is able to deliver a Th signal to the infected B cell in the absence of Th cells. The parallel between CD40-mediated Th and LMP1 signaling extends to the ability of LMP1 to drive immunoglobulin class switching (He et al. 2003; Uchida et al. 1999). LMP1, like CD40, also turns off expression of bcl-6 (Panagopoulos et al. 2004) and turns on bcl-2 (Henderson et al. 1991). Through its ability to regulate bcl-2 and bcl-6, LMP1 (Carbone et al. 1998) almost certainly plays a role in driving the latently infected B cell to leave the GC and differentiate into a memory cell (Fig. 7).

LMP2 is also a membrane protein, but it delivers a constitutive, ligand-independent BCR signal (Caldwell et al. 1998). For a detailed discussion of LMP2, see the chapter authored by Richard Longnecker. LMP2a contains the same signaling

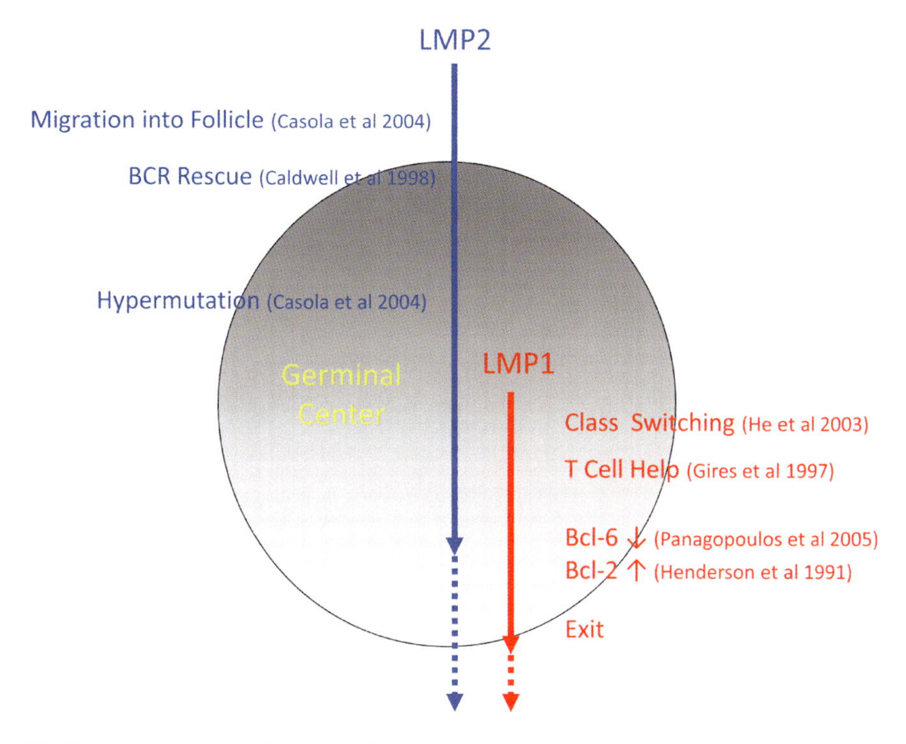

Fig. 7 A summary of the functions of LMP1 and LMP2a demonstrated in vitro or in vivo with transgenic mice that could contribute to the GC processing of a latently infected B cell

motifs (ITAMs) (Beaufils et al. 1993), as the α- and β-chains of the BCR. These motifs allow it to engage signaling molecules employed by the BCR (Miller et al. 1995; Kurosaki 1999). The BCR produces two types of signals (MacLennan 1998): One (tonic) is required to ensure the survival of resting B cells (Lam et al. 1997; Maruyama et al. 2000), while the other (activating) leads to cellular activation, proliferation, and ultimately differentiation into immunoglobulin-secreting plasma cells (MacLennan 1994; Liu and Arpin 1997). LMP2a is able to provide the tonic but not the activating signal (Caldwell et al. 1998) and in the absence of a BCR is able to drive GC formation in mucosal tissue where the cells show evidence of having undergone mutation of their immunoglobulin genes (Casola et al. 2004b). Thus, LMP2a almost certainly plays a role in driving the latently infected B cell into and through the GC (Fig. 7).

In sum, LMP1 and LMP2a have the capacity to provide the latently infected B cell with a whole range of signals associated with GC development (Fig. 7).

3.2.7 Does EBV Do It All—The Conundrum of LMP1 and LMP2

One critical remaining question is: does EBV do it all? The signaling properties of LMP1 and LMP2 imply that together they could potentially provide all the signals necessary to rescue a latently infected B cell from the GC into memory, bypassing the normal mechanisms of antigen selection. If so, the immunoglobulin genes of latently infected memory B cells should either be unmutated or show an unselected pattern of mutations. However, the expressed immunoglobulins in latently infected memory B cells from the blood have undergone CSR, have no stop codons, and display the SHM pattern expected for antigen-selected memory cells (Souza et al. 2005, 2007). Thus, it seems that the expression of LMP1 and LMP2 has little discernible impact on the selection process as EBV-infected cells transit the GC into memory.

Experiments involving the expression of either LMP1 or LMP2 in the B cell compartment of transgenic mice indicate that alone these molecules can have devastating physiologic effects. In such studies, LMP1 could exclude B cells from the GC (Uchida et al. 1999) and even drive the development of B cell lymphomas (Kulwichit et al. 1998). LMP2 on the other hand was shown to replace the BCR allowing BCR-negative B cells to survive and enter the periphery (Caldwell et al. 1998) (a particularly relevant observation for Hodgkin's lymphoma see below) and in some models break tolerance allowing autoreactive cells to survive in the periphery (Chang et al. 2012; Swanson-Mungerson and Longnecker 2007; Swanson-Mungerson et al. 2005). These observations suggest that deregulated expression of LMP1 or LMP2 may play an important role in the pathogenesis of lymphoma and autoimmune disease development but seemed strangely at odds with the striking lack of B cell lymphoma and autoimmunity in the vast population of EBV-infected people. However, in humans, LMP1 and LMP2 are usually expressed together and a follow-up study on double transgenic mice revealed that now the mice did not develop lymphoma or autoimmune disease and their B cells were able to comfortably transit the GC, undergo affinity maturation, and enter the memory compartment (Vrazo et al. 2012).

Thus, it seems that LMP1 and LMP2, when coexpressed in vivo, can modulate each other's signaling. For example, in vitro, LMP1 when expressed alone can downregulate bcl-6 (Panagopoulos et al. 2004) and upregulate bcl-2, yet in the GC, LMP1 expression, in the presence of LMP2, is fully compatible with bcl-6 expression and is not associated with the upregulation of bcl-2 (Roughan and Thorley-Lawson 2009). What then is the role of these proteins in the GC? Because their functions are so tuned to the requirements of the GC and they are specifically expressed there, it seems certain that they must play some important role. What could this be? A clue comes from the analysis of the small subset of bcl-2-positive cells in the GC, those about to leave, which revealed that they only express LMP1, not LMP2, i.e., LMP1 expression in vivo, just as in vitro, is associated with upregulation of bcl-2 but only in the absence of LMP2. It seems likely therefore that the expression of LMP1 or LMP2 alone in the GC is strictly regulated to occur only at specific moments to achieve specific ends. Based on what we know so far, LMP2 expression alone in latently infected cells would ensure that the cells form GCs in mucosal epithelium; LMP1 and LMP2a together drive CSR and SHM and provide the requisite survival signals, and LMP1 alone ensures exit from the GC and terminal differentiation by switching off bcl-6 and switching on bcl-2 (Fig. 7). To test this hypothesis will require careful dissection of infected GC populations. Previous attempts at this showed no differences (Babcock et al. 2000; Roughan and Thorley-Lawson 2009), but were based on the now discredited marker CD77 (Victora et al. 2010) and were therefore artifacts. Recently, an accurate phenotype for GC subsets has been described (Victora and Nussenzweig 2012; Victora et al. 2010), making these studies now feasible.

3.3 EBV Persistence in the Peripheral Memory B Cell Compartment

How the transition from GC to resting long-lived memory B cell is achieved for any cell is not fully understood, but we may assume that once the mechanism is uncovered, we will find that the virus exploits it to gain access to the memory compartment.

3.3.1 The Resting Memory B cell

EBV, in the peripheral blood, is found only in B cells (Miyashita et al. 1995) that have the phenotype expected of a latently infected, long-lived, GC-derived, resting, memory B cell, i.e., classical memory B cells (Table 2) (Babcock et al. 1998; Decker et al. 1996; Joseph et al. 2000b; Miyashita et al. 1997). Persistence in memory B cells, first demonstrated for EBV, may be a common strategy for all B lymphotropic gammaherpesviruses (Barton et al. 2011). Restriction of EBV in the periphery to the GC-derived memory compartment is so tight that less than 1 in 10,000 latently infected cells in the blood are in the naïve compartment (Hochberg et al. 2004). They have the phenotypic hallmarks of classical GC-derived memory B cells being CD27+ (Joseph et al. 2000b; Klein et al. 1998) and having undergone CSR and SHM (Babcock et al. 1998; Joseph et al. 2000b; Souza et al. 2007). They are also CD23− and CD80− (B cell activation markers) (Miyashita et al. 1995), and >90 % are in the G0 stage of the cell cycle (Miyashita et al. 1997; Hochberg et al. 2004) all characteristics of resting B cells.

The latently infected cells occupy a skewed niche within the memory compartment, being excluded from the IgD+ memory subset, but otherwise are evenly distributed among B cells carrying the different immunoglobulin isotypes. This

Table 2 Phenotype of EBV Infected Cells in the Blood

Phenotype	Implication
CD19+, CD20+, CD3-	B Cell
CD23-, CD80-, Ki67-, G0 stage of cell cycle	Resting Cell
CD27+, Ig genes hypermutated and class switched	GC Derived Memory Cell
IgD-	Not marginal zone B cells
CD5-	Not B1 cells
Episomal viral genomes, no linear form	Latently Infected

suggests that once they enter memory, the EBV-infected cells cannot be distinguished from uninfected cells by host homeostasis mechanisms. The pattern of SHM (Souza et al. 2005) they display is that expected for antigen-selected memory cells (Souza et al. 2005). They tend to accumulate more mutations than their uninfected counterparts and actually showed a reduced proclivity to be self-reactive (Tracy et al. 2012). However, these differences were modest and may simply reflect differences between mucosal (EBV+) and splenic (peripheral) derived memory B cells. What is apparent though is that EBV does not significantly disrupt the normal processing of latently infected cells into memory. Deviations from normal B cell biology are not tolerated in these cells despite the potentially potent signaling capacities of LMP1 and LMP2.

EBV is not found in the CD5+ B1 subset (Joseph et al. 2000b), nor in circulating IgD+/IgM+/CD27+ marginal zone memory cells (Joseph et al. 2000b; Souza et al. 2007). These are both long-lived compartments of B cells (Youinou et al. 1999; Kantor 1991) that frequently have specificity for polyantigens such as bacterial cell wall components (Hardy 2008) but neither of which develop through GCs. The absence of EBV from these subsets provides further support for the conclusion that transit of the GC is required for the production of memory B cells latently infected with EBV. Studies claiming to find EBV preferentially in IgA-bearing B cells (Ehlin-Henriksson et al. 1999) or in IgD+ memory cells (Chaganti et al. 2009) were technically flawed and have not been substantiated [for a detailed discussion of the issues, see Joseph et al. (2000b) and Thorley-Lawson et al. (2013), respectively].

Memory cells latently infected with EBV in the peripheral blood do not express any of the known latent proteins (Hochberg et al. 2003a; Hochberg and Thorley-Lawson 2005). This is an important point to stress. Several studies have identified EBV latent gene expression in the peripheral blood based on RT-PCR analysis. However, these were not quantitative studies and were performed on bulk preparations of B cells (Babcock et al. 1999; Chen et al. 1995; Qu and Rowe 1992; Tierney et al. 1994). Because the assays used are so sensitive and variable in their sensitivity, it is impossible to know whether the signals are from rare infected cells expressing the transcript or are representative of the whole infected population of cells. It turns out that the former is true. By performing a limiting dilution RT-PCR analysis (Hochberg et al. 2003a; Hochberg and Thorley-Lawson 2005), it was possible to show that >99 % of the infected cells do not express transcripts for any of the known latent proteins. Indeed, the single-cell analysis afforded by this approach revealed that when latent gene transcripts were found, they were not part of any known transcription programs, indicating that they almost certainly are residual transcripts of no biological significance.

We may conclude therefore that the memory B cell is the site of long-term viral persistence. Here, it can remain for the lifetime of the host because immunological memory is for life, but the virus is no longer pathogenic to the host because the genes that drive cellular proliferation and threaten neoplastic disease are turned off. Similarly, the virus is safe from immunosurveillance because no viral proteins are expressed to act as targets of the immune system. The transcription program used in these cells, where no viral proteins are expressed, is called the latency program (Hochberg et al. 2003a and Table 1) reflecting its role at the site of latent persistence.

The frequency of infected memory B cells for an individual healthy carrier is very stable over time (Hadinoto et al. 2009; Khan et al. 1996). However, the level of infected cells in a population ranges widely from 5 to 3000 for every 10^7 memory B cells both in the peripheral blood (mean $110/10^7$) and in Waldeyer's ring (mean $175/10^7$—the virus is evenly distributed throughout the ring) (Laichalk et al. 2002). The level of infected cells is similar between peripheral blood and Waldeyer's ring but at least 20-fold lower in the other lymphoid tissue tested (spleen and mesenteric lymph node) (Laichalk et al. 2002), suggesting preferential homing to the lymphoepithelium of Waldeyer's ring. Based on these measurements, the total body load calculates to 10^4–10^7 (mean 0.5×10^6) infected memory B cells per person representing a small, stable, and, most critically, "safe" pool of infected cells that guarantees long-term persistence. Only ~1 % of these cells reside in the peripheral blood.

3.3.2 Memory B Cell Homeostasis—The Maintenance of Long-Term Memory and Persistent Infection

The survival of memory B cells requires a tonic signal from the BCR (Maruyama et al. 2000), and the number of cells is controlled by homeostasis mediated by cytokines such as BAFF (Mackay and Schneider 2009; Stadanlick and Cancro 2008). The tonic BCR signal can be completely replaced by LMP2 (Caldwell et al. 1998), raising the possibility that persistently infected cells could be BCR independent. However, this is not the case, infected cells in the periphery do not express LMP2 (Hochberg et al. 2003a), and as already noted, they express a functional, possibly, antigen-selected BCR. A number of independent lines of evidence suggest that memory B cells latently infected with EBV are also maintained by homeostasis:

1. EBV-infected memory B cells in the periphery of adult humans are >90 % in a resting state, but at any given time, around 2–3 % of the cells are undergoing cell division (Miyashita et al. 1997; Hochberg et al. 2004). This is exactly the same rate as has been reported (2.7 %) for normal memory B cells (Hochberg et al. 2004; Macallan et al. 2005; Miyashita et al. 1997).
2. The half-life of both EBV-infected and EBV-uninfected memory B cells is virtually identical –7.5 ± 3.7 days (Hadinoto et al. 2008) and 11 ± 4 days (Macallan et al. 2005), respectively.
3. Latently infected memory cells in the periphery express no viral latent proteins. Therefore, when they divide, it must be driven by normal homeostasis signals.

We may conclude therefore that the pool of latently infected memory B cells is indistinguishable to the host from normal memory B cells.

When EBV-infected cells divide, they must express EBNA1 because the viral genome cannot replicate in its absence (Yates et al. 1985). Predictably, therefore, latently infected memory cells in the periphery express EBNA1 when they undergo cell division (Hochberg et al. 2003a). This is the third example we will encounter of a latent gene transcription program used in lymphoma, being found in a normal infected B cell counterpart in vivo. In this case, the EBNA1 only program, which is used in Burkitt's lymphoma (BL), is found in dividing, latently infected memory B cells in the blood (Table 1, Fig. 5). EBNA1 expression during cell division is the only potential point of attack for the immune system against the pool of latently infected memory cells. It is perhaps not surprising therefore that EBNA1 has evolved so as not to be processed and presented efficiently to the immune system [Levitskaya et al. (1995, 1997) reviewed recently in Daskalogianni et al. (2014)], thus minimizing the risk of attack.

3.4 Viral Replication—Plasma Cell Differentiation, Stress, and the Role of Epithelial Cells

3.4.1 Terminal Differentiation—Maintenance of Stable Antibody Production and Viral Shedding

The last component of persistent infection to be discussed is that infectious virus is continuously shed into the saliva (Golden et al. 1973; Hadinoto et al. 2009). Unlike the level of latently infected memory cells, which is strikingly stable over long time periods, virus shedding fluctuates dramatically. The level of shedding is relatively stable over short periods (hours–days) but varies through 3.5–5.5 orders of magnitude over longer periods (Hadinoto et al. 2009). This variation means, contrary to what is generally believed, that the definition of high and low shedder is not so much a function of variation between individuals but within individuals over time. Also an important but simple insight, that had gone unrealized in the field, was that EBV shedding into saliva must be continuous and rapid. This is because the virus must be replaced ~2 min which is how frequently, on average, a normal individual swallows. Thus, the mouth is not, as often cited, a reservoir of virus but a conduit through which a continuous flow stream of virus passes in saliva (Fig. 8). Consequently, virus is being shed at a much higher rate than is generally appreciated.

Memory B cells, transiting the nasopharyngeal lymphoid tissue, presumably must occasionally initiate virus replication and release the virus. From cell surface phenotyping of fractionated tonsil cells, it is clear that the B cells replicating the virus in the lymphoepithelium of the tonsils are plasma cells (CD38hi, CD10−, CD19−, CD20lo, sIg−, and cIg+) (Laichalk and Thorley-Lawson 2005), a conclusion consistent with histological observations (Niedobitek et al. 2000; Anagnostopoulos et al. 1995). Quantitative estimates suggest that somewhere in the region of ~250 cells are undergoing replication in Waldeyer's ring at any one time (Hawkins et al. 2013; Laichalk and Thorley-Lawson 2005). However, sequentially fewer cells express the immediate early, early, and then late antigens of the lytic cycle such that only ~10 % of the cells complete the replicative cycle. Thus, only a handful of B cells are actually releasing virus in Waldeyer's ring at any given time. This sequential diminution in the numbers of cells replicating the virus as they proceed through the cycle may indicate that replication is frequently abortive or may be the result of aggressive immunosurveillance by CTL

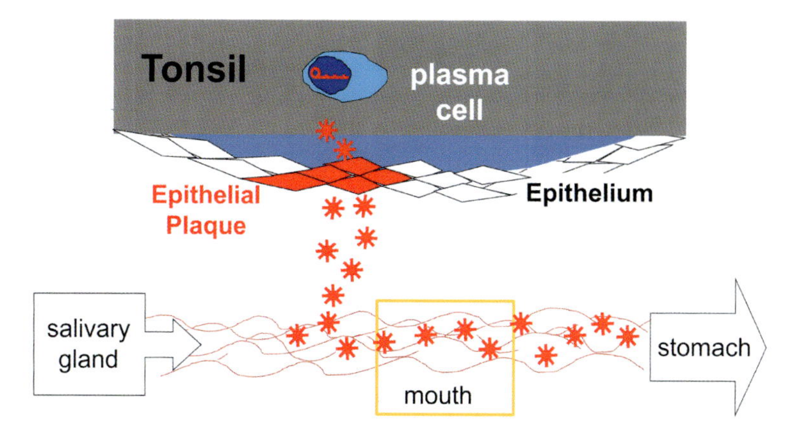

Fig. 8 A model of EBV reactivation and shedding. The known data fit a model where a single latently infected memory B cell in the tonsil occasionally differentiates into a plasma cell and releases virus that infects epithelial cells. The infection spreads exponentially through the epithelium, resulting in the shedding of virus. The plaque is eventually eliminated by the immune response. Meanwhile, another plaque initiates elsewhere in the Waldeyer's ring. The data are consistent with their being no more than three such plaques in Waldeyer's ring at any one time. Virus is continuously shed into the mouth where it mingles with saliva for about 2 min before being swallowed. Thus, the mouth is a flow stream of EBV not a static reservoir

(Callan et al. 1998b). This despite mechanisms that the virus employs during its lytic cycle to reduce CTL surveillance (Ressing et al. 2008). For a detailed discussion of immune evasion by EBV, see the chapter authored by Emmanuel Wietz.

It has been shown that differentiation into plasma cells, and not the signals that induce differentiation, initiates viral replication (Laichalk and Thorley-Lawson 2005). Again, the biology of the virus is intimately tailored and responsive to normal B cell biology. This was confirmed by in vitro studies in cells showing that the promoter for BZLF1, the gene that begins viral replication, becomes active only after memory cells differentiate into plasma cells, that it is active in plasma cell lines and is activated by the plasma cell-associated transcription factors XBP-1 and Blimp1. The molecular mechanism behind this activation process has been comprehensively reviewed recently (Kenney and Mertz 2014). For a detailed discussion, see the chapter authored by Ayman El-Guindy.

The signal that causes latently infected memory B cells to undergo terminal differentiation is unclear. It has been suggested that immunological B cell memory may be sustained through bystander T cell help (Bernasconi et al. 2002) such that a memory B cell transiting through a lymph node will, when it encounters bystander T cell help, undergo a cell division that will generate one memory cell and one plasma cell. This ensures the stability of the memory pool, while a continuous supply of plasma cells is produced that will guarantee stable production of antibody. Applied to EBV, this could explain how the population of latently infected memory cells could be maintained for years, while, through the generation of plasma cells, virus can also be continuously produced.

An alternate hypothesis is that the generation of plasma cells replicating EBV is stimulated by cognate antigen and T cell help. This hypothesis has the attractive feature that latency would need to be established in antigen-specific memory cells in the tonsil. These cells would then enter the peripheral circulation where they would maintain persistent infection. As these cells reenter secondary lymphoid tissue, the site where they would most likely reencounter cognate antigen would be the tonsil. This would provide a mechanism for preferential homing and reactivation of the latently infected memory cells in the tonsil compared to other lymph nodes. Although it is difficult to conceive of a mechanism by which the virus could access antigen-specific naive B cells with a high enough probability and frequency to be feasible, this model is very consistent with the observation that the latently infected memory cells appear to bear antigen-selected BCRs.

3.4.2 Stress—An Alternate Pathway to Viral Replication

Indications of a second pathway to viral replication come from in vitro studies that a number of stress-inducing agents including TGFb and chemotherapy agents, BCR cross-linking and hypoxia can also initiate viral replication in cell lines (Kenney and Mertz 2014). In these systems, however, there is acute activation of the BZLF1 promoter within minutes of receiving the stimulus, and not surprisingly, the cells do not undergo plasma cell differentiation prior to viral replication. In a similar fashion, explanted infected peripheral memory cells will acutely undergo spontaneous reactivation (Rickinson et al. 1977), presumably in response to the stress induced upon being placed in culture. What these systems have in common is the induction of apoptosis in the B cells in response to the stress signal (Inman et al. 2001). However, EBV encodes a homologue of the antiapoptotic gene bcl-2 that is expressed during viral replication in vitro (Henderson et al. 1993) and this protects these cells from stress-induced death and apoptosis, while the virus replicates (Inman et al. 2001). It is known that B cells are particularly prone to apoptosis. It seems therefore that in addition to replication in plasma cells located in the epithelium of the tonsil for infectious spread, the virus has developed an escape hatch that allows it to exit any infected B cell that may begin to die by apoptosis.

3.4.3 Replication in Epithelial Cells

Although we lack a direct demonstration that EBV replicates in epithelial cells in vivo, the indirect evidence is compelling:

1. The strongest evidence comes from numerical arguments. Put simply, there are not enough B cells replicating the virus in Waldeyer's ring to account for either the amount or extreme variability of EBV shedding in saliva. For a detailed discussion of the numbers, see Hadinoto et al. (2009). The dynamics of virus shedding is most simply explained by single B cells sporadically releasing virus that infects neighboring epithelial cells (Fig. 8). (This mechanism is

analogous to the neurotropic herpesviruses (HSV and VZV) that persist silently in ganglia but when reactivated travel down the neurons to replicate in fibroblasts.) Epithelial infection by EBV spreads at an exponential rate and is terminated randomly, resulting in infected plaques of epithelial cells ranging in size from 1 to 10^5 cells, more than sufficient to account for the observed rate of shedding. At any one time, there would be a very small number (≤ 3) of such infected epithelial plaques in the entire Waldeyer's ring that would be transient and usually small, explaining why they have previously gone undetected.

2. Cell cultures of primary epithelial cells from tonsils carry already infected cells that are both latently infected and replicating the virus (Pegtel et al. 2004).
3. EBV is found in oral hairy leukoplakia which represents a lesion where EBV is actively replicating in the epithelium of the tongue (Greenspan et al. 1985). This indicates that EBV can replicate in epithelial cells in vivo.
4. The glycoprotein patterns on the virus differ depending on whether the virus emerges from a B cell or an epithelial cell (Borza and Hutt-Fletcher 2002). This happens in such a way that the virus bears an epithelial tropic pattern of viral glycoproteins when it emerges from B cells and a B lymphotropic pattern when it emerges from epithelial cells. These results imply that the virus has evolved to efficiently shuttle back and forth between epithelial and B cells.
5. A unique receptor, $\alpha 5\beta 1$ integrin, for EBV is expressed on epithelial cells that allows infection only on the basolateral surface (Tugizov et al. 2003). As with the glycoprotein patterns, this implies that epithelial cell infection by EBV is only used in one direction, in this case specifically restricted to exit by the virus.

Taken together, this evidence presents a strong circumstantial argument that tonsillar epithelium is actively infected with replicating EBV as an ongoing part of normal viral persistence and provides an explanation for the presence of the virus in the associated diseases of epithelial cells.

4 The Cyclic Pathogen Refinement of GCM

If a biological model is correct, then it should be logically rigorous and able to be expressed mathematically. Mathematical modeling is not really that different from how biology has always been done, it is just a more rigorous way to organize data and a more logical way to make testable predictions based on hypotheses. However, most biological systems are not well enough characterized quantitatively to be amenable to this type of analysis. This is under appreciated by biologists who tend to see the failing of modeling (despite its obvious utility in other more quantitative sciences such as physics and engineering) as a consequence of the limitations of modeling itself rather than the lack of rigor in understanding the biological system being studied. Persistent EBV infection is an exception.

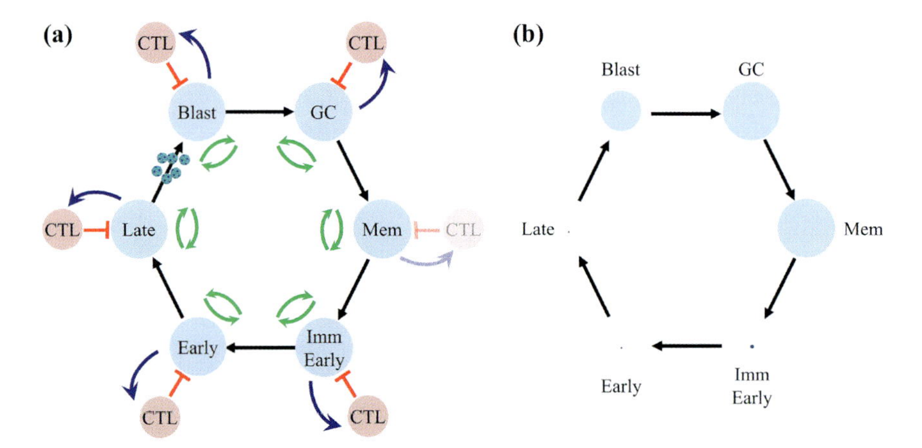

Fig. 9 The cyclic pathogen model (CPM). **a** CPM is a mathematical description of the GCM. It consists of a cycle of 6 infected stages (*blue circles* based on the biological GCM illustrated in Fig. 4). These are blast, GC, memory and immediate early, early and late lytically infected B cells, each of which is potentially controlled by the immune response (*red circles*). The single lytic stage in the GCM is broken down into three discrete stages which are known to be recognized independently by the immune response. Biologically, there is never a CTL response against the memory stage; however, the model allows analysis of theoretical conditions such as the memory compartment being regulated by CTL. This model can be described by a system of differential equations employing rate constants for the stimulation of CTL (*blue arrows*), killing of CTL targets (*red arrows*), and the proliferation and death of each stage (*green arrows*). For this system, there is one and only one mathematical solution that is stable and biologically credible. This solution accurately describes biologically persistent infection. **b** Shows the infected populations as *circles* whose area is proportional to their frequency within all tonsils ($1:5:1.5.10^2:10^4:10^4:0.5.10^4$, Late:Early:ImmEarly:Memory:GC:Blast). This highlights the very large range in the sizes of these populations

The GCM, as generalized in Fig. 9, can be described by a system of differential equations—the cyclic pathogen model (CPM) (Delgado-Eckert and Shapiro 2011) for which there is one and only one solution that is stable and biologically credible. We have sufficient quantitative information to be able to know, derive precisley, or estimate approximately values for all the parameters (rate constants) governing these equations. When solved with this parameter set, the model very precisely replicates the actual dynamics of the infection (Hawkins et al. 2013). This includes predicting which and to what extent each infected stage is recognized by CTL and even precisely predicting the expected sizes of the infected memory and GC populations and the extent to which they vary between infected individuals. Furthermore, when marginally non-biological values are assigned to parameters, the model fails to replicate infection. This is an important result that seems to have gone unappreciated in the biological community. The chances that one could randomly pluck a complex model such as the one shown in Fig. 9 and have it predict correctly when and only when biological values are applied are vanishingly small. The fact that this model works so well is a convincing argument for the biological accuracy of the GCM in explaining EBV persistence.

The mathematical description of the cyclic pathogen model and its subsequent analysis also provided important new insights including:

1. There are two possible mechanisms for EBV persistence in B cell memory. In one, the virus persists through homeostasis independently of new infection and addition to that compartment. In the second (predicted by the CPM), it is the cycle of infected states that accounts for persistence. Aggressive intervention with antivirals should distinguish these since they should have no impact on the memory compartment if the former is true but will reduce the overall level of viral infection if the latter is true. Indeed, long-term treatment with antivirals, which dramatically reduce viral shedding, produced a parallel decline in the level of infected memory B cells (Hoshino et al. 2009). This confirms the prediction from the CPM. For a detailed discussion of antiviral interventions, see the chapter authored by Richard Ambinder.
2. Based on the same arguments, CPM predicts that an effective vaccine against primary EBV infection will also be effective over time in reducing and eventually eliminating persistent infection because it will interdict the cycle of infection required for long-term persistence. For a detailed discussion of vaccine strategies, see the chapter authored by Rajiv Khanna.
3. To a biological eye, it is apparent that EBV persists because it can attain latent infection of resting memory B cells that are invisible to the immune system. However, the CPM provides a different interpretation, fully compatible with all the biological data, namely that it is the cycle of infection that allows persistence. Persistence is possible even if the memory compartment were highly immunogenic; however, the overall structure and dynamics of the persistent infection would look nothing like what is actually observed (in passing, it is worth noting the utility of modeling in allowing such biologically impossible experiments to be performed mathematically). This is a further validation of the accuracy of the mathematical model. Thus, access to the immunologically protected memory compartment defines the overall pattern, features, and dynamics of persistent infection but alone does not account for it.
4. It explains how infection can be stable at a very low level. This is crucial for both the host and the virus because it imposes the minimum burden on the host within which EBV wants to persist for life. For an average person, there is ~1 infected cell per 5 ml of blood. Such a low level of infection leaves the virus vulnerable to extinction through stochastic variation, yet the value only varies by a factor of perhaps ± 25 % over many years (Hadinoto et al. 2009). This is because the cycle of infection ensures that an obliterated population can be rapidly repopulated returning the system to the same equilibrium as before.
5. The absolute levels of infection are defined by the level of the immune response against viral proteins. This includes cytotoxic responses to infected cells expressing latent and lytic proteins and neutralizing antibody against infectious virus. The prediction that the immune system only moderates the overall viral load, not the form of persistence, is confirmed in studies of immunosuppressed individuals. Here, in the presence of a minimally effective

immune response, the levels of virus-infected memory B cells increase on average 50-fold (Babcock et al. 1999). However, the regulation of viral persistence is intact and the virus in the blood remains restricted to resting memory B cells. This means that the immune response per se plays no role in regulating the mechanisms of viral persistence, but it only regulates absolute levels of the infection.

6. That the system is a simple circle is amply demonstrated by studies on acutely infected individuals. Here, the system is allowed to run unchecked until the immune response is activated. In this case, as many as a staggering 50 % of all memory cells may become latently infected with EBV until the immune system begins to reduce the overall load of infection (Hadinoto et al. 2008; Hochberg et al. 2004). Impressively, the regulation still holds and the virus remains restricted to resting memory cells in the blood again highlighting that the immune system only functions to regulate the level not the form of the infection.

5 The Model of Persistence—A Summary

In summary, persistent infection by EBV can be seen as a self-perpetuating circle of infection, differentiation, persistent infection, reactivation, and reinfection (Figs. 4 and 9) that exploits virtually every aspect of mature B cell biology. The expansion of the virus is counterbalanced by the immune response. It is this cycle of infection together with the quiescent infection of peripheral memory B cells that allows the virus to be maintained at the extremely low and stable levels characteristic of persistent infection. In doing so, EBV does not disrupt the normal processing of latently infected cells into memory, and in so far as the presence of the virus may cause deviations from normal B cell biology, they are not detectable by the time the cells enter the memory compartment.

6 Disease Pathogenesis—Insights from the GCM

The GCM explains that EBV needs to transit the GC to access the resting memory compartment. EBV-infected and GC B cells are tightly regulated because they both proliferate rapidly—a risk factor for cancer. GC cells also actively undergo DNA breakage and mutagenesis during CSR and SHM, additional risk factors for tumor development and the production of autoreactive B cells. Furthermore, in the GC, EBV expresses LMP1, a growth-promoting potential oncogene, and LMP2, a pro-survival molecule able to rescue autoreactive B cells. Thus, the presence of EBV in GC B cells presents a nexus for disease risk, especially cancer (EBV-positive Hodgkin's disease and Burkitt's lymphoma both arise from EBV-infected GC cells) and autoimmunity. It is not surprising, therefore, that EBV has been linked with a number of such diseases.

6.1 Infectious Mononucleosis—Acute Infection (AIM)

Delayed infection by EBV can cause infectious mononucleosis (AIM). Why adolescence and adults get AIM is not clear. It is likely immunopathologic in nature, meaning the disease symptoms are caused by the inflammatory response of the immune system rather than the virus itself. For a detailed discussion of AIM, see the chapter authored by Kristin Hogquist. The intensity of the disease varies but can last for weeks or months before finally resolving (Hoagland 1967). IM is characterized by a lymphocytosis (Wood and Frenkel 1967) due to the appearance of large numbers of "atypical" lymphocytes which are predominantly CD8+ T cells, representing a vigorous CTL response to the virus (Strang and Rickinson 1987; Callan et al. 1998b).

6.1.1 AIM and the GCM

Virologically and immunologically, we know nothing about what is happening in the newly infected host until they arrive at the clinic with symptoms some 5 weeks into the infection (Hoagland 1964). We may assume though that when the virus initially infects, there is nothing to control the cycle of infection, latency, reactivation, and reinfection, shown in Fig. 4. Consequently, the memory compartment begins to fill up with latently infected B cells. A staggering level of infection is achieved that can reach ≥ 50 % of all memory B cells (Hochberg et al. 2004). Despite this overwhelming invasion of the B cell compartment by EBV no cells expressing the lymphoblastoid form of latency/infection are detected in the periphery, the virus remains restricted to resting memory B cells. This is fully consistent with the GCM which predicts that the lymphoblastoid form of latency is restricted to the lymphoid tissue and tightly regulated such that the cells rapidly transit into the GC to become memory cells before entering the circulation.

By the time patients experience symptoms and arrive at the clinic, the infection is always resolving (Hadinoto et al. 2008). Viral shedding and the levels of newly infected B cells are all falling. All that is left is the massive level of infection in the memory compartment. Since these cells are not seen by the immune response, their levels decrease simply by attrition as they initiate viral replication and are immediately killed by CTLs that recognize the immediate early lytic antigens. Consequently, at this time, as many as half of all the CTLs in the body are directed against EBV-infected cells expressing these targets (Callan et al. 1998b). It is most likely that this destruction of large numbers of infected B cells is responsible for the inflammatory response leading to the fever and malaise characteristic of IM.

There then ensues a parallel decrease in the number of latently infected memory B cells (dying as they enter viral replication and are destroyed by CTL) and the number of CTLs that they stimulate (Catalina et al. 2001; Hadinoto et al. 2008). For the next few weeks, there is an exponential decrease in the levels of latently infected memory B cells (half-life ~7.5 days) and CTL against the IE proteins (half-life ~73 days). Eventually, the level of infected memory B cells drops to a point where the rate of attrition is matched by the steady-state low-level production of newly infected memory B cells. At this time, the level of infected memory cells and CTL against IE proteins begins to stabilize. Thus, the acute infection is eventually limited by the immune response but at an excruciatingly slow rate. The ensuing events are strikingly different from infection with most other viruses. Shedding of EBV does not stop, but continues for life (Hadinoto et al. 2009; Yao et al. 1985). B cells expressing the growth and default programs persist in Waldeyer's ring (Babcock et al. 2000), and latently infected memory cells, expressing the latency program, remain in the blood for life (Babcock et al. 1998; Khan et al. 1996). At the same time, levels of neutralizing antibodies (Henle and Henle 1979) and CTL (Callan et al. 1998a; Steven et al. 1996) also continue at significant and stable levels for the lifetime of the host. For a detailed discussion of the immune responses that regulate EBV, see the chapters authored by David Nadal, Martin Rowe, Jaap Middeldorp, and Andrew Hislop et al. It is apparent therefore that the immune response ameliorates and counterbalances the infection but never clears it. An equilibrium is established between the immune response and the various states of viral latency allowing the virus to persist at stable levels without causing significant impairment to the host. The CPM states that these countervailing forces are responsible for maintaining the stable fixed point that mathematically describes EBV persistence.

6.1.2 Why Do Adolescents Get AIM?

The age dependence of symptoms has led to the suggestion that IM is a disease of a mature immune response. What this means mechanistically is less clear. It has also been suggested that the tonsil is an immune-privileged site during AIM, since the majority of circulating CTLs at this time lack the requisite mucosal homing receptors to enter the tonsil (Hislop et al. 2005). However, a much simpler and more likely explanation for this observation is that the majority of EBV-infected cells reside in the spleen during acute infection as evidenced by the well-documented symptom of splenomegaly. Thus, most of the CTL should be homing to this non-mucosal site until the infection resolves.

A more compelling explanation for AIM is the theory of heterologous immunity (Clute et al. 2010; Selin et al. 1998). In this theory, the CTL response to new infections early in life is by naive T cells which produce high-affinity CTLs that efficiently and rapidly clear the virus infection and then become memory CTL. As the organism ages, memory CTLs accumulate to a variety of pathogens. Exposure to a novel infection later in life is more likely to trigger cross-reacting memory CTL than naive CTL. These memory CTLs are of lower affinity and may produce an ineffectual response allowing extensive viral replication and spread that triggers a massive inflammatory response that lasts until the virus is finally brought under control.

6.2 Autoimmune Disease

EBV has variously been linked with a number of autoimmune diseases including SLE (James et al. 1997), rheumatoid arthritis (Lotz and Roudier 1989), Sjorgen's syndrome (Fox et al. 1987), and most recently and aggressively with multiple sclerosis [reviewed in Ascherio and Munger (2010)]. For a detailed discussion of EBV and autoimmune disease, see the chapter authored by Alberto Ascherio. The driving concept behind these associations is the knowledge that EBV can cause the activation and proliferation of infected B cells in an antigen-independent fashion and that such cells are immortal in tissue culture. This raises the possibility that, by infecting them, EBV infection could allow the survival of autoreactive clones of B cells. However, EBV does not persist by immortalizing B cells in vivo, but by differentiating the infected cells into a resting memory state. Consequently, the fundamental rationale for the association must be modified to suggest that the EBV latent genes may rescue a forbidden clone from a GC into the memory compartment. Such a latently infected cell itself could not produce antibodies because upon plasma cell differentiation it would produce infectious virus and die. However, theoretically, such a cell could present autoimmune antigens and break tolerance. There is evidence to support this idea in that LMP2, which is expressed in the GC, is capable of breaking tolerance in a mouse model of autoreactivity (Swanson-Mungerson and Longnecker 2007; Swanson-Mungerson et al. 2005), and has even been shown to exacerbate disease in a mouse model of MS (Chang et al. 2012). Such behavior must be anomalous, however, because the expressed immunoglobulins of latently infected memory cells in healthy humans are, if anything, skewed away from self-reactivity (Tracy et al. 2012).

Demonstrating experimentally a causative role for EBV in autoimmune disease is difficult because infection usually occurs early in life and by adulthood >90 % of the population is infected. Analysis is further complicated by the fact that EBV is carried in the peripheral circulation by infected memory B cells, so sensitive tests will detect EBV in any inflamed tissue, regardless of the virus's role in causing the inflammation. The GCM/CPM adds another layer of complication which is that EBV uses virtually every aspect of mature B cell biology to establish and maintain persistent infection and virus shedding which is counterbalanced by the immune response to produce a defined stable level of infection. The corollary is that EBV is exquisitely sensitive to changes in the immune system. Any disease that affects the immune system will have an impact on the regulation of EBV persistence. This could result in an increase in the numbers of infected cells in the blood (peripheral blood burden) and/or an increase in virus shedding. Thus, changes in virological or immunological parameters of EBV infection associated with an autoimmune disease are most likely an indirect effect of a compromised immune system caused by the disease as is the case with SLE (Gross et al. 2005) rather than a cause of the disease.

6.3 Cancer

The motivating force behind associating EBV with cancer is obvious. It is well established that EBV has latent proteins that can drive cellular proliferation, at least in B lymphocytes, and it is highly likely that inappropriate or deregulated expression of these genes could play a causative role in tumor development. EBV-associated cancers fall into three discrete groups.

1. Tumors for which there are claims that remain to be substantiated. These would include, but not be limited to, breast (Bonnet et al. 1999) and hepatocellular carcinoma (Sugawara et al. 1999). In these cases, the doubt usually exists through the inability of investigators to reproducibly detect the virus in the tumor cells. Since the assays used are usually based on either PCR or immunohistochemistry, they are subject to the vagaries of those techniques which include a high level of false positives and dependence on technical skill to perform well-controlled studies. Thus, it becomes difficult to resolve whether conflicting results are caused by false-positive artifacts or technical inconsistencies.
2. Tumors for which there is strong supportive evidence, but the tumors arise in cell types for which no latently infected biological equivalent has been established. Such tumors may arise through accidental infection leading to inappropriate viral latent gene expression. This includes such tumors as nasopharyngeal (Raab-Traub 2002), gastric (Shibata et al. 1991; Shibata and Weiss 1992), and salivary gland (Raab-Traub et al. 1991) carcinomas. For a detailed discussion of EBV-positive carcinoma, see the chapter authored by Nancy Raab-Traub. Also included are leiomyosarcoma (van Gelder et al. 1995; Timmons et al. 1995) and T and NK lymphomas (Chiang et al. 1996; Tao et al. 1995). In these cases, there is a high degree of correlation between the disease and EBV [e.g., 100 % of undifferentiated NPC contains EBV (Andersson-Anvret et al. 1977)] and reproducible detection of the virus in the tumor cells. Since the viral episome is lost from cells absent some selective pressure for its retention (Kirchmaier and Sugden 1995), the consistent presence of viral DNA in any tumor is *prime facie* evidence that the virus is playing a role in the growth/survival of the tumor cells (Vereide and Sugden 2009). These tumors also frequently express the latent gene LMP1 which is known to be highly oncogenic when constitutively expressed (Baichwal and Sugden 1988; Moorthy and Thorley-Lawson 1992; Nicholson et al. 1997; Uchida et al. 1999; Wang et al. 1985). This type of tumor is relevant to the GCM which posits that EBV latent gene expression patterns have evolved to be regulated in concert with normal B cell activation and differentiation, with the ultimate goal of establishing persistent infection in a memory B cell where the latent genes are no longer expressed. It follows that if EBV fortuitously gains access to a cell type which is not a natural target of infection, i.e., a non-B cell, this could lead to aberrant latent gene expression that would not be regulated appropriately. This could result in constitutive expression of LMP1 for example.

3. Tumors for which there is good evidence linking EBV. These are the lymphomas IL, HD, and BL. There is convincing epidemiological, serological, and molecular biological evidence associating EBV with these tumors. The GCM provided the first and, to date, only explanation for the origin of these lymphomas and the reason they express restricted patterns of latent proteins (Thorley-Lawson and Gross 2004). Indeed, it is supportive of the GCM and cannot be a coincidence that tumors arise from each of the three proliferative stages of EBV infection predicted by the model (Fig. 5). These are IL from cells expressing the growth program (new infection), HD from cells expressing the default program (GC cells), and BL from cells expressing EBNA1 only (late GC cell).

6.3.1 Lymphoma in the Immunosuppressed—IL

Patients who are immunosuppressed are at risk for diseases such as post-transplant lymphoproliferative disease (PTLD) in organ transplant patients and immunoblastic lymphoma in AIDS patients. These are a heterogeneous collection of B cell disorders [reviewed in Hopwood and Crawford (2000)] that usually carry the virus and express the growth program (Thomas et al. 1990). The obvious explanation for PTLD is that suppression of the immune response allows uninhibited growth of EBV-infected cells; however, it is not that simple. If EBV was able to freely drive cell growth in the absence of an immune response, there would be several consequences. First, all immunosuppressed patients who are EBV-infected should develop the disease. Second, it should be a polyclonal and disseminated disease since EBV would drive the growth of many infected cells throughout the body. Finally, the lymphomas would be expected to arise most frequently in the places where infected cells are known to express the growth program—Waldeyer's ring (Joseph et al. 2000a). In reality, the disease has none of these features. Only a small fraction of patients (0.1–10 % depending on the setting) develop the disease, it is usually oligoclonal arising from one or a few infected cells, and it often occurs in extranodal sites such as the brain and gut (Penn 1998; Hopwood and Crawford 2000). This indicates that the disease is not simply EBV-driven growth but involves a rare event where EBV infection has gone wrong.

The GCM states that the growth program of EBV is used specifically to activate newly infected naïve B cells in Waldeyer's ring, so they can then differentiate into resting memory B cells. It follows that for a lymphoblastoid cell, proliferating due to the growth program, to survive and evolve into a lymphoma, the cell must be unable to exit the cell cycle. This could occur if infection of the wrong B cell type or in the wrong location occurs (inappropriate infection).

IL in the immunosuppressed is therefore a consequence of two events. The rare specific event is the expression of the growth program in a B cell that cannot exit the cell cycle. The global event is immunosuppression that prevents the elimination of these rare cells. At this stage of the disease, the tumor cells are still susceptible to immunosurveillance and regression can be achieved by reducing immunosuppression (Starzl et al. 1984) or by treatment with autologous CTL (Rooney et al. 1995). For a detailed discussion of the application of adoptive transfer for treatment of EBV tumors, see the chapter authored by Stephen Gottschalk and Cliona Rooney. However, in the absence of T cell immunity, the proliferating cells acquire additional genetic damage and more malignant clones arise (Knowles et al. 1995). These cells ultimately become unresponsive to reduced immunosuppression or immunotherapy and are usually fatal.

6.3.2 Hodgkin's Disease (HD)

For a detailed discussion of HD, see the chapter authored by Paul Murray and Andy Bell. HD is a tumor of germinal center cells (Kuppers 2012; Kuppers and Rajewsky 1998) and is characterized by the unusual Hodgkin's Reed–Sternberg (HRS) tumor cells. AIM and elevated antibody titers to EBV are both risk factors for HD (Ambinder 2007; Henle and Henle 1979; Hjalgrim et al. 2007), and up to 40 % of the tumors contain EBV (Glaser et al. 1997). The virus in the tumors is clonal and expresses the default transcription program (Oudejans et al. 1996; Deacon et al. 1993; Herbst et al. 1991; Niedobitek et al. 1997), the same transcription program used by latently infected GC B cells (Babcock et al. 2000). Thus, the cell origin and the viral gene expression data agree that HD arises from an EBV-infected GC B cell expressing the default program (Fig. 5). In effect, the viral gene expression pattern in HD is not created within and selected by the tumor, but is a natural consequence of the cellular origin of the tumor. The presence of EBV in ~40 % of the tumors would seem to rule out a chance association of the virus with the tumor. But this does not take into account that the levels of EBV-infected B cells reach extremely high levels during AIM, frequently 10–50 % (Hochberg et al. 2003b), so there is a very high probability (as high as 50 %) that the premalignant GC cell will have EBV in it by chance. Therefore, it remains a possibility that it is the immunological disruption of AIM which is the risk factor and EBV is simply a passenger that plays no role in tumor development.

However, retention of the virus in HD strongly argues that it must be contributing something to tumor cell survival/growth (Vereide and Sugden 2009). One specific contribution has been identified for the subset of HD tumors that express immunoglobulin genes crippled by mutation. These cases are almost universally EBV positive (Bechtel et al. 2005) and express LMP2 which has been shown independently to replace the missing BCR-derived tonic signal necessary for the survival of B cells with crippled BCRs (Mancao and Hammerschmidt 2007).

6.3.3 Burkitt's Lymphoma (BL)

For a detailed discussion of BL and diffuse large B cell lymphoma, see the chapters authored by Ann Moorman and Rosemary Rochford and Sandeep Dave. BL has the pedigree of being the tumor in which EBV was originally discovered, but its contribution to BL still remains enigmatic. The defining genetic lesion in BL is deregulated activation of the c-myc oncogene due to reciprocal translocation with one of the immunoglobulin genes (Klein 1983; Leder 1985; Manolov and Manolova 1972). BL can occur without EBV, and expression in transgenic mice of c-myc, in the context of the immunoglobulin translocation, has been shown to be sufficient to produce Burkitt's lymphoma-like tumors (Kovalchuk et al. 2000), suggesting that deregulated c-myc is sufficient to produce the tumors. This raises the question as to what role EBV may play. The most compelling evidence of EBV's involvement in BL is the retention of the genome by the tumors (Vereide and Sugden 2009) and the high frequency of tumors carrying the virus (de-Thé 1985) in the endemic (eBL) regions of Africa (>95 % contain EBV DNA). The frequency is lower (15–85 %) in the sporadic form of the tumor (sBL). The presence of clonal EBV in the tumors (Gulley et al. 1992) has also been interpreted as evidence of EBV's role, but in actuality, this only means that EBV was present prior to the last event that produced the tumor. Curiously, none of the viral growth-promoting latent genes are expressed in the tumor cells, the only latent protein present being EBNA1 (Gregory et al. 1990), along with the non-coding small RNAs EBERs and the microRNAs. Transcriptionally, this looks like a situation where the virus is just along for the ride since EBNA1 has to be expressed to allow duplication of the viral genome. However, there is evidence for all the EBV genes expressed in BL that they may contribute to pathogenesis, usually by limiting sensitivity to apoptosis (Iwakiri 2014; Kennedy et al. 2003; Vereide et al. 2014; Wilson et al. 1996).

What is required for understanding how EBV may predispose to BL is an explanation for why only EBNA1 is expressed. In the case of HD, we have shown that the default program is expressed because the tumor derives from an EBV-infected GC cell and the default program is what the virus naturally expresses in a GC cell. Applying this thinking to BL, there is currently only one way known to produce the "EBNA1-only" phenotype of BL in a non-tumor cell. This is when a latently infected GC cell that becomes a memory cell expressing the latency program, i.e., no latent proteins, divides, as part of normal B cell homeostasis (Figs. 3 and 4). At this time, the virus turns on expression of "EBNA1 only" to ensure replication of the viral DNA with the cell. BL has the phenotypic (Gregory et al. 1987) and gene expression profile of a light zone (LZ) GC cell (Victora et al. 2012). The LZ is where GC cells express c-myc (Dominguez-Sola et al. 2012) before they begin to proliferate again and also where GC cells reside prior to exit. Thus, this would be the location where viral gene expression would be expected to shut down prior to exiting the GC. Lastly, although BL has the characteristics of a GC cell, the tumor actually grows in extrafollicular locations (Klein et al. 1995). Therefore, a consistent scenario is that BL is derived from a LZ GC cell that has left the follicle to become a resting memory cell but cannot achieve this because it continues to proliferate due to an activated c-myc and therefore constitutively expresses EBNA1 only. This scenario also accounts for the presence of clonal EBV in the tumors because the virus would already be present when the major transformation event, c-myc translocation, occurs.

There are two major infectious players in the predisposition to eBL: malaria and EBV. Recently, experimental evidence has been presented to account for this based on the GCM of EBV and the notion that BL arises from a latently infected GC B cell (Thorley-Lawson and Allday 2008; Torgbor et al. 2014). It is known that expression of the growth program that occurs prior to entry into the GC includes the epigenetic silencing of proapoptotic functions, including bim, an important regulator of myc-induced apoptosis (Allday 2009, 2013). This, together with the antiapoptotic activities associated with the EBNA1, EBERs and microRNAs expressed in latently infected GC cells leave these cells more resistant to apoptosis induced by a translocated/deregulated myc gene. The myc translocation itself is believed to be mediated by the enzyme AID which is uniquely expressed in the GC (Ramiro et al. 2004; Robbiani et al. 2008). Infection by *P. falciparum* malaria has two consequences. First, it increases the viral burden of EBV, resulting in higher numbers of latently infected cells transiting the GC and able to resist apoptosis. Second, it drives deregulated expression of AID in B cells, potentially increasing the frequency of translocation events (Torgbor et al. 2014). Taken together, the increased levels of virus-infected cells and rate of myc translocations in the GC induced by malaria can account for the close association of eBL with malaria and EBV.

6.3.4 X-linked Lymphoproliferative Disease—XLP

For a detailed discussion of EBV infection in primary immunodeficiencies, see the chapter authored by Jeffrey Cohen. XLP is a rare X-linked immunodeficiency (Purtilo et al. 1975; Seemayer et al. 1995) which frequently results in lymphoma or fulminating AIM [reviewed in Bassiri et al. (2008), Purtilo et al. (1975), Seemayer et al. (1995)]. Responses to other virus infections are typically normal, but ~75 % of the boys typically succumb within one month of primary EBV infection. Death is due to the accumulation of EBV-infected B cells expressing the growth program in tissues such as the liver or subsequently by the widespread tissue damage associated with a pronounced virus-associated hemophagocytic syndrome. Surviving boys typically have severely disrupted immune systems, resulting in varying degrees of hypogammaglobulinemia. The XLP gene itself, SH2D1A or SAP (Sayos et al. 1998; Coffey et al. 1998), encodes for a small signaling molecule of 128 amino acids that consists essentially of a single SH2 domain with a small C-terminal extension that is an important regulator of T and NK cell interactions and activation. It has been suggested that the inefficient recognition of SAP-deficient B cells, the target cell for EBV-driven growth, accounts for the disease (Dupre et al. 2005). However, studies in SAP-deficient mice and humans have demonstrated defects in long-term B cell memory (Crotty et al. 2003; Ma et al. 2006) due to their inability to develop functional GCs. This suggests an alternative scenario based on the GCM, namely that these patients may be unable to process latently infected blasts into memory because of their defective GCs. This would result in the infected cells being permanently stuck in the proliferative phase driven by the growth program which, together with the defective T cell response, could lead to uncontrolled proliferation and death.

7 Other Sites of EBV Persistence

7.1 The Epithelium

The role of the epithelium in persistence as a site of viral replication is now broadly accepted. Whether it itself is an independent site of persistent infection is less clear. Two early studies suggested that it is not. Patients undergoing complete bone marrow ablation as part of a bone marrow transplant lost their EBV (Gratama et al. 1988), and patients with XLA, an X-linked genetic disorder where the patients lack B cells, showed no signs of being infected (Faulkner et al. 1999). However, the technical aspects of these papers leave much to be desired and there is a need to reproduce them using modern sensitive, quantitative techniques. This is especially true in light of recent work from the laboratory of Katherine Luzuriaga (Renzette et al. 2014). They have presented intriguing evidence from deep sequencing of EBV within infected individuals that variants are primarily generated over time in saliva not in the blood. This is completely consistent with the GCM concept that the site of latency is a resting memory B cell in the blood which divides only rarely and therefore would accumulate variants extremely slowly, whereas the virus replicates in the epithelium, providing a site for the more rapid accumulation of mutations. However, since (1) the variants would arise as single virions and need to be amplified by reinfection to be detected and (2) the variants are retained over time, this is striking evidence suggesting that the epithelium may be a site where the virus can persist through continuous reinfection and replication in new epithelial plaques.

7.2 The Tonsil Intraepithelial (Marginal Zone) B Cell—A Second Route to Persistence?

The study of persistent infection by EBV has been driven from the start by the property that EBV is able to establish latent persistent infection in vitro by driving newly infected B cells to become latently infected proliferating lymphoblasts expressing the growth program. As a consequence, it was initially assumed that proliferating latently infected lymphoblasts represented the mechanism by which the virus persisted in vivo. We now know from the GCM that this is incorrect. Rather EBV uses lymphoblastoid activation of newly infected naïve B cells transiently in vivo to gain access to the resting memory compartment—the actual site of viral persistence. However, lymphoblastoid activation by EBV infection in vitro is not transient; it results in indefinite proliferation. The question remains therefore: Does extended lymphoblastoid proliferation driven by EBV have a biologically relevant role in vivo? Is this an in vitro counterpart of an in vivo infected, proliferating cell where the virus might persist or is it an artifact of selection for growth in culture?

For a proliferating blast to persist for a long period of time in vivo, it would need to evade CTL, but it turns out that is not so hard to do. One can envision a scenario where infection occurs in vivo, driving the expansion of latently infected lymphoblasts that subsequently stimulate a robust CTL response. The CTLs begin to kill the blasts reducing their numbers, and therefore the antigenic load, leading in turn to attrition of the CTLs. Thus, both populations will be reduced till they reach a point where the time it takes for a CTL to find its target, the lymphoblastoid cell expressing the growth program, exactly equals the time it takes for that cell to die and an equilibrium will have been established (Hawkins et al. 2013). Besides avoiding CTL, long-term proliferation of lymphoblastoid cells also requires that latent proteins have functions specifically evolved to override the cellular mechanisms that normally limit proliferation, i.e., the virus would have to fundamentally change the nature of the B cell (Allday 2013; Price and Luftig 2014). It is now clear that in vitro at least this is indeed the case since EBNA3A and 3C specifically act to override cell cycle checkpoints (Allday 2013). Moreover, this activity is not required for the initial phase of rapid proliferation only coming into play as late as 7 days post-infection. Clearly, this activity must exist to sustain long-term proliferation. Confirmation that these arguments are correct requires that such cells must be demonstrated to exist in vivo. What type of cell might this be?

When naïve B cells are infected in culture, they become activated proliferating blasts that express AID and the memory cell marker CD27 and undergo SHM (Siemer et al. 2008). However, they are unable to undergo CSR (Heath et al. 2012), remain IgD+, and do not express bcl-6 (Siemer et al. 2008), and hence, they would be unable to enter a GC (Kitano et al. 2011). Thus, the phenotype of a lymphoblastoid cell derived from an infected naïve B cell in vitro is IgD+, CD27+, AID+, and bcl-6−, with Ig genes that are somatically mutated but not class-switched [for a complete gene expression profile, see White et al. (2010) and http://www.epstein-barrvirus.org/]. This is reminiscent of resident, tonsil intraepithelial (marginal zone) B cells (Dono et al. 2003; Spencer et al. 1985; Weill et al. 2009; Xu et al. 2007) and completely distinct from GC cells, which are bcl-6+ and undergoing Ig class switching, and GC-derived memory cells, which are IgD− and AID−, and have class-switched Ig genes (Table 3). Can we find such cells in tonsils? The answer is tentatively yes. We have found a population of IgD+ CD27+ B cells in the tonsil that express the growth program and have proliferated extensively (Torgbor and Thorley-Lawson unpublished observations). If it can be confirmed that they are also AID+ and bcl-6− this will be compelling evidence that EBV is able to enter into and persist for some period of time in the resident tonsil intraepithelial (marginal zone) memory B cell compartment.

Table 3 Lymphoblasts transformed in vitro by EBV most closely resemble activated marginal zone memory B cells

Marker	Lymphoblasts	Marginal Zone B Cells	Memory B Cells	GC B Cells
AID	+	+	-	+
bcl-6	-	-	-	+
CD27	+	+	+	+/-
SHM	+	+	+	+
CSR	-	-	+	+

It is important to reiterate that what we are discussing here is the resident, marginal zone-like, intraepithelial, memory compartment of the tonsils. These are thought to be distinct from circulating marginal zone memory cells (Weill et al. 2009; Weller et al. 2004) which have been shown in repeated studies to lack EBV (Joseph et al. 2000b; Souza et al. 2007). The potential presence of EBV in the tonsil subset and absence from the circulating subset support a separate origin for these two types of cells.

There remains much work to be done to investigate this hypothesis not the least of which is whether lymphoblastoid proliferation is a form of long- or short-term persistence, do the cells somehow transition to a resting state, why do they not usually enter the periphery, and how does the virus get back out again from these cells. But the central question remains: What is the biological significance, if any, of long-term proliferation driven by EBV in vivo?

7.3 GC-Independent Maturation of Infected Naïve Blasts

As noted above, direct infection of naïve B cells leads them to become blasts that have many characteristics of memory cells including somatically mutated Ig genes and expression of CD27. This has led to the suggestion that EBV could drive the differentiation of infected naïve B cells all the way to a memory phenotype without the need to access the GC (Heath et al. 2012). This model does not provide a mechanism for the cells to leave the cell cycle and does not account for the different viral latency programs nor the origin of the different lymphomas. More critically, the cells produced in vitro did not undergo CSR and presumably do not express antigen-selected patterns of SHM, two well-known characteristics of the latently infected memory B cells seen in vivo. Rather than contradicting the GCM, these studies actually provide an elegant refinement because the authors reported that the cells would undergo CSR if provided exogenous T cell help which can only be found in the GC (Victora and Nussenzweig 2012). Thus, these studies suggest that infection of naïve B cells in vivo can initiate the GC process, but the cells need to migrate into and through a GC to emerge as class-switched memory B cells with antigen-selected patterns of SHM.

7.4 Two Pathways to Persistence?

The results discussed in the two previous sections raise the intriguing possibility that a newly infected naïve B cell in vivo in Waldeyer's ring may have two routes to persistence (Fig. 10). The key may lie in the observation that upon initial infection in vitro, cells undergo a brief period (~3 days) of rapid proliferation before transitioning to a stable slower proliferative state that goes on indefinitely—the lymphoblastoid cell line [Nikitin et al. (2010), Thorley-Lawson and Strominger (1978) and see Sect. 3.2.3]. If this occurs in vivo, then the initial phase of rapid proliferation may indicate the initiation of the GC reaction; hence, the cells express AID and begin SHM in the absence of bcl-6 (see preceding sections). Because these cells lack bcl-6, they likely will be unable to physically enter the GC (Kitano et al. 2011). Instead, they will arrest at the T cell/B cell boundary of

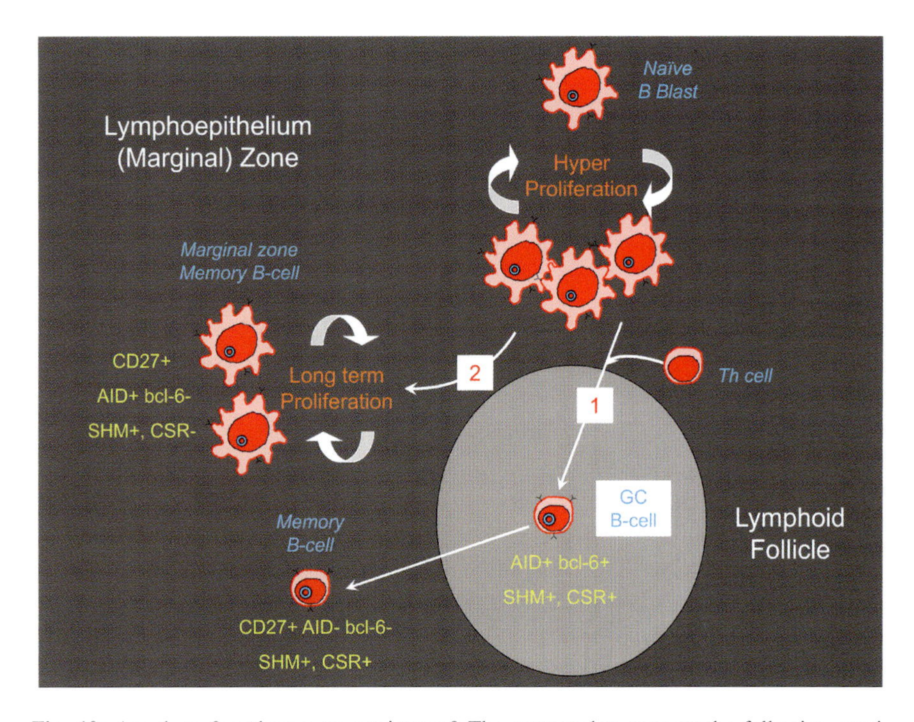

Fig. 10 Are there 2 pathways to persistence? The current data suggest the following possible hypothetical model. Infected naive blasts will migrate to the follicle because they express the chemokine receptor EBI2. They express AID and undergo SHM but will not enter the GC because they are bcl-6 negative. If they receive the necessary signals (cytokines/T cell help), they will enter the follicle switch on bcl-6, undergo CSR, and eventually leave as resting memory B cells as described by the GCM (Route 1). If, however, the cells do not receive the necessary signal to turn on bcl-6, they will continue to proliferate as marginal zone memory B cells (Route 2). The ultimate fate of such cells is unclear. For example, to be biologically relevant, they would need to release infectious virus at some point. What is clear is that they appear capable of extensive proliferation despite the presence of CTL

the GC. The speculation is that if they access T cell help and/or other signals at this time, they would become bcl-6+, proceed into the GC, switch to the default program, and undergo CSR and some version of affinity maturation. It is interesting to note here that one of the latent genes expressed in the growth program, EBNA3B, specifically activates the expression of cytokines that would attract Th cells (White et al. 2012). Eventually, they exit into the periphery as a latently infected GC-derived memory cell as described by the GCM. If, however, after 2–3 days of hyperproliferation, the cells cannot access the necessary signals at the T/B cell boundary, they would transition to the phase of slower, long-term proliferation, remain bcl-6 negative, fail to enter the GC, and instead remain in the interfollicular lymphoepithelium as latently infected marginal zone B cells.

7.5 Direct Infection of Memory Cells

Direct infection of memory B cells was first raised as a possibility over 15 years ago (Babcock et al. 1998) and was subsequently proposed by Rajewsky and coworkers (Kurth et al. 2000, 2003). However, no further evidence for or explanation of a mechanism behind this idea has been produced. Problems with the model include the following:

1. In repeated experiments, we have never detected evidence for the presence of directly infected memory B cells in the tonsil.
2. It fails to provide an explanation for the different latency transcription programs and especially why EBV would have a program (the default program) specifically designed to allow the survival of GC B cells.
3. It has failed to provide evidence or a mechanism for how the directly infected memory B cells transit to a resting state.
4. It does not explain why EBV in the periphery is restricted only to GC-derived memory B cells.
5. Predictions made by the model were incorrect when tested experimentally, instead supporting the GCM. Thus, infected GC B cells express the viral default transcription program in vivo (Babcock et al. 2000; Roughan and Thorley-Lawson 2009) (as predicted by the GCM), not the growth program [as predicted by the direct infection model (Siemer et al. 2008)], and in a transgenic mouse model, one of the EBV latent proteins expressed in the GC (LMP2a) was shown to drive B cells to form GCs in the absence of antigen as required by the GCM and contrary to the idea that EBV directly infects memory cells (Casola et al. 2004a).

8 Conclusions

The GC model of EBV infection demonstrates that persistent infection by EBV is a self-renewing circle of infection, differentiation, persistent infection, reactivation, and reinfection (Fig. 4) that elegantly exploits virtually every aspect of mature B cell biology to:

1. Establish persistent infection (B cell activation with the growth program and GC differentiation with the default program);
2. Maintain persistent infection (latency in the long-lived memory pool, maintained and regulated through the processes of homeostasis);
3. Replicate for reinfection and infectious spread (reactivation of viral replication in response to terminal differentiation into plasma cells).

It is this cycle of infection together with the quiescent infection in memory that allows the virus to be maintained at the extremely low and stable level of infection observed. In doing so, EBV does not detectably disrupt the normal processing of latently infected cells into memory.

This remains the only model, consistent with experimental observation that provides a framework for uniting and understanding the disparate behaviors of EBV, for example:

Why does EBV drive the activation and proliferation of B cells which put the host at risk for neoplastic disease? Because the latently infected resting naïve B cell has to become activated so that it can subsequently differentiate through the GC to become a resting memory B cell where it can persist in a state that is no longer a pathogenic risk to the host.

EBV uses a different transcription program in different forms of lymphoma (IL, HD, and BL). Why? The pattern of genes expressed by the different lymphomas is indicative of the infected cell of origin. The existence of lymphomas expressing all three of the transcription programs associated with the proliferation of infected B cells—growth program (IL), default program (HD), and EBNA1 only (BL)—suggests that each of these stages in the EBV life cycle is vulnerable to deregulation leading to lymphoma.

EBV infects and persists in >90 % of the adult human population almost always benignly despite its ability to make cells grow. The proliferating cells are short-lived and not normally a pathogenic threat because the virus is programmed to ensure that they rapidly differentiate into resting memory cells.

LMP1 and LMP2 have signaling properties analogous to T cell help and the BCR, respectively. Why would two EBV latent proteins mimic B cell survival and differentiation signals? LMP1 and LMP2 have these properties because they are replicating the signals that are normally used to rescue and differentiate normal GC B cells into memory. Hence, LMP1 and LMP2 are the only viral regulatory proteins expressed in infected GC cells.

The epitopes on the latent proteins recognized by cytotoxic T cells are conserved (*Khanna* et al. 1997). *Why would the virus do this?* Once the virus has colonized the memory compartment, any infected cell that continues to express the growth program is a threat to the host. The virus ensures that any cell population that continues to expand due to the growth program will be eliminated by conserving the targets of EBV-specific CTL.

The GCM has also provided insights into the behavior of memory B cells. Notably, it has overturned the belief that memory cells do not recirculate (Gray et al. 1982) because latently infected memory B cells clearly do recirculate (Laichalk et al. 2002). In addition, the restriction of EBV to the isotype-switched GC-derived memory pool and absence from the marginal zone memory pool in the periphery support the view that these marginal zone memory B cells arise independently of the GC (Weill et al. 2009; Weller et al. 2004). Lastly, the presence of EBV in tonsil intraepithelial (marginal) zone B cells supports the idea that this subset is functionally distinct from the circulating/splenic marginal zone B cell.

9 To Be Continued

Although the details of the GCM are likely to change and much is still to be learned, it seems certain that an ultimate understanding of EBV infection will involve a model, whereby EBV uses the normal biology of mature B lymphocytes to establish and maintain persistent infection. The most interesting unanswered questions that remain about EBV persistence are as follows:

1. Is/are there GC-independent mechanisms/sites of persistent infection?
2. What, if any, is the biological significance in vivo of the in vitro phenomenon of long-term lymphoblastoid proliferation?
3. Why does the virus encode for a BCR surrogate if it is persisting in B cells with an apparently normal BCR?
4. What is the relative contribution of viral latent proteins (especially LMP1 and LMP2a) and physiologic signals (Th and BCR) to the production of latently infected memory cells? Could the requirement for both be providing us new insights into the complexities involved in producing and maintaining immunological memory?

10 Final Thought—EBV Is Not As Safe As You Might Think!

EBV seems like a pretty safe virus. It infects virtually every human being for life, and the infection is almost always benign. However, XLP arises during acute EBV infection and almost always results in death. It is caused by mutations in the SH2D1A gene (Sayos et al. 1998; Coffey et al. 1998). So all that stands between EBV switching from a benign lifetime persistent infection to a life-threatening acute disease is a single point mutation in the XLP gene. Put another way, the reader of this chapter would likely have expired and not be around to read this if it was not for that single mutation.

Acknowledgments The work described here is in large part the consequence of research carried out by a number of graduate students in my own laboratory too numerous to mention individually but hopefully appropriately referenced in the text. I would also like to express my thanks to Michael Lawson for a very careful and thorough editing of the text. To the extent that this chapter is comprehendible, it is due to him. Finally, I would like to acknowledge NIH, who have supported my laboratory continuously through Public Health Service grants R01 CA65883 and R01 AI18757.

References

Adams A, Lindahl T (1975) Epstein-Barr virus genomes with properties of circular DNA molecules in carrier cells. Proc Natl Acad Sci USA 72:1477–1481

Allday MJ (2009) How does Epstein-Barr virus (EBV) complement the activation of Myc in the pathogenesis of Burkitt's lymphoma? Semin Cancer Biol 19:366–376

Allday MJ (2013) EBV finds a polycomb-mediated, epigenetic solution to the problem of oncogenic stress responses triggered by infection. Front Genet 4:212

Allday MJ, Crawford DH, Griffin BE (1989) Epstein-Barr virus latent gene expression during the initiation of B cell immortalization. J Gen Virol 70:1755–1764

Allday MJ, Sinclair A, Parker G, Crawford DH, Farrell PJ (1995) Epstein-Barr virus efficiently immortalizes human B cells without neutralizing the function of p53. EMBO J 14:1382–1391

Allen CD, Ansel KM, Low C, Lesley R, Tamamura H, Fujii N, Cyster JG (2004) Germinal center dark and light zone organization is mediated by CXCR4 and CXCR5. Nat Immunol 5:943–952

Ambinder RF (2007) Epstein-barr virus and hodgkin lymphoma. Hematol Am Soc Hematol Educ Program 2007:204–209

Anagnostopoulos I, Hummel M, Kreschel C, Stein H (1995) Morphology, immunophenotype, and distribution of latently and/or productively Epstein-Barr virus-infected cells in acute infectious mononucleosis: implications for the interindividual infection route of Epstein-Barr virus. Blood 85:744–750

Andersson-Anvret M, Forsby N, Klein G, Henle W (1977) Relationship between the Epstein-Barr virus and undifferentiated nasopharyngeal carcinoma: correlated nucleic acid hybridization and histopathological examination. Int J Cancer 20:486–494

Arrand JJ, Rymo L (1982) Characterisation of the major Epstein-Barr virus specific RNA in Burkitt lymphoma derived cells. J Virol 41:376–389

Artavanis-Tsakonas S, Matsuno K, Fortini ME (1995) Notch signaling. Science 268:225–232

Ascherio A, Munger KL (2010) 99th Dahlem conference on infection, inflammation and chronic inflammatory disorders: Epstein-Barr virus and multiple sclerosis: epidemiological evidence. Clin Exp Immunol 160:120–124

Babcock GJ, Decker LL, Volk M, Thorley-Lawson DA (1998) EBV persistence in memory B cells in vivo. Immunity 9:395–404

Babcock GJ, Decker LL, Freeman RB, Thorley-Lawson DA (1999) Epstein-barr virus-infected resting memory B cells, not proliferating lymphoblasts, accumulate in the peripheral blood of immunosuppressed patients. J Exp Med 190:567–576

Babcock GJ, Hochberg D, Thorley-Lawson AD (2000) The expression pattern of Epstein-Barr virus latent genes in vivo is dependent upon the differentiation stage of the infected B cell. Immunity 13:497–506

Baichwal VR, Sugden B (1988) Transformation of Balb 3T3 cells by the BNLF-1 gene of Epstein-Barr virus. Oncogene 2:461–467

Banchereau J, Bazan F, Blanchard D, Briere F, Galizzi JP, Van KC, Liu YJ, Rousset F, Saeland S (1994) The CD40 antigen and its ligand. Annu Rev Immunol 12:881–922

Barton E, Mandal P, Speck SH (2011) Pathogenesis and host control of gammaherpesviruses: lessons from the mouse. Annu Rev Immunol 29:351–397

Bassiri H, Janice YEO WC, Rothman J, Koretzky GA, Nichols KE (2008) X-linked lymphoproliferative disease (XLP): a model of impaired anti-viral, anti-tumor and humoral immune responses. Immunol Res 42:145–159

Basso K, Dalla-Favera R (2010) BCL6: master regulator of the germinal center reaction and key oncogene in B cell lymphomagenesis. Adv Immunol 105:193–210

Beaufils P, Choquet D, Mamoun RZ, Malissen B (1993) The (YXXL/I)2 signalling motif found in the cytoplasmic segments of the bovine leukaemia virus envelope protein and Epstein-Barr virus latent membrane protein 2A can elicit early and late lymphocyte activation events. EMBO J 12:5105–5112

Bechtel D, Kurth J, Unkel C, Kuppers R (2005) Transformation of BCR-deficient germinal-center B cells by EBV supports a major role of the virus in the pathogenesis of Hodgkin and posttransplantation lymphomas. Blood 106:4345–4350

Bernasconi NL, Traggiai E, Lanzavecchia A (2002) Maintenance of serological memory by polyclonal activation of human memory B cells. Science 298:2199–2202

Birkenbach M, Josefsen K, Yalamanchili R, Lenoir G, Kieff E (1993) Epstein-Barr virus-induced genes: first lymphocyte-specific G protein-coupled peptide receptors. J Virol 67:2209–2220

Bonnet M, Guinebretiere JM, Kremmer E, Grunewald V, Benhamou E, Contesso G, Joab I (1999) Detection of Epstein-Barr virus in invasive breast cancers. J Natl Cancer Inst 91:1376–1381

Borza CM, Hutt-Fletcher LM (2002) Alternate replication in B cells and epithelial cells switches tropism of Epstein-Barr virus. Nat Med 8:594–599

Brandtzaeg P, Baekkevold ES, Farstad IN, Jahnsen FL, Johansen FE, Nilsen EM, Yamanaka T (1999a) Regional specialization in the mucosal immune system: what happens in the microcompartments? Immunol Today 20:141–151

Brandtzaeg P, Farstad IN, Haraldsen G (1999b) Regional specialization in the mucosal immune system: primed cells do not always home along the same track. Immunol Today 20:267–277

Calame KL, Lin KI, Tunyaplin C (2003) Regulatory mechanisms that determine the development and function of plasma cells. Annu Rev Immunol 21:205–230 Epub 2001 Dec 19

Caldwell RG, Wilson JB, Anderson SJ, Longnecker R (1998) Epstein-Barr virus LMP2A drives B cell development and survival in the absence of normal B cell receptor signals. Immunity 9:405–411

Callan MF, Annels N, Steven N, Tan L, Wilson J, McMichael AJ, Rickinson AB (1998a) T cell selection during the evolution of CD8+ T cell memory in vivo. Eur J Immunol 28:4382–4390

Callan MF, Tan L, Annels N, Ogg GS, Wilson JD, O'Callaghan CA, Steven N, McMichael AJ, Rickinson AB (1998b) Direct visualization of antigen-specific CD8+ T cells during the primary immune response to Epstein-Barr virus In vivo. J Exp Med 187:1395–1402

Carbone A, Gaidano G, Gloghini A, Larocca LM, Capello D, Canzonieri V, Antinori A, Tirelli U, Falini B, Dalla-Favera R (1998) Differential expression of BCL-6, CD138/syndecan-1, and Epstein-Barr virus-encoded latent membrane protein-1 identifies distinct histogenetic subsets of acquired immunodeficiency syndrome-related non-Hodgkin's lymphomas. Blood 91:747–755

Casola S, Otipoby KL, Alimzhanov M, Humme S, Uyttersprot N, Kutok JL, Carroll MC, Rajewsky K (2004a) B cell receptor signal strength determines B cell fate. Nat Immunol 5:317–327

Casola S, Otipoby KL, Alimzhanov M, Humme S, Uyttersprot N, Kutok JL, Carroll MC, Rajewsky K (2004b) B cell receptor signal strength determines B cell fate. Nat Immunol 5:317–327

Catalina MD, Sullivan JL, Bak KR, Luzuriaga K (2001) Differential evolution and stability of epitope-specific CD8(+) T cell responses in EBV infection. J Immunol 167:4450–4457

Cattoretti G, Chang CC, Cechova K, Zhang J, Ye BH, Falini B, Louie DC, Offit K, Chaganti RS, Dalla-Favera R (1995) BCL-6 protein is expressed in germinal-center B cells. Blood 86:45–53

Chaganti S, Heath EM, Bergler W, Kuo M, Buettner M, Niedobitek G, Rickinson AB, Bell AI (2009) Epstein-Barr virus colonization of tonsillar and peripheral blood B-cell subsets in primary infection and persistence. Blood 113:6372–6381

Chang RA, Miller SD, Longnecker R (2012) Epstein-Barr virus latent membrane protein 2A exacerbates experimental autoimmune encephalomyelitis and enhances antigen presentation function. Sci Rep 2:353

Chatterjee B, Leung CS, Munz C (2014) Animal models of Epstein Barr virus infection. J Immunol Methods 40:80–87

Chen F, Zou JZ, Di RL, Winberg G, Hu LF, Klein E, Klein G, Ernberg I (1995) A subpopulation of normal B cells latently infected with Epstein-Barr virus resembles Burkitt lymphoma cells in expressing EBNA-1 but not EBNA-2 or LMP1. J Virol 69:3752–3758

Chiang AK, Tao Q, Srivastava G, Ho FC (1996) Nasal NK- and T-cell lymphomas share the same type of Epstein-Barr virus latency as nasopharyngeal carcinoma and Hodgkin's disease. Int J Cancer 68:285–290

Clute SC, Naumov YN, Watkin LB, Aslan N, Sullivan JL, Thorley-Lawson DA, Luzuriaga K, Welsh RM, Puzone R, Celada F, Selin LK (2010) Broad cross-reactive TCR repertoires recognizing dissimilar Epstein-Barr and influenza A virus epitopes. J Immunol 185:6753–6764

Coffey AJ, Brooksbank RA, Brandau O, Oohashi T, Howell GR, Bye JM, Cahn AP, Durham J, Heath P, Wray P, Pavitt R, Wilkinson J, Leversha M, Huckle E, Shaw-Smith CJ, Dunham A, Rhodes S, Schuster V, Porta G, Yin L, Serafini P, Sylla B, Zollo M, Franco B, Bentley DR et al (1998) Host response to EBV infection in X-linked lymphoproliferative disease results from mutations in an SH2-domain encoding gene. Nat Genet 20:129–135

Collins CM, Speck SH (2014) Expansion of murine gammaherpesvirus latently infected B cells requires T follicular help. PLoS Pathog 10:e1004106

Crotty S, Kersh EN, Cannons J, Schwartzberg PL, Ahmed R (2003) SAP is required for generating long-term humoral immunity. Nature 421:282–287

Daskalogianni C, Pyndiah S, Apcher S, Mazars A, Manoury B, Ammari N, Nylander K, Voisset C, Blondel M, Fahraeus R (2014) Epstein-Barr virus-encoded EBNA1 and ZEBRA: targets for therapeutic strategies against EBV-carrying cancers. J Pathol 235:334–341

Deacon EM, Pallesen G, Niedobitek G, Crocker J, Brooks L, Rickinson AB, Young LS (1993) Epstein-Barr virus and Hodgkin's disease: transcriptional analysis of virus latency in the malignant cells. J Exp Med 177:339–349

Decker LL, Klaman LD, Thorley-Lawson DA (1996) Detection of the latent form of Epstein-Barr virus DNA in the peripheral blood of healthy individuals. J Virol 70:3286–3289

Decker LL, Babcock GJ, Thorley-Lawson DA (2001) Detection and discrimination of latent and replicative herpesvirus infection at the single cell level in vivo. Methods Mol Biol 174:111–116

Delecluse HJ, Hammerschmidt W (2000) The genetic approach to the Epstein-Barr virus: from basic virology to gene therapy. Mol Pathol 53:270–279

Delgado-Eckert E, Shapiro M (2011) A model of host response to a multi-stage pathogen. J Math Biol 63:201–227

de-Thé G (1985) Epstein-Barr virus and Burkitt's lymphoma worldwide: the causal relationship revisited. In: Olweny CLM, Lenoir GM, O'conor GT (eds) Burkitt's Lymphoma a human cancer model. Oxford University Press, New York

Dominguez-Sola D, Victora GD, Ying CY, Phan RT, Saito M, Nussenzweig MC, Dalla-Favera R (2012) The proto-oncogene MYC is required for selection in the germinal center and cyclic reentry. Nat Immunol 13:1083–1091

Dono M, Zupo S, Colombo M, Massara R, Gaidano G, Taborelli G, Ceppa P, Burgio VL, Chiorazzi N, Ferrarini M (2003) The human marginal zone B cell. Ann N Y Acad Sci 987:117–124
Dupre L, Andolfi G, Tangye SG, Clementi R, Locatelli F, Arico M, Aiuti A, Roncarolo MG (2005) SAP controls the cytolytic activity of CD8+ T cells against EBV-infected cells. Blood 105:4383–4389
Edson CM, Thorley-Lawson DA (1981) Epstein-Barr virus membrane antigens: characterization, distribution, and strain differences. J Virol 39:172–184
Ehlin-Henriksson B, Zou JZ, Klein G, Ernberg I (1999) Epstein-Barr virus genomes are found predominantly in IgA-positive B cells in the blood of healthy carriers. Int J Cancer 83:50–54
Faulkner GC, Burrows SR, Khanna R, Moss DJ, Bird AG, Crawford DH (1999) X-Linked agammaglobulinemia patients are not infected with Epstein-Barr virus: implications for the biology of the virus. J Virol 73:1555–1564
Fingeroth JD, Weis JJ, Tedder TF, Strominger JL, Biro PA, Fearon DT (1984) Epstein-Barr virus receptor of human B lymphocytes is the C3d receptor CR2. Proc Natl Acad Sci USA 81:4510–4514
Fox RI, Chilton T, Scott S, Benton L, Howell FV, Vaughan JH (1987) Potential role of Epstein-Barr virus in Sjogren's syndrome. Rheum Dis Clin North Am 13:275–292
Gatto D, Brink R (2013) B cell localization: regulation by EBI2 and its oxysterol ligand. Trends Immunol 34:336–341
Gires O, Zimber-Strobl U, Gonnella R, Ueffing M, Marschall G, Zeidler R, Pich D, Hammerschmidt W (1997) Latent membrane protein 1 of Epstein-Barr virus mimics a constitutively active receptor molecule. EMBO J 16:6131–6140
Glaser SL, Lin RJ, Stewart SL, Ambinder RF, Jarrett RF, Brousset P, Pallesen G, Gulley ML, Khan G, O'Grady J, Hummel M, Preciado MV, Knecht H, Chan JK, Claviez A (1997) Epstein-Barr virus-associated Hodgkin's disease: epidemiologic characteristics in international data. Int J Cancer 70:375–382
Golden HD, Chang RS, Prescott W, Simpson E, Cooper TY (1973) Leukocyte-transforming agent: prolonged excretion by patients with mononucleosis and excretion by normal individuals. J Infect Dis 127:471–473
Gratama JW, Oosterveer MA, Zwaan FE, Lepoutre J, Klein G, Ernberg I (1988) Eradication of Epstein-Barr virus by allogeneic bone marrow transplantation: implications for sites of viral latency. Proc Natl Acad Sci USA 85:8693–8696
Gray D, Maclennan IC, Bazin H, Khan M (1982) Migrant mu+ delta+ and static mu+ delta-B lymphocyte subsets. Eur J Immunol 12:564–569
Greenspan JS, Greenspan D, Lennette ET, Abrams DI, Conant MA, Petersen V, Freese UK (1985) Replication of Epstein-Barr virus within the epithelial cells of oral "hairy" leukoplakia, an AIDS-associated lesion. N Engl J Med 313:1564–1571
Gregory CD, Tursz T, Edwards CF, Tetaud C, Talbot M, Caillou B, Rickinson AB, Lipinski M (1987) Identification of a subset of normal B cells with a Burkitt's lymphoma (BL)-like phenotype. J Immunol 139:313–318
Gregory CD, Rowe M, Rickinson AB (1990) Different Epstein-Barr virus-B cell interactions in phenotypically distinct clones of a Burkitt's lymphoma cell line. J Gen Virol 71:1481–1495
Gross AJ, Hochberg D, Rand WM, Thorley-Lawson DA (2005) EBV and systemic lupus erythematosus: a new perspective. J Immunol 174:6599–6607
Gulley ML, Raphael M, Lutz CT, Ross DW, Raab-Traub N (1992) Epstein-Barr virus integration in human lymphomas and lymphoid cell lines. Cancer 70:185–191
Hadinoto V, Shapiro M, Greenough TC, Sullivan JL, Luzuriaga K, Thorley-Lawson DA (2008) On the dynamics of acute EBV infection and the pathogenesis of infectious mononucleosis. Blood 111:1420–1427
Hadinoto V, Shapiro M, Sun CC, Thorley-Lawson DA (2009) The dynamics of EBV shedding implicate a central role for epithelial cells in amplifying viral output. PLoS Pathog 5:e1000496

Hardy R (2008) Chapter 7: B Lymphocyte Development and Biology. Fundamental Immunology, 6th edn. Lippincott Williams and Wilkins, Philadelphia

Hawkins JB, Delgado-Eckert E, Thorley-Lawson DA, Shapiro M (2013) The cycle of EBV infection explains persistence, the sizes of the infected cell populations and which come under CTL regulation. PLoS Pathog 9:e1003685

He B, Raab-Traub N, Casali P, Cerutti A (2003) EBV-encoded latent membrane protein 1 cooperates with BAFF/BLyS and APRIL to induce T cell-independent Ig heavy chain class switching. J Immunol 171:5215–5224

Heath E, Begue-Pastor N, Chaganti S, Croom-Carter D, Shannon-Lowe C, Kube D, Feederle R, Delecluse HJ, Rickinson AB, Bell AI (2012) Epstein-Barr virus infection of naive B cells in vitro frequently selects clones with mutated immunoglobulin genotypes: implications for virus biology. PLoS Pathog 8:e1002697

Henderson S, Rowe M, Gregory C, Croom-Carter D, Wang F, Longnecker R, Kieff E, Rickinson A (1991) Induction of bcl-2 expression by Epstein-Barr virus latent membrane protein 1 protects infected B cells from programmed cell death. Cell 65:1107–1115

Henderson S, Huen D, Rowe M, Dawson C, Johnson G, Rickinson A (1993) Epstein Barr virus-coded BHRF1 protein, a viral homologue of Bcl-2, protects human B cells from programmed cell death. Proc Natl Acad Sci USA 90:8479–8483

Henle W, Henle G (1979) Seroepidemiology of the virus. In: Epstein MA, Achong BG (eds) The Epstein-Barr virus. Springer, Berlin

Henle W, Diehl V, Kohn G, Zur Hausen H, Henle G (1967) Herpes-type virus and chromosome marker in normal leukocytes after growth with irradiated Burkitt cells. Science 157:1064–1065

Herbst H, Dallenbach F, Hummel M, Niedobitek G, Pileri S, Muller LN, Stein H (1991) Epstein-Barr virus latent membrane protein expression in Hodgkin and Reed-Sternberg cells. Proc Natl Acad Sci USA 88:4766–4770

Hickabottom M, Parker GA, Freemont P, Crook T, Allday MJ (2002) Two nonconsensus sites in the Epstein-Barr virus oncoprotein EBNA3A cooperate to bind the co-repressor carboxyl-terminal-binding protein (CtBP). J Biol Chem 277:47197–47204 (Epub 2002 Oct 7)

Hislop AD, Kuo M, Drake-Lee AB, Akbar AN, Bergler W, Hammerschmitt N, Khan N, Palendira U, Leese AM, Timms JM, Bell AI, Buckley CD, Rickinson AB (2005) Tonsillar homing of Epstein-Barr virus-specific CD8+ T cells and the virus-host balance. J Clin Invest 115:2546–2555

Hjalgrim H, Smedby KE, Rostgaard K, Molin D, Hamilton-Dutoit S, Chang ET, Ralfkiaer E, Sundstrom C, Adami HO, Glimelius B, Melbye M (2007) Infectious mononucleosis, childhood social environment, and risk of Hodgkin lymphoma. Cancer Res 67:2382–2388

Hoagland RJ (1955) The transmission of infectious mononucleosis. Am J Med Sci 229:262–272

Hoagland RJ (1964) The incubation period of infectious mononucleosis. Am J Public Health Nations Health 54:1699–1705

Hoagland RJ (1967) Infectious mononucleosis. Infectious mononucleosis. Grune and Stratton Inc, New York/London

Hochberg DR, Thorley-Lawson DA (2005) Quantitative detection of viral gene expression in populations of Epstein-Barr virus-infected cells in vivo. Methods Mol Biol 292:39–56

Hochberg D, Middeldorp JM, Catalina M, Sullivan JL, Luzuriaga K, Thorley-Lawson DA (2003a) Demonstration of the Burkitt's Lymphoma Epstein-Barr virus phenotype in dividing latently infected memory cells in vivo. Proc Natl Acad Sci USA 101:239–244

Hochberg D, Vorobyova T, Catalina M, Sullivan JS, Luzuriaga K, Thorley-Lawson DA (2003b) Acute infection with Epstein-Bar virus targets and overwhelms the memory B cell compartment with latently infected cells (Submitted)

Hochberg D, Souza T, Catalina M, Sullivan JL, Luzuriaga K, Thorley-Lawson DA (2004) Acute infection with Epstein-Bar virus targets and overwhelms the peripheral memory B cell compartment with resting, latently infected cells. J Virol (in press)

Hopwood P, Crawford DH (2000) The role of EBV in post-transplant malignancies: a review. J Clin Pathol 53:248–254

Hoshino Y, Katano H, Zou P, Hohman P, Marques A, Tyring SK, Follmann D, Cohen JI (2009) Long-term administration of valacyclovir reduces the number of Epstein-Barr virus (EBV)-infected B cells but not the number of EBV DNA copies per B cell in healthy volunteers. J Virol 83:11857–11861

Hurley EA, Thorley-Lawson DA (1988) B cell activation and the establishment of Epstein-Barr virus latency. J Exp Med 168:2059–2075

Inman GJ, Binne UK, Parker GA, Farrell PJ, Allday MJ (2001) Activators of the Epstein-Barr virus lytic program concomitantly induce apoptosis, but lytic gene expression protects from cell death. J Virol 75:2400–2410

Iwakiri D (2014) Epstein-Barr virus-encoded RNAs: key molecules in viral pathogenesis. Cancers (Basel) 6:1615–1630

Izumi KM, Kieff ED (1997) The Epstein-Barr virus oncogene product latent membrane protein 1 engages the tumor necrosis factor receptor-associated death domain protein to mediate B lymphocyte growth transformation and activate NF- kappaB. Proc Natl Acad Sci USA 94:12592–12597

James JA, Kaufman KM, Farris AD, Taylor-Albert E, Lehman TJ, Harley JB (1997) An increased prevalence of Epstein-Barr virus infection in young patients suggests a possible etiology for systemic lupus erythematosus. J Clin Invest 100:3019–3026

Joseph AM, Babcock GJ, Thorley-Lawson DA (2000a) Cells expressing the Epstein-Barr virus growth program are present in and restricted to the naive B-cell subset of healthy tonsils. J Virol 74:9964–9971

Joseph AM, Babcock GJ, Thorley-Lawson DA (2000b) EBV persistence involves strict selection of latently infected B cells. J Immunol 165:2975–2981

Kaiser C, Laux G, Eick D, Jochner N, Bornkamm GW, Kempkes B (1999) The proto-oncogene c-myc is a direct target gene of Epstein-Barr virus nuclear antigen 2. J Virol 73:4481–4484

Kantor AB (1991) The development and repertoire of B-1 cells (CD5 B cells). Immunol Today 12:389–391

Kempkes B, Spitkovsky D, Jansen-Durr P, Ellwart JW, Kremmer E, Delecluse HJ, Rottenberger C, Bornkamm GW, Hammerschmidt W (1995) B-cell proliferation and induction of early G1-regulating proteins by Epstein-Barr virus mutants conditional for EBNA2. EMBO J 14:88–96

Kennedy G, Komano J, Sugden B (2003) Epstein-Barr virus provides a survival factor to Burkitt's lymphomas. Proc Natl Acad Sci USA 100:14269–14274

Kenney SC, Mertz JE (2014) Regulation of the latent-lytic switch in Epstein-Barr virus. Semin Cancer Biol 26:60–68

Khan G, Miyashita EM, Yang B, Babcock GJ, Thorley-Lawson DA (1996) Is EBV persistence in vivo a model for B cell homeostasis? Immunity 5:173–179

Khanna R, Burrows SR, Neisig A, Neefjes J, Moss DJ, Silins SL (1997) Hierarchy of Epstein-Barr virus-specific cytotoxic T-cell responses in individuals carrying different subtypes of an HLA allele: implications for epitope-based antiviral vaccines. J Virol 71:7429–7435

Kieff E, Rickinson AB (2007) Epstein-Barr virus and its replication. In: Knipe DM, Howley PM (eds) Fields virology, 5th edn. Lippincott Williams & Wilkins, Philadelphia

Kirchmaier AL, Sugden B (1995) Plasmid maintenance of derivatives of oriP of Epstein-Barr virus. J Virol 69:1280–1283

Kis LL, Takahara M, Nagy N, Klein G, Klein E (2006) Cytokine mediated induction of the major Epstein-Barr virus (EBV)-encoded transforming protein, LMP-1. Immunol Lett 104:83–88

Kis LL, Salamon D, Persson EK, Nagy N, Scheeren FA, Spits H, Klein G, Klein E (2010) IL-21 imposes a type II EBV gene expression on type III and type I B cells by the repression of C- and activation of LMP-1-promoter. Proc Natl Acad Sci USA 107:872–877

Kitano M, Moriyama S, Ando Y, Hikida M, Mori Y, Kurosaki T, Okada T (2011) Bcl6 protein expression shapes pre-germinal center B cell dynamics and follicular helper T cell heterogeneity. Immunity 34:961–972
Klein G (1983) Specific chromosomal translocations and the genesis of B-cell-derived tumors in mice and men. Cell 32:311–315
Klein U, Klein G, Ehlin-Henriksson B, Rajewsky K, Kuppers R (1995) Burkitt's lymphoma is a malignancy of mature B cells expressing somatically mutated V region genes. Mol Med 1:495–505
Klein U, Rajewsky K, Kuppers R (1998) Human immunoglobulin (Ig)M+IgD+ peripheral blood B cells expressing the CD27 cell surface antigen carry somatically mutated variable region genes: CD27 as a general marker for somatically mutated (memory) B cells. J Exp Med 188:1679–1689
Knight JS, Lan K, Subramanian C, Robertson ES (2003) Epstein-Barr virus nuclear antigen 3C recruits histone deacetylase activity and associates with the corepressors mSin3A and NCoR in human B-cell lines. J Virol 77:4261–4272
Knowles DM, Cesarman E, Chadburn A, Frizzera G, Chen J, Rose EA, Michler RE (1995) Correlative morphologic and molecular genetic analysis demonstrates three distinct categories of posttransplantation lymphoproliferative disorders. Blood 85:552–565
Kovalchuk AL, Qi CF, Torrey TA, Taddesse-Heath L, Feigenbaum L, Park SS, Gerbitz A, Klobeck G, Hoertnagel K, Polack A, Bornkamm GW, Janz S, III Morse HC (2000) Burkitt lymphoma in the mouse. J Exp Med 192:1183–1190
Kulwichit W, Edwards RH, Davenport EM, Baskar JF, Godfrey V, Raab-Traub N (1998) Expression of the Epstein-Barr virus latent membrane protein 1 induces B cell lymphoma in transgenic mice. Proc Natl Acad Sci USA 95:11963–11968
Kuppers R (2012) New insights in the biology of Hodgkin lymphoma. Hematology Am Soc Hematol Educ Program 2012:328–334
Kuppers R, Rajewsky K (1998) The origin of Hodgkin and Reed/Sternberg cells in Hodgkin's disease. Annu Rev Immunol 16:471–493
Kurosaki T (1999) Genetic analysis of B cell antigen receptor signaling. Annu Rev Immunol 17:555–592
Kurth J, Spieker T, Wustrow J, Strickler GJ, Hansmann LM, Rajewsky K, Kuppers R (2000) EBV-infected B cells in infectious mononucleosis: viral strategies for spreading in the B cell compartment and establishing latency. Immunity 13:485–495
Kurth J, Hansmann ML, Rajewsky K, Kuppers R (2003) Epstein-Barr virus-infected B cells expanding in germinal centers of infectious mononucleosis patients do not participate in the germinal center reaction. Proc Natl Acad Sci USA 100:4730–4735
Kuzembayeva M, Hayes M, Sugden B (2014) Multiple functions are mediated by the miRNAs of Epstein-Barr virus. Curr Opin Virol 7C:61–65
Laichalk LL, Thorley-Lawson DA (2005) Terminal differentiation into plasma cells initiates the replicative cycle of Epstein-Barr virus in vivo. J Virol 79:1296–1307
Laichalk LL, Hochberg D, Babcock GJ, Freeman RB, Thorley-Lawson DA (2002) The dispersal of mucosal memory B cells: evidence from persistent EBV infection. Immunity 16:745–754
Lam N, Sugden B (2003) CD40 and its viral mimic, LMP1: similar means to different ends. Cell Signal 15:9–16
Lam KP, Kuhn R, Rajewsky K (1997) In vivo ablation of surface immunoglobulin on mature B cells by inducible gene targeting results in rapid cell death. Cell 90:1073–1083
Leder P (1985) Translocations among antibody genes in human cancer. In: Lenoir GM, O'conor GT, Olweny CLM (eds) Burkitt's Lymphomaa human cancer model. Oxford University Press, New York
Levitskaya J, Coram M, Levitsky V, Imreh S, Steigerwald MP, Klein G, Kurilla MG, Masucci MG (1995) Inhibition of antigen processing by the internal repeat region of the Epstein-Barr virus nuclear antigen-1. Nature 375:685–688

Levitskaya J, Sharipo A, Leonchiks A, Ciechanover A, Masucci MG (1997) Inhibition of ubiquitin/proteasome-dependent protein degradation by the Gly-Ala repeat domain of the Epstein-Barr virus nuclear antigen 1. Proc Natl Acad Sci USA 94:12616–12621
Li Q, Spriggs MK, Kovats S, Turk SM, Comeau MR, Nepom B, Hutt-Fletcher LM (1997) Epstein-Barr virus uses HLA class II as a cofactor for infection of B lymphocytes. J Virol 71:4657–4662
Liu YJ, Arpin C (1997) Germinal center development. Immunol Rev 156:111–126
Lotz M, Roudier J (1989) Epstein-Barr virus and rheumatoid arthritis: cellular and molecular aspects. Rheumatol Int 9:147–152
Ma CS, Pittaluga S, Avery DT, Hare NJ, Maric I, Klion AD, Nichols KE, Tangye SG (2006) Selective generation of functional somatically mutated IgM+ CD27+, but not Ig isotype-switched, memory B cells in X-linked lymphoproliferative disease. J Clin Invest 116:322–333
Macallan DC, Wallace DL, Zhang Y, Ghattas H, Asquith B, de Lara C, Worth A, Panayiotakopoulos G, Griffin GE, Tough DF, Beverley PC (2005) B-cell kinetics in humans: rapid turnover of peripheral blood memory cells. Blood 105:3633–3640
Mackay F, Schneider P (2009) Cracking the BAFF code. Nat Rev Immunol 9:491–502
Maclennan IC (1994) Germinal centers. Annu Rev Immunol 12:117–139
Maclennan IC (1998) B-cell receptor regulation of peripheral B cells. Curr Opin Immunol 10:220–225
Mancao C, Hammerschmidt W (2007) Epstein-Barr virus latent membrane protein 2A is a B-cell receptor mimic and essential for B-cell survival. Blood 110:3715–3721
Manolov G, Manolova Y (1972) Marker band in one chromosome 14 from Burkitt lymphomas. Nature 237:33–34
Martinez-Valdez H, Guret C, de Bouteiller O, Fugier I, Banchereau J, Liu YJ (1996) Human germinal center B cells express the apoptosis-inducing genes Fas, c-myc, P53, and Bax but not the survival gene bcl-2. J Exp Med 183:971–977
Maruyama M, Lam KP, Rajewsky K (2000) Memory B-cell persistence is independent of persisting immunizing antigen. Nature 407:636–642
Matsumoto AK, Martin DR, Carter RH, Klickstein LB, Ahearn JM, Fearon DT (1993) Functional dissection of the CD21/CD19/TAPA-1/Leu-13 complex of B lymphocytes. J Exp Med 178:1407–1417
Miller CL, Burkhardt AL, Lee JH, Stealey B, Longnecker R, Bolen JB, Kieff E (1995) Integral membrane protein 2 of Epstein-Barr virus regulates reactivation from latency through dominant negative effects on protein-tyrosine kinases. Immunity 2:155–166
Miyashita EM, Yang B, Lam KM, Crawford DH, Thorley-Lawson DA (1995) A novel form of Epstein-Barr virus latency in normal B cells in vivo. Cell 80:593–601
Miyashita EM, Yang B, Babcock GJ, Thorley-Lawson DA (1997) Identification of the site of Epstein-Barr virus persistence in vivo as a resting B cell. J Virol 71:4882–4891
Moorthy R, Thorley-Lawson DA (1992) Mutational analysis of the transforming function of the EBV encoded LMP-1. Curr Top Microbiol Immunol 182:359–365
Mosialos G, Birkenbach M, Yalamanchili R, Vanarsdale T, Ware C, Kieff E (1995) The Epstein-Barr virus transforming protein LMP1 engages signaling proteins for the tumor necrosis factor receptor family. Cell 80:389–399
Muramatsu M, Nagaoka H, Shinkura R, Begum NA, Honjo T (2007) Discovery of activation-induced cytidine deaminase, the engraver of antibody memory. Adv Immunol 94:1–36
Nanbo A, Sugden A, Sugden B (2007) The coupling of synthesis and partitioning of EBV's plasmid replicon is revealed in live cells. EMBO J 26:4252–4262
Nemerow GR, Wolfert R, McNaughton ME, Cooper NR (1985) Identification and characterization of the Epstein-Barr virus receptor on human B lymphocytes and its relationship to the C3d complement receptor (CR2). J Virol 55:347–351

Nicholson LJ, Hopwood P, Johannessen I, Salisbury JR, Codd J, Thorley-Lawson D, Crawford DH (1997) Epstein-Barr virus latent membrane protein does not inhibit differentiation and induces tumorigenicity of human epithelial cells. Oncogene 15:275–283
Niedobitek G, Kremmer E, Herbst H, Whitehead L, Dawson CW, Niedobitek E, von Ostau C, Rooney N, Grasser FA, Young LS (1997) Immunohistochemical detection of the Epstein-Barr virus-encoded latent membrane protein 2A in Hodgkin's disease and infectious mononucleosis. Blood 90:1664–1672
Niedobitek G, Agathanggelou A, Steven N, Young LS (2000) Epstein-Barr virus (EBV) in infectious mononucleosis: detection of the virus in tonsillar B lymphocytes but not in desquamated oropharyngeal epithelial cells. Mol Pathol 53:37–42
Nikitin PA, Yan CM, Forte E, Bocedi A, Tourigny JP, White RE, Allday MJ, Patel A, Dave SS, Kim W, Hu K, Guo J, Tainter D, Rusyn E, Luftig MA (2010) An ATM/Chk2-mediated DNA damage-responsive signaling pathway suppresses Epstein-Barr virus transformation of primary human B cells. Cell Host Microbe 8:510–522
Nilsson K (1979) The nature of lymphoid cell lines and their relationship to the virus. In: Epstein MA, Achong BG (eds) The Epstein-Barr virus. Springer, Berlin
Nonkwelo C, Skinner J, Bell A, Rickinson A, Sample J (1996) Transcription start sites downstream of the Epstein-Barr virus (EBV) Fp promoter in early-passage Burkitt lymphoma cells define a fourth promoter for expression of the EBV EBNA-1 protein. J Virol 70:623–627
O'Nions J, Allday MJ (2003) Epstein-Barr virus can inhibit genotoxin-induced G1 arrest downstream of p53 by preventing the inactivation of CDK2. Oncogene 22:7181–7191
O'nions J, Allday MJ (2004) Deregulation of the Cell Cycle by the Epstein-Barr Virus. In: Vande Woude GF, Klein G (eds) Advances in Cancer Research. Elsevier, New York
Oudejans JJ, Dukers DF, Jiwa NM, van den Brule AJ, Grasser FA, de Bruin PC, Horstman A, Vos W, van Gorp J, Middeldorp JM, Meijer CJ (1996) Expression of epstein-barr virus encoded nuclear antigen 1 in benign and malignant tissues harbouring EBV. J Clin Pathol 49:897–902
Panagopoulos D, Victoratos P, Alexiou M, Kollias G, Mosialos G (2004) Comparative analysis of signal transduction by CD40 and the Epstein-Barr virus oncoprotein LMP1 in vivo. J Virol 78:13253–13261
Parker GA, Crook T, Bain M, Sara EA, Farrell PJ, Allday MJ (1996) Epstein-Barr virus nuclear antigen (EBNA)3C is an immortalizing oncoprotein with similar properties to adenovirus E1A and papillomavirus E7. Oncogene 13:2541–2549
Pegtel DM, Middeldorp J, Thorley-Lawson DA (2004) Epstein-Barr virus infection in ex vivo tonsil epithelial cell cultures of asymptomatic carriers. J Virol 78:12613–12624
Penn I (1998) The role of immunosuppression in lymphoma formation. Springer Semin Immunopathol 20:343–355
Perry ME (1994) The specialised structure of crypt epithelium in the human palatine tonsil and its functional significance. J Anat 185(Pt 1):111–127
Perry M, Whyte A (1998) Immunology of the tonsils. Immunol Today 19:414–421
Polack A, Hortnagel K, Pajic A, Christoph B, Baier B, Falk M, Mautner J, Geltinger C, Bornkamm GW, Kempkes B (1996) c-myc activation renders proliferation of Epstein-Barr virus (EBV)- transformed cells independent of EBV nuclear antigen 2 and latent membrane protein 1. Proc Natl Acad Sci U S A 93:10411–10416
Pope JH, Horne MK, Scott W (1968) Transformation of foetal human keukocytes in vitro by filtrates of a human leukaemic cell line containing herpes-like virus. Int J Cancer 3:857–866
Price AM, Luftig MA (2014) Dynamic Epstein-Barr virus gene expression on the path to B-cell transformation. Adv Virus Res 88:279–313
Purtilo DT, Cassel CK, Yang JP, Harper R (1975) X-linked recessive progressive combined variable immunodeficiency (Duncan's disease). Lancet 1:935–940
Qu L, Rowe DT (1992) Epstein-Barr virus latent gene expression in uncultured peripheral blood lymphocytes. J Virol 66:3715–3724

Quigley MF, Gonzalez VD, Granath A, Andersson J, Sandberg JK (2007) CXCR5+ CCR7− CD8 T cells are early effector memory cells that infiltrate tonsil B cell follicles. Eur J Immunol 37:3352–3362
Raab-Traub N (2002) Epstein-Barr virus in the pathogenesis of NPC. Semin Cancer Biol 12:431–441
Raab-Traub N, Dambaugh T, Kieff E (1980) DNA of Epstein-Barr virus VIII: B95-8, the previous prototype, is an unusual deletion derivative. Cell 22:257–267
Raab-Traub N, Rajadurai P, Flynn K, Lanier AP (1991) Epstein-Barr virus infection in carcinoma of the salivary gland. J Virol 65:7032–7036
Radkov SA, Touitou R, Brehm A, Rowe M, West M, Kouzarides T, Allday MJ (1999) Epstein-Barr virus nuclear antigen 3C interacts with histone deacetylase to repress transcription. J Virol 73:5688–5697
Ramiro AR, Jankovic M, Eisenreich T, Difilippantonio S, Chen-Kiang S, Muramatsu M, Honjo T, Nussenzweig A, Nussenzweig MC (2004) AID is required for c-myc/IgH chromosome translocations in vivo. Cell 118:431–438
Renzette N, Somasundaran M, Brewster F, Coderre J, Weiss ER, McManus M, Greenough T, Tabak B, Garber M, Kowalik TF, Luzuriaga K (2014) Epstein-Barr virus latent membrane protein 1 genetic variability in peripheral blood B cells and oropharyngeal fluids. J Virol 88:3744–3755
Ressing ME, Horst D, Griffin BD, Tellam J, Zuo J, Khanna R, Rowe M, Wiertz EJ (2008) Epstein-Barr virus evasion of CD8(+) and CD4(+) T cell immunity via concerted actions of multiple gene products. Semin Cancer Biol 18:397–408
Rickinson AB, Finerty S, Epstein MA (1977) Mechanism of the establishment of Epstein-Barr virus genome-containing lymphoid cell lines from infectious mononucleosis patients: studies with phosphonoacetate. Int J Cancer 20:861–868
Robbiani DF, Bothmer A, Callen E, Reina-San-martin B, Dorsett Y, Difilippantonio S, Bolland DJ, Chen HT, Corcoran AE, Nussenzweig A, Nussenzweig MC (2008) AID is required for the chromosomal breaks in c-myc that lead to c-myc/IgH translocations. Cell 135:1028–1038
Rooney C, Howe JG, Speck SH, Miller G (1989) Influence of Burkitt's lymphoma and primary B cells on latent gene expression by the nonimmortalizing P3J-HR-1 strain of Epstein-Barr virus. J Virol 63:1531–1539
Rooney CM, Smith CA, Ng CY, Loftin S, Li C, Krance RA, Brenner MK, Heslop HE (1995) Use of gene-modified virus-specific T lymphocytes to control Epstein-Barr-virus-related lymphoproliferation. Lancet 345:9–13
Roughan JE, Thorley-Lawson DA (2009) The intersection of Epstein-Barr virus with the germinal center. J Virol 83:3968–3976
Roughan JE, Torgbor C, Thorley-Lawson DA (2010) Germinal center B cells latently infected with Epstein-Barr virus proliferate extensively but do not increase in number. J Virol 84:1158–1168
Salamon D, Adori M, Ujvari D, Wu L, Kis LL, Madapura HS, Nagy N, Klein G, Klein E (2012) Latency type-dependent modulation of Epstein-Barr virus-encoded latent membrane protein 1 expression by type I interferons in B cells. J Virol 86:4701–4707
Sayos J, Wu C, Morra M, Wang N, Zhang X, Allen D, van Schaik S, Notarangelo L, Geha R, Roncarolo MG, Oettgen H, de Vries JE, Aversa G, Terhorst C (1998) The X-linked lymphoproliferative-disease gene product SAP regulates signals induced through the co-receptor SLAM. Nature 395:462–469
Schaefer BC, Strominger JL, Speck SH (1995) Redefining the Epstein-Barr virus-encoded nuclear antigen EBNA-1 gene promoter and transcription initiation site in group I Burkitt lymphoma cell lines. Proc Natl Acad Sci USA 92:10565–10569
Schwickert TA, Victora GD, Fooksman DR, Kamphorst AO, Mugnier MR, Gitlin AD, Dustin ML, Nussenzweig MC (2011) A dynamic T cell-limited checkpoint regulates affinity-dependent B cell entry into the germinal center. J Exp Med 208:1243–1252

Seemayer TA, Gross TG, Egeler RM, Pirruccello SJ, Davis JR, Kelly CM, Okano M, Lanyi A, Sumegi J (1995) X-linked lymphoproliferative disease: twenty-five years after the discovery. Pediatr Res 38:471–478

Selin LK, Varga SM, Wong IC, Welsh RM (1998) Protective heterologous antiviral immunity and enhanced immunopathogenesis mediated by memory T cell populations. J Exp Med 188:1705–1715

Shibata D, Weiss LM (1992) Epstein-Barr virus-associated gastric adenocarcinoma. Am J Pathol 140:769–774

Shibata D, Tokunaga M, Uemura Y, Sato E, Tanaka S, Weiss LM (1991) Association of Epstein-Barr virus with undifferentiated gastric carcinomas with intense lymphoid infiltration. Lymphoepithelioma-like carcinoma. Am J Pathol 139:469–474

Siemer D, Kurth J, Lang S, Lehnerdt G, Stanelle J, Kuppers R (2008) EBV transformation overrides gene expression patterns of B cell differentiation stages. Mol Immunol 45:3133–3141

Sinclair AJ, Farrell PJ (1995) Host cell requirements for efficient infection of quiescent primary B lymphocytes by Epstein-Barr virus. J Virol 69:5461–5468

Sinclair AJ, Palmero I, Peters G, Farrell PJ (1994) EBNA-2 and EBNA-LP cooperate to cause G0 to G1 transition during immortalization of resting human B lymphocytes by Epstein-Barr virus. EMBO J 13:3321–3328

Skalsky RL, Corcoran DL, Gottwein E, Frank CL, Kang D, Hafner M, Nusbaum JD, Feederle R, Delecluse HJ, Luftig MA, Tuschl T, Ohler U, Cullen BR (2012) The viral and cellular microRNA targetome in lymphoblastoid cell lines. PLoS Pathog 8:e1002484

Souza TA, Stollar BD, Sullivan JL, Luzuriaga K, Thorley-Lawson DA (2005) Peripheral B cells latently infected with Epstein-Barr virus display molecular hallmarks of classical antigen-selected memory B cells. Proc Natl Acad Sci USA 102:18093–18098

Souza TA, Stollar BD, Sullivan JL, Luzuriaga K, Thorley-Lawson DA (2007) Influence of EBV on the peripheral blood memory B cell compartment. J Immunol 179:3153–3160

Speck SH (2002) EBV framed in Burkitt lymphoma. Nat Med 8:1086–1087

Spencer J, Finn T, Pulford KA, Mason DY, Isaacson PG (1985) The human gut contains a novel population of B lymphocytes which resemble marginal zone cells. Clin Exp Immunol 62:607–612

Spencer J, Perry ME, Dunn-Walters DK (1998) Human marginal-zone B cells. Immunol Today 19:421–426

Stadanlick JE, Cancro MP (2008) BAFF and the plasticity of peripheral B cell tolerance. Curr Opin Immunol 20:158–161

Starzl TE, Nalesnik MA, Porter KA, Ho M, Iwatsuki S, Griffith BP, Rosenthal JT, Hakala TR, Jr Shaw BW, Hardesty RL et al (1984) Reversibility of lymphomas and lymphoproliferative lesions developing under cyclosporin-steroid therapy. Lancet 1:583–587

Steven NM, Leese AM, Annels NE, Lee SP, Rickinson AB (1996) Epitope focusing in the primary cytotoxic T cell response to Epstein-Barr virus and its relationship to T cell memory. J Exp Med 184:1801–1813

Strang G, Rickinson AB (1987) Multiple HLA class I-dependent cytotoxicities constitute the "non-HLA-restricted" response in infectious mononucleosis. Eur J Immunol 17:1007–1013

Sugawara Y, Mizugaki Y, Uchida T, Torii T, Imai S, Makuuchi M, Takada K (1999) Detection of Epstein-Barr virus (EBV) in hepatocellular carcinoma tissue: a novel EBV latency characterized by the absence of EBV-encoded small RNA expression. Virology 256:196–202

Swanson-Mungerson M, Longnecker R (2007) Epstein-Barr virus latent membrane protein 2A and autoimmunity. Trends Immunol 28:213–218

Swanson-Mungerson MA, Caldwell RG, Bultema R, Longnecker R (2005) Epstein-Barr virus LMP2A alters in vivo and in vitro models of B-cell anergy, but not deletion, in response to autoantigen. J Virol 79:7355–7362

Tang X, Hori S, Osamura RY, Tsutsumi Y (1995) Reticular crypt epithelium and intra-epithelial lymphoid cells in the hyperplastic human palatine tonsil: an immunohistochemical analysis. Pathol Int 45:34–44

Tao Q, Ho FC, Loke SL, Srivastava G (1995) Epstein-Barr virus is localized in the tumour cells of nasal lymphomas of NK, T or B cell type. Int J Cancer 60:315–320

Thomas JA, Hotchin NA, Allday MJ, Amlot P, Rose M, Yacoub M, Crawford DH (1990) Immunohistology of Epstein-Barr virus-associated antigens in B cell disorders from immunocompromised individuals. Transplantation 49:944–953

Thorley-Lawson DA (2001) Epstein-Barr virus: exploiting the immune system. Nat Rev Immunol 1:75–82

Thorley-Lawson DA (2005) EBV persistence and latent infection in vivo. In: Robertson ES (ed) Epstein-Barr Virus. Caister Academic Press, Norfolk

Thorley-Lawson DA, Allday MJ (2008) The curious case of the tumour virus: 50 years of Burkitt's lymphoma. Nat Rev Microbiol 6:913–924

Thorley-Lawson DA, Babcock GJ (1999) A model for persistent infection with Epstein-Barr virus: the stealth virus of human B cells. Life Sci 65:1433–1453

Thorley-Lawson DA, Gross A (2004) Persistence of the Epstein-Barr virus and the origins of associated lymphomas. N Engl J Med 350:1328–1337

Thorley-Lawson DA, Mann KP (1985) Early events in Epstein-Barr virus infection provide a model for B cell activation. J Exp Med 162:45–59

Thorley-Lawson DA, Poodry CA (1982) Identification and isolation of the main component (gp350–gp220) of Epstein-Barr virus responsible for generating neutralizing antibodies in vivo. J Virol 43:730–736

Thorley-Lawson DA, Strominger JL (1978) Reversible inhibition by phosphonoacetic acid of human B lymphocyte transformation by Epstein-Barr virus. Virology 86:423–431

Thorley-Lawson DA, Schooley RT, Bhan AK, Nadler LM (1982) Epstein-Barr virus superinduces a new human B cell differentiation antigen (B-LAST 1) expressed on transformed lymphoblasts. Cell 30:415–425

Thorley-Lawson DA, Nadler LM, Bhan AK, Schooley RT (1985) BLAST-2 [EBVCS], an early cell surface marker of human B cell activation, is superinduced by Epstein Barr virus. Journal of Immunology 134:3007–3012

Thorley-Lawson DA, Hawkins JB, Tracy SI, Shapiro M (2013) The pathogenesis of Epstein-Barr virus persistent infection. Curr Opin Virol 3:227–232

Tierney RJ, Steven N, Young LS, Rickinson AB (1994) Epstein-Barr virus latency in blood mononuclear cells: analysis of viral gene transcription during primary infection and in the carrier state. J Virol 68:7374–7385

Timmons CF, Dawson DB, Richards CS, Andrews WS, Katz JA (1995) Epstein-Barr virus-associated leiomyosarcomas in liver transplantation recipients. Origin from either donor or recipient tissue. Cancer 76:1481–1489

Torgbor C, Awuah P, Deitsch K, Kalantari P, Duca KA, Thorley-Lawson DA (2014) A multifactorial role for *P. falciparum malaria* in endemic Burkitt's lymphoma pathogenesis. PLoS Pathog 10:e1004170

Touitou R, Hickabottom M, Parker G, Crook T, Allday MJ (2001) Physical and functional interactions between the corepressor CtBP and the Epstein-Barr virus nuclear antigen EBNA3C. J Virol 75:7749–7755

Tracy SI, Kakalacheva K, Lunemann JD, Luzuriaga K, Middeldorp J, Thorley-Lawson DA (2012) Persistence of Epstein-Barr virus in self-reactive memory B cells. J Virol 86:12330–12340

Tsai CN, Liu ST, Chang YS (1995) Identification of a novel promoter located within the Bam HI Q region of the Epstein-Barr virus genome for the EBNA 1 gene. DNA Cell Biol 14:767–776

Tugizov SM, Berline JW, Palefsky JM (2003) Epstein-Barr virus infection of polarized tongue and nasopharyngeal epithelial cells. Nat Med 9:307–314

Tugizov SM, Herrera R, Palefsky JM (2013) Epstein-Barr virus transcytosis through polarized oral epithelial cells. J Virol 87:8179–8194

Uchida J, Yasui T, Takaoka-Shichijo Y, Muraoka M, Kulwichit W, Raab-Traub N, Kikutani H (1999) Mimicry of CD40 signals by Epstein-Barr virus LMP1 in B lymphocyte responses. Science 286:300–303

van Gelder T, Vuzevski VD, Weimar W (1995) Epstein-Barr virus in smooth-muscle tumors. N Engl J Med 332:1719

Vereide D, Sugden B (2009) Proof for EBV's sustaining role in Burkitt's lymphomas. Semin Cancer Biol 19:389–393

Vereide DT, Seto E, Chiu YF, Hayes M, Tagawa T, Grundhoff A, Hammerschmidt W, Sugden B (2014) Epstein-Barr virus maintains lymphomas via its miRNAs. Oncogene 33:1258–1264

Victora GD, Nussenzweig MC (2012) Germinal centers. Annu Rev Immunol 30:429–457

Victora GD, Schwickert TA, Fooksman DR, Kamphorst AO, Meyer-Hermann M, Dustin ML, Nussenzweig MC (2010) Germinal center dynamics revealed by multiphoton microscopy with a photoactivatable fluorescent reporter. Cell 143:592–605

Victora GD, Dominguez-Sola D, Holmes AB, Deroubaix S, Dalla-Favera R, Nussenzweig MC (2012) Identification of human germinal center light and dark zone cells and their relationship to human B-cell lymphomas. Blood 120:2240–2248

Vrazo AC, Chauchard M, Raab-Traub N, Longnecker R (2012) Epstein-Barr virus LMP2A reduces hyperactivation induced by LMP1 to restore normal B cell phenotype in transgenic mice. PLoS Pathog 8:e1002662

Wade M, Allday MJ (2000) Epstein-Barr virus suppresses a G(2)/M checkpoint activated by genotoxins. Mol Cell Biol 20:1344–1360

Wang F (2013) Nonhuman primate models for Epstein-Barr virus infection. Curr Opin Virol 3:233–237

Wang D, Liebowitz D, Kieff E (1985) An EBV membrane protein expressed in immortalized lymphocytes transforms established rodent cells. Cell 43:831–840

Weill JC, Weller S, Reynaud CA (2009) Human marginal zone B cells. Annu Rev Immunol 27:267–285

Weller S, Braun MC, Tan BK, Rosenwald A, Cordier C, Conley ME, Plebani A, Kumararatne DS, Bonnet D, Tournilhac O, Tchernia G, Steiniger B, Staudt LM, Casanova JL, Reynaud CA, Weill JC (2004) Human blood IgM "memory" B cells are circulating splenic marginal zone B cells harboring a prediversified immunoglobulin repertoire. Blood 104:3647–3654

White RE, Groves IJ, Turro E, Yee J, Kremmer E, Allday MJ (2010) Extensive co-operation between the Epstein-Barr virus EBNA3 proteins in the manipulation of host gene expression and epigenetic chromatin modification. PLoS ONE 5:e13979

White RE, Ramer PC, Naresh KN, Meixlsperger S, Pinaud L, Rooney C, Savoldo B, Coutinho R, Bodor C, Gribben J, Ibrahim HA, Bower M, Nourse JP, Gandhi MK, Middeldorp J, Cader FZ, Murray P, Munz C, Allday MJ (2012) EBNA3B-deficient EBV promotes B cell lymphomagenesis in humanized mice and is found in human tumors. J Clin Invest 122:1487–1502

Wilson JB, Bell JL, Levine AJ (1996) Expression of Epstein-Barr virus nuclear antigen-1 induces B cell neoplasia in transgenic mice. EMBO J 15:3117–3126

Woisetschlaeger M, Yandava CN, Furmanski LA, Strominger JL, Speck SH (1990) Promoter switching in Epstein-Barr virus during the initial stages of infection of B lymphocytes. Proc Natl Acad Sci USA 87:1725–1729

Wood TA, Frenkel EP (1967) The atypical lymphocyte. Am J Med 42:923–936

Xu W, He B, Chiu A, Chadburn A, Shan M, Buldys M, Ding A, Knowles DM, Santini PA, Cerutti A (2007) Epithelial cells trigger frontline immunoglobulin class switching through a pathway regulated by the inhibitor SLPI. Nat Immunol 8:294–303

Yao QY, Rickinson AB, Epstein MA (1985) A re-examination of the Epstein-Barr virus carrier state in healthy seropositive individuals. Int J Cancer 35:35–42

Yates JL, Warren N, Sugden B (1985) Stable replication of plasmids derived from Epstein-Barr virus in various mammalian cells. Nature 313:812–815

Ye BH, Cattoretti G, Shen Q, Zhang J, Hawe N, de Waard R, Leung C, Nouri-Shirazi M, Orazi A, Chaganti RS, Rothman P, Stall AM, Pandolfi PP, Dalla-Favera R (1997) The BCL-6 proto-oncogene controls germinal-centre formation and Th2-type inflammation. Nat Genet 16:161–170
Youinou P, Jamin C, Lydyard PM (1999) CD5 expression in human B-cell populations. Immunol Today 20:312–316
Zimber-Strobl U, Strobl LJ (2001) EBNA2 and Notch signalling in Epstein-Barr virus mediated immortalization of B lymphocytes. Semin Cancer Biol 11:423–434

Infectious Mononucleosis

Samantha K. Dunmire, Kristin A. Hogquist and Henry H. Balfour

Abstract Infectious mononucleosis is a clinical entity characterized by sore throat, cervical lymph node enlargement, fatigue, and fever most often seen in adolescents and young adults and lasting several weeks. It can be caused by a number of pathogens, but this chapter only discusses infectious mononucleosis due to primary Epstein–Barr virus (EBV) infection. EBV is a γ-herpesvirus that infects at least 90 % of the population worldwide. The virus is spread by intimate oral contact among teenagers and young adults. How preadolescents acquire the virus is not known. A typical clinical picture with a positive heterophile test is usually sufficient to make the diagnosis, but heterophile antibodies are not specific and do not develop in some patients. EBV-specific antibody profiles are the best choice for staging EBV infection. In addition to causing acute illness, there can also be long-term consequences as the result of acquisition of the virus. Several EBV-related illnesses occur including certain cancers and autoimmune diseases, as well as complications of primary immunodeficiency in persons with the certain genetic mutations. A major obstacle to understanding these sequelae has been the lack of an efficient animal model for EBV infection, although progress in primate and mouse models has recently been made. Key future challenges are to develop protective vaccines and effective treatment regimens.

S.K. Dunmire (✉) · K.A. Hogquist
Center for Immunology, University of Minnesota, Minneapolis, MN 55455, USA
e-mail: dunmi002@umn.edu

K.A. Hogquist
e-mail: hogqu001@umn.edu

H.H. Balfour
Department of Laboratory Medicine and Pathology, Department of Pediatrics, University of Minnesota, University of Minnesota Medical School, Minneapolis, MN 55455, USA
e-mail: balfo001@umn.edu

C. Münz (ed.), *Epstein Barr Virus Volume 1*, Current Topics in Microbiology and Immunology 390, DOI 10.1007/978-3-319-22822-8_9

Contents

1 Introduction 213
2 Epidemiology of Primary EBV Infection 213
  2.1 Age-Specific Prevalence of EBV Antibodies 213
  2.2 Routes of Transmission of Primary EBV Infection 214
3 Clinical Manifestations of Primary EBV Infection 215
  3.1 Acute Illness 215
  3.2 Complications of the Acute Illness 216
  3.3 EBV-Associated Diseases 218
4 Virus–Host Interactions During Primary EBV Infection 219
  4.1 Incubation Period 219
  4.2 Acute Infection 220
  4.3 Convalescence 224
5 Diagnosis of Infcctious Mononuclcosis Duc to EBV 224
6 Genetic Susceptibility 226
7 Prevention of Primary EBV Infection 229
8 Treatment 230
9 Animal Models of Infectious Mononucleosis 230
  9.1 Humanized Mice 230
  9.2 Rabbits 231
  9.3 Non-human Primates 232
10 Summary and Outlook 232
References 233

Abbreviations

CAEBV Chronic active Epstein–Barr virus
DC Dendritic cells
EBNA Epstein–Barr nuclear antigen
EBV Epstein–Barr virus
eBL Endemic Burkitt's lymphoma
EIA Enzyme immunoassay
HL Hodgkin's lymphoma
HLA Human leukocyte antigen
HLH Hemophagocytic lymphohistiocytosis
IFN Interferon
LCV Lymphocryptovirus
MHC Major histocompatibility complex
MS Multiple sclerosis
NK Natural killer
NHANES National Health and Nutrition Examination Survey
NIH National Institutes of Health
NPC Nasopharyngeal carcinoma
SAP Signaling lymphocytic activation molecule-associated protein
VCA Viral capsid antigen
XLP X-linked lymphoproliferative disease

1 Introduction

Infectious mononucleosis is a clinical entity characterized by sore throat, cervical lymph node enlargement, fatigue, and fever. It can be caused by a number of pathogens, but this chapter considers it as disease resulting from primary Epstein–Barr virus (EBV) infection and is focused on the immunocompetent host. Infectious mononucleosis was the name coined by Sprunt and Evans (1920) to describe a syndrome that resembled an acute infectious disease accompanied by atypical large peripheral blood lymphocytes. These atypical lymphocytes, also known as Downey cells (Downey and McKinlay 1923), are actually activated CD8 T lymphocytes, most of which are responding to EBV-infected cells. Infectious mononucleosis is medically important because of the severity and duration of the acute illness and also because of its long-term consequences especially the development of certain cancers and autoimmune disorders.

2 Epidemiology of Primary EBV Infection

2.1 Age-Specific Prevalence of EBV Antibodies

EBV infection is extremely common worldwide and approximately 90 % of adults become antibody-positive before the age of 30 (de-The et al. 1975; Venkitaraman et al. 1985; Levin et al. 2010). A recent example is that 1037 (90 %) of 1148 subjects 18 and 19 years old participating in the US National Health and Nutrition Examination Surveys (NHANES) between 2003 and 2010 had IgG antibodies against EBV viral capsid (VCA) antigen, indicative of prior infection (Balfour et al. 2013a, b).

The prevalence of EBV antibodies in preadolescent children is lower, varying from 20 to 80 % depending on age and geographic location. Factors clearly related to early acquisition of primary EBV infection include geographic region (reviewed in Hjalgrim et al. 2007), and race/ethnicity (Balfour et al. 2013a, b; Condon et al. 2014). Other factors implicated are socioeconomic status (Henle et al. 1969; Hesse et al. 1983; Crowcroft et al. 1998), crowding or sharing a bedroom (Sumaya et al. 1975; Crowcroft et al. 1998), maternal education (Figueira-Silva and Pereira 2004), day care attendance (Hesse et al. 1983), and school catchment area (Crowcroft et al. 1998).

Regarding race/ethnicity, it was recently shown that antibody prevalence across all age groups of USA children 6–19 years old enrolled in NHANES between 2003 and 2010 was substantially higher in non-Hispanic blacks and Mexican Americans than non-Hispanic whites (Balfour et al. 2013a, b). The greatest disparity in antibody prevalence was among the younger children, especially the 6- to 8-year-olds. Interestingly, the difference in antibody prevalence between whites and non-whites diminished during the teenage years. Thus, family environment and/or social practices may differ among white and non-white families, which could account for this disparity in antibody prevalence in younger children. Within each race/ethnicity group, older age, lack of health insurance, and lower household education and income were statistically significantly associated with higher antibody prevalence.

These NHANES findings were confirmed (Condon et al. 2014) and extended to include younger children (18 months to 6 years of age) living in the Minneapolis-St. Paul metropolitan area. The Twin Cities study showed that the divergence in age-specific antibody prevalence between blacks and whites was clearly apparent by the age of 5 years.

The age at which primary EBV infection is acquired may be increasing in developed countries (Morris and Edmunds 2002; Takeuchi et al. 2006; Balfour et al. 2013a, b). This is important to monitor because there is a complex interplay between age of acquisition, symptomatic versus asymptomatic infection, and the subsequent risk of EBV-associated cancers or autoimmune diseases. For example, younger age at the time of primary EBV infection among Kenyan infants was associated with elevated levels of EBV viremia throughout infancy, leading the investigators to postulate that these infants were at higher risk for endemic Burkitt's lymphoma (Piriou et al. 2012; Slyker et al. 2013). Another study found that Greenland Eskimo children acquired primary EBV infection at an earlier age and had higher titers of IgG antibody against VCA than age-matched Danish children (Melbye et al. 1984). The authors speculated that early infection with "a large inoculum of EBV" explained why Eskimos were at high risk for nasopharyngeal carcinoma versus Danes who were not. Nevertheless, late acquisition of primary EBV infection is also detrimental in several contexts. Adolescents and young adults are more likely to experience infectious mononucleosis during primary infection than children (Krabbe et al. 1981). Furthermore, multiple sclerosis (MS) is an inflammatory autoimmune disease that involves EBV infection and risk of MS is higher among individuals who have experienced infectious mononucleosis (Ascherio and Munger 2010). Infectious mononucleosis also increases the risk of Hodgkin's lymphoma (Hjalgrim et al. 2000). Thus, since age of primary EBV infection is an important factor in infectious mononucleosis, it is an important consideration for EBV-related diseases.

2.2 Routes of Transmission of Primary EBV Infection

Kissing is the major route of transmission of primary EBV infection among adolescents and young adults. This was elegantly documented by Hoagland's careful clinical observations (Hoagland 1955) and confirmed many decades later by a prospective study at the University of Minnesota (Balfour et al. 2013a, b). Penetrative sexual intercourse has been postulated to enhance transmission (Crawford et al. 2006), but we have found that subjects reporting deep kissing with or without coitus had the same risk of primary EBV infection throughout their undergraduate years (Balfour et al. 2013a, b).

The incubation period of infectious mononucleosis is approximately 6 weeks. Hoagland's clinical records suggested an incubation period of 32–49 days based on the dates of kissing episodes until the onset of infectious mononucleosis (Hoagland 1955). A well-documented case was reported by Svedmyr et al. (1984) in which the kissing event occurred 38 days prior to onset of symptoms. Behavioral data from our medical history questionnaires collected during prospective studies are consistent with an incubation period of 42 days (Balfour et al. Unpublished observations).

Besides deep kissing, primary EBV infection can also be transmitted by blood transfusion (Gerber et al. 1969), solid organ transplantation (Hanto et al. 1981), or hematopoietic cell transplantation (Shapiro et al. 1988), but these routes account for relatively few cases overall. Alfieri et al. (1996) used polymorphisms in the EBV *BAM*HI-K fragment length and size polymorphisms in EBV nuclear antigen EBNA-1, EBNA-2 and EBNA-3 proteins to identify the specific blood donor responsible for transmitting EBV to a 16-year-old liver transplant recipient who subsequently developed infectious mononucleosis.

The way young children contract EBV is unknown. A reasonable supposition is that they are infected by their parents or siblings who are "carriers" of the virus and who intermittently shed it in their oral secretions (Sumaya and Ench 1986). An extreme example of this is the very early acquisition of EBV among three distinct Melanesian populations whose infants have multiple caregivers that premasticate the baby's food (Lang et al. 1977).

3 Clinical Manifestations of Primary EBV Infection

3.1 Acute Illness

Our prospective studies have determined that 75 % of young adults between the ages of 18 and 22 develop typical infectious mononucleosis after primary EBV infection. Approximately 15 % have atypical symptoms and 10 % are completely asymptomatic (Balfour et al. 2013a, b); Balfour et al. Unpublished observations). There are two common presentations among symptomatic patients. The first is the abrupt onset of sore throat, which many patients say is the worst sore throat they have ever had. Patients may also notice a swollen neck that results from cervical lymph node enlargement. Parenthetically, anterior and posterior cervical nodes are usually equally enlarged. The second common presentation is the gradual onset of malaise, myalgia ("body aches") and fatigue. Table 1 shows the frequency of signs and their median duration in 72 undergraduate students studied prospectively. Most findings have a median duration of 10 days or less, but fatigue and cervical lymphadenopathy persist for a median of 3 weeks. Other findings, seen in fewer than 20 % of cases in our experience, include abdominal pain, hepatomegaly, splenomegaly, nausea, vomiting, palatal petechiae, periorbital and eyelid edema, and rash. Rash is seen more often in patients given penicillin derivatives, which is most likely due to transient penicillin hypersensitivity (Balfour et al. 1972).

Table 1 Clinical features of primary EBV infections in 72 undergraduate students studied prospectively (48 women, 24 men; age range, 18–22 years)

Feature	No. of subjects (%)	Median duration (days)
Sore throat	68 (94 %)	10
Cervical lymphadenopathy	58 (81 %)	21
Fatigue	52 (72 %)	20
Upper respiratory symptoms	46 (64 %)	4.5
Headache	38 (53 %)	9.5
Decreased appetite	38 (53 %)	9.5
Feels feverish	34 (47 %)	4
Myalgia (body aches)	33 (46 %)	3

Subclinical hepatitis documented by elevated levels of alanine aminotransferase occurs in approximately 75 % of prospectively followed patients and, in some cases (5–10 %), overt hepatitis develops with tender hepatomegaly and jaundice (Balfour et al. Unpublished observations).

Primary EBV infection in preadolescents has not been thoroughly investigated most likely because prospective studies in young children are logistically difficult to conduct. The assumption has been that the majority of primary EBV infections in children before puberty are asymptomatic but that is not necessarily so. Young children, especially those under the age of 4 years, may not develop a positive heterophile antibody response during primary EBV infection (Horwitz et al. 1981), and unless specific EBV assays are performed, the diagnosis will be missed.

3.2 Complications of the Acute Illness

Fortunately, serious complications during the acute phase of primary EBV infection are rare. Table 2 shows reported complications divided into those estimated to occur in at least 1 % of patients and those that are seen in fewer than 1 % of cases (Hoagland and Henson 1957; White and Karofsky 1985; Robinson 1988; Connelly and DeWitt 1994; Jenson 2000). Splenic rupture is the most feared complication, which has kept many athletes out of competition for weeks. Current consensus is that athletes may return to contact sports 3 weeks after onset of infectious mononucleosis provided they are afebrile, their energy has returned to normal, and they have no other abnormalities associated with primary EBV infection (Putukian et al. 2008).

Although most symptoms associated with infectious mononucleosis resolve in a matter of months, there can be severe and lasting disease that develops following primary EBV infection. One of these complications may manifest in the form of chronic active EBV (CAEBV). Patients presenting with CAEBV generally exhibit signs that can occur during infectious mononucleosis such as fever, lymphadenopathy, splenomegaly, and hepatitis and show markedly elevated levels of EBV DNA

Table 2 Complications during acute primary EBV infection

Frequency of complications	Complication
≥1 %	Airway obstruction due to oropharyngeal inflammation
	Meningoencephalitis
	Hemolytic anemia
	Streptococcal pharyngitis
	Thrombocytopenia
<1 %	Conjunctivitis
	Hemophagocytic syndrome
	Myocarditis
	Neurologic disorders (other than meningoencephalitis)
	Neutropenia
	Pancreatitis
	Parotitis
	Pericarditis
	Pneumonitis
	Psychological disorders
	Splenic rupture

in the blood (Kimura et al. 2001). Less frequently, patients may also present with lymphoma or hemophagocytic disease, a complication of EBV that is discussed in greater detail below (Kimura et al. 2003). Interestingly, in many cases of CAEBV outside of the USA, particularly in Japan, EBV is reported to infect T or NK cells rather than its usual reservoir of B cells (Quintanilla-Martinez et al. 2000; Kimura et al. 2001). Several instances of B cell tropic CAEBV have also been reported, but these fall into the minority of documented incidences (Schooley et al. 1986; Kimura et al. 2003). While some patients have T or NK cell dysfunction, none of the subjects in a National Institutes of Health (NIH) study had mutations typically associated with EBV-related immunodeficiencies and thus the disease observed in that study was considered to be largely insidious. The most successful treatment for CAEBV has been hematopoietic stem cell transplant. In the same NIH study, all but one of the patients who presented with CAEBV died within an average of six years unless a transplant was received. Those that survived all subsequently became negative for EBV DNA in the blood (Cohen et al. 2011a, b).

EBV may also cause hemophagocytic disease, which is alternately referred to in the literature as EBV-associated hemophagocytic syndrome or EBV-associated hemophagocytic lymphohistiocytosis (HLH). The relative rarity of any form of HLH stands as a barrier to diagnosis, and thus, cases of EBV-HLH are even more uncommon. The disease is characterized by fever, splenomegaly, and cytopenias, though the key laboratory signs are high levels of ferritin and soluble CD25 (Jordan et al. 2011; Janka 2012). The distribution of EBV-HLH seems to be similar to CAEBV, focused mainly in Asian populations and infecting T or NK cells in those groups (Kawaguchi et al. 1993) though B cells are also infected in other populations (Beutel et al. 2009). EBV-HLH may be related to one of several primary immunodeficiencies discussed below, but EBV infection may be a triggering event even in the absence of an identified genetic condition. Interestingly, transcriptome profiling studies showed that the peripheral blood gene expression signature observable during infectious mononucleosis highly resembles that of HLH (Dunmire et al. 2014). This reinforces links between primary EBV infections as a trigger in the initiation of HLH.

X-linked lymphoproliferative syndrome (XLP) is a disease characterized by anemia, hypergammaglobulinemia, and lymphohistiocytosis. Generally, the boys who present with this disease exhibit massive cellular responses to primary EBV infection that result in hemophagocytic pathology, even though they are simultaneously unable to control EBV-transformed B cells (Cannons et al. 2011). It was discovered that the main deficiency involved with XLP is in the signaling lymphocytic activation molecule-associated protein (SAP), which is encoded by the human gene *SH2D1A*. Mutations in this gene disrupt the ability of T cells and NK cells to interact with B cells, resulting in a lack of immunoglobulin class switching and meaning that T cells and NK cells cannot efficiently recognize B cell targets to induce death (Hislop et al. 2010; Zhao et al. 2012).

3.3 EBV-Associated Diseases

EBV has been shown to be the causative agent of about 1 % of the worldwide human cancer burden. In particular, EBV infection is associated with neoplasia of lymphoid and epithelial origins including endemic Burkitt's lymphoma (eBL) and Hodgkin's lymphoma (HL) in the case of the former, as well as nasopharyngeal carcinoma and gastric carcinoma in the case of the latter. EBV is considered the etiologic agent in 95 % of cases of eBL, which occur in regions where malaria is common (Brady et al. 2007). Likewise, EBV can be detected in a high proportion of HL cases in underdeveloped nations, but accounts for less than half of cases in Western countries (Flavell and Murray 2000). It is important to note that incidence of infectious mononucleosis is exceptionally low in Southeast Asia and equatorial Africa where EBV infection during childhood is nearly ubiquitous; thus, it might be extrapolated that infectious mononucleosis does not have a strong correlation with either eBL or HL in these areas (de-The et al. 1978). Emerging evidence suggests that previous presentation with infectious mononucleosis can increase the risk of HL (Hjalgrim et al. 2000). While associations between infectious mononucleosis and epithelial carcinomas have not been explored, the presence of EBV in tumors from nasopharyngeal and gastric carcinoma patients is well documented. About 10 % of human gastric carcinomas are EBV-positive (Iizasa et al. 2012). Like eBL and HL seen in underdeveloped nations, nasopharyngeal carcinomas from endemic regions are virtually all positive for EBV DNA (Raab-Traub 2002), with these tumors thought to be derived from a single EBV-infected epithelial cell (Raab-Traub and Flynn 1986; Pathmanathan et al. 1995).

It is possible that achieving a very high viral titer in the blood at any point in life predisposes individuals to subsequent EBV-related cancers. For example, patients who present with eBL and endemic nasopharyngeal carcinomas live in malaria endemic areas. Evidence shows that patients being treated for malarial disease can have extremely high titers of EBV in the blood (Nijie 2009). Titers of this magnitude are seen exclusively in patients presenting with infectious mononucleosis in developed countries. Thus, it may be possible to reduce occurrences of cancer with prophylactic or therapeutic vaccines aimed at preventing primary EBV infection or at the very least reducing the set point at which the virus is maintained in these individuals.

In recent years, infectious diseases have been emerging as possible triggers for autoimmune disorders. EBV infection in particular has come to be highly associated with occurrence of MS. EBV as a causation factor in MS was first proposed over thirty years ago (Warner and Carp 1981). Many correlative observations for this trend exist, including a low incidence of infectious mononucleosis and MS in developing countries, and MS usually first manifests after the adolescent years during which EBV would be acquired, increasing at a rate of 11 % per year following primary EBV (Ascherio and Munger 2010). There is also a high association between patients who recall having infectious mononucleosis and subsequent development of MS (Alotaibi et al. 2004; Pohl et al. 2006; Banwell et al. 2007). Furthermore, MS in EBV-negative individuals occurs very infrequently (Levin et al. 2010; Pakpoor et al. 2013).

A causative role for EBV was supported by examination of the antibody profiles of patients with MS, scrutinizing the viral loads, epitope specificity, and quantity of antibodies, especially those against EBNA. The risk of MS increases positively with levels of circulating anti-EBNA antibodies (Ascherio et al. 2001; DeLorenze et al. 2006; Levin et al. 2005; Sundstrom et al. 2004). The ability to discriminate MS cases and controls was substantially enhanced by the inclusion of quantitative measures of the anti-EBNA-1 response to EBV infection (Strautins et al. 2014). Interestingly, the strong genetic association of MS with particular human leukocyte antigen (HLA) alleles primarily reflects the association with anti-EBV responses (Rubicz et al. 2013). Researchers recently treated an MS patient with autologous T cells expanded by exposure to EBV antigens. Transfer of the EBV-specific CD8 T cells resulted in a decrease in anti-EBV antibody as well as the size and number of MS related lesions in the brain (Pender et al. 2014).

4 Virus–Host Interactions During Primary EBV Infection

4.1 Incubation Period

The long incubation period of EBV continues to be poorly understood due to a lack of samples obtained between the time of infection and presentation with EBV-related symptoms. During primary infection, viral replication is first detected in the oral cavity (Balfour et al. 2013a, b). The virus infects tonsillar epithelial cells as well as B cells in the parenchyma of the tonsil (Wang et al. 1998). There may be a cyclic nature to the pattern of infection in the oral cavity as it has been shown in vitro that virus derived from epithelial cells has a much higher entry efficiency for infecting B cells and vice versa, resulting a switched viral tropism depending on the cell type in which the virus replicates (Borza and Hutt-Fletcher 2002). At some point during the incubation period, the virus moves from the oral cavity to the blood. Little is known about the kinetics or means of this transition. A type I interferon response was detected by gene expression profiling approximately 2 weeks prior to symptom onset in some subjects experiencing primary EBV infection (Dunmire et al. 2014). Viral genomes can sometimes be detected in the peripheral blood as early as three weeks prior to symptom onset and consistently at least one week prior to illness (Dunmire et al. unpublished observations), where it is probably maintained latently in resting memory phenotype B cells (Hadinoto et al. 2008).

4.2 Acute Infection

The kinetics of the EBV viral loads and EBV-specific antibody responses during primary EBV infection are illustrated in Figs. 1 and 2. High viral loads in both the

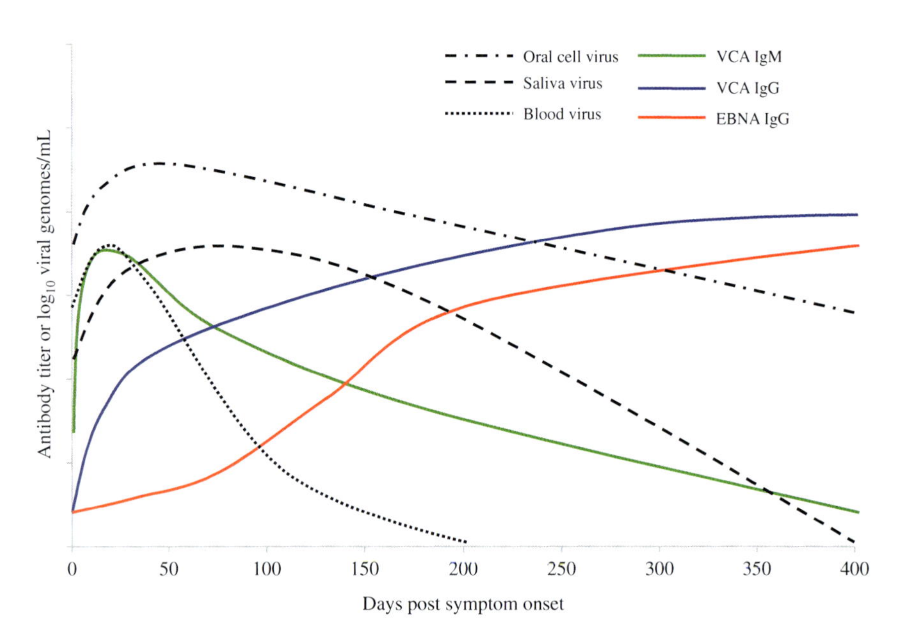

Fig. 1 Kinetics of EBV viral load and antibody responses in subjects with primary EBV infection. Depicted are viral loads in whole blood, saliva, and oral cell pellets (*black lines*) as well as IgM and IgG antibodies to VCA and IgG to EBNA1 (*colored lines*). *Note* the limit of detection of the EBV viral genomes in blood was 200 copies per mL of whole blood

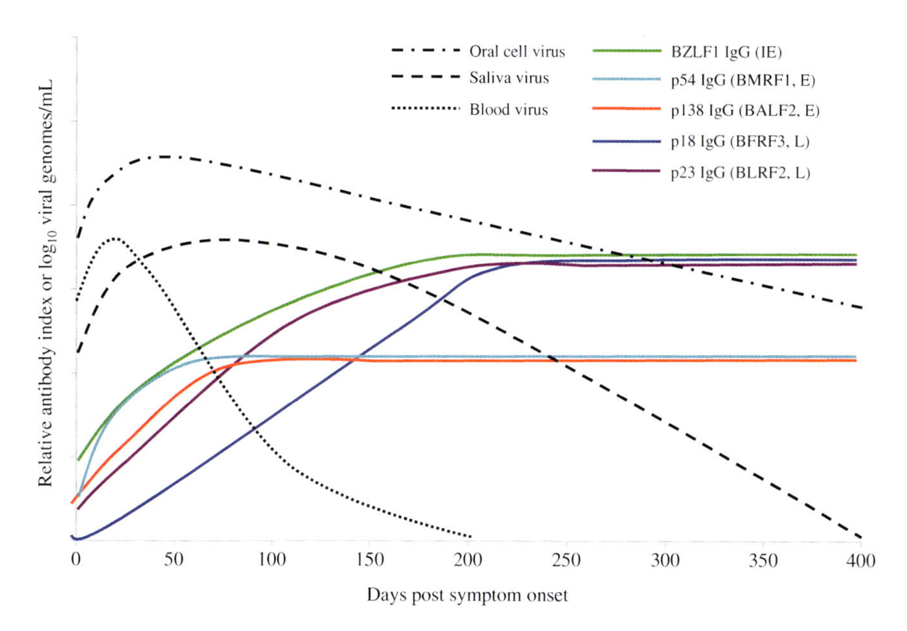

Fig. 2 Kinetics of antibody responses to additional EBV antigens as determined by immunoblot. Depicted are viral loads in whole blood, saliva, and oral cell pellets (*black lines*) as well as the relative kinetics of antibodies generated against the EBV antigens p18, p23, BZLF1, p138, and p54 (*colored lines*). The gene name of each antigen and the stage of expression (*IE* immediate early, *E* early, or *L* late) are indicated in *parentheses*

oral cavity and blood are detected around the time of symptom onset in infectious mononucleosis and accompanied by production of IgM antibodies to EBV VCA and an extraordinary expansion of CD8 T lymphocytes (Balfour et al. 2013a, b). While CD8 T cell responses during primary EBV infection have been thoroughly discussed elsewhere (Hislop et al. 2007; Odumade et al. 2011), a brief discussion of CD8 T cell and B cell responses follows. Of particular interest is the response of cytotoxic CD8 T cells, which have been shown to be important in the control of EBV-infected B cells as evidenced by the disease that occurs in patients who lack elements of CD8 T cell function such as the ability to interact with and kill infected B cells (Rigaud et al. 2006; Palendira et al. 2011).

During infectious mononucleosis when there are very high numbers of circulating CD8 T cells, many of these cells are EBV-specific and directed toward lytically expressed proteins from the immediate early and early stages of the lytic cycle with particular predilection for the immediate early. Cells specific for some late antigens tend to emerge only after the patient has been infected for some time as discovered by generating T cell clones from infectious mononucleosis patients (Abbott et al. 2013). The case of latency, however, is rather different and is often dependent on the relevant HLA type of the EBV-infected individual in question. Immunodominant epitopes for the most prevalent HLA types generally include those derived from latently expressed proteins, especially EBNA-2 and EBNA-3, although some patients develop a strong population of CD8 T cells specific for less readily expressed antigens such as EBNA-1 (Blake et al. 2000; Hislop et al. 2007).

Both CD8 and CD4 T cells require cell-to-cell contact in order to become activated and perform related functions (Adhikary et al. 2007; Merlo et al. 2010). Although total CD4 numbers do not increase appreciably during infectious mononucleosis, evidence exists to support the concept that CD4 T cells are activated and help control infected B cells. Using major histocompatibility complex (MHC) II tetramers, it was shown that several lytic antigens are recognized by CD4 T cells during acute infection and that these cells are maintained at low levels in the blood of EBV-infected hosts (Long et al. 2013). That study also revealed that the response to different antigens varies: CD4(+) T cell responses to EBNA1 did not develop until much later, which likely explains the delay in EBNA1 IgG antibody responses.

Natural killer (NK) cells are also emerging as important players during infectious mononucleosis. Several immunodeficiencies involving T and NK cells and/or their cytolysis pathways result in severe EBV-related outcomes (Menasche et al. 2005; Parvaneh et al. 2013). These include familial hemophagocytic lymphohistiocytosis 2 (FHL2), x-linked lymphoproliferative syndrome (XLP), XIAP deficiency, and x-linked immunodeficiency with Mg+2 defect (XMEN) disorders and are discussed in detail in another chapter of this book. The value of NK cells specifically was suggested by the observation that NK cells preferentially killed EBV-infected cells as the virus entered the lytic cycle (Pappworth et al. 2007). The role of NK cells in vivo was investigated using a humanized mouse model, where NOD-scid $\gamma c-/-$(NSG) mice were reconstituted with CD34+ lin-hematopoietic stem cells (Strowig et al. 2010; Ramer et al. 2011) and infected with the B95.8 strain of EBV and monitored for signs of infectious mononucleosis such as CD8 lymphocytosis and viremia. Animals depleted of NK cells experienced more severe EBV-related disease (Chijioke et al. 2013). It is interesting to note that NK depletion after an established EBV infection did not have significant effects, in contrast to the effect of depletion before infection. Given the gap in the robustness of responses observed between tonsillar and peripheral blood NK cells, it is possible that peripheral NK cells during the systemic infectious mononucleosis phase are less important than those during early infection in the oropharynx (Strowig et al. 2008). Indeed, studies in humans have disagreed about the state of NK cells in peripheral blood during infectious mononucleosis. Work from Williams et al. showed an inverse correlation between NK cell numbers in the periphery and virus in the blood (Williams et al. 2005). In contrast, a larger prospective study found a positive correlation (Balfour et al. 2013a, b). Thus, the interplay between NK cells and blood virus in human subjects needs further study. Of course, the specific type or subset of NK cells may be more important than total numbers. NKG2C+ NK cells were shown to specifically respond to and play a crucial role in immunity to cytomegalovirus (Lopez-Verges et al. 2011). However, NKG2C+ NK cells do not expand upon EBV infection (Hendricks et al. 2014). Rather NK cells expressing

NKG2A and CD54 could be found in higher numbers in the tonsils of EBV carriers (Lunemann et al. 2013) and in the peripheral blood during acute infection (Hendricks et al. 2014). In a study of pediatric infectious mononucleosis patients, it was recently shown that CD56dim NKG2A+ KIR-NK cells preferentially proliferate in response to EBV-infected cells during acute infection (Azzi et al. 2014).

EBV also seems to have evolved mechanisms to interfere with the activation of NK cells during viral replication. The protein encoded by the EBV open reading frame BILF1 downregulates expression of HLA-A and HLA-B on the surface of infected cells but not HLA-C, which is inhibitory to NK cells (Griffin et al. 2013). Several specific populations of NK cells have been implicated in limiting the transformation of B cells by EBV in vitro. When exposed to dendritic cells (DCs) prepared with EBV, CD56bright CD16-NK cells were preferentially primed and were able to limit B cell transformation in vitro in an interferon (IFN) γ-dependent manner. Interestingly, tonsillar NK cells were much more efficient than NK cells derived from peripheral blood (Ferlazzo et al. 2004; Strowig et al. 2008). Further understanding of NK cell recognition of EBV-infected cells and their responses during human infection are needed at this point.

Despite our growing understanding of the innate and adaptive immune response to EBV, it remains unclear why primary EBV infection leads to infectious mononucleosis in adolescents yet is most often asymptomatic in young children. It is possible that adolescents receive a larger initial virus inoculum when transmission occurs through deep kissing. Our data did not find any correlation between virus copy number in the oral cavity and severity of illness (Balfour et al. 2013a, b); however, peak virus copy number in the oral cavity may not directly reflect the initial virus inoculum, nor is it possible to measure the initial virus inoculum with natural infection. Another idea put forward is that infectious mononucleosis in adolescents may reflect the response of cross-reactive memory CD8 T cells. For example, influenza-specific CD8+ T cells might cross-react with EBV (Clute et al. 2005), and as adolescents are presumably more likely to have high numbers of influenza-specific CD8+ T cells, they could react more strongly with EBV. However, we have not seen evidence of influenza–EBV dual-specific CD8+ T cells in our cohort (Odumade et al. 2012) and it remains debatable whether preexisting (cross-reactive) CD8+ T cell immunity to EBV would increase or decrease infectious mononucleosis. A CD8+ T cell peptide epitope vaccine was effective in generating EBV-specific CD8+ T cell responses, and there was no incidence of infectious mononucleosis in the vaccinated group, although the study was small (Burrows et al. 1990). An exciting proposition has arisen from recent data that implicate an NKG2A+ NK cell as important in EBV control (Azzi et al. 2014; Hendricks et al. 2014). Azzi and colleagues showed that CD56dim NKG1A+ KIR-NK cells were found at significantly lower levels in the peripheral blood of adolescents and adults compared to children, suggesting that decreased NK-mediated immune control of EBV could explain why adolescents and adults experience infectious mononucleosis more frequently than children.

4.3 Convalescence

During the convalescent period of infectious mononucleosis (3–6 months postinfection), the number of CD8+ T cells declines to normal levels (Balfour et al. 2013a, b). Previous work in mouse suggested that infection with herpesviruses may cause long-term changes to the "readiness" of host immune cells, thus priming them for subsequent bacterial infections (Barton et al. 2007). These effects were later shown to be transient in that model (Yager et al. 2009), but work in human subjects showed that there were no long-term gene expression changes observable in peripheral blood mononuclear cells following acquisition of EBV (Dunmire et al. 2014). The virus is maintained in resting memory-like B cells. Probably, the most interesting phenomenon that occurs during this phase of infection is related to the levels of antibody that are produced and maintained by latently infected hosts.

Some peculiar trends exist in the antibody response to EBV. For example, antibodies against EBNA-1 display an unusually long delay between virus acquisition and the presence of anti-EBNA-1 IgG, as it generally appears only after a patient with infectious mononucleosis has convalesced (Henle et al. 1987; Hille et al. 1993). This is especially odd given the high levels of class-switched antibody toward EBV antigens that can be measured in many patients presenting with infectious mononucleosis, including responses to latently expressed gene products of EBNA-2 and EBNA-3 (Long et al. 2013) (Figs. 1 and 2). The delayed kinetic might be associated with alanine–glycine-rich regions within the structure of EBNA-1's protein product, which has been shown to inhibit proteosomal processing of relevant epitopes (Levitskaya et al. 1997). That this is the case, however, is not clear and may have to do more with accumulation of protein released from cells for cross-presentation. In contrast, high levels of antibody specific for the immediate early antigen BZLF1 are maintained for the life of the host. This is likely because BZLF1 is expressed by cells undergoing viral reactivation and thus is more frequently presented to B cells in latently infected hosts in comparison with proteins that are coded for later in the viral replication process (Massa et al. 2007).

5 Diagnosis of Infectious Mononucleosis Due to EBV

Infectious mononucleosis due to EBV should be suspected in patients, especially teenagers and young adults, who present with an acute illness characterized by sore throat, cervical lymphadenopathy, fever, and fatigue. Clinical signs that make the diagnosis more likely are exudative pharyngitis with swelling of the uvula and tonsils; periorbital and eyelid edema; and symmetrical cervical and postauricular lymphadenopathy.

A heterophile test using one of the numbers of commercially available antibody kits is most often done to support the clinical diagnosis (Table 3). Heterophile tests are relatively inexpensive and easy to perform. However, heterophile antibodies

Table 3 Staging EBV infection by enzyme immunoassay patterns

Stage of infection	VCA IgM	VCA IgG	EBNA-1 IgG
Naïve	Negative	Negative	Negative
Acute primary[a]	1—2+	Negative—1+	Negative
Subacute[b]	3—4+	2—4+	Negative—1+
Convalescent[c]	Negative—3+	3—4+	Negative—2+
Past[d] (90–95 % of cases)	Negative	3+	3—4+
Past[d] (5–10 % of cases)	Negative	3+	Negative

[a]0–3 weeks after onset of illness
[b]3 weeks to 3 months after onset of illness
[c]3–6 months after onset of illness
[d]>6 months after onset of illness

by definition are not specific. They are IgM class antibodies directed against mammalian erythrocytes. False-positive heterophile tests have been reported in a myriad of conditions including other acute infections, autoimmune disease, and cancer (Sadoff and Goldsmith 1971; Phillips 1972; Horwitz et al. 1979; Hendry and Longmore 1982; Fisher and Bhalara 2004). Although heterophile tests are most commonly used to diagnose infectious mononucleosis, the US Centers for Disease Control and Prevention has recently advised against them “for general use” because of their non-specificity and the possibility of false-negative results especially in young children (Centers for Disease Control and Prevention 2014). In our experience, however, if the clinical picture is typical of infectious mononucleosis and the heterophile antibody test is positive, no additional diagnostic procedures are necessary.

No single antibody test is perfect for confirming the diagnosis of primary EBV infection. Most patients (75 %) will have VCA IgM antibodies at the onset of clinical illness and 95 % eventually make them (Table 3). In prospective studies of EBV-naïve college students, we detected VCA IgM by EIA as early as 8 days before onset of symptoms. The median first day of detection was 2 days after onset of illness (Fig. 1). However, a problem with IgM class antibody tests in general is cross-reactivity with related pathogens. In the case of VCA IgM antibodies, false-positive results have been reported especially with cytomegalovirus infections (Guerrero-Ramos et al. 2014).

Depending on the assay platform and antigen used in the assay, VCA IgG antibodies are first detected during the first month of illness. IgG class antibody against the p18 component of VCA develops later than against the p23 component (see Fig. 2). Using an EIA with p18 as the antigen, we found in prospective studies that the median first day of detection after onset of illness was 31 with a very wide range of 1–118 days. Everyone who experiences a primary EBV infection develops IgG antibodies to VCA (Balfour et al. 2013a, b), so this is the best single test to verify a previous EBV infection. It is superior to EBNA-1 antibody tests because 5–10 % of patients never make EBNA-1 antibodies. Trends in viral load and antibody titer are shown in Figs. 1 and 2.

Antibodies against EBNA-1 are slow to appear with a median first day of detection of 91 days (Fig. 1). Because of this, the presence of EBNA-1 antibodies during an acute illness rules out primary EBV infection. In general, the vast majority of EBV infections in immunocompetent patients can be staged by assaying a blood sample for VCA IgM, VCA IgG, and EBNA-1 antibodies and interpreting the results as shown in Table 4. Western blots or immunoblots can be used to confirm results of screening tests and also stage EBV infections (Schubert et al. 1998; Bauer 2001).

Measurement of IgG antibodies against EBV early antigen (EA) is not useful for the diagnosis of primary infection because only 60–80 % of acutely ill patients are positive and EA antibodies can be found for years in 20 % of healthy individuals (Hess 2004; Centers for Disease Control and Prevention 2014). When the available antibody data do not distinguish the stage of an EBV infection, IgG avidity assays may be useful (Schubert et al. 1998; Nystad and Myrmel 2007). The principle is that during the course of infection, antibodies with high binding strength to their target are selected. In other words, IgG antibodies during the acute phase of infection do not bind to their target as tightly as antibodies produced during convalescence. Low-avidity antibodies can be dissociated from their target by exposure to urea or another chaotropic reagent. Avidity is determined by comparing the amount of antibody detected with and without urea treatment.

The best test for diagnosing and monitoring EBV infections in the immunocompromised host is the blood viral load (or quantitative EBV DNAemia assay) usually performed on a PCR platform (Holman et al. 2012). Most of these infections are not primary, but there are a few that are and result in classic infectious mononucleosis. The monitoring concept is to anticipate the risk of impending EBV disease based on sequential changes in blood viral load. A threshold amount, which varies from center to center, is established that triggers intervention with changes in immunosuppressive and/or antiviral drug regimens.

6 Genetic Susceptibility

Given the disparities in antibody prevalence among populations of different racial backgrounds observed in surveys of children in the USA, there may exist variances in the genetic susceptibility of certain race/ethnicities to EBV. This theory is supported by the incidence of infectious mononucleosis in family groups. A study examining infectious mononucleosis concordance in twins from the California Twin Program found that monozygotic twins were twice as likely to both develop infectious mononucleosis than dizygotic twin pairs. When the analysis was limited to same-sex dizygotic twins, the risk in those groups was higher (Hwang et al. 2012). These findings were then further expanded to include first-degree relatives in a large study of Danish families surveyed by governmental registries. The rate

Table 4 Diagnostic tests for EBV infections

Test	Advantages	Disadvantages	Proportion of patients positive[a]	Median day detected after onset of illness (range of days until detection)[a]
Heterophile antibody	Inexpensive, easy to perform, becomes negative 3–12 months postinfection	Non-specific (false positives due to acute infections and autoimmune diseases); may be negative during first week of illness and persistently negative in young children[b]	85 % (72 % the first week)	0 (6–31)
VCA IgM antibody	Specific, becomes negative 3–12 months postinfection	Not usually performed at point of care sites	95 % (85 % the first week)	2 (21–20)
VCA IgG antibody	Best test for diagnosis of past EBV infection	Not usually performed at point of care sites	100 %	31 (1–118)
EBNA-1 IgG antibody	Best test to distinguish acute from convalescent EBV infection	Not usually performed at point of care sites; 5–10 % of patients never become positive	90–95 %	91 (6–479)
EA antibody	A marker of acute infection	Absent in 20–40 % of acute illnesses; persists for years in ~20 % of cases	Not tested	Not tested
Immunoblot antibodies	Can be used to stage infection (acute, convalescent, past)	Relatively expensive	100 %[c]	2 (25–60)
Blood viral load	Correlates with severity of illness; best test to monitor infection in the immuncompromised host	Viremia is short-lived and may be missed in immunocompetent patients	80 %	4 (8–38)

(continued)

Table 4 (continued)

Test	Advantages	Disadvantages	Proportion of patients positive[a]	Median day detected after onset of illness (range of days until detection)[a]
Oral viral load	Noninvasive, confirms past infection	Cannot be used to diagnose acute infection because virtually all antibody-positive adults shed oral virus intermittently	100 %	−4 (21–31)

[a]Based on prospective studies of EBV-naive college students who developed primary EBV infections (Balfour et al. 2005, 2013a; Balfour et al. Unpublished observations)

[b]Although heterophile tests are most commonly used to diagnose infectious mononucleosis, the dendritic cell (DC):conventional (cDC) has recently advised against them "for general use" (Centers for Disease Control and Prevention 2014)

[c]To one or more antigens

ratios for same-sex twins were highest, followed by groups of siblings (Rostgaard et al. 2014). It is important to note that siblings tend to have similar environments and behaviors, which may explain why the degree of concordance is so high.

Further evidence lends credence to a genetic basis for susceptibility to infectious mononucleosis. Recently, the HLA locus was identified as a major factor influencing antibodies to EBNA-1 in large Mexican American families (Rubicz et al. 2013). Although the authors interpreted this genetic influence on susceptibility to EBV infection, not all individuals who are infected with EBV make a strong antibody response to EBNA-1. Therefore, it is possible that the HLA locus influences the response to infection rather than the infection rate per se. In regard to infectious mononucleosis, the effect on adaptive response to primary infection may be sufficient to achieve a change in whether or not primary infection is symptomatic. Class II MHC is also required for viral entry into the cell, although whether or not there are alterations in viral entry efficiency between HLAs prevalent in different racial groups has not been explored.

7 Prevention of Primary EBV Infection

Given the disease burden associated with acute and chronic EBV diseases, development of an EBV vaccine has long been a priority for researchers in the field. The National Cancer Institute recommended that more clinical trials be conducted to test the safety and efficacy of a vaccine to prevent infectious mononucleosis and cancers caused by EBV (Cohen et al. 2011a, b). Although the first phase 1 trial for a prophylactic EBV vaccine occurred almost twenty years ago (Gu et al. 1995), there has been relatively little progress since. In total, three prophylactic vaccines have been tested in humans, and although all proved at least moderately immunogenic, none provided sterilizing immunity (Balfour 2014). However, sterilizing immunity is probably not necessary to impact symptomatic disease caused by primary EBV infection. For example, a phase 2 trial in Belgium showed that vaccination with a gp350 subunit adjuvant vaccine could reduce the number of cases of infectious mononucleosis (Sokal et al. 2007).

Whether or not a vaccine exclusively targeting gp350 is sufficient to prevent EBV-related disease, however, is unknown. In the case of epithelial neoplasia, it seems less convincing on the grounds that gp350 is not strictly required for viral entry into epithelial cells and can be achieved via the viral proteins gH and gL, albeit less efficiently (Hutt-Fletcher 2007). Increasing the range of the vaccine to include other proteins necessary for this entry such as the aforementioned gH or gL, especially given what is known concerning the switch tropism of EBV between B cells and epithelial cells, might greatly improve the efficacy of a vaccine with the goal of preventing EBV-positive lymphomas and carcinomas.

8 Treatment

There is no currently accepted specific treatment for infectious mononucleosis. While it is clear that acyclovir and valacyclovir have an antiviral effect in vivo, a clinical benefit has not been convincingly demonstrated to date (Tynell et al. 1996). Ganciclovir and valganciclovir have been used to treat EBV infections in immunocompromised hosts, but there are no controlled trials demonstrating clinical efficacy. Corticosteroids are often prescribed to treat inflammatory complications such as airway obstruction or autoimmune phenomena such as anemia and thrombocytopenia, but the value of these drugs is controversial, and they may impair clearance of the viral load (Luzuriaga and Sullivan 2010).

9 Animal Models of Infectious Mononucleosis

9.1 Humanized Mice

One of the major barriers to studying many human viruses is the lack of a small animal model. EBV belongs in this group because it only infects primates. Although studies have been performed with murine gammaherpesvirus-68 (MHV-68), there are important genetic differences between that virus and EBV. In order to directly evaluate the effects of EBV on various lymphoid compartments in vivo, efforts toward developing feasible methods for modeling human infections have resulted in the creation of humanized mice. Humanized mice have only become a viable option for studying human diseases within the last ten years or so due to low engraftment of human cells even in animals with severe combined immunodeficiency (SCID) or knockout of one of the recombinase activating genes (RAG). With the advent of common gamma chain (a receptor component for IL-2, 4, 7, 9, 15, and 21) knockouts, engraftment improved dramatically.

The two types of mice most commonly employed are the previously mentioned NSG and the BALB/c RAG−/− γc−/−. Cells derived from either human fetal liver, human fetal thymus, or CD34+ hematopoietic stem cells are then transferred into neonatal mice to reconstitute their immune systems (Leung et al. 2013). Of these two mouse strains, the NSG mouse has more complete reconstitution of the T and NK compartments and could be maintained for 22 weeks with latent virus detectable and without developing tumors or other EBV-related pathology (Strowig et al. 2009). This model has also been shown to give an approximation of human immune components (Strowig et al. 2010; Ramer et al. 2011). Mice are bred onto a transgenic HLA background. This then allows for thymic selection of the human-derived cells and later specific responses during infection of these animals with EBV (Yajima et al. 2008; Shultz et al. 2010).

It is important to note, however, that study of EBV within this context may neglect important aspects of interplay between EBV and epithelial cells, which have been shown to be important during the replication of EBV, chiefly within the oropharynx (Borza and Hutt-Fletcher 2002). Nevertheless, humanized mice represent a significant step forward to investigating cellular responses in vivo.

In particular, the roles of certain subsets of cells have been interrogated through the depletion of these populations prior to infection with EBV. A significant gap in our understanding of the early innate response to EBV existed and had only been addressed in vitro prior to the advent of this model. Using the NSG mouse and virus obtained from the B95-8 cell line, investigators can now effectively emulate events that occur early during infection in the peripheral blood. Humanized mice exhibit classical features of infectious mononucleosis such as elevated CD8 counts and high levels of IFNγ (Chijioke et al. 2013). Specific deletion of subsets of NK cells could then be performed to examine which were most important during the response to primary EBV infection. Furthermore, the roles of adaptive immune cells may be examined as well. Investigators looking at CD4 and CD8 T cells found EBV-specific T cells were HLA restricted and responded to autologous lymphoblastoid cell lines. When either CD4 or CD8 T cells were depleted from these mice, they developed EBV-related pathologies (Strowig et al. 2009). These results are not particularly surprising given the same might be observed in primary human immunodeficiencies as previously described, but provided important evidence that the humanized mouse can be an appropriate tool for asking questions about the response to EBV. Mouse models, however, neglect nearly all aspects of initial infection as events in the oral cavity cannot occur as EBV does not have tropism for murine epithelial cells. In order to examine infection of the tonsil and oral epithelium, primate models might be preferred as there is established infection in the oral cavity similar to EBV in humans.

9.2 Rabbits

Studies from Japan have suggested that EBV infection may be modeled in rabbits (Takashima et al. 2008; Okuno et al. 2010). Animals in these studies were alternately inoculated intravenously, intranasally, and perorally with EBV derived from the B95-8 cell line. Most animals had only transient levels of virus detectable in the blood, but two had consistent viral titers. Both T and B cells appeared to be infected in these incidences (Takashima et al. 2008; Okuno et al. 2010). It is important to note that while early antigen IgG titers were maintained, VCA IgG antibodies were transient even when very high quantities of virus were used (Okuno et al. 2010). Though not a model of infection with a great deal of similarity to infectious mononucleosis, rabbits may still provide interesting insight with regard to the kinetics and magnitude of antibody responses to EBV, which has implications for the development and testing of humoral component vaccines such as the gp350 subunit vaccine.

9.3 Non-human Primates

Non-human primates are the other major option for investigating EBV infection in vivo. The gamma herpesvirus lymphocryptovirus (LCV) exists in two types: that which infects old world and new world primates. The LCV that infects old world primates has higher genetic similarity to EBV than new world LCV. The open reading frames of rhesus LCV have 28–98 % amino acid identity with EBV (Wang 2013). Thus, rhesus LCV is used to model EBV in rhesus macaques. The symptoms of EBV and LCV are very similar, and LCV can be given orally to emulate natural infection routes in humans. One difference is that the incubation period of LCV is generally shorter, lasting about three weeks rather than six.

LCV can be manipulated using a bacterial artificial chromosome system to allow for mutation of the LCV genome. This enables researchers to examine the possible effects of EBV homologues in vivo. For example, BARF-1 has been shown to bind and inhibit the signaling of colony stimulating factor 1 (Elegheert et al. 2012; Shim et al. 2012), which can promote the maturation and maintenance of type I interferon producing plasmacytoid DCs (Fancke et al. 2008). Knocking out BARF-1 in LCV resulted in more favorable outcomes and lower viral loads in infected rhesus macaques (Ohashi et al. 2012).

The LCV model can also be used to test potential prophylactic or therapeutic vaccines. An LCV gp350 subunit vaccine protected against infection and reduced the viral set point in rhesus macaques (Sashihara et al. 2011). More recently, Leskowitz and colleagues showed that an adenovirus-based vaccine encoding LCV EBNA-1 induced expansion of CD4+ and CD8+ T cells specific for EBNA-1 in rhesus macaques with natural persistent LCV infection (Leskowitz et al. 2014).

10 Summary and Outlook

EBV is one of the most important human pathogens. Although this virus was discovered more than 50 years ago and infects more than 90 % of the worldwide population, there are large gaps in our knowledge of its epidemiology and pathogenesis. Our future challenge is to focus research on the following five gaps.

1. We don't know how EBV is transmitted to young children.
2. We don't know why some adolescents and young adults become very ill from a primary EBV infection, while others remain completely asymptomatic.
3. We don't have an approved specific treatment for EBV infections.
4. We don't have an approved EBV vaccine.
5. Finally, we don't know the mechanism by which EBV induces malignancies or autoimmune diseases. In terms of EBV-associated cancer, we do know a reasonable amount about how this virus infects and transforms lymphocytes and epithelial cells. What we don't understand is how these cells escape immune recognition.

References

Abbott RJ, Quinn LL, Leese AM, Scholes HM, Pachnio A, Rickinson AB (2013) CD8+ T cell responses to lytic EBV infection: late antigen specificities as subdominant components of the total response. J Immunol 191:5398–5409. doi:10.4049/jimmunol.1301629

Adhikary D, Behrends U, Boerschmann H, Pfunder A, Burdach S, Moosmann A, Witter K, Bornkamm GW, Mautner J (2007) Immunodominance of lytic cycle antigens in Epstein-Barr virus-specific CD4+ T cell preparations for therapy. PLoS One 2:e583. doi:10.1371/journal.pone.0000583

Alfieri C, Tanner J, Carpentier L, Perpete C, Savoie A, Paradis K, Delage G, and Joncas J (1996) Epstein-Barr virus transmission from a blood donor to an organ transplant recipient with recovery of the same virus strain from the recipient's blood and oropharynx. Blood 87:812–817

Alotaibi S, Kennedy J, Tellier R, Stephens D, Banwell B (2004) Epstein-Barr virus in pediatric multiple sclerosis. JAMA 291:1875–1879. doi:10.1001/jama.291.15.1875

Ascherio A, Munger KL, Lennette ET, Spiegelman D, Hernan MA, Olek MJ, Hankinson SE, Hunter DJ (2001) Epstein-Barr virus antibodies and risk of multiple sclerosis: a prospective study. JAMA 286:3083–3088. doi:10.1001/jama.286.24.3083

Ascherio A, Munger KL (2010) Epstein-barr virus infection and multiple sclerosis: a review. J Neuroimmune Pharmacol 5:271–277. doi:10.1007/s11481-010-9201-3

Azzi T, Lunemann A, Murer A, Ueda S, Beziat V, Malmberg KJ, Staubli G, Gysin C, Berger C, Munz C, Chijioke O, Nadal D (2014) Role for early-differentiated natural killer cells in infectious mononucleosis. Blood 124:2533–2543. doi:10.1182/blood-2014-01-553024

Balfour HH Jr (2014) Progress, prospects, and problems in Epstein-Barr virus vaccine development. Curr Opin Virol 6C:1–5. doi:10.1016/j.coviro.2014.02.005

Balfour HH, Jr, Holman CJ, Hokanson KM, Lelonek MM, Giesbrecht JE, White DR, Schmeling DO, Webb CH, Cavert W, Wang DH et al. (2005). A prospective clinical study of Epstein-Barr virus and host interactions during acute infectious mononucleosis. J Infect Dis 192: 1505–1512. doi:10.1086/491740

Balfour HH Jr, Forte FA, Simpson RB, Zolov DM (1972) Penicillin-related exanthems in infectious mononucleosis identical to those associated with ampicillin. Clin Pediatr 11:417–421

Balfour HH Jr, Odumade OA, Schmeling DO, Mullan BD, Ed JA, Knight JA, Vezina HE, Thomas W, Hogquist KA (2013a) Behavioral, virologic, and immunologic factors associated with acquisition and severity of primary Epstein-Barr virus infection in university students. J Infect Dis 207:80–88. doi:10.1093/infdis/jis646

Balfour HH Jr, Sifakis F, Sliman JA, Knight JA, Schmeling DO, Thomas W (2013b) Age-specific prevalence of Epstein-Barr virus infection among individuals aged 6–19 years in the United States and factors affecting its acquisition. J Infect Dis 208:1286–1293. doi:10.1093/infdis/jit321

Banwell B, Krupp L, Kennedy J, Tellier R, Tenembaum S, Ness J, Belman A, Boiko A, Bykova O, Waubant E, Mah JK, Stoian C, Kremenchutzky M, Bardini MR, Ruggieri M, Rensel M, Hahn J, Weinstock-Guttman B, Yeh EA, Farrell K, Freedman M, Iivanainen M, Sevon M, Bhan V, Dilenge ME, Stephens D, Bar-Or A (2007) Clinical features and viral serologies in children with multiple sclerosis: a multinational observational study. Lancet Neurol 6:773–781. doi:10.1016/S1474-4422(07)70196-5

Barton ES, White DW, Cathelyn JS, Brett-McClellan KA, Engle M, Diamond MS, Miller VL, Virgin HW (2007) Herpesvirus latency confers symbiotic protection from bacterial infection. Nature 447:326–329. doi:10.1038/nature05762

Bauer G (2001) Simplicity through complexity: immunoblot with recombinant antigens as the new gold standard in Epstein-Barr virus serology. Clin Lab 47:223–230

Beutel K, Gross-Wieltsch U, Wiesel T, Stadt UZ, Janka G, Wagner HJ (2009) Infection of T lymphocytes in Epstein-Barr virus-associated hemophagocytic lymphohistiocytosis in children of non-Asian origin. Pediatr Blood Cancer 53:184–190. doi:10.1002/pbc.22037

Blake N, Haigh T, Shaka'a G, Croom-Carter D, Rickinson A (2000) The importance of exogenous antigen in priming the human CD8+ T cell response: lessons from the EBV nuclear antigen EBNA1. J Immunol 165:7078–7087

Borza CM, Hutt-Fletcher LM (2002) Alternate replication in B cells and epithelial cells switches tropism of Epstein-Barr virus. Nat Med 8:594–599. doi:10.1038/nm0602-594

Brady G, MacArthur GJ, Farrell PJ (2007) Epstein-Barr virus and Burkitt lymphoma. J Clin Pathol 60:1397–1402. doi:10.1136/jcp.2007.047977

Burrows SR, Sculley TB, Misko IS, Schmidt C, Moss DJ (1990) An Epstein-Barr virus-specific cytotoxic T cell epitope in EBV nuclear antigen 3 (EBNA 3). J Exp Med 171:345–349

Cannons JL, Tangye SG, Schwartzberg PL (2011) SLAM family receptors and SAP adaptors in immunity. Annu Rev Immunol 29:665–705. doi:10.1146/annurev-immunol-030409-101302

Centers for Disease Control and Prevention (2014) Epstein-Barr virus and infectious mononucleosis. http://www.cdc.gov/epstein-barr/laboratory-testing.html. Accessed 08 Aug 2014

Chijioke O, Muller A, Feederle R, Barros MH, Krieg C, Emmel V, Marcenaro E, Leung CS, Antsiferova O, Landtwing V, Bossart W, Moretta A, Hassan R, Boyman O, Niedobitek G, Delecluse HJ, Capaul R, Munz C (2013) Human natural killer cells prevent infectious mononucleosis features by targeting lytic Epstein-Barr virus infection. Cell Rep 5:1489–1498. doi:10.1016/j.celrep.2013.11.041

Clute SC, Watkin LB, Cornberg M, Naumov YN, Sullivan JL, Luzuriaga K, Welsh RM, Selin LK (2005) Cross-reactive influenza virus-specific CD8+ T cells contribute to lymphoproliferation in Epstein-Barr virus-associated infectious mononucleosis. J Clin Invest 115:3602–3612. doi:10.1172/JCI25078

Cohen JI, Fauci AS, Varmus H, Nabel GJ (2011) Epstein-Barr virus: an important vaccine target for cancer prevention. Sci Transl Med 3:107fs107. doi:10.1126/scitranslmed.3002878

Cohen JI, Jaffe ES, Dale JK, Pittaluga S, Heslop HE, Rooney CM, Gottschalk S, Bollard CM, Rao VK, Marques A, Burbelo PD, Turk SP, Fulton R, Wayne AS, Little RF, Cairo MS, El-Mallawany NK, Fowler D, Sportes C, Bishop MR, Wilson W, Straus SE (2011b) Characterization and treatment of chronic active Epstein-Barr virus disease: a 28-year experience in the United States. Blood 117:5835–5849. doi:10.1182/blood-2010-11-316745

Condon LM, Cederberg LE, Rabinovitch MD, Liebo RV, Go JC, Delaney AS, Schmeling DO, Thomas W, Balfour HH Jr (2014) Age-specific prevalence of Epstein-Barr virus infection among minnesota children effects of race/ethnicity and family environment. Clin Infect Dis. doi:10.1093/cid/ciu342

Connelly KP, DeWitt LD (1994) Neurologic complications of infectious mononucleosis. Pediatr Neurol 10:181–184

Crawford DH, Macsween KF, Higgins CD, Thomas R, McAulay K, Williams H, Harrison N, Reid S, Conacher M, Douglas J, Swerdlow AJ (2006) A cohort study among university students: identification of risk factors for Epstein-Barr virus seroconversion and infectious mononucleosis. Clin Infect Dis 43:276–282. doi:10.1086/505400

Crowcroft NS, Vyse A, Brown DW, Strachan DP (1998) Epidemiology of Epstein-Barr virus infection in pre-adolescent children: application of a new salivary method in Edinburgh, Scotland. J Epidemiol Community Health 52:101–104

DeLorenze GN, Munger KL, Lennette ET, Orentreich N, Vogelman JH, Ascherio A (2006) Epstein-Barr virus and multiple sclerosis: evidence of association from a prospective study with long-term follow-up. Arch Neurol 63:839–844. doi:10.1001/archneur.63.6.noc50328

de-The G, Day NE, Geser A, Lavoue MF, Ho JH, Simons MJ, Sohier R, Tukei P, Vonka V, Zavadova H (1975) Sero-epidemiology of the Epstein-Barr virus: preliminary analysis of an international study—a review. IARC scientific publications, pp 3–16

de-The G, Geser A, Day NE, Tukei PM, Williams EH, Beri DP, Smith PG, Dean AG, Bronkamm GW, Feorino P, Henle W (1978) Epidemiological evidence for causal relationship between Epstein-Barr virus and Burkitt's lymphoma from Ugandan prospective study. Nature 274:756–761

Downey HM, McKinlay CA (1923) Acute lymphadenosis compared with acute lymphatic leukemia. Arch Intern Med 32:82–112

Dunmire SK, Odumade OA, Porter JL, Reyes-Genere J, Schmeling DO, Bilgic H, Fan D, Baechler EC, Balfour HH Jr, Hogquist KA (2014) Primary EBV infection induces an expression profile distinct from other viruses but similar to hemophagocytic syndromes. PLoS One 9:e85422. doi:10.1371/journal.pone.0085422

Elegheert J, Bracke N, Pouliot P, Gutsche I, Shkumatov AV, Tarbouriech N, Verstraete K, Bekaert A, Burmeister WP, Svergun DI, Lambrecht BN, Vergauwen B, Savvides SN (2012) Allosteric competitive inactivation of hematopoietic CSF-1 signaling by the viral decoy receptor BARF1. Nat Struct Mol Biol 19:938–947. doi:10.1038/nsmb.2367

Fancke B, Suter M, Hochrein H, O'Keeffe M (2008) M-CSF: a novel plasmacytoid and conventional dendritic cell poietin. Blood 111:150–159. doi:10.1182/blood-2007-05-089292

Ferlazzo G, Pack M, Thomas D, Paludan C, Schmid D, Strowig T, Bougras G, Muller WA, Moretta L, Munz C (2004) Distinct roles of IL-12 and IL-15 in human natural killer cell activation by dendritic cells from secondary lymphoid organs. Proc Natl Acad Sci USA 101:16606–16611. doi:10.1073/pnas.0407522101

Figueira-Silva CM, Pereira FE (2004) Prevalence of Epstein-Barr virus antibodies in healthy children and adolescents in Vitoria, State of Espirito Santo, Brazil. Rev Soc Bras Med Trop 37:409–412. doi:10.1590/S0037-86822004000500008

Fisher BA, Bhalara S (2004) False-positive result provided by rapid heterophile antibody test in a case of acute infection with hepatitis E virus. J Clin Microbiol 42:4411. doi:10.1128/JCM.42.9.4411.2004

Flavell KJ, Murray PG (2000) Hodgkin's disease and the Epstein-Barr virus. Mol Pathol 53:262–269

Gerber P, Walsh JH, Rosenblum EN, Purcell RH (1969) Association of EB-virus infection with the post-perfusion syndrome. Lancet 1:593–595

Griffin BD, Gram AM, Mulder A, Van Leeuwen D, Claas FH, Wang F, Ressing ME, Wiertz E (2013) EBV BILF1 evolved to downregulate cell surface display of a wide range of HLA class I molecules through their cytoplasmic tail. J Immunol 190:1672–1684. doi:10.4049/jimmunol.1102462

Gu SY, Huang TM, Ruan L, Miao YH, Lu H, Chu CM, Motz M, Wolf H (1995) First EBV vaccine trial in humans using recombinant vaccinia virus expressing the major membrane antigen. Dev Biol Stand 84:171–177

Guerrero-Ramos A, Patel M, Kadakia K, Haque T (2014) Performance of the architect EBV antibody panel for determination of Epstein-Barr virus infection stage in immunocompetent adolescents and young adults with clinical suspicion of infectious mononucleosis. Clin Vaccine Immunol 21:817–823. doi:10.1128/CVI.00754-13

Hadinoto V, Shapiro M, Greenough TC, Sullivan JL, Luzuriaga K, Thorley-Lawson DA (2008) On the dynamics of acute EBV infection and the pathogenesis of infectious mononucleosis. Blood 111:1420–1427. doi:10.1182/blood-2007-06-093278

Hanto DW, Frizzera G, Purtilo DT, Sakamoto K, Sullivan JL, Saemundsen AK, Klein G, Simmons RL, Najarian JS (1981) Clinical spectrum of lymphoproliferative disorders in renal transplant recipients and evidence for the role of Epstein-Barr virus. Cancer Res 41:4253–4261

Hendricks DW, Balfour HH Jr, Dunmire SK, Schmeling DO, Hogquist KA, Lanier LL (2014) Cutting edge: NKG2ChiCD57+ NK cells respond specifically to acute infection with cytomegalovirus and not Epstein-Barr virus. J Immunol. doi:10.4049/jimmunol.1303211

Hendry BM, Longmore JM (1982) Systemic lupus erythematosus presenting as infectious mononucleosis with a false positive monospot test. Lancet 1:455

Henle G, Henle W, Clifford P, Diehl V, Kafuko GW, Kirya BG, Klein G, Morrow RH, Munube GM, Pike P, Tukei PM, Ziegler JL (1969) Antibodies to Epstein-Barr virus in Burkitt's lymphoma and control groups. J Natl Cancer Inst 43:1147–1157

Henle W, Henle G, Andersson J, Ernberg I, Klein G, Horwitz CA, Marklund G, Rymo L, Wellinder C, Straus SE (1987) Antibody responses to Epstein-Barr virus-determined nuclear antigen (EBNA)-1 and EBNA-2 in acute and chronic Epstein-Barr virus infection. Proc Natl Acad Sci USA 84:570–574

Hess RD (2004) Routine Epstein-Barr virus diagnostics from the laboratory perspective: still challenging after 35 years. J Clin Microbiol 42:3381–3387. doi:10.1128/JCM.42.8.3381-3387.2004

Hesse J, Ibsen KK, Krabbe S, Uldall P (1983) Prevalence of antibodies to Epstein-Barr virus (EBV) in childhood and adolescence in Denmark. Scand J Infect Dis 15:335–338

Hille A, Klein K, Baumler S, Grasser FA, Mueller-Lantzsch N (1993) Expression of Epstein-Barr virus nuclear antigen 1,2A and 2B in the baculovirus expression system: serological evaluation of human antibodies to these proteins. J Med Virol 39:233–241
Hislop AD, Palendira U, Leese AM, Arkwright PD, Rohrlich PS, Tangye SG, Gaspar HB, Lankester AC, Moretta A, Rickinson AB (2010) Impaired Epstein-Barr virus-specific CD8+ T-cell function in X-linked lymphoproliferative disease is restricted to SLAM family-positive B-cell targets. Blood 116:3249–3257. doi:10.1182/blood-2009-09-238832
Hislop AD, Taylor GS, Sauce D, Rickinson AB (2007) Cellular responses to viral infection in humans: lessons from Epstein-Barr virus. Annu Rev Immunol 25:587–617. doi:10.1146/annurev.immunol.25.022106.141553
Hjalgrim H, Askling J, Sorensen P, Madsen M, Rosdahl N, Storm HH, Hamilton-Dutoit S, Eriksen LS, Frisch M, Ekbom A, Melbye M (2000) Risk of Hodgkin's disease and other cancers after infectious mononucleosis. J Natl Cancer Inst 92:1522–1528
Hjalgrim H, Friborg J, Melbye M (2007) The epidemiology of EBV and its association with malignant disease. In: Arvin A et al. (eds) Human herpesviruses: biology, therapy, and immunoprophylaxis. Cambridge
Hoagland RJ (1955) The transmission of infectious mononucleosis. Am J Med Sci 229:262–272
Hoagland RJ, Henson HM (1957) Splenic rupture in infectious mononucleosis. Ann Intern Med 46:1184–1191
Holman CJ, Karger AB, Mullan BD, Brundage RC, Balfour HH Jr (2012) Quantitative Epstein-Barr virus shedding and its correlation with the risk of post-transplant lymphoproliferative disorder. Clin Transplant 26:741–747. doi:10.1111/j.1399-0012.2012.01608.x
Horwitz CA, Henle W, Henle G, Goldfarb M, Kubic P, Gehrz RC, Balfour HH Jr, Fleisher GR, Krivit W (1981) Clinical and laboratory evaluation of infants and children with Epstein-Barr virus-induced infectious mononucleosis: report of 32 patients (aged 10–48 months). Blood 57:933–938
Horwitz CA, Henle W, Henle G, Penn G, Hoffman N, Ward PC (1979) Persistent falsely positive rapid tests for infectious mononucleosis. Report of five cases with four-six-year follow-up data. Am J Clin Pathol 72:807–811
Hutt-Fletcher LM (2007) Epstein-Barr virus entry. J Virol 81:7825–7832. doi:10.1128/JVI.00445-07
Hwang AE, Hamilton AS, Cockburn MG, Ambinder R, Zadnick J, Brown EE, Mack TM, Cozen W (2012) Evidence of genetic susceptibility to infectious mononucleosis: a twin study. Epidemiol Infect 140:2089–2095. doi:10.1017/S0950268811002457
Iizasa H, Nanbo A, Nishikawa J, Jinushi M, Yoshiyama H (2012) Epstein-Barr virus (EBV)-associated gastric carcinoma. Viruses 4:3420–3439
Janka GE (2012) Familial and acquired hemophagocytic lymphohistiocytosis. Annu Rev Med 63:233–246. doi:10.1146/annurev-med-041610-134208
Jenson HB (2000) Acute complications of Epstein-Barr virus infectious mononucleosis. Curr Opin Pediatr 12:263–268
Jordan MB, Allen CE, Weitzman S, Filipovich AH, McClain KL (2011) How i treat hemophagocytic lymphohistiocytosis. Blood 118:4041–4052. doi:10.1182/blood-2011-03-278127
Kawaguchi H, Miyashita T, Herbst H, Niedobitek G, Asada M, Tsuchida M, Hanada R, Kinoshita A, Sakurai M, Kobayashi N et al (1993) Epstein-Barr virus-infected T lymphocytes in Epstein-Barr virus-associated hemophagocytic syndrome. J Clin Invest 92:1444–1450. doi:10.1172/JCI116721
Kimura H, Hoshino Y, Kanegane H, Tsuge I, Okamura T, Kawa K, Morishima T (2001) Clinical and virologic characteristics of chronic active Epstein-Barr virus infection. Blood 98:280–286
Kimura H, Morishima T, Kanegane H, Ohga S, Hoshino Y, Maeda A, Imai S, Okano M, Morio T, Yokota S, Tsuchiya S, Yachie A, Imashuku S, Kawa K, Wakiguchi H, Japanese Association for Research on Epstein-Barr V, Related D (2003) Prognostic factors for chronic active Epstein-Barr virus infection. J Infect Dis 187:527–533. doi:10.1086/367988
Krabbe S, Hesse J, Uldall P (1981) Primary Epstein-Barr virus infection in early childhood. Arch Dis Child 56:49–52

Lang DJ, Garruto RM, Gajdusek DC (1977) Early acquisition of cytomegalovirus and Epstein-Barr virus antibody in several isolated Melanesian populations. Am J Epidemiol 105:480–487

Leskowitz R, Fogg MH, Zhou XY, Kaur A, Silveira EL, Villinger F, Lieberman PM, Wang F, Ertl HC (2014) Adenovirus-based vaccines against rhesus lymphocryptovirus EBNA-1 induce expansion of specific CD8+ and CD4+ T cells in persistently infected rhesus macaques. J Virol 88:4721–4735. doi:10.1128/JVI.03744-13

Leung C, Chijioke O, Gujer C, Chatterjee B, Antsiferova O, Landtwing V, McHugh D, Raykova A, Munz C (2013) Infectious diseases in humanized mice. Eur J Immunol 43:2246–2254. doi:10.1002/eji.201343815

Levin LI, Munger KL, Rubertone MV, Peck CA, Lennette ET, Spiegelman D, Ascherio A. (2005). Temporal relationship between elevation of epstein-barr virus antibody titers and initial onset of neurological symptoms in multiple sclerosis. JAMA 293: 2496–2500. doi:10.1001/jama.293.20.2496

Levin LI, Munger KL, O'Reilly EJ, Falk KI, Ascherio A (2010) Primary infection with the Epstein-Barr virus and risk of multiple sclerosis. Ann Neurol 67:824–830. doi:10.1002/ana.21978

Levitskaya J, Sharipo A, Leonchiks A, Ciechanover A, Masucci MG (1997) Inhibition of ubiquitin/proteasome-dependent protein degradation by the Gly-Ala repeat domain of the Epstein-Barr virus nuclear antigen 1. Proc Natl Acad Sci USA 94:12616–12621

Long HM, Chagoury OL, Leese AM, Ryan GB, James E, Morton LT, Abbott RJ, Sabbah S, Kwok W, Rickinson AB (2013) MHC II tetramers visualize human CD4+ T cell responses to Epstein-Barr virus infection and demonstrate atypical kinetics of the nuclear antigen EBNA1 response. J Exp Med 210:933–949. doi:10.1084/jem.20121437

Lopez-Verges S, Milush JM, Schwartz BS, Pando MJ, Jarjoura J, York VA, Houchins JP, Miller S, Kang SM, Norris PJ, Nixon DF, Lanier LL (2011) Expansion of a unique CD57(+)NKG2Chi natural killer cell subset during acute human cytomegalovirus infection. Proc Natl Acad Sci USA 108:14725–14732. doi:10.1073/pnas.1110900108

Lunemann A, Vanoaica LD, Azzi T, Nadal D, Munz C (2013) A distinct subpopulation of human NK cells restricts B cell transformation by EBV. J Immunol 191:4989–4995. doi:10.4049/jimmunol.1301046

Luzuriaga K, Sullivan JL (2010) Infectious mononucleosis. N Engl J Med 362:1993–2000. doi:10.1056/NEJMcp1001116

Massa J, Munger KL, O'Reilly EJ, Falk KI, Ascherio A (2007) Plasma titers of antibodies against Epstein-Barr virus BZLF1 and risk of multiple sclerosis. Neuroepidemiology 28:214–215. doi:10.1159/000108113

Melbye M, Ebbesen P, Levine PH, Bennike T (1984) Early primary infection and high Epstein-Barr virus antibody titers in Greenland Eskimos at high risk for nasopharyngeal carcinoma. Int J Cancer 34:619–623

Menasche G, Feldmann J, Fischer A, de Saint Basile G (2005) Primary hemophagocytic syndromes point to a direct link between lymphocyte cytotoxicity and homeostasis. Immunol Rev 203:165–179. doi:10.1111/j.0105-2896.2005.00224.x

Merlo A, Turrini R, Trento C, Zanovello P, Dolcetti R, Rosato A (2010) Impact of gamma-chain cytokines on EBV-specific T cell cultures. J Transl Med 8:121. doi:10.1186/1479-5876-8-121

Morris MC, Edmunds WJ (2002) The changing epidemiology of infectious mononucleosis? J Infect 45:107–109. doi:10.1053/jinf.2002.1022

Nijie R, Bell AI, Jia H, Croom-Carter D, Chaganti S, Hislop AD, Whittle H, Rickinson AB (2009). The effects of acute malaria on Epstein-Barr virus (EBV) load and EBV-specific T cell immunity in Gambian children. J Infect Dis 199:31–38. doi:10.1086/594373

Nystad TW, Myrmel H (2007) Prevalence of primary versus reactivated Epstein-Barr virus infection in patients with VCA IgG-, VCA IgM- and EBNA-1-antibodies and suspected infectious mononucleosis. J Clin Virol 38:292–297. doi:10.1016/j.jcv.2007.01.006

Odumade OA, Hogquist KA, Balfour HH Jr (2011) Progress and problems in understanding and managing primary Epstein-Barr virus infections. Clin Microbiol Rev 24:193–209. doi:10.1128/cmr.00044-10

Odumade OA, Knight JA, Schmeling DO, Masopust D, Balfour HH Jr, Hogquist KA (2012) Primary Epstein-Barr virus infection does not erode preexisting CD8(+) T cell memory in humans. J Exp Med 209:471–478. doi:10.1084/jem.20112401

Ohashi M, Fogg MH, Orlova N, Quink C, Wang F (2012) An Epstein-Barr virus encoded inhibitor of colony stimulating factor-1 signaling is an important determinant for acute and persistent EBV infection. PLoS Pathog 8:e1003095. doi:10.1371/journal.ppat.1003095

Okuno K, Takashima K, Kanai K, Ohashi M, Hyuga R, Sugihara H, Kuwamoto S, Kato M, Sano H, Sairenji T, Kanzaki S, Hayashi K (2010) Epstein-Barr virus can infect rabbits by the intranasal or peroral route: an animal model for natural primary EBV infection in humans. J Med Virol 82:977–986. doi:10.1002/jmv.21597

Pakpoor J, Disanto G, Gerber JE, Dobson R, Meier UC, Giovannoni G, Ramagopalan SV (2013) The risk of developing multiple sclerosis in individuals seronegative for Epstein-Barr virus: a meta-analysis. Mult Scler 19:162–166. doi:10.1177/1352458512449682

Palendira U, Low C, Chan A, Hislop AD, Ho E, Phan TG, Deenick E, Cook MC, Riminton DS, Choo S, Loh R, Alvaro F, Booth C, Gaspar HB, Moretta A, Khanna R, Rickinson AB, Tangye SG (2011) Molecular pathogenesis of EBV susceptibility in XLP as revealed by analysis of female carriers with heterozygous expression of SAP. PLoS Biol 9:e1001187. doi:10.1371/journal.pbio.1001187

Pappworth IY, Wang EC, Rowe M (2007) The switch from latent to productive infection in epstein-barr virus-infected B cells is associated with sensitization to NK cell killing. J Virol 81:474–482. doi:10.1128/JVI.01777-06

Parvaneh N, Filipovich AH, Borkhardt A (2013) Primary immunodeficiencies predisposed to Epstein-Barr virus-driven haematological diseases. Br J Haematol 162:573–586. doi:10.1111/bjh.12422

Pathmanathan R, Prasad U, Sadler R, Flynn K, Raab-Traub N (1995) Clonal proliferations of cells infected with Epstein-Barr virus in preinvasive lesions related to nasopharyngeal carcinoma. N Engl J Med 333:693–698. doi:10.1056/NEJM199509143331103

Pender MP, Csurhes PA, Smith C, Beagley L, Hooper KD, Raj M, Coulthard A, Burrows SR, Khanna R (2014) Epstein-Barr virus-specific adoptive immunotherapy for progressive multiple sclerosis. Mult Scler. doi:10.1177/1352458514521888

Phillips GM (1972) False-positive monospot test result in rubella. JAMA 222:585

Piriou E, Asito AS, Sumba PO, Fiore N, Middeldorp JM, Moormann AM, Ploutz-Snyder R, Rochford R (2012) Early age at time of primary Epstein-Barr virus infection results in poorly controlled viral infection in infants from Western Kenya: clues to the etiology of endemic Burkitt lymphoma. J Infect Dis 205:906–913. doi:10.1093/infdis/jir872

Pohl D, Krone B, Rostasy K, Kahler E, Brunner E, Lehnert M, Wagner HJ, Gartner J, Hanefeld F (2006) High seroprevalence of Epstein-Barr virus in children with multiple sclerosis. Neurology 67:2063–2065. doi:10.1212/01.wnl.0000247665.94088.8d

Putukian M, O'Connor FG, Stricker P, McGrew C, Hosey RG, Gordon SM, Kinderknecht J, Kriss V, Landry G (2008) Mononucleosis and athletic participation: an evidence-based subject review. Clin J Sport Med 18:309–315. doi:10.1097/JSM.0b013e31817e34f8

Quintanilla-Martinez L, Kumar S, Fend F, Reyes E, Teruya-Feldstein J, Kingma DW, Sorbara L, Raffeld M, Straus SE, Jaffe ES (2000) Fulminant EBV(+) T-cell lymphoproliferative disorder following acute/chronic EBV infection: a distinct clinicopathologic syndrome. Blood 96:443–451

Raab-Traub N (2002) Epstein-Barr virus in the pathogenesis of NPC. Semin Cancer Biol 12:431–441

Raab-Traub N, Flynn K (1986) The structure of the termini of the Epstein-Barr virus as a marker of clonal cellular proliferation. Cell 47:883–889

Ramer PC, Chijioke O, Meixlsperger S, Leung CS, Munz C (2011) Mice with human immune system components as in vivo models for infections with human pathogens. Immunol Cell Biol 89:408–416. doi:10.1038/icb.2010.151

Rigaud S, Fondaneche MC, Lambert N, Pasquier B, Mateo V, Soulas P, Galicier L, Le Deist F, Rieux-Laucat F, Revy P, Fischer A, de Saint Basile G, Latour S (2006) XIAP deficiency in humans causes an X-linked lymphoproliferative syndrome. Nature 444:110–114. doi:10.1038/nature05257

Robinson RG (1988) Abdominal complications of infectious mononucleosis. J Am Board Fam Pract 1:207–210

Rostgaard K, Wohlfahrt J, Hjalgrim H (2014) A genetic basis for infectious mononucleosis: evidence from a family study of hospitalized cases in Denmark. Clin Infect Dis. doi:10.1093/cid/ciu204

Rubicz R, Yolken R, Drigalenko E, Carless MA, Dyer TD, Bauman L, Melton PE, Kent JW Jr, Harley JB, Curran JE, Johnson MP, Cole SA, Almasy L, Moses EK, Dhurandhar NV, Kraig E, Blangero J, Leach CT, Goring HH (2013) A genome-wide integrative genomic study localizes genetic factors influencing antibodies against Epstein-Barr virus nuclear antigen 1 (EBNA-1). PLoS Genet 9:e1003147. doi:10.1371/journal.pgen.1003147

Sadoff L, Goldsmith O (1971) False-positive infectious mononucleosis spot test in pancreatic carcinoma. JAMA 218:1297–1298

Sashihara J, Hoshino Y, Bowman JJ, Krogmann T, Burbelo PD, Coffield VM, Kamrud K, Cohen JI (2011) Soluble rhesus lymphocryptovirus gp350 protects against infection and reduces viral loads in animals that become infected with virus after challenge. PLoS Pathog 7:e1002308. doi:10.1371/journal.ppat.1002308

Schooley RT, Carey RW, Miller G, Henle W, Eastman R, Mark EJ, Kenyon K, Wheeler EO, Rubin RH (1986) Chronic Epstein-Barr virus infection associated with fever and interstitial pneumonitis. Clinical and serologic features and response to antiviral chemotherapy. Ann Intern Med 104:636–643

Schubert J, Zens W, Weissbrich B (1998) Comparative evaluation of the use of immunoblots and of IgG avidity assays as confirmatory tests for the diagnosis of acute EBV infections. J Clin Virol 11:161–172

Shapiro RS, McClain K, Frizzera G, Gajl-Peczalska KJ, Kersey JH, Blazar BR, Arthur DC, Patton DF, Greenberg JS, Burke B et al (1988) Epstein-Barr virus associated B cell lymphoproliferative disorders following bone marrow transplantation. Blood 71:1234–1243

Shim AH, Chang RA, Chen X, Longnecker R, He X (2012) Multipronged attenuation of macrophage-colony stimulating factor signaling by Epstein-Barr virus BARF1. Proc Natl Acad Sci USA 109:12962–12967. doi:10.1073/pnas.1205309109

Shultz LD, Saito Y, Najima Y, Tanaka S, Ochi T, Tomizawa M, Doi T, Sone A, Suzuki N, Fujiwara H, Yasukawa M, Ishikawa F (2010) Generation of functional human T-cell subsets with HLA-restricted immune responses in HLA class I expressing NOD/SCID/IL2r gamma(null) humanized mice. Proc Natl Acad Sci USA 107:13022–13027. doi:10.1073/pnas.1000475107

Slyker JA, Casper C, Tapia K, Richardson B, Bunts L, Huang ML, Maleche-Obimbo E, Nduati R, John-Stewart G (2013) Clinical and virologic manifestations of primary Epstein-Barr virus (EBV) infection in Kenyan infants born to HIV-infected women. J Infect Dis 207:1798–1806. doi:10.1093/infdis/jit093

Sokal EM, Hoppenbrouwers K, Vandermeulen C, Moutschen M, Leonard P, Moreels A, Haumont M, Bollen A, Smets F, Denis M (2007) Recombinant gp350 vaccine for infectious mononucleosis: a phase 2, randomized, double-blind, placebo-controlled trial to evaluate the safety, immunogenicity, and efficacy of an Epstein-Barr virus vaccine in healthy young adults. J Infect Dis 196:1749–1753. doi:10.1086/523813

Sprunt TP, Evans FA (1920) Mononuclear leucocytosis in reaction to acute infections ("infectious mononucleosis"). Johns Hopkins Hosp Bull 31:410–417

Strautins K, Tschochner M, James I, Choo L, Dunn DS, Pedrini M, Kermode A, Carroll W, Nolan D (2014) Combining HLA-DR risk alleles and anti-Epstein-Barr virus antibody profiles to stratify multiple sclerosis risk. Mult Scler 20:286–294. doi:10.1177/1352458513498829

Strowig T, Brilot F, Arrey F, Bougras G, Thomas D, Muller WA, Munz C (2008) Tonsilar NK cells restrict B cell transformation by the Epstein-Barr virus via IFN-gamma. PLoS Pathog 4:e27. doi:10.1371/journal.ppat.0040027

Strowig T, Chijioke O, Carrega P, Arrey F, Meixlsperger S, Ramer PC, Ferlazzo G, Munz C (2010) Human NK cells of mice with reconstituted human immune system components require preactivation to acquire functional competence. Blood 116:4158–4167. doi:10.1182/blood-2010-02-270678

Strowig T, Gurer C, Ploss A, Liu YF, Arrey F, Sashihara J, Koo G, Rice CM, Young JW, Chadburn A, Cohen JI, Munz C (2009) Priming of protective T cell responses against virus-induced tumors in mice with human immune system components. J Exp Med 206:1423–1434. doi:10.1084/jem.20081720

Sumaya CV, Ench Y (1986) Epstein-Barr virus infections in families: the role of children with infectious mononucleosis. J Infect Dis 154:842–850

Sumaya CV, Henle W, Henle G, Smith MH, LeBlanc D (1975) Seroepidemiologic study of Epstein-Barr virus infections in a rural community. J Infect Dis 131:403–408

Sundstrom P, Juto P, Wadell G, Hallmans G, Svenningsson A, Nystrom L, Dillner J, Forsgren L (2004) An altered immune response to Epstein-Barr virus in multiple sclerosis: a prospective study. Neurology 62:2277–2282

Svedmyr E, Ernberg I, Seeley J, Weiland O, Masucci G, Tsukuda K, Szigeti R, Masucci MG, Blomogren H, Berthold W (1984) Virologic, immunologic, and clinical observations on a patient during the incubation, acute, and convalescent phases of infectious mononucleosis. Clin Immunol Immunopathol 30:437–450

Takashima K, Ohashi M, Kitamura Y, Ando K, Nagashima K, Sugihara H, Okuno K, Sairenji T, Hayashi K (2008) A new animal model for primary and persistent Epstein-Barr virus infection: human EBV-infected rabbit characteristics determined using sequential imaging and pathological analysis. J Med Virol 80:455–466. doi:10.1002/jmv.21102

Takeuchi K, Tanaka-Taya K, Kazuyama Y, Ito YM, Hashimoto S, Fukayama M, Mori S (2006) Prevalence of Epstein-Barr virus in Japan: trends and future prediction. Pathol Int 56:112–116. doi:10.1111/j.1440-1827.2006.01936.x

Tynell E, Aurelius E, Brandell A, Julander I, Wood M, Yao QY, Rickinson A, Akerlund B, Andersson J (1996) Acyclovir and prednisolone treatment of acute infectious mononucleosis: a multicenter, double-blind, placebo-controlled study. J Infect Dis 174:324–331

Venkitaraman AR, Lenoir GM, John TJ (1985) The seroepidemiology of infection due to Epstein-Barr virus in southern India. J Med Virol 15:11–16

Wang F (2013) Nonhuman primate models for Epstein-Barr virus infection. Curr Opin Virol 3:233–237. doi:10.1016/j.coviro.2013.03.003

Wang X, Kenyon WJ, Li Q, Mullberg J, Hutt-Fletcher LM (1998) Epstein-Barr virus uses different complexes of glycoproteins gH and gL to infect B lymphocytes and epithelial cells. J Virol 72:5552–5558

Warner HB, Carp RI (1981) Multiple sclerosis and Epstein-Barr virus. Lancet 2:1290

White LR, Karofsky PS (1985) Review of the clinical manifestations, laboratory findings, and complications of infectious mononucleosis. Wis Med J 84:19–25

Williams H, McAulay K, Macsween KF, Gallacher NJ, Higgins CD, Harrison N, Swerdlow AJ, Crawford DH (2005) The immune response to primary EBV infection: a role for natural killer cells. Br J Haematol 129:266–274. doi:10.1111/j.1365-2141.2005.05452.x

Yager EJ, Szaba FM, Kummer LW, Lanzer KG, Burkum CE, Smiley ST, Blackman MA (2009) Gamma-herpesvirus-induced protection against bacterial infection is transient. Viral Immunol 22:67–72. doi:10.1089/vim.2008.0086

Yajima M, Imadome K, Nakagawa A, Watanabe S, Terashima K, Nakamura H, Ito M, Shimizu N, Honda M, Yamamoto N, Fujiwara S (2008) A new humanized mouse model of Epstein-Barr virus infection that reproduces persistent infection, lymphoproliferative disorder, and cell-mediated and humoral immune responses. J Infect Dis 198:673–682. doi:10.1086/590502

Zhao F, Cannons JL, Dutta M, Griffiths GM, Schwartzberg PL (2012) Positive and negative signaling through SLAM receptors regulate synapse organization and thresholds of cytolysis. Immunity 36:1003–1016. doi:10.1016/j.immuni.2012.05.017

Primary Immunodeficiencies Associated with EBV Disease

Jeffrey I. Cohen

Abstract Epstein-Barr virus (EBV) infects nearly all humans and usually is asymptomatic, or in the case of adolescents and young adults, it can result in infectious mononucleosis. EBV-infected B cells are controlled primarily by NK cells, iNKT cells, CD4 T cells, and CD8 T cells. While mutations in proteins important for B cell function can affect EBV infection of these cells, these mutations do not result in severe EBV infection. Some genetic disorders affecting T and NK cell function result in failure to control EBV infection, but do not result in increased susceptibility to other virus infections. These include mutations in *SH2D1A, BIRC4, ITK, CD27, MAGT1, CORO1A,* and *LRBA*. Since EBV is the only virus that induces proliferation of B cells, the study of these diseases has helped to identify proteins critical for interactions of T and/or NK cells with B cells. Mutations in three genes associated with hemophagocytic lymphohistiocytosis, *PRF1, STXBP2, and UNC13D*, can also predispose to severe chronic active EBV disease. Severe EBV infection can be associated with immunodeficiencies that also predispose to other viral infections and in some cases other bacterial and fungal infections. These include diseases due to mutations in *PIK3CD, PIK3R1, CTPS1, STK4, GATA2, MCM4, FCGR3A, CARD11, ATM, and WAS*. In addition, patients with severe combined immunodeficiency, which can be due to mutations in a number of different genes, are at high risk for various infections as well as EBV B cell lymphomas. Identification of proteins important for control of EBV may help to identify new targets for immunosuppressive therapies.

J.I. Cohen (✉)
Medical Virology Section, Laboratory of Infectious Diseases, National Institutes of Health, 50 South Drive, MSC 8007, Bethesda, MD 20892, USA
e-mail: jcohen@niaid.nih.gov

C. Münz (ed.), *Epstein Barr Virus Volume 1*, Current Topics in Microbiology and Immunology 390, DOI 10.1007/978-3-319-22822-8_10

Contents

1 Introduction 243
2 Immunodeficiencies Specific for EBV Disease 244
  2.1 X-Linked Lymphoproliferative Disease 1 (XLP1) 244
  2.2 X-Linked Lymphoproliferative Disease 2 (XLP2) 251
  2.3 IL-2-Inducible T Cell Kinase (ITK) 252
  2.4 CD27 252
  2.5 Magnesium Transporter 1 (MagT1) Protein 253
  2.6 Coronin Actin Binding Protein 1A 254
  2.7 LPS-Responsive Beige-Like Anchor (LRBA) Protein 254
3 Proteins Associated with EBV Disease and Familial Hemophagocytic Lymphohistiocytosis (FHL) 254
  3.1 Perforin 254
  3.2 Munc13-4 255
  3.3 Munc18-2 255
4 Genes Associated with EBV and Other Infections 256
  4.1 Phosphatidylinositol-3-OH Kinase (PI3K) Catalytic Subunit P110δ 256
  4.2 Cytidine 5′-Triphosphate Synthase 1 (CTPS1) 258
  4.3 Serine/Threonine Kinase 4 (STK4) 258
  4.4 GATA Binding Protein 2 (GATA2) 259
  4.5 Minichromosome Maintenance Complex Component 4 (MCM4) 259
  4.6 Fcγ Receptor 3A (CD16a) 259
  4.7 Caspase Recruitment Domain-Containing Protein 11 (CARD11): Gain of Function Mutations 260
  4.8 Other Immunodeficiencies Associated with Multiple Infections 260
References 260

Abbreviations

APDS	Activated PI3Kδ syndrome
CAEBV	Chronic active EBV disease
CARD11	Caspase recruitment domain family, member 11
CTL	Cytotoxic T lymphocyte
CTPS1	Cytidine 5′-triphosphate synthase
EBV	Epstein-Barr virus
FHL	Familial hemophagocytic lymphohistiocytosis
GATA2	GATA binding protein 2
HLH	Hemophagocytic lymphohistiocytosis
HPV	Human papillomavirus
HSCT	Hematopoietic stem cell transplantation
iNKT	Invariant NKT
ITK	IL-2-inducible T cell kinase
LRBA	LPS-responsive beige-like anchor
MagT1	Magnesium transporter 1
MCM4	Minichromosome maintenance complex component 4
MST1	Mammalian sterile 20-like protein

MTOR	Mammalian target of rapamycin
NK	Natural killer
PASLI	P110δ-activating mutation causing senescent T cells, lymphadenopathy, and immune deficiency
PI3K	Phosphatidylinositol-3-OH kinase
PIP2	Phosphatidylinositol-(4,5)-biphosphate
PIP3	Phosphatidylinositol-(3,4,5)-triphosphate
PKC	Protein kinase C
PLC	Phospholipase C
PML	Progressive multifocal leukoencephalopathy
SAP	SLAM-associated protein
SCID	Severe combined immunodeficiency
SH2	Src homology 2
SM	Sec1/munc18
SNARE	Soluble NSF attachment protein receptor
STK4	Serine/threonine kinase 4
XIAP	X-linked inhibitor of apoptosis
XLP	X-linked lymphoproliferative disease
XMEN	X-linked immunodeficiency with magnesium defect, EBV infection, and neoplasia

1 Introduction

Epstein-Barr virus (EBV) infects 95 % of the human population, and most individuals are infected asymptomatically and are unaware that they shed virus from the oropharynx throughout their life. Infection of adolescents and young adults often results in infectious mononucleosis with fever, sore throat, lymphadenopathy, and splenomegaly. EBV infection of B cells can result in a lytic infection with death of the cells. More often, virus infection of B cells results in a latent infection with expression of a very limited number of viral proteins. Certain B cell abnormalities affect the interaction of EBV with the host. Patients with Bruton's agammaglobulinemia lack mature B cells, and their B cells cannot be infected with EBV (Faulkner et al. 1999). Patients with the hyperIgM syndrome have mutations in CD40L, and their cells are impaired for transformation by EBV in vitro (Imadome et al. 2003). B cells from patients with mutations in STAT3 have impaired proliferation in response to virus infection (Koganti et al. 2014). However, no B cell defects have been described to date that result in more severe EBV infection.

Control of EBV-infected cells is mediated by natural killer (NK cells), invariant NKT (iNKT) cells, CD4 T cells, and CD8 T cells (Chung et al. 2013; Hislop et al. 2007). Impairment in the function of these cell types can result in failure to adequately control EBV infection. This can lead to persistent EBV viremia, lymphoproliferation, and ultimately EBV lymphomas.

Surprisingly, many of the genes important for controlling EBV infection are not critical for controlling other pathogens including other members of the herpesvirus family. This indicates a unique characteristic of EBV relative to other pathogens. EBV is the only human virus that induces growth proliferation of B cells. Therefore, host cell proteins important for interactions of T or NK cells with B cells are often critical for control of EBV. Identification of cellular gene mutations associated with severe EBV infection helps us to identify proteins important for the B cell–T cell or B cell–NK cell immunologic synapse. These cellular proteins may also serve as targets for new immunosuppressive medications.

2 Immunodeficiencies Specific for EBV Disease

See Tables 1 and 2.

2.1 X-Linked Lymphoproliferative Disease 1 (XLP1)

The first genetic disorder linked with severe and often fatal EBV disease was XLP1, also known as Duncan's disease. Purtillo and colleagues described an X-linked lymphoproliferative disease (XLP) in boys that often presented with fulminant infectious mononucleosis with lymphocytic and histiocytic infiltration of the bone marrow, central nervous system, and other organs (Purtilo et al. 1975). Female carriers are not affected. If untreated, patients often die from bone marrow failure with bleeding and infection, or liver failure. Subsequent studies showed that the disease has a variety of phenotypes including fatal infectious mononucleosis, aplastic anemia, or hypogammaglobulinemia after primary EBV infection, immunoblastic B cell lymphoma, Burkitt lymphoma, or plasmacytoma and that the underlying cause is an abnormal immune response to infection with EBV (Purtilo 1976; Purtilo et al. 1977). More recent studies show that the presenting symptoms of XLP1 include hemophagocytic lymphohistiocytosis (HLH) in 31.9 % of persons, dysgammaglobulinemia in 22 %, a family history of XLP1 alone in 16.5 %, lymphoma in 14.3 %, fulminant infectious mononucleosis in 7.7 %, and other symptoms in 7.7 % (Booth et al. 2011). These other symptoms include aplastic anemia, lymphomatoid granulomatosis, and vasculitis. Patients have impaired antibody responses to vaccination. Most lymphomas are EBV-positive non-Hodgkin's B lymphomas, although T cell lymphomas and EBV-negative B cell lymphomas have also been reported. In most cases, there is not a clear correlation between specific mutations and clinical phenotype (Booth et al. 2011; Filipovich et al. 2010). Within families, different members can have different severity of presentation despite the same mutation.

The gene responsible for XLP1 was identified in 1988 and is referred to as *SH2D1A* (Coffey et al. 1998), *DSHP* (Nichols et al. 1998), and *SAP* (Sayos et al.

Table 1 Clinical features of immunodeficiencies predisposing to EBV disease

Protein	Gene	Name of syndrome	Transmission	EBV-associated symptoms	Non-infectious diseases in the absence of EBV infection
SLAM-associated protein	*SH2D1A*	XLP1	X-linked	Fulminant infectious mononucleosis, B cell lymphoma, HLH, lymphomatoid granulomatosis	EBV-negative B cell lymphoma, aplastic anemia, vasculitis
X-linked inhibitor of apoptosis	*BIRC4*	XLP2	X-linked	Fulminant infectious mononucleosis, HLH, splenomegaly, cytopenias	Colitis, HLH, inflammatory bowel disease
IL-2-inducible T cell kinase	*ITK*	EBV-associated autosomal lymphoproliferative syndrome	Autosomal recessive	Lymphoproliferation, Hodgkin's lymphoma, HLH, hepatosplenomegaly, lung disease, lymphomatoid granulomatosis	Autoimmune kidney disease
CD27	*CD27*	CD27 deficiency	Autosomal recessive	Lymphoproliferative disease, HLH, lymphoma, aplastic anemia	None
Magnesium transporter 1 protein	*MAGT1*	XMEN	X-linked	B cell lymphoma	Autoimmune cytopenias
Coronin actin binding protein 1A	*CORO1A*	Coronin 1A deficiency	Autosomal recessive	B cell lymphoma, lymphoproliferative disease	Neurocognitive impairment
LPS-responsive beige-like anchor protein	LRBA		Autosomal recessive	B cell lymphoproliferative disease, EBV viremia	Inflammatory bowel disease, chronic diarrhea, autoimmune cytopenias
Perforin	*PRF1*	FHL2	Autosomal recessive	HLH, CAEBV, splenomegaly	HLH
Munc13-4	*UNC13D*	FHL3	Autosomal recessive	CAEBV, vasculitis, hepatitis, splenomegaly	HLH
Munc18-2	*STXBP2*	FHL5	Autosomal recessive	CABEV, lymphoma, splenomegaly	HLH, colitis, bleeding
PI3K catalytic subunit 110δ: gain of function mutations	*PIK3CD*	PASLI, APDS	Autosomal dominant	Lymphoma	Lymphoid nodules in upper airway and gastrointestinal tract, autoimmune cytopenias

(continued)

Table 1 (continued)

Protein	Gene	Name of syndrome	Transmission	EBV-associated symptoms	Non-infectious diseases in the absence of EBV infection
PI3K catalytic subunit 110δ: loss of function mutation	*PIK3CD*		Autosomal recessive	EBV viremia	Lymphadenopathy, hepatosplenomegaly, autoantibodies
CTP synthase 1	*CTPS1*	CTP synthase 1 deficiency	Autosomal recessive	Severe infectious mononucleosis; lymphoproliferative disease, lymphoma	None
Serine threonine kinase 4	*STK4*	STK4 deficiency	Autosomal recessive	B cell lymphoma, lymphoproliferative disease, autoimmune hemolytic anemia	Dermatitis, autoimmune cytopenias
GATA binding protein 2	*GATA2*	MonoMac	Autosomal dominant	Severe infectious mononucleosis, EBV-positive smooth muscle tumors, CAEBV, HLH	Myeloid malignancies, myelodysplastic syndrome, autoimmune disease, pulmonary alveolar proteinosis, primary lymphedema.
Minichromosome maintenance complex component 4	*MCM4*	Classical NK cell deficiency type 2	Autosomal recessive	EBV lymphoproliferative disease, lymphoma	Adrenal insufficiency, growth retardation
Fcγ receptor 3A (CD16a)	*FCGR3A*		Autosomal recessive	EBV-positive Castleman's disease	None
Caspase recruitment domain-containing protein 11: gain of function mutations	*CARD11*		Autosomal dominant	EBV viremia	Autoimmune neutropenia

Abbreviations: *XLP* X-linked lymphoproliferative disease; *HLH* hemophagocytotic lymphohistiocytosis; *XMEN* X-linked immunodeficiency with magnesium defect; *EBV* infection, and neoplasia; *CAEBV* chronic active EBV disease; *FHL* familial hemophagocytic lymphohistiocytosis; *PI3K* phosphoinositide 3-kinase; *PASLI* p110δ-activating mutation causing senescent T cells, lymphadenopathy, and immune deficiency; *APDS* activated PI3Kδ syndrome; *CTP* cytidine 5′-triphosphate

Table 2 Immunologic and infectious features of immunodeficiencies predisposing to EBV disease

Protein	Humoral immune findings	Cellular immune findings	Chronic EBV viremia	Other infections	References
SLAM-associated protein	Low IgG, increased IgA, increased IgM, reduced antibody response to vaccinations and infections	Absent iNKT cells, normal numbers of B, T cells; reduced memory B cells; impaired T and NK cell killing	No	None	Coffey et al. (1998); Nichols et al. (1998); Sayos et al. (1998)
X-linked inhibitor of apoptosis	Low IgG	Normal or low numbers of iNKT cells; normal numbers of B, T, NK cells	No	None	Rigaud et al. (2006); Speckmann et al. (2013)
IL-2-inducible T cell kinase	Low or normal IgG	Absent iNKT cells; low numbers of WBC, CD4 T cells; normal numbers of B cells	Yes	None	Huck et al. (2009); Linka et al. (2012)
CD27	Low or normal IgG	Normal or reduced iNKT and memory B cells	Yes	None	van Montfrans et al. (2012); Salzer et al. (2013)
Magnesium transporter 1 protein	Low or normal IgG	Low numbers of CD4 cells; normal numbers of iNKT cells; reduced levels of NKG2D on NK and T cells; impaired T cell killing of EBV-transformed B cells; impaired NK cell function	Yes	Bacterial sinusitis, chronic diarrhea	Li et al. (2011, 2014); Chaigne-Delalande et al. (2013)
Coronin actin binding protein 1A	Normal or low normal IgG, normal or low IgM	Reduced numbers of CD4 T, CD8 T, naïve T cells, and B cells; low or absent iNKT cells	Yes	Recurrent ear, nose, throat, and upper respiratory tract infections	Moshous et al. (2013)
LPS-responsive beige-like anchor protein	Normal or low IgG	Normal or low B cell numbers; normal T, NK cell numbers	Yes	Usually none, may have recurrent otitis media and pneumonia	Alangari et al. (2012)

(continued)

Table 2 (continued)

Protein	Humoral immune findings	Cellular immune findings	Chronic EBV viremia	Other infections	References
Perforin	Normal IgG	Low numbers of B cells, CD4 cells, and neutrophils; impaired NK cell and CTL killing	No	None	Katano et al. (2004)
Munc13-4	Low IgG	Low numbers of neutrophils, impaired NK cell and CTL killing	No	Recurrent upper airway infections	Rohr et al. (2010)
Munc18-2	Low IgG	Low numbers of neutrophils, impaired NK cell and CTL killing	Yes	Recurrent upper airway infections	Rohr et al. (2010); Cohen et al. (2015)
PI3K catalytic subunit 110δ: gain of function mutations	Low, normal, or high IgG	Low numbers of CD4 cells, memory T cells, and naïve CD4 T cells; increased CD8 cells and senescent effector CD8 T cells	Yes	Sinus and pulmonary infections, CMV viremia	Lucas et al. (2014); Angulo et al. (2013)
PI3K catalytic subunit 110δ: loss of function mutation	High IgG and IgM	Increased numbers of B cells, reduced memory B cell, and diminished NK cell killing	Yes	Recurrent otitis media and sinusitis	Kuehn et al. (2013)
CTP synthase 1	Normal or increased IgG	Lymphopenia, low CD4/CD8 ratios, low naïve CD4 T cells, low CD27 memory B cells; absent iNKT cells; increased effector memory T cells	Yes	Herpesviruses, encapsulated bacteria	Martin et al. (2014)
Serine threonine kinase 4	High IgG	Low numbers of CD4 cells, naïve T cells, B cells, and neutrophils	Yes	Severe HPV and molluscum contagiosum infection, candida, recurrent bacterial and virus infections	Nehme et al. (2012); Abdollahpour et al. (2012)

(continued)

Table 2 (continued)

Protein	Humoral immune findings	Cellular immune findings	Chronic EBV viremia	Other infections	References
GATA binding protein 2	Normal IgG	Low numbers of B cells, CD4 T cells, NK cells, dendritic cells, and monocytes	Yes	Severe HSV, VZV, CMV, HPV, non-tuberculous mycobacteria, fungal infections	Spinner et al. (2014)
Minichromosome maintenance complex component 4	Normal IgG	Low numbers of NK cells, absent CD56^{dim} NK cells	Unknown	Respiratory infections, recurrent herpesvirus infections	Gineau et al. (2012); Eidenschenk et al. (2006)
Fcγ receptor 3A	Normal IgG	Variable numbers of NK cells, but impaired function	Unknown	Respiratory infections, severe herpesvirus infections	deVries et al. (1996); Grier et al. (2012)
Caspase recruitment domain-containing protein 11: gain of function mutations	Normal or slightly increased IgG, low IgM, reduced antibody response to polysaccharide-based vaccines	B cell lymphocytosis, normal T cell numbers	Yes	Respiratory infections, sinusitis, otitis media, molluscum contagiosum	Snow et al. (2012)

Abbreviations: *iNKT* invariant NKT; *HPV* human papillomavirus; *PML* progressive multifocal leukoencephalopathy; *CTL* cytotoxic T lymphocyte; *CTP* cytidine 5′-triphosphate

1998) and is expressed on T, NK, and iNKT cells. Mutations in SAP account for 60–70 % of cases of XLP. The protein encoded by the gene, SAP (SLAM-associated protein), is an adapter molecule that is expressed on T, NK, and iNKT cells. SAP consists of a single Src homology 2 (SH2) domain and interacts with several proteins including SLAM, 2B4, NTB-A, CD84, Ly108, and Ly9 (Cannons et al. 2010). The interaction of SAP with SLAM reduces production of IFN-γ (Latour et al. 2001) and T cell killing of virus-infected cells (Dupré et al. 2005). The interaction of SAP with 2B4 (Parolini et al. 2000) and NTB-A (Bottino et al. 2001) increases NK cell cytotoxicity. SAP-deficient cells are impaired for formation of immunologic synapses and killing of B cells, but not dendritic cells (Qi et al. 2008; Zhao et al. 2012). This results in inefficient recruitment and retention of T cells to germinal centers. CD84 and Ly108 are critical for T and B cell contacts, and CD84 is required for germinal center formation; in the absence of SAP, germinal centers are defective (Cannons et al. 2011). SAP is also important for control of T cell proliferation and apoptosis during antigen stimulation; in the absence of SAP, T cells are resistant to apoptosis mediated by T cell receptor restimulation (Snow et al. 2009). Taken together, these effects result in impaired T and NK cell cytotoxicity, with massive proliferation of T and NK cells, excessive cytokine production, and HLH.

XLP1 patients have impaired controlled of EBV, but not other virus infections or bacteria. These patients lack class-switched memory B cells (Chaganti et al. 2008). SAP knockout mice lack virus-specific memory B cells and long-lived plasma cells, due to a defect in CD4 T cells (Crotty et al. 2003). Patients who survived EBV infection were found to have impaired recognition of SLAM-ligand EBV-transformed B cells expressing EBV protein, but were able to recognize SLAM-ligand-negative EBV-infected B cells (Hislop et al. 2010). Somatic reversion of SAP-mutated cells in patients who survived XLP1 occurred solely in memory CD8 T cells, and these T cells proliferated and were cytotoxic for EBV-infected B cells (Palendira et al. 2012). A study of female XLP carriers (who have SAP^+/SAP^- alleles) showed that memory CD8 T cells specific for influenza and CMV were present in both SAP^+ and SAP^- cells; however, EBV-specific T cells were only present in SAP^+ cells (Palendira et al. 2011). These differences are due to the failure of SAP^- cells to respond to antigens presented by B cells, and since EBV is the only human virus that latently infects B cells, these results help to explain why XLP1 is a disease confined to EBV and does not predispose to infections by other pathogens. Blocking NTB-A and 2B4, both of which bind to SAP, restores the ability of SAP^- T cells to respond to antigen presentation by B cells. While patients with XLP1 have normal numbers of T and B cells, they lack iNKT cells (Nichols et al. 2005) which are critical for T cell receptor-induced cellular cytotoxicity (Das et al. 2013). These patients also have impaired T cell production of IL-10 (Ma et al. 2005). Thus, patients with XLP1 have impaired recognition of antigens presented by B cells, absent iNKT cells, impaired T cell cytotoxicity, and reduced expression of IL-10 by T cells.

While intravenous immunoglobulin, which contains neutralizing antibody to EBV, has been used to try to prevent EBV infection in patients with XLP1, breakthrough infections have occurred resulting in death. Rituximab, an anti-CD20 monoclonal antibody, was reported to reverse fulminant infectious mononucleosis in two patients with XLP1 (Milone et al. 2005). Reduced intensity conditioning hematopoietic stem cell transplantation (HSCT) resulted in an 80 % one-year survival rate in patients presenting with XLP1, regardless of whether they had a history of HLH (Marsh et al. 2014). In the future, genetic therapy correcting the SAP gene directly may be possible, and this has been demonstrated using retrovirus-mediated gene transfer in a mouse model of XLP1 (Rivat et al. 2013).

2.2 X-Linked Lymphoproliferative Disease 2 (XLP2)

XLP2, due to a mutation in *BIRC4* which encodes the X-linked inhibitor of apoptosis (XIAP), was initially described in patients from three families who presented with HLH, often but not always, associated with EBV, splenomegaly, hypogammaglobulinemia, and colitis (Rigaud et al. 2006). XIAP is expressed in B, T, NK, and dendritic cells as well as non-hematopoietic cells. Mutations in XIAP account for about 20–30 % of cases of XLP. While the initial paper reported that patients had low numbers of iNKT cells, a later report indicated that patients had normal numbers of these cells (Marsh et al. 2009). These patients have normal numbers of B, T, and NK cells. A review of 25 cases of XLP2 found that only 8 presented with HLH; other patients presented with colitis, severe infectious mononucleosis, or splenomegaly; iNKT cells were not reduced (Speckmann et al. 2013). There was no clear correlation of specific mutations with the severity of disease (Filipovich et al. 2010; Speckmann et al. 2013). Patients who underwent reduced conditioning HSCT had a better long-term survival rate (55 %) than those who had ablative conditioning (14 %) (Marsh et al. 2013).

Mutations in XIAP result in enhanced T cell apoptosis in response to stimulation with anti-Fas antibody, anti-CD3 antibody, and trimeric TRAIL (Rigaud et al. 2006); however, a subsequent report did not find enhanced apoptosis with anti-Fas antibody (Marsh et al. 2010). T cells from patients with XIAP have enhanced T cell reactivation-induced cell death. XIAP is also important for NOD2-mediated signaling (Krieg et al. 2009) as well as activation of NF-κB. Thus, patients with XLP2 have enhanced T cell apoptosis to various stimuli including T cell receptor restimulation.

A comparison of patients with XLP1 and XLP2 found that HLH was more common in XLP2 (76 %) versus XLP1 (55 %), but was more likely to be fatal in patients with XLP1 (61 %) versus XLP2 (23 %) (Pachlopnik Schmid et al. 2011). Infection with EBV triggered the onset of HLH in 92 % of persons with XLP1 and 83 % with XLP2. Significantly, more patients with XLP1 had hypogammaglobulinemia (67 %) and lymphoma (30 %) than those with XLP2 (33 and 0 %, respectively). In contrast, significantly more patients with XLP2 had cytopenias (52 %) and splenomegaly (87 %) in the absence of HLH, and hemorrhagic colitis (17 %)

than XLP1 (12, 7, 0 %, respectively). The large number of differences between XLP1 and XLP2 has led some authors to propose that XLP2 should be reclassified instead as an X-linked familial HLH disorder (Filipovich et al. 2010).

2.3 IL-2-Inducible T Cell Kinase (ITK)

IL-2-inducible T cell kinase (ITK) deficiency was first reported in two girls with homozygous mutations who died with B cell proliferation due to EBV (Huck et al. 2009). The disease has been called EBV-associated autosomal lymphoproliferative syndrome. ITK is a member of the TEC family of kinases and has a critical role in T cell receptor signaling. It is important for T cell proliferation and differentiation. Both girls had high levels of eomesodermin in CD8 T cells, and iNKT cells were absent from the one girl who was tested. A review of seven patients (four girls and three boys) with ITK deficiency from 4 families found that all patients presented with fever, lymphadenopathy, and all of whom were tested had markedly elevated EBV DNA in the peripheral blood (Linka et al. 2012). Most had hepatosplenomegaly, pulmonary disease, hypogammaglobulinemia, leukopenia, and CD4 lymphopenia. Pathology showed Hodgkin's lymphoma in four patients, B cell lymphoproliferative disease in two patients, and both large B cell lymphoma and lymphomatoid granulomatosis in one patient. Two patients underwent HSCT, one survived, and the other died of complications associated with graft-versus-host disease. All patients that were tested had low numbers of iNKT cells and an impaired calcium flux in T cells after stimulation of the T cell receptor with anti-CD3 antibody.

2.4 CD27

Two brothers were reported with homozygous mutations in CD27 and persistent EBV viremia: One had aplastic anemia and the other had hypogammaglobulinemia (van Montfrans et al. 2012). Both patients had undetectable CD27 on all lymphocytes, but normal numbers of lymphocyte subsets. T cell proliferative responses to CD27-dependent mitogens (CD2 and pokeweed mitogen) were reduced, and T cell-dependent B cell responses to vaccines were impaired. CD27 is a member of the tumor necrosis receptor family and binds to its ligand, CD70, and provides costimulatory signaling to activate B, T, and NK cells. CD27 is a marker for memory B cells and enhances B cell differentiation, and T and NK cell function.

Eight patients in three additional families with CD27 deficiency were subsequently reported: Three patients had asymptomatic deficiencies in memory B cells, three had EBV HLH and lymphoproliferative disease, and two had lymphoma (Salzer et al. 2013). Three patients developed hypogammaglobulinemia after primary EBV infection. Two with severe disease had reduced NK cell function and diminished numbers of iNKT cells. One patient received repeated courses of rituximab and two underwent HSCT.

2.5 Magnesium Transporter 1 (MagT1) Protein

MAGT1 encodes a magnesium transporter protein located in the plasma cell membrane. MagT1 protein allows an influx of magnesium into cells after stimulation of the T cell receptor which results in activation of T cells (Li et al. 2011). The influx of magnesium results in increased calcium signaling and activation of PLC (phospholipase C)-γ1, PKC (protein kinase C)-θ, and NF-κB. Thus, magnesium, like calcium, can act as an intracellular second messenger to couple events on the cell surface with changes in the cytoplasm and nucleus.

Seven patients have been reported with mutations in *MAGT1* who had markedly elevated levels of EBV DNA in the blood; these patients ranged in age from 3 to 45 years old with a mean age of 16 years (Chaigne-Delalande et al. 2013). Four patients had B lymphomas, three of whom were tested for EBV and were positive. Stimulation of the T cell receptor in PBMCs from the patients resulted in impaired calcium signaling and reduced activation of PLC γ1, PKC-θ, and NF-κB.

Patients with mutations in *MAGT1* often have low CD4 cell counts with an inverted CD4:CD8 ratio, reduced NKG2D (an NK cell activating receptor) on NK cells and cytotoxic T lymphocytes (CTLs), and impaired T cell activation. Their CTLs showed reduced killing of autologous EBV-transformed B cells, and their NK cells were impaired for killing other target cells. In addition to elevated levels of EBV in the blood, some patients had hypogammaglobulinemia, sinusitis, and chronic diarrhea. Two patients had autoimmune cytopenias. All patients had splenomegaly; one had hemophagocytosis. Two patients underwent HSCT and both died of complications related to the transplant. The disease has been termed XMEN (X-linked immunodeficiency with magnesium defect, EBV infection, and neoplasia).

Patients with mutations in *MAGT1* had lower levels of intracellular magnesium, which suggested that supplemental magnesium might improve their immune responses. In vitro supplemental magnesium of PBMCs from patients resulted in increased levels of intracellular magnesium, NKG2D, and improved NK cell and CTL cytotoxicity. Treatment of two patients with magnesium supplementation resulted in increased levels of intracellular magnesium, increased expression of NKG2D on CTLs, improved CTL activity against autologous EBV-transformed B cells, and a reduction in the percentage of EBV-infected cells in the blood.

2.6 Coronin Actin Binding Protein 1A

Three siblings in one family that presented with EBV B cell lymphoproliferative disease in early childhood were found to have homozygous mutations in *CORO1A* which encodes coronin actin binding protein 1A (Moshous et al. 2013). One patient had an EBV-positive lymphoproliferative process and two had EBV lymphomas. Two of the patients in whom EBV DNA levels in the blood were measured had elevated levels. All three patients had recurrent ear, nose, and throat as well as upper respiratory tract infections. Two of the patients died: one in preparation for HSCT and one from graft-versus-host disease after HSCT. Coronin actin binding protein 1A binds to actin-related protein 2/3 and is important for T cell synapse formation and T cell receptor signaling. The patients had reduced numbers of CD4, CD8, CD19, and naïve T cells, a reduced T cell repertoire, low or no iNKT cells, and few mucosal-associated invariant T cells.

2.7 LPS-Responsive Beige-Like Anchor (LRBA) Protein

Patients with mutations in *LRBA* present with inflammatory bowel disease, chronic diarrhea, and autoimmune cytopenias (Alangari et al. 2012). One patient presented with EBV lymphoproliferative disease, elevated EBV DNA in the blood, and autoimmune pancytopenia. The LRBA protein has domains that are conserved with the Chediak–Higashi syndrome protein and is important for endocytosis of ligand-activated receptors; however, its role in immunity is not well understood.

3 Proteins Associated with EBV Disease and Familial Hemophagocytic Lymphohistiocytosis (FHL)

See Tables 1 and 2.

3.1 Perforin

Familial hemophagocytic lymphohistiocytosis (FHL) is a group of diseases due to mutations in proteins important for maturation or release of cytotoxic granules from CTLs and NK cells, or for entry of cytotoxic proteins from these granules into target cells. Four genes have been identified in which mutations cause FLH-*PRF1, UNC13D, STX11,* and *STXBP2*, which are responsible for FHL2, FHL3, FHL4, and FHL5, respectively.

Perforin is encoded by *PRF1* and is expressed in cytotoxic granules of CTLs and NK cells. When foreign antigens are expressed on antigen-presenting cells, CTLs become activated and granules containing perforin and granzymes dock on the plasma membrane and are released. Perforin oligomerizes to form pores in target cells which allows entry of granzymes into these cells resulting in activation of caspases and death of the cells. Mutations in perforin result in an autosomal recessive disorder known as FHL2. Perforin mutations result in impaired killing of target cells by CTLs and NK cells.

We described a boy who presented with EBV-positive infectious mononucleosis followed by persistent splenomegaly and lymphadenopathy and was diagnosed with chronic active EBV disease (CAEBV) and HLH (Katano et al. 2004; Cohen et al. 2011). The patient had different mutations in the two alleles of perforin which resulted in reduced expression of the native form of the protein. The patient only expressed the immature form of perforin, since his perforin was not cleaved at the carboxyl terminus to yield the active form of the protein. Accordingly, his T cells were impaired for killing target cells.

3.2 *Munc13-4*

Munc13-4 is encoded by *UNC13D*. Munc13-4 interacts with syntaxin 11 to change the conformation of syntaxin from a closed to an open conformation; a soluble NSF attachment protein receptor (SNARE) complex is formed between v-SNARE on cytotoxic granules and the target membrane t-SNARE syntaxin 11. This allows priming of cytotoxic granules and ultimately results in fusion of the granules with the membrane of the cell with exocytosis of granules. Mutations in munc13-4 result in an autosomal recessive disease referred to as FHL3 with impaired NK and T cell cytotoxicity.

Mutations in munc13-4 were reported in one patient with CAEBV who had cerebral vasculitis, hypogammaglobulinemia, chronic hepatitis, splenomegaly, and recurrent respiratory infections (Rohr et al. 2010). The patient was a compound heterozygote for munc13-4 mutations. The patient was initially EBV seropositive and then developed CAEBV with a high viral load and HLH and died of the disease.

3.3 *Munc18-2*

Munc18-2 is encoded by *STXBP2*, a member of the sec1/munc18 (SM) family of proteins that are important for SNARE-mediated membrane fusion. Munc18-2 binds to syntaxin 11, on the plasma membrane of NK cells, and to v-SNARE, on cytotoxic granules. Thus, munc18-2 forms a bridge assisting in the docking of cytotoxic granules to the plasma membrane of CTLs or NK cells. Mutations in munc18-2 result in an autosomal recessive disease referred to as FHL5. Deficiency in munc18-2 results in impaired binding of munc18-2 to syntaxin 11, reduced stability of both proteins, and impaired exocytosis of cytotoxic granules from CTLs or NK cells (Côte et al. 2009; zur Stadt et al. 2009). Mutations in munc18-2 affect folding of the protein which impair its binding activity (Hackmann et al. 2013). These patients have impaired NK and T cell killing of target cells.

Mutations in munc18-2 were reported in four patients with CABEV (Rohr et al. 2010). Three patients had homozygous mutations and one was a compound heterozygote. Two patients developed HLH after primary EBV infection: one presented with HLH-like symptoms and then severe HLH after primary EBV infection, and one was initially EBV seropositive and then developed CAEBV with a high viral load and HLH. All four patients had hypogammaglobulinemia, three had persistent splenomegaly, two had recurrent infections, and one had Hodgkin's lymphoma. Three underwent HSCT at ages 6, 16, and 16 and one survived; the fourth patient with lymphoma had a HSCT and remains alive and well. Another patient with compound heterozygous mutations in munc18-2 presented with late onset CAEBV and did well after HSCT (Cohen et al. 2015).

4 Genes Associated with EBV and Other Infections

See Tables 1 and 2.

4.1 Phosphatidylinositol-3-OH Kinase (PI3K) Catalytic Subunit P110δ

4.1.1 PI3K P110δ Gain of Function Mutations

Thirty-one patients with gain of function mutations in the p110δ catalytic subunit of phosphatidylinositol-3-OH kinase (PI3K) have been reported who had impaired control of EBV (Lucas et al. 2014; Angulo et al. 2013). PI3K is activated in T cells after ligand binding to the T cell receptor (Okkenhaug and Vanhaesebroeck 2003). This results in binding of the p85 regulatory domain of PI3K to phosphorylated tyrosine residues on proteins and its dissociation from the p110 catalytic subunit of PI3K. p110δ is found exclusively in lymphocytes. Free p110δ is then recruited to the plasma membrane, and it phosphorylates PIP2 (phosphatidylinositol-(4,5)-biphosphate) to PIP3 (phosphatidylinositol-(3,4,5)-triphosphate). This results in phosphorylation of Akt (also known as protein kinase B) which phosphorylates mammalian target of rapamycin (mTOR). The mTOR complex (composed of mTOR, raptor, and GβL) phosphorylates 4E-BP1 (a protein translation initiation inhibitor) and p70S6 kinase (which promotes protein translation). Phosphorylation of the former inhibits its ability to block eukaryotic translation initiation factor eiF4E, while phosphorylation of the latter activates the S6 ribosomal protein to increase protein translation. This results in increased protein synthesis, cell growth, proliferation, differentiation, and survival.

Heterozygous mutations (N334K, E525K, and E1021K) in p110δ result in gain of function mutations (Lucas et al. 2014; Angulo et al. 2013). These likely block the interaction of p110δ with p85 (to allow unbridled activity of p110δ) or promote the association of p110δ with the cell membrane. This results in increased activation (phosphorylation) of PI3K either in the presence or in the absence of T cell receptor stimulation. Stimulation of peripheral blood mononuclear cells from these patients with antibody to CD3 and CD28 results in reduced IL-2 secretion and decreased proliferation compared with controls. Surprisingly, these patients have increased numbers of EBV-specific T cells based on tetramer staining and increased EBV-specific effector memory cells. Patients with mutations in p110δ have reduced memory CD8 T cells, reduced naive CD4 T cells, increased senescent effector CD8 T cells, reduced class-switched IgG and IgA cells, and increased activation-induced cell death.

Patients present early in childhood with sinus and pulmonary infections, persistent EBV and/or CMV viremia, lymphoproliferative disease with lymphoid nodules in mucosa that can obstruct the lungs and gastrointestinal tract, and autoimmune cytopenias. Patients have normal or elevated IgM, variable levels of IgG, reduced IgA, and impaired antibody production after vaccination. Patients have decreased CD4 cells and increased CD8 cells with an inverted CD4/CD8 ratio, and a progressive B and T cell immunodeficiency. Two patients developed EBV-positive B cell lymphomas and one patient had a marginal zone lymphoma. The impaired ability to control EBV may be due to the low numbers of CD4 cells and/or the reduction in memory CD8 T cells.

This disease has been termed PASLI (p110δ-activating mutation causing senescent T cells, lymphadenopathy, and immune deficiency) (Lucas et al. 2014) or APDS (activated PI3K*δ* syndrome) (Angulo et al. 2013). Treatment of one patient with rapamycin, an mTOR inhibitor, resulted in reduced CD8 T cell numbers, increased IL-2 secretion, and increased T cell proliferation after stimulation with anti-CD3 and anti-CD28 antibody in vitro (Lucas et al. 2014). The patient had a reduction in the size of his lymph nodes, liver, and spleen. Treatment of cells in vitro from patients with mutations in p110δ with specific inhibitors of p110δ reduced the activity of the protein (Angulo et al. 2013).

A similar disease due to gain of function mutations in the p85 subunit of PI3K (*PIK3R1*) has been associated with EBV viremia (Deau et al. 2014).

4.1.2 PI3K P110δ Loss of Function Mutation

One patient has been reported who initially presented with recurrent otitis media and sinusitis, generalized lymphadenopathy, hepatosplenomegaly, B cell lymphocytosis, and persistent EBV viremia (Kuehn et al. 2013). The patient's serum had autoantibodies to several cellular proteins, and his NK cells had diminished cytotoxicity.

4.2 Cytidine 5′-Triphosphate Synthase 1 (CTPS1)

Eight patients from 5 families were reported with mutations in cytidine 5′-triphosphate synthase (CTPS1): Four patients presented with severe infectious mononucleosis (three of which had chronic EBV viremia), three with lymphoproliferative disease involving the central nervous system (two of whom had EBV-positive non-Hodgkin's lymphoma), and one with asymptomatic chronic EBV viremia (Martin et al. 2014). All of the patients had other severe herpesvirus infections during childhood and infections with encapsulated bacteria. One had Streptococcus pneumoniae sepsis and meningitis, and one had Neisseria meningitidis meningitis. Six of the eight underwent HSCT and four survived; one died of graft-versus-host disease and one from disseminated varicella-zoster virus.

Most patients with CTPS1 had lymphopenia with low CD4:CD8 ratios during infections. One patient who was studied more intensively had low numbers of naïve CD4 T cells, increased effector memory T cells, low numbers of CD27 memory B cells, and absent iNKT cells. Proliferation and DNA synthesis of T cells in response to anti-CD3 antibody and proliferation of B cells in response to anti-B cell receptor and CpG were impaired. Reduced levels of CTP were present in stimulated T and B cells from the patients. CTPS1 expression is normally increased with T cell activation; therefore, deficiency of the protein presumably limits the ability of T cells to proliferate and control virus and bacterial infections.

4.3 Serine/Threonine Kinase 4 (STK4)

Three patients from two families with mutations in serine/threonine kinase 4 (STK4) were reported with high levels of EBV DNA in the blood. One patient developed an EBV-positive Hodgkin's lymphoma and survived, a second patient developed disseminated EBV B-cell lymphoproliferative lesions and died after HSCT from graft-versus host disease and infectious complications, and a third patient with autoimmune hemolytic anemia underwent HSCT and also died from graft-versus-host disease and infection (Nehme et al. 2012). The patients had a history of recurrent bacterial and viral infections, dermatitis, CD4 lymphopenia, reductions in the numbers of naïve T cells, impaired T cell proliferative responses to phytohemagglutinin and to anti-CD3 antibody, reduced T cell receptor repertoires, and increased levels of IgG. Increased Fas expression was present on the surface of the cells, which showed increased sensitivity to Fas-induced apoptosis.

Another study reported three patients from one family with STK4 mutations; one patient with generalized lymphadenopathy had a biopsy which showed a monoclonal EBV lymphoproliferative process that was reported to resemble a lymphoplasmacytic lymphoma (Abdollahpour et al. 2012). These patients had bacterial and viral infections (including extensive warts and molluscum contagiosum), mucocutaneous candidiasis, neutropenia, CD4 lymphopenia, and B cell lymphopenia, and most had elevated levels of IgG. STK4 is also referred to as mammalian sterile 20-like protein (MST1) and is involved in signaling pathways important for cell proliferation and apoptosis; STK4 is cleaved by caspases and is thought to be a pro-apoptotic protein.

4.4 GATA Binding Protein 2 (GATA2)

Patients with mutations in GATA binding protein 2 (GATA2) can have various signs and symptoms including acute myeloid leukemia, myelodysplastic syndrome, autoimmune disease, pulmonary alveolar proteinosis, and primary lymphedema. Patients with GATA2 mutations have presented with chronic active EBV disease, EBV-positive smooth muscle tumors, and persistent EBV viremia (Hsu et al. 2011; Spinner et al. 2014). In addition, these patients also are susceptible to other severe herpesvirus infections as well as severe human papillomavirus, fungal, and non-tuberculous mycobacterial infections. GATA2 encodes a transcription factor important for hematopoiesis; accordingly, patients with mutations in GATA2 often have low numbers of B cells, CD4 T cells, NK cells, dendritic cells, red blood cells, neutrophils, monocytes, and platelets.

4.5 Minichromosome Maintenance Complex Component 4 (MCM4)

Patients with mutations in minichromosome maintenance complex component 4 (MCM4) present with adrenal insufficiency, growth retardation, low numbers of NK cells, and absent $CD56^{dim}$ NK cells (Gineau et al. 2012). These latter cells are cytotoxic and produce cytokines after recognition of target cells. One patient developed an EBV lymphoma (Eidenschenk et al. 2006). MCM4 is a DNA helicase that is important for DNA replication.

4.6 Fcγ Receptor 3A (CD16a)

Two patients with mutations in Fcγ receptor 3A (CD16) were described with EBV diseases; one patient had a prolonged illness with fever and malaise associated with EBV infection (deVries et al. 1996) and the second was reported to have recurrent lymphadenopathy due to EBV-positive Castleman's disease (Grier et al. 2012). The former patient also had severe infections with Bacille Calmette–Guerin and varicella-zoster virus, while the latter patient also had severe HPV infections and deficient NK cell cytotoxicity. Fcγ receptor 3A is expressed on NK cells and neutrophils; mutations in this receptor are responsible for classical NK cell deficiency.

4.7 Caspase Recruitment Domain-Containing Protein 11 (CARD11): Gain of Function Mutations

Patients with germ line gain of function mutations in *CARD11* present with B cell lymphocytosis, splenomegaly, lymphadenopathy with florid follicular hyperplasia, recurrent sinusitis, and otitis media (Snow et al. 2012). One patient presented with persistently elevated EBV DNA in the blood as well as splenomegaly, lymphadenopathy, bronchiectasis, recurrent otitis media, and molluscum contagiosum. The CARD11 protein is required for activation of NF-κB by antigen receptor in B and T cells. Somatic gain of function mutations of CARD11 are present in many diffuse large B cell lymphomas.

4.8 Other Immunodeficiencies Associated with Multiple Infections

Other primary immunodeficiency diseases associated with multiple infections can present with EBV lymphoproliferative disease or lymphomas. Ataxia telangiectasia is an autosomal recessive disease due to a mutation in *ATM* which encodes a serine/threonine kinase that is important for DNA repair. In addition to neurologic and skin disease, these patients often develop sinopulmonary infections, interstitial lung disease, and are at increased risk for malignancies. These patients often have increased levels of EBV DNA in the blood and can develop EBV lymphomas.

Wiskott–Aldrich syndrome is an X-linked disorder due to mutations in *WAS*. The Wiskott–Aldrich syndrome protein is important for the formation of the immunologic synapse which is the site of interaction between antigen-presenting cells and T cells. In addition to increased propensity of infections, these patients have thrombocytopenia, eczema, and autoimmune disease. Patients may develop EBV lymphomas.

Patients with severe combined immunodeficiency (SCID) can have mutations in a number of different genes; these result in impaired B cell and T cell immunity. In addition to increased infections, these patients often have chronic diarrhea and failure to thrive. These patients are susceptible to EBV-positive B cell lymphomas.

References

Abdollahpour H, Appaswamy G, Kotlarz D, Diestelhorst J, Beier R, Schäffer AA, Gertz EM, Schambach A, Kreipe HH, Pfeifer D, Engelhardt KR, Rezaei N, Grimbacher B, Lohrmann S, Sherkat R, Klein C (2012) The phenotype of human STK4 deficiency. Blood 119(15):3450–3457

Alangari A, Alsultan A, Adly N, Massaad MJ, Kiani IS, Aljebreen A, Raddaoui E, Almomen AK, Al-Muhsen S, Geha RS, Alkuraya FS (2012) LPS-responsive beige-like anchor (LRBA) gene mutation in a family with inflammatory bowel disease and combined immunodeficiency. J Allergy Clin Immunol 130(2):481–488

Angulo I, Vadas O, Garçon F, Banham-Hall E, Plagnol V, Leahy TR, Baxendale H, Coulter T, Curtis J, Wu C, Blake-Palmer K, Perisic O, Smyth D, Maes M, Fiddler C, Juss J, Cilliers D, Markelj G, Chandra A, Farmer G, Kielkowska A, Clark J, Kracker S, Debré M, Picard C, Pellier I, Jabado N, Morris JA, Barcenas-Morales G, Fischer A, Stephens L, Hawkins P, Barrett JC, Abinun M, Clatworthy M, Durandy A, Doffinger R, Chilvers ER, Cant AJ, Kumararatne D, Okkenhaug K, Williams RL, Condliffe A, Nejentsev S (2013) Phosphoinositide 3-kinase δ gene mutation predisposes to respiratory infection and airway damage. Science 342(6160):866–871

Booth C, Gilmour KC, Veys P, Gennery AR, Slatter MA, Chapel H, Heath PT, Steward CG, Smith O, O'Meara A, Kerrigan H, Mahlaoui N, Cavazzana-Calvo M, Fischer A, Moshous D, Blanche S, Pachlopnik Schmid J, Latour S, de Saint-Basile G, Albert M, Notheis G, Rieber N, Strahm B, Ritterbusch H, Lankester A, Hartwig NG, Meyts I, Plebani A, Soresina A, Finocchi A, Pignata C, Cirillo E, Bonanomi S, Peters C, Kalwak K, Pasic S, Sedlacek P, Jazbec J, Kanegane H, Nichols KE, Hanson IC, Kapoor N, Haddad E, Cowan M, Choo S, Smart J, Arkwright PD, Gaspar HB (2011) X-linked lymphoproliferative disease due to SAP/SH2D1A deficiency: a multicenter study on the manifestations, management and outcome of the disease. Blood 117(1):53–62

Bottino C, Falco M, Parolini S, Marcenaro E, Augugliaro R, Sivori S, Landi E, Biassoni R, Notarangelo LD, Moretta L, Moretta A (2001) NTB-A [correction of GNTB-A], a novel SH2D1A-associated surface molecule contributing to the inability of natural killer cells to kill Epstein-Barr virus-infected B cells in X-linked lymphoproliferative disease. J Exp Med 194(3):235–246

Cannons JL, Qi H, Lu KT, Dutta M, Gomez-Rodriguez J, Cheng J, Wakeland EK, Germain RN, Schwartzberg PL (2010) Optimal germinal center responses require a multistage T cell: B cell adhesion process involving integrins, SLAM-associated protein, and CD84. Immunity 32(2):253–265

Cannons JL, Tangye SG, Schwartzberg PL (2011) SLAM family receptors and SAP adaptors in immunity. Annu Rev Immunol 29:665–705

Chaganti S, Ma CS, Bell AI, Croom-Carter D, Hislop AD, Tangye SG, Rickinson AB (2008) Epstein-Barr virus persistence in the absence of conventional memory B cells: IgM^{+} IgD^{+} $CD27^{+}$ B cells harbor the virus in X-linked lymphoproliferative disease patients. Blood 112(3):672–679

Chaigne-Delalande B, Li FY, O'Connor GM, Lukacs MJ, Jiang P, Zheng L, Shatzer A, Biancalana M, Pittaluga S, Matthews HF, Jancel TJ, Bleesing JJ, Marsh RA, Kuijpers TW, Nichols KE, Lucas CL, Nagpal S, Mehmet H, Su HC, Cohen JI, Uzel G, Lenardo MJ (2013) Mg^{2+} regulates cytotoxic functions of NK and CD8 T cells in chronic EBV infection through NKG2D. Science 341(6142):186–191

Chung BK, Tsai K, Allan LL, Zheng DJ, Nie JC, Biggs CM, Hasan MR, Kozak FK, van den Elzen P, Priatel JJ, Tan R (2013) Innate immune control of EBV-infected B cells by invariant natural killer T cells. Blood 122(15):2600–2608

Coffey AJ, Brooksbank RA, Brandau O, Oohashi T, Howell GR, Bye JM, Cahn AP, Durham J, Heath P, Wray P, Pavitt R, Wilkinson J, Leversha M, Huckle E, Shaw-Smith CJ, Dunham A, Rhodes S, Schuster V, Porta G, Yin L, Serafini P, Sylla B, Zollo M, Franco B, Bolino A, Seri M, Lanyi A, Davis JR, Webster D, Harris A, Lenoir G, de St Basile G, Jones A, Behloradsky BH, Achatz H, Murken J, Fassler R, Sumegi J, Romeo G, Vaudin M, Ross MT, Meindl A, Bentley DR (1998) Host response to EBV infection in X-linked lymphoproliferative disease results from mutations in an SH2-domain encoding gene. Nat Genet 20(2):129–135

Cohen JI, Jaffe ES, Dale JK, Pittaluga S, Heslop HE, Rooney CM, Gottschalk S, Bollard CM, Rao VK, Marques A, Burbelo PD, Turk SP, Fulton R, Wayne AS, Little RF, Cairo MS, El-Mallawany NK, Fowler D, Sportes C, Bishop MR, Wilson W, Straus SE (2011) Characterization and treatment of chronic active Epstein-Barr virus disease: a 28-year experience in the United States. Blood 117(22):5835–5849

Cohen JI, Niemela JE, Stoddard JL, Pittaluga S, Heslop H, Jaffe ES, Dowdell K (2015) Late-onset severe chronic active EBV in a patient for five years with mutations in STXBP2 (MUNC18-2) and PRF1 (Perforin 1). J Clin Immunol 35(5):445–448

Côte M, Ménager MM, Burgess A, Mahlaoui N, Picard C, Schaffner C, Al-Manjomi F, Al-Harbi M, Alangari A, Le Deist F, Gennery AR, Prince N, Cariou A, Nitschke P, Blank U, El-Ghazali G, Ménasché G, Latour S, Fischer A, de Saint Basile G (2009) Munc18-2 deficiency causes familial hemophagocytic lymphohistiocytosis type 5 and impairs cytotoxic granule exocytosis in patient NK cells. J Clin Invest 119(12):3765–3773

Crotty S, Kersh EN, Cannons J, Schwartzberg PL, Ahmed R (2003) SAP is required for generating long-term humoral immunity. Nature 421(6920):282–287

Das R, Bassiri H, Guan P, Wiener S, Banerjee PP, Zhong MC, Veillette A, Orange JS, Nichols KE (2013) The adaptor molecule SAP plays essential roles during invariant NKT cell cytotoxicity and lytic synapse formation. Blood 121(17):3386–3395

de Vries E, Koene HR, Vossen JM, Gratama JW, von dem Borne AE, Waaijer JL, Haraldsson A, de Haas M, van Tol MJ (1996) Identification of an unusual Fc gamma receptor IIIa (CD16) on natural killer cells in a patient with recurrent infections. Blood 88(8):3022–3027

Deau MC, Heurtier L, Frange P, Suarez F, Bole-Feysot C, Nitschke P, Cavazzana M, Picard C, Durandy A, Fischer A, Kracker S (2014) A human immunodeficiency caused by mutations in the PIK3R1 gene. J Clin Invest 124(9):3923–3928

Dupré L, Andolfi G, Tangye SG, Clementi R, Locatelli F, Aricò M, Aiuti A, Roncarolo MG (2005) SAP controls the cytolytic activity of $CD8^{+}$ T cells against EBV-infected cells. Blood 105(11):4383–4389

Eidenschenk C, Dunne J, Jouanguy E, Fourlinnie C, Gineau L, Bacq D, McMahon C, Smith O, Casanova JL, Abel L, Feighery C (2006) A novel primary immunodeficiency with specific natural-killer cell deficiency maps to the centromeric region of chromosome 8. Am J Hum Genet 78(4):721–727

Faulkner GC, Burrows SR, Khanna R, Moss DJ, Bird AG, Crawford DH (1999) X-Linked agammaglobulinemia patients are not infected with Epstein-Barr virus: implications for the biology of the virus. J Virol 73(2):1555–1564

Filipovich AH, Zhang K, Snow AL, Marsh RA (2010) X-linked lymphoproliferative syndromes: brothers or distant cousins? Blood 116(18):3398–3408

Gineau L, Cognet C, Kara N, Lach FP, Dunne J, Veturi U, Picard C, Trouillet C, Eidenschenk C, Aoufouchi S, Alcaïs A, Smith O, Geissmann F, Feighery C, Abel L, Smogorzewska A, Stillman B, Vivier E, Casanova JL, Jouanguy E (2012) Partial MCM4 deficiency in patients with growth retardation, adrenal insufficiency, and natural killer cell deficiency. J Clin Invest 122(3):821–832

Grier JT, Forbes LR, Monaco-Shawver L, Oshinsky J, Atkinson TP, Moody C, Pandey R, Campbell KS, Orange JS (2012) Human immunodeficiency-causing mutation defines CD16 in spontaneous NK cell cytotoxicity. J Clin Invest 122(10):3769–3780

Hackmann Y, Graham SC, Ehl S, Höning S, Lehmberg K, Aricò M, Owen DJ, Griffiths GM (2013) Syntaxin binding mechanism and disease-causing mutations in Munc18-2. Proc Natl Acad Sci USA 110(47):E4482–E4491

Hislop AD, Taylor GS, Sauce D, Rickinson AB (2007) Cellular responses to viral infection in humans: lessons from Epstein-Barr virus. Annu Rev Immunol 25:587–617

Hislop AD, Palendira U, Leese AM, Arkwright PD, Rohrlich PS, Tangye SG, Gaspar HB, Lankester AC, Moretta A, Rickinson AB (2010) Impaired Epstein-Barr virus-specific $CD8^{+}$ T-cell function in X-linked lymphoproliferative disease is restricted to SLAM family-positive B-cell targets. Blood 116(17):3249–3257

Hsu AP, Sampaio EP, Khan J, Calvo KR, Lemieux JE, Patel SY, Frucht DM, Vinh DC, Auth RD, Freeman AF, Olivier KN, Uzel G, Zerbe CS, Spalding C, Pittaluga S, Raffeld M, Kuhns DB, Ding L, Paulson ML, Marciano BE, Gea-Banacloche JC, Orange JS, Cuellar-Rodriguez J, Hickstein DD, Holland SM (2011) Mutations in GATA2 are associated with the autosomal dominant and sporadic monocytopenia and mycobacterial infection (MonoMAC) syndrome. Blood 118(10):2653–2655

Huck K, Feyen O, Niehues T, Rüschendorf F, Hübner N, Laws HJ, Telieps T, Knapp S, Wacker HH, Meindl A, Jumaa H, Borkhardt A (2009) Girls homozygous for an IL-2-inducible T cell kinase mutation that leads to protein deficiency develop fatal EBV-associated lymphoproliferation. J Clin Invest 119(5):1350–1358

Imadome K, Shirakata M, Shimizu N, Nonoyama S, Yamanashi Y (2003) CD40 ligand is a critical effector of Epstein-Barr virus in host cell survival and transformation. Proc Natl Acad Sci USA 100(13):7836–7840

Katano H, Ali MA, Patera AC, Catalfamo M, Jaffe ES, Kimura H, Dale JK, Straus SE, Cohen JI (2004) Chronic active Epstein-Barr virus infection associated with mutations in perforin that impair its maturation. Blood 103(4):1244–1252

Koganti S, de la Paz A, Freeman AF, Bhaduri-McIntosh S (2014) B lymphocytes from patients with a hypomorphic mutation in STAT3 resist Epstein-Barr virus-driven cell proliferation. J Virol 88(1):516–524

Krieg A, Correa RG, Garrison JB, Le Negrate G, Welsh K, Huang Z, Knoefel WT, Reed JC (2009) XIAP mediates NOD signaling via interaction with RIP2. Proc Natl Acad Sci USA 106(34):14524–14529

Kuehn HS, Niemela JE, Rangel-Santos A, Zhang M, Pittaluga S, Stoddard JL, Hussey AA, Evbuomwan MO, Priel DA, Kuhns DB, Park CL, Fleisher TA, Uzel G, Oliveira JB (2013) Loss-of-function of the protein kinase C δ (PKCδ) causes a B-cell lymphoproliferative syndrome in humans. Blood 121(16):3117–3125

Latour S, Gish G, Helgason CD, Humphries RK, Pawson T, Veillette A (2001) Regulation of SLAM-mediated signal transduction by SAP, the X-linked lymphoproliferative gene product. Nat Immunol 2(8):681–690

Li FY, Chaigne-Delalande B, Kanellopoulou C, Davis JC, Matthews HF, Douek DC, Cohen JI, Uzel G, Su HC, Lenardo MJ (2011) Second messenger role for Mg^{2+} revealed by human T-cell immunodeficiency. Nature 475(7357):471–476

Li FY, Chaigne-Delalande B, Su H, Uzel G, Matthews H, Lenardo MJ (2014) XMEN disease: a new primary immunodeficiency affecting Mg^{2+} regulation of immunity against Epstein-Barr virus. Blood 123(14):2148–2152

Linka RM, Risse SL, Bienemann K, Werner M, Linka Y, Krux F, Synaeve C, Deenen R, Ginzel S, Dvorsky R, Gombert M, Halenius A, Hartig R, Helminen M, Fischer A, Stepensky P, Vettenranta K, Köhrer K, Ahmadian MR, Laws HJ, Fleckenstein B, Jumaa H, Latour S, Schraven B, Borkhardt A (2012) Loss-of-function mutations within the IL-2 inducible kinase ITK in patients with EBV-associated lymphoproliferative diseases. Leukemia 26(5):963–971

Lucas CL, Kuehn HS, Zhao F, Niemela JE, Deenick EK, Palendira U, Avery DT, Moens L, Cannons JL, Biancalana M, Stoddard J, Ouyang W, Frucht DM, Rao VK, Atkinson TP, Agharahimi A, Hussey AA, Folio LR, Olivier KN, Fleisher TA, Pittaluga S, Holland SM, Cohen JI, Oliveira JB, Tangye SG, Schwartzberg PL, Lenardo MJ, Uzel G (2014) Dominant-activating germline mutations in the gene encoding the PI(3)K catalytic subunit p110δ result in T cell senescence and human immunodeficiency. Nat Immunol 15(1):88–97

Ma CS, Hare NJ, Nichols KE, Dupré L, Andolfi G, Roncarolo MG, Adelstein S, Hodgkin PD, Tangye SG (2005) Impaired humoral immunity in X-linked lymphoproliferative disease is associated with defective IL-10 production by $CD4^+$ T cells. J Clin Invest 115(4):1049–1059

Marsh RA, Villanueva J, Kim MO, Zhang K, Marmer D, Risma KA, Jordan MB, Bleesing JJ, Filipovich AH (2009) Patients with X-linked lymphoproliferative disease due to BIRC4 mutation have normal invariant natural killer T-cell populations. Clin Immunol 132(1):116–123

Marsh RA, Madden L, Kitchen BJ, Mody R, McClimon B, Jordan MB, Bleesing JJ, Zhang K, Filipovich AH (2010) XIAP deficiency: a unique primary immunodeficiency best classified as X-linked familial hemophagocytic lymphohistiocytosis and not as X-linked lymphoproliferative disease. Blood 116(7):1079–1082

Marsh RA, Rao K, Satwani P, Lehmberg K, Müller I, Li D, Kim MO, Fischer A, Latour S, Sedlacek P, Barlogis V, Hamamoto K, Kanegane H, Milanovich S, Margolis DA, Dimmock D, Casper J, Douglas DN, Amrolia PJ, Veys P, Kumar AR, Jordan MB, Bleesing JJ, Filipovich AH (2013) Allogeneic hematopoietic cell transplantation for XIAP deficiency: an international survey reveals poor outcomes. Blood 121(6):877–883

Marsh RA, Bleesing JJ, Chandrakasan S, Jordan MB, Davies SM, Filipovich AH (2014) Reduced-intensity conditioning hematopoietic cell transplantation is an effective treatment for patients with SLAM-associated protein deficiency/X-linked lymphoproliferative disease type 1. Biol Blood Marrow Transplant. pii:S1083-8791(14)00350-4

Martin E, Palmic N, Sanquer S, Lenoir C, Hauck F, Mongellaz C, Fabrega S, Nitschké P, Esposti MD, Schwartzentruber J, Taylor N, Majewski J, Jabado N, Wynn RF, Picard C, Fischer A, Arkwright PD, Latour S (2014) CTP synthase 1 deficiency in humans reveals its central role in lymphocyte proliferation. Nature 510(7504):288–292

Milone MC, Tsai DE, Hodinka RL, Silverman LB, Malbran A, Wasik MA, Nichols KE (2005) Treatment of primary Epstein-Barr virus infection in patients with X-linked lymphoproliferative disease using B-cell-directed therapy. Blood 105(3):994–996

Moshous D, Martin E, Carpentier W, Lim A, Callebaut I, Canioni D, Hauck F, Majewski J, Schwartzentruber J, Nitschke P, Sirvent N, Frange P, Picard C, Blanche S, Revy P, Fischer A, Latour S, Jabado N, de Villartay JP (2013) Whole-exome sequencing identifies Coronin-1A deficiency in 3 siblings with immunodeficiency and EBV-associated B-cell lymphoproliferation. J Allergy Clin Immunol 131(6):1594–1603

Nehme NT, Pachlopnik Schmid J, Debeurme F, André-Schmutz I, Lim A, Nitschke P, Rieux-Laucat F, Lutz P, Picard C, Mahlaoui N, Fischer A, de Saint Basile G (2012) MST1 mutations in autosomal recessive primary immunodeficiency characterized by defective naive T-cell survival. Blood 119(15):3458–3468

Nichols KE, Harkin DP, Levitz S, Krainer M, Kolquist KA, Genovese C, Bernard A, Ferguson M, Zuo L, Snyder E, Buckler AJ, Wise C, Ashley J, Lovett M, Valentine MB, Look AT, Gerald W, Housman DE, Haber DA (1998) Inactivating mutations in an SH2 domain-encoding gene in X-linked lymphoproliferative syndrome. Proc Natl Acad Sci USA 95(23):13765–13770

Nichols KE, Hom J, Gong SY, Ganguly A, Ma CS, Cannons JL, Tangye SG, Schwartzberg PL, Koretzky GA, Stein PL (2005) Regulation of NKT cell development by SAP, the protein defective in XLP. Nat Med 11(3):340–345

Okkenhaug K, Vanhaesebroeck B (2003) PI3K in lymphocyte development, differentiation and activation. Nat Rev Immunol 3(4):317–330

Pachlopnik Schmid J, Canioni D, Moshous D, Touzot F, Mahlaoui N, Hauck F, Kanegane H, Lopez-Granados E, Mejstrikova E, Pellier I, Galicier L, Galambrun C, Barlogis V, Bordigoni P, Fourmaintraux A, Hamidou M, Dabadie A, Le Deist F, Haerynck F, Ouachée-Chardin M, Rohrlich P, Stephan JL, Lenoir C, Rigaud S, Lambert N, Milili M, Schiff C, Chapel H, Picard C, de Saint Basile G, Blanche S, Fischer A, Latour S (2011) Clinical similarities and differences of patients with X-linked lymphoproliferative syndrome type 1 (XLP-1/SAP deficiency) versus type 2 (XLP-2/XIAP deficiency). Blood 117(5):1522–1529

Palendira U, Low C, Chan A, Hislop AD, Ho E, Phan TG, Deenick E, Cook MC, Riminton DS, Choo S, Loh R, Alvaro F, Booth C, Gaspar HB, Moretta A, Khanna R, Rickinson AB, Tangye SG (2011) Molecular pathogenesis of EBV susceptibility in XLP as revealed by analysis of female carriers with heterozygous expression of SAP. PLoS Biol 9(11):e1001187

Palendira U, Low C, Bell AI, Ma CS, Abbott RJ, Phan TG, Riminton DS, Choo S, Smart JM, Lougaris V, Giliani S, Buckley RH, Grimbacher B, Alvaro F, Klion AD, Nichols KE, Adelstein S, Rickinson AB, Tangye SG (2012) Expansion of somatically reverted memory $CD8^+$ T cells in patients with X-linked lymphoproliferative disease caused by selective pressure from Epstein-Barr virus. J Exp Med 209(5):913–924

Parolini S, Bottino C, Falco M, Augugliaro R, Giliani S, Franceschini R, Ochs HD, Wolf H, Bonnefoy JY, Biassoni R, Moretta L, Notarangelo LD, Moretta A (2000) X-linked lymphoproliferative disease. 2B4 molecules displaying inhibitory rather than activating function are responsible for the inability of natural killer cells to kill Epstein-Barr virus-infected cells. J Exp Med 192(3):337–346

Purtilo DT (1976) Pathogenesis and phenotypes of an X-linked recessive lymphoproliferative syndrome. Lancet 2(7991):882–885

Purtilo DT, Cassel CK, Yang JP, Harper R (1975) X-linked recessive progressive combined variable immunodeficiency (Duncan's disease). Lancet 1(7913):935–940

Purtilo DT, DeFlorio D Jr, Hutt LM, Bhawan J, Yang JP, Otto R, Edwards W (1977) Variable phenotypic expression of an X-linked recessive lymphoproliferative syndrome. N Engl J Med 297(20):1077–1080

Qi H, Cannons JL, Klauschen F, Schwartzberg PL, Germain RN (2008) SAP-controlled T-B cell interactions underlie germinal centre formation. Nature 455(7214):764–769

Rigaud S, Fondanèche MC, Lambert N, Pasquier B, Mateo V, Soulas P, Galicier L, Le Deist F, Rieux-Laucat F, Revy P, Fischer A, de Saint Basile G, Latour S (2006) XIAP deficiency in humans causes an X-linked lymphoproliferative syndrome. Nature 444(7115):110–114

Rivat C, Booth C, Alonso-Ferrero M, Blundell M, Sebire NJ, Thrasher AJ, Gaspar HB (2013) SAP gene transfer restores cellular and humoral immune function in a murine model of X-linked lymphoproliferative disease. Blood 121(7):1073–1076

Rohr J, Beutel K, Maul-Pavicic A, Vraetz T, Thiel J, Warnatz K, Bondzio I, Gross-Wieltsch U, Schündeln M, Schütz B, Woessmann W, Groll AH, Strahm B, Pagel J, Speckmann C, Janka G, Griffiths G, Schwarz K, zur Stadt U, Ehl S (2010) Atypical familial hemophagocytic lymphohistiocytosis due to mutations in UNC13D and STXBP2 overlaps with primary immunodeficiency diseases. Haematologica 95(12):2080–2087

Salzer E, Daschkey S, Choo S, Gombert M, Santos-Valente E, Ginzel S, Schwendinger M, Haas OA, Fritsch G, Pickl WF, Förster-Waldl E, Borkhardt A, Boztug K, Bienemann K, Seidel MG (2013) Combined immunodeficiency with life-threatening EBV-associated lymphoproliferative disorder in patients lacking functional CD27. Haematologica 98(3):473–478

Sayos J, Wu C, Morra M, Wang N, Zhang X, Allen D, van Schaik S, Notarangelo L, Geha R, Roncarolo MG, Oettgen H, De Vries JE, Aversa G, Terhorst C (1998) The X-linked lymphoproliferative-disease gene product SAP regulates signals induced through the co-receptor SLAM. Nature 395(6701):462–469

Snow AL, Marsh RA, Krummey SM, Roehrs P, Young LR, Zhang K, van Hoff J, Dhar D, Nichols KE, Filipovich AH, Su HC, Bleesing JJ, Lenardo MJ (2009) Restimulation-induced apoptosis of T cells is impaired in patients with X-linked lymphoproliferative disease caused by SAP deficiency. J Clin Invest 119(10):2976–2989

Snow AL, Xiao W, Stinson JR, Lu W, Chaigne-Delalande B, Zheng L, Pittaluga S, Matthews HF, Schmitz R, Jhavar S, Kuchen S, Kardava L, Wang W, Lamborn IT, Jing H, Raffeld M, Moir S, Fleisher TA, Staudt LM, Su HC, Lenardo MJ (2012) Congenital B cell lymphocytosis explained by novel germline CARD11 mutations. J Exp Med 209(12):2247–2261

Speckmann C, Lehmberg K, Albert MH, Damgaard RB, Fritsch M, Gyrd-Hansen M, Rensing-Ehl A, Vraetz T, Grimbacher B, Salzer U, Fuchs I, Ufheil H, Belohradsky BH, Hassan A, Cale CM, Elawad M, Strahm B, Schibli S, Lauten M, Kohl M, Meerpohl JJ, Rodeck B, Kolb R, Eberl W, Soerensen J, von Bernuth H, Lorenz M, Schwarz K, Zur Stadt U, Ehl S (2013) X-linked inhibitor of apoptosis (XIAP) deficiency: the spectrum of presenting manifestations beyond hemophagocytic lymphohistiocytosis. Clin Immunol 149(1):133–141

Spinner MA, Sanchez LA, Hsu AP, Shaw PA, Zerbe CS, Calvo KR, Arthur DC, Gu W, Gould CM, Brewer CC, Cowen EW, Freeman AF, Olivier KN, Uzel G, Zelazny AM, Daub JR, Spalding CD, Claypool RJ, Giri NK, Alter BP, Mace EM, Orange JS, Cuellar-Rodriguez J, Hickstein DD, Holland SM (2014) GATA2 deficiency: a protean disorder of hematopoiesis, lymphatics, and immunity. Blood 123(6):809–821

van Montfrans JM, Hoepelman AI, Otto S, van Gijn M, van de Corput L, de Weger RA, Monaco-Shawver L, Banerjee PP, Sanders EA, Jol-van der Zijde CM, Betts MR, Orange JS, Bloem AC, Tesselaar K (2012) CD27 deficiency is associated with combined immunodeficiency and persistent symptomatic EBV viremia. J Allergy Clin Immunol 129(3):787–793

Zhao F, Cannons JL, Dutta M, Griffiths GM, Schwartzberg PL (2012) Positive and negative signaling through SLAM receptors regulate synapse organization and thresholds of cytolysis. Immunity 36(6):1003–1016

zur Stadt U, Rohr J, Seifert W, Koch F, Grieve S, Pagel J, Strauss J, Kasper B, Nürnberg G, Becker C, Maul-Pavicic A, Beutel K, Janka G, Griffiths G, Ehl S, Hennies HC (2009) Familial hemophagocytic lymphohistiocytosis type 5 (FHL-5) is caused by mutations in Munc18-2 and impaired binding to syntaxin 11. Am J Hum Genet 85(4):482–492

Burkitt's Lymphoma

Rosemary Rochford and Ann M. Moormann

Abstract Endemic Burkitt's lymphoma (BL) remains the most prevalent pediatric cancer in sub-Saharan Africa even though it was the first human cancer with a viral etiology described over 50 years ago. Epstein–Barr virus (EBV) was discovered in a BL tumor in 1964 and has since been implicated in other malignancies. The etiology of endemic BL has been linked to EBV and *Plasmodium falciparum* malaria co-infection. While epidemiologic studies have yielded insight into EBV infection and the etiology of endemic BL, the modulation of viral persistence in children by malaria and deficits in EBV immunosurveillance has more recently been reified. Renewed efforts to design prophylactic and therapeutic EBV vaccines provide hope of preventing EBV-associated BL as well as increasing the ability to cure this cancer.

Contents

1 The Early Work ... 268
2 The Role of EBV in Endemic BL ... 271
3 The Role of Malaria on EBV ... 273
4 EBV-Specific T-Cell Immunosurveillance ... 276
5 Other Potential Co-factors in Endemic BL ... 278
6 Future Directions ... 279
References ... 280

R. Rochford
Department of Microbiology and Immunology, SUNY Upstate Medical University, Syracuse, NY, USA
e-mail: rochforr@upstate.edu

A.M. Moormann (✉)
Program in Molecular Medicine, University of Massachusetts Medical School, 373 Plantation Street, Worcester, MA 01605, USA
e-mail: ann.moormann@umassmed.edu

C. Münz (ed.), *Epstein Barr Virus Volume 1*, Current Topics in Microbiology and Immunology 390, DOI 10.1007/978-3-319-22822-8_11

1 The Early Work

Endemic Burkitt's lymphoma was first described by Denis Burkitt, a British surgeon living in Uganda in the late 1950s. In his initial report (Burkitt 1958), he described a malignant tumor which was occurring in young children between the ages of 2 and 14 years. The striking feature of this cancer was the involvement of the jaw, which led Burkitt to classify this as a sarcoma. Additionally, Burkitt noted the involvement of other anatomical sites in many of these children, including the abdomen. Following histological review, Burkitt and O'Conor (1961) reported that these jaw and abdominal tumors were identical and that this was a distinct pathological entity. Upon further examination, it was concluded that the tumor was in fact a malignant lymphoma, not a sarcoma. Strangely, this lymphoma seemed to only involve extranodal sites. Referring to this disease generally as malignant lymphoma, it was reported to be very common in Uganda and accounted for approximately half of the cases of cancer that occurred in children. The authors examined 106 cases and reported an age distribution similar to what was previous described by Burkitt. Interestingly, it was noted that there was approximately a 2:1 male-to-female sex ratio in the distribution of these cases. By collecting information through personal contact and questionnaire, it was reported that malignant lymphomas were seen in a relatively limited geographic area across most of central Africa, which was referred to as the "lymphoma belt."

In order to elucidate the cause of this tumor, Burkitt began examining its geographic occurrence in Uganda (Burkitt 1962a) and traveled on a series of "tumor safaris" across much of central and east Africa, attempting to visit as many hospitals as possible along the "edge" of this belt (Burkitt 1962a, b). He further traveled to areas of West Africa in order to examine the occurrence of the cancer there. Burkitt found the only factor that seemed to be related to the incidence of this lymphoma was altitude, whereby he noted that this cancer only appeared in populations residing in areas with an altitude of less than 5000 feet and near the equator. When moving geographically away from the equator, this altitude cut-off decreased along with the incidence of BL. By mapping the occurrence of BL along the edge of the lymphoma belt, it was determined that altitude was a proxy measure of the minimum temperature. That is, this lymphoma only occurred in areas where the minimum temperature during the coldest season of the year was higher than 60 °F. This hypothesis was supported by the fact that the only area of Africa that met these temperature requirements was a belt across central Africa. This belt was strikingly similar geographically to Burkitt's so-called lymphoma belt. At a time when no infectious agents were known to cause malignancies in humans, Burkitt suggested that the temperature dependency indicated that a vector may be responsible for transmission of the disease and hypothesized the transmitted agent to be a virus. After examining rainfall levels in West Africa, Burkitt further hypothesized that the occurrence of this cancer was also humidity dependent, suggesting that this was perhaps an indication of vegetation dependence (Burkitt 1962a).

The geographic restriction of this cancer, which was now referred to as Burkitt's lymphoma (BL), was examined in more detail by Haddow (1963). He noted that the temperature and rainfall requirements described by Burkitt corresponded closely with the distribution of certain species of mosquitoes and that at temperatures between 60 and 65 °F, viruses, such as the yellow fever virus and dengue virus, fail to develop in a mosquito vector. Because many mosquito species rely on an annual rainfall of at least 20 inches, it was hypothesized that BL would only be seen in areas where the annual rainfall was greater than 20 inches a year. To test this hypothesis, a map of the areas in Africa which met these geographic requirements was obtained and the location of all known cases of BL was overlaid on this map. Upon examination, 95 % of geographic locations which had reported cases of BL occurred in these areas. It was concluded that BL was geographically associated with a mean average temperature of greater than 60 °F (15.5 °C) for the coolest month and an annual rainfall of at least 20 inches (500 mm).

Therefore, Haddow made the following deductions (Haddow 1964):

1. If an arbovirus was involved in the generation of BL, the number of cases occurring was far too small to permit maintenance of the disease by direct case-to-case transmission.
2. If this was a zoonotic agent that occasionally escapes into humans, one would expect definite outbreaks to occur in limited areas. There was yet no evidence of this. In addition, the curious age distribution would remain unexplained.
3. Thus, the infection, if it did occur essentially in humans, must be an exceedingly common and widespread one in order to permit continuous transmission, with the tumor being a rare and extreme manifestation.

In order to explore the arbovirus hypothesis, Haddow analyzed the age distribution of 363 cases collected by Burkitt from Uganda, Tanganyika (now Tanzania), Mozambique, and Nigeria. The age distribution showed a peak between the ages of 6 and 7 years and only 2 % of cases occurring in individuals above the age of 20 years. From this, it was concluded that the risk of responding to the postulated infection by development of the tumor was eliminated by age 20, suggesting a very common viral infection. Furthermore, the fact that the tumor did not occur in extremely young children suggested that they were protected by maternal passive immunity following birth, consistent with the hypothesis that this agent was common.

In a report publish in 1966 (Burkitt and Wright 1966), Burkitt et al. examined the occurrence of 450 histologically determined cases of BL over an eight year time period. He found the occurrence of the cancer to be closely correlated geographically with temperature. High rates, ranging from 8.7 to 13.4 per 100,000 children per year, were seen in lowland areas along the Nile, while rates were almost 20 times lower in the mountainous area of southwestern Uganda. Consistent with the previously reported sex bias, the overall incidence was 2.3 males per female, even though the admission rates to pediatric wards were equal for sexes in the hospitals from which records were obtained. In areas with high rates of BL, the average age of tumor occurrence was 8.1 years old, similar to previous reports. Interestingly, in areas with a low rate of BL, the average age of incidence was 16.2 years, markedly older. Furthermore, among immigrants who moved to areas of high BL incidence, similar rates of disease to native residents were seen. However, BL occurred among older age groups for those migrating to high incident areas with immigrants accounting for nearly 50 % of patients over the age of 15 year in these areas. Using this evidence, the authors suggested that in highly endemic areas, exposure to the "mysterious" arbovirus occurred more commonly early in life, causing the tumor to develop soon thereafter, whereas in areas where BL was less common exposure tended to occur later in life. The authors reported that all tribes living in lowland areas had similar rates of BL and that the ratio of African to Asian cases of BL was consistent with the population ratio of these two groups [also see (Burkitt 1962c)], suggesting that geographic differences in the occurrence of this cancer could not be explained by human genetic factors.

While visiting the territories of Papua and New Guinea in 1960, Ten Seldam observed 2 cases of lymphosarcoma in young boys and was struck by the similarity to the BL patients which had been described in Africa. Ten Seldam et al. (1966) examined 35 cases of childhood lymphoma that occurred in these territories between 1958 and 1963. Of these, approximately one-third had the typical jaw involvement described by Burkitt which was about half the ratio of tumors with jaw involvement found in Africa. However, this group reported an extraordinarily uniform histological picture in the appearance of these tumors with those described in Africa. Geographically, these cases occurred in areas which met the temperature and rainfall requirements previously described. Interestingly, it was noted that *Plasmodium falciparum* malaria was also holoendemic in both areas of Africa and New Guinea, where repeated malarial infections were nearly universal during the first years of life (Dalldorf et al. 1964).

Following this lead, Burkitt collected and examined evidence suggesting that malaria could play an etiologic role in BL and in 1969 published an extensive review that used a number of arguments to suggest that malaria was involved in the generation of BL (Burkitt 1969). In this review, Burkitt discussed in detail epidemiologic evidence for the association between malaria and BL, most importantly the fact that holoendemic malaria was only known to occur in Africa and New Guinea, as well as pathological evidence, such as the possibility that the occurrence of BL in extranodal sites could be explained by hyperreactivity of the reticuloendothelial system caused by chronic malaria infections, also known as tropical splenomegaly syndrome (Fakunle and Greenwood 1976). This review further discussed the evidence for the involvement of a virus, such as the virus isolated by Epstein and colleagues from BL tissue culture cells. With increasing evidence pointing toward the role of an infectious agent in the etiology of BL, yet no conclusive results to date, Burkitt stated, "Some of the strongest evidence implicating an infective agent as partly responsible for the occurrence of Burkitt's lymphoma is the demonstration of case clustering in space and time. This would be consistent with the postulation of a viral infection, either as a vectored agent primarily responsible for the tumor or as a virus not necessarily vectored but promoting actual tumor formation."

In 1964, EBV was discovered in a BL tumor from a Ugandan patient when Burkitt sent a biopsy specimen to Anthony Epstein, Yvonne Barr, and Bert Achong who were able to visualize the virus within this B-cell tumor using electron microscopy (Epstein et al. 1964). Details of this discovery of EBV can be found in the chapter by Dr. Epstein. Since then, EBV has been associated with nearly all endemic (African) BL tumors, with clear evidence that EBV infection precedes B-cell clonal expansion (Neri et al. 1991). Yet BL can present without detectable EBV within the tumor even though all BL tumors have the *c-myc* translocation (Dalla-Favera et al. 1982). Other forms of BL have been classified as sporadic BL in which only 20 % of the patients have detectable EBV within their tumors, immune deficiency (AIDS) associated BL with EBV in situ in 30–40 % of the patients' tumors, and an "intermediate" form of BL found in less developed countries such as Brazil, Middle East, and other areas that do not have malaria. It is tempting to speculate that EBV triggered these forms of BL but was lost during rapid cell replication hijacked by the myc-oncogene. For the purposes of this review, we will limit our discussion to the EBV-associated form of BL.

2 The Role of EBV in Endemic BL

Despite inconsistencies between studies investigating space-time clustering, seasonality, sex ratios, etc. (Biggar and Nkrumah 1979; Morrow et al. 1977; Rainey et al. 2007; Siemiatycki et al. 1980), these early descriptive studies provided a great deal of evidence implicating EBV and holoendemic malaria as etiologic agents in the generation of BL. The advent of a serologic test to diagnose primary EBV infection and follow the stages of infection heralded a means to investigate the temporal role of EBV in BL etiology, and EBV gene expression patterns are discussed in more detail in the chapter on EBV serology. In brief, Gertrude and Werner Henle developed a diagnostic assay using immunoglobulin G (IgG) and IgM antibodies to EBV following the hierarchy of antigen expression (Henle et al. 1969). Antibody titers to viral capsid antigen (VCA) , early antigen-diffuse (EA-D) and Epstein–Barr virus nuclear antigen (EBNA-1) are used to determine the stage of infection. There is some debate as to the interpretation of EBV serology patterns, however, IgM to VCA indicates early-acute primary infection, whereas elevated IgG to VCA and EA-D indicate viral reactivation. The development of IgG to EBNA-1 marks the recovery phase and is an indicator of remote infection.

Serological surveys conducted by Guy de Thé in the 1970s demonstrated that nearly all African children were EBV seropositive by three years of age (de-The et al. 1978), in contrast to children in developed countries who seroconvert later in life with ~30 % of the population remaining seronegative until adolescence (Balfour et al. 2013). This suggested that EBV may be necessary but not sufficient to cause BL tumorigenesis and raised the question as to whether changes in serological profiles to EBV antigens could be predictive of BL. The first serosurvey study, reported in 1978, was conducted by de Thé et al. (1978) in the West

Nile district of Uganda from 1971 to 1976. Serum samples were taken from 42,000 children below the age of 8 years. Over the course of the study, 14 children enrolled were clinically diagnosed with BL. Serum antibodies from these cases were then compared to the following 3 controls: (1) serum from a neighbor of the same age and sex, (2) 4 control sera selected from the sera bank of the same age, sex, and approximate locality, and (3) sera from a random sample of the surveyed population. Whereas no difference was seen in pre-BL sera antibodies to EBV EA and EBNA antigens for patients in comparison with matched controls, higher titers of antibodies to VCA were present in pre-BL sera of cases, with an average geometric mean titer (GMT) approximately 3.4 times higher. When GMTs to VCA were compared to a random sample of the population, all but 2 cases were higher than the mean for children in that corresponding age group in the normal population, indicating that the age, sex, and geographically adjusted risk of developing BL was approximately 30 times higher for children with a VCA titer 2 dilutions above the mean than for children with average sera titers. Antibody levels to herpes simplex virus and cytomegalovirus showed no difference between cases and controls, indicating that EBV plays a specific role in the development of BL, not just that elevated VCA titers were a marker for overall immune activation. Further support for the role of EBV in the etiology of endemic BL came from studies showing that the virus was clonal in the tumors suggesting that infection preceded the development of malignancy (Neri et al. 1991).

In 1977, Guy de Thé proposed that perinatal infection with EBV was a risk factor for BL (de-The 1977) but not until 2012, there was evidence to support this hypothesis (Piriou et al. 2012). In a prospective study by Piriou et al. based in Kenya, two groups of infants were followed from 1 month of age through 2 years of age. The first group of 68 children was from a malaria endemic region and the second group of 82 children was from an area with low malaria transmission. EBV infection was detected either by evidence of EBV DNA in the blood or by levels of IgG and IgM to VCA and of IgG to EBNA-1. Several observations were made. First, infants born in a malaria holoendemic region were infected with EBV earlier in life than infants born in a region with low malaria transmission (mean age 7.28 months compared to 8.39 months, respectively). Second, ~35 % of infants from the malaria holoendemic region were infected before 6 months of age. And finally, infection with EBV early in life predicted higher viral load over time suggesting that early age of infection resulted in poor control of the virus.

At a molecular level, several lines of evidence also point to a role for the virus in driving tumorigenesis. These include as follows: the maintenance of the viral genome in BL tumors (zur Hausen et al. 1970), the viral genome is present in all cells within the tumors (zur Hausen et al. 1970), the virus is clonal within the tumors (Neri et al. 1991), and the viral protein EBNA-1 (Lindahl et al. 1974) along with viral non-coding BART microRNAs (Tao et al. 1998; Xue et al. 2002) are consistently expressed within the tumors. In addition, EBV infection within the context of type I latency program promoted BL cell growth by inhibiting c-myc-induced apoptosis through the upregulation of Bcl-2 and a commensurate decrease in c-myc expression (Ruf et al. 2001). The consequences of EBV gene expression patterns that diverge from restricted Qp promotor latency I may have implications for responsiveness to chemotherapy. Studies conducted by Griffin et al. found that BL tumors from Malawian patients expressed an array of lytic genes and expression of EBV BZLF1 replication activator intermediate early promotor (ZEBRA) correlated with responsiveness to cyclophosphamide (Labrecque et al. 1999). Studies of BL cell lines concluded that Wp-restricted tumors were more aggressive and resistant to apoptosis (Kelly et al. 2013). In addition, the transformation capacity of EBV also highlights the virus' oncogenic potential. EBNA-1 induces lymphomas in a transgenic mouse model (Wilson et al. 1996) supporting a potential direct role for this EBV latent protein in oncogenesis. An alternative model for the role of EBV in oncogenesis is that EBV infection of a B cell increases the risk for other cellular changes. These could be either through epigenetic modifications (Kaneda et al. 2012) or by promoting genetic instability (Gruhne et al. 2009).

3 The Role of Malaria on EBV

In an attempt to retrospectively determine malaria infection history of BL patients, a case–control study was conducted in Uganda from 1994 to 1999 (Carpenter et al. 2008). Carpenter et al. found that children diagnosed with BL received more frequent treatments for malaria in the preceding year and were less likely to live in a household using insecticide-treated bednets compared to controls. IgG VCA titers were measured on 126 BL patients and 70 controls. Consistent with the earlier studies, IgG VCA titers were significantly higher in the BL patients surveyed compared to children without BL. Antibodies to malaria were measured by indirect immunofluorescence assay (Sulzer et al. 1969). The odds ratio (OR) estimates for high antimalarial antibody titers verged on significance, however, when the cases and controls were matched according to residential district, antimalarial antibody titers were no longer a risk factor, whereas elevated antibody levels to VCA remained highly significant. A replication case–control study was conducted by Mutalima et al. in Malawi in 2005–2006 (Mutalima et al. 2008). IgG titers to VCA and malaria parasite schizont extract (Verra et al. 2007) were higher in 137 BL patients compared to 91 controls, OR = 14.8 (95 % CI 5.8–38.5) and OR = 2.4 (1.2–4.4), respectively, supporting an interaction between malaria and EBV in increasing the risk for BL.

Studies conducted by Piriou et al. (2009) in western Kenya compared EBV serological profiles in 67 children residing in a malaria holoendemic area to 102 children residing in an area with sporadic malaria transmission. The cumulative effect of chronic and/or repeated *P. falciparum* malaria infections resulted in elevated IgG antibody titers to VCA, EBNA-1, EA-D, and another immediate early protein, Z trans-activation antigen, Zta (also known as ZEBRA), which is typically not induced in healthy individuals (Fachiroh et al. 2006). These data suggest that over time, children co-infected with malaria experienced more EBV reactivation than children not infected with malaria. A complementary study went on to compare EBV and malaria serological profiles in 32 children with BL and 25 controls who were frequency matched for age and sex (Asito et al. 2010). This study used a panel of liver- and blood-stage malaria antigens in a multiplex serological assay to generated IgG antibody titers and found no difference in median antibody levels to specific antimalarial proteins between BL cases and age-, sex- and geographically matched controls. In contrast, BL cases had elevated antibody titers to Zta and VCA compared to controls, consistent with the Ugandan study (Carpenter et al. 2008) but contradicting the Malawian study (Mutalima et al. 2008). In a more recent study in Uganda by Orem et al. (2014), 46 BL patients and 50 children with other forms of non-Hodgkin's lymphoma had significantly elevated IgG levels to EA-D, yet IgG titers to VCA did not differ between cases and controls. Activation of EBV could be due to poor general health that may affect control over EBV latency and may be an effect modifier in children residing in malaria holoendemic areas.

The first intervention study attempting to examine the causative role of malaria in BL was conducted by Geser et al. whereby they attempted to prevent malaria in the entire childhood population of the North Mara lowlands of Tanzania from 1977 to 1982 and compare the rates of BL during this study period to previous and subsequent years (Geser and Brubaker 1985; Geser et al. 1989). The group carried out baseline malaria surveys in 1974, 1975, and 1976 as well as continuous surveys from 1978 to 1982, while chloroquine tablets were given twice monthly to all children in the region from 1977 to 1982. Prior to the study period, from 1964 to 1976, the annual incidence of BL averaged 4.3 per 100,000 children, ranging from 2.6 to 6.9 per 100,000 children. Immediately following the start of the intervention, the incidence of BL began dropping rapidly, hitting a minimum of 0.5 per 100,000 children in 1980 and 1981. Subsequently, from 1982 onward the rate rose sharply, hitting a peak of 7.1 per 100,000 children in 1984. Using simple linear regression, the overall drop in incidence from 1964 to 1982 was highly significant, whereas the slight decline in the rate prior to intervention (1964–1976) was only marginally significant. Upon preliminary analysis, this drop seemed to correspond with the effects of the intervention. However, following a large drop in the prevalence of malaria parasitemia in the population (48 % in 1976 to 11 % in 1977), the levels of parasitemia subsequently rose during the study period and returned to pre-intervention levels by 1981, primarily due to inefficiency of drug distribution. The fact that this intervention was only marginally successful in preventing malaria makes it difficult to determine whether the drop in BL was actually due to the antimalarial intervention. Additionally, the researchers found upon closer scrutiny that BL incidence in the North Mara district may have started dropping in 1972, five years prior to the initiation of this intervention. Serological work from this study did show that neither the prevalence of EBV nor the geometric mean titer to EBV antibodies in children varied throughout the study, indicating declining EBV prevalence was not associated with the decline in BL. Furthermore, it was found from the malaria surveys that *P. falciparum* was responsible for about 90 % of malaria infections; consistent with the observation from other areas that *P. falciparum* is the dominant parasite species in areas with a high incidence of BL.

Transient lymphopenia and selective immune suppression has long been recognized as a complication of acute malaria infections (Greenwood et al. 1972; Williamson and Greenwood 1978). Acute malaria has also been shown to induce pathophysiological changes in B-cell homeostasis (Asito et al. 2008) as well as long-term changes in memory B-cell subsets (Portugal et al. 2012; Weiss et al. 2009) leading to a possibility that malaria could alter EBV persistence. A direct effect of *P. falciparum* on EBV-infected B cells was shown by Chêne et al. (2007). BL cell lines were incubated with the cysteine-rich inter-domain region 1 α (CIDRα) of the *P. falciparum* erythrocyte membrane protein which resulted in lytic cycle activation. In support of this as a potential mechanism for altering EBV persistence, children from Uganda with acute malaria had EBV viremia (Donati et al. 2006) and healthy children from a malaria endemic region had evidence of viral reactivation as indicated by viremia (Rasti et al. 2005). Serologic evidence for increased viral reactivation in children living in malaria holoendemic regions was also shown in a study by Piriou et al. (2009) in Kenya. That there might also be an increase in latently infected cells was first shown by Lam et al. (1991). A separate study by Nije et al. (2009) of Gambian children with acute malaria confirmed this observation by demonstrating an elevated viral load in peripheral blood mononuclear cells from children with acute malaria.

More recent studies to elucidate a mechanism by which malaria could play a role in BL pathogenesis have focused on the enzyme activation-induced cytidine deaminase (AID) which is highly expressed in germinal center B cells (Muramatsu et al. 2000; Ramiro et al. 2004, 2006) and is responsible for somatic hypermutation and class switch recombination. In a mouse model, AID was also found to be required for the *c-myc* translocation (Robbiani et al. 2008). Incubation of human B cells with *P. falciparum* extracts induces AID and class switch recombination (Potup et al. 2009). Subsequently, a study by Torgbor et al. (2014) using B-cell lines and palatine tonsil tissue obtained from 12 Ghanaian children compared to one North American, malaria-naïve control demonstrated that *P. falciparum* malaria, more specifically the malaria-derived hemozoin/DNA complex that is a TLR9 agonist (Parroche et al. 2007), deregulates AID expression. Another study by Wilmore et al. of healthy Kenyan children with divergent malaria exposure examined AID expression in peripheral blood mononuclear cells and demonstrated that malaria-exposed children with detectable EBV circulating in their peripheral blood had higher AID expression compared to EBV-seropositive children with no detectable virus by PCR (Wilmore et al. 2014). This study is important for solidifying the link between malaria and EBV co-infection in endemic BL etiology in that malaria-non-exposed children with measurable EBV load had similar AID levels compared to EBV PCR negative children. Repeated or chronic malaria-induced upregulation of AID could in turn increase EBV load due to the preference of EBV to infect B cells with mutated immunoglobulin (Heath et al. 2012). This mechanism may explain in part the synergy between malaria and EBV leading to BL and is compatible with malaria-induced alternations in T-cell immunosurveillance.

4 EBV-Specific T-Cell Immunosurveillance

T-cell responses that control persistent EBV infections in immune competent individuals and T-cell responses in individuals suffering from acute infectious mononucleosis and X-linked lympho-proliferative disorder are covered in more detail in other chapters of this book and in a recent review by Rickinson et al. (2014). Suffice it to mention, studies of EBV-specific T-cell immunity developed by children infected within the first few years of life and in children diagnosed with BL are few due to challenges in gaining access to pediatric patients and appropriately matched controls as well as limitations in blood volumes ethically obtainable from children.

Early studies to assess the effects of malaria infection on EBV-specific T-cell immunity were conducted by Moss et al. (1983) using blood samples from adults residing in two areas of Papua New Guinea: a lowland coastal region with holoendemic malaria and a high incidence of BL, and a highland area where malaria was not endemic and the incidence of BL was low. Antibody titers to both *P. falciparum* and EBV antigens were determined for both study groups as well as with a group of Caucasian controls. Levels of EBV-specific T-cell-mediated immunity were determined by regression assay. Antibody titers to malaria were higher in the lowland group in comparison with the other two groups and there was no significant difference in anti-VCA and anti-EBNA between the three groups. Regression assays indicated a highly significant difference in the levels of EBV-specific T-cell-mediated immunity associated with holoendemic malaria. Additionally, spontaneous B-cell transformation was seen in 14 of 55 cultures from peripheral blood samples from individuals residing the area with holoendemic malaria, whereas no transformation was seen in any of the cultures from the other two study groups. This study concluded that adults in areas with holoendemic malaria had decreased T-cell-mediated immunity to EBV. It should be noted that no other viruses were examined in this study; therefore, it was not possible to determine whether this was a generalized T-cell immunosuppression or if it was specific to only EBV. After stratifying by study group, it was found that there was no significant correlation between regression end-points and antibody levels to EBNA or VCA, consistent with the results from the study by Biggar et al. (1981) in Ghana that found no significant differences in EBV antibody titers between African communities in malarious and non-malarious regions.

A similar study examining T-cell-mediated immunity to EBV was conducted by Whittle et al. (1984) in the Gambia. In this study, blood samples were taken from nine children ranging in age from 5 to 18 years during and 3 weeks following an acute infection with *P. falciparum* malaria. Regression assays for T-cell control of EBV showed significantly higher regression indices during acute infection, which fell to normal levels following recovery, implying that a transient loss of T-cell-mediated immunity to EBV occurs only during acute malarial infection. Analysis of the cell types in the blood indicated that during acute malaria infection, children had a reduced total number of T cells, as well as reduced ratio of

T-helper to T-suppressor cells. Furthermore, there was an increased ratio of B cells to T-suppressor cells. Using this information, the group hypothesized the following model: In children destined to develop BL, a large number of their B cells are infected by EBV during initial infection [based of the work of de Thé et al. (1978)]. Repeated attacks with malaria result in the loss of T-cell control of EBV, allowing EBV-infected B cells to proliferate and increase in number, which increases the likelihood that chromosomal translocation and malignancy will occur.

It was not until years later that this model could be put to the test. Studies conducted by Moormann et al. (2005) in Kenya measured EBV loads by real-time quantitative PCR in 104 children residing in a malaria holoendemic area and compared them to 127 children from a highland area that experienced sporadic malaria transmission. EBV load was highest in the malaria-exposed children (1–4 years of age) prior to the peak age of BL onset. This study went on to investigate the specificity of the T-cell immune deficiency described by earlier studies by comparing IFN-γ ELISPOT responses to pools of EBV lytic and latent antigen peptides specific to CD8+ T cell (Moormann et al. 2007). Immune responses to EBV were contrasted to those against malaria peptides. Children were stratified by age group: 1–4 years of age, when malaria morbidity is the highest; 5–9 years of age, when premunition to malaria is being developed characterized by semi-protective immunity permissive of asymptomatic parasitemias and coinciding with the peak age-incidence for BL; and 10–14 years of age, when children have developed anti-disease immunity to malaria (Riley and Stewart 2013) and have not succumbed to BL. This study demonstrated deficient IFN-γ responses to EBV lytic and latent antigen peptides restricted to 5- to 9-year-old children who had been residing in a malaria holoendemic area compared to age-matched control children from an area with sporadic malaria transmission (Moormann et al. 2007). Longitudinal studies conducted by Snider et al. (2012) suggest that T-cell immunity to EBV lytic antigens may diminish prior to those against EBV latent antigens. This would be consistent with a model whereby malaria induces lytic reactivation, and lytic-specific T cells become exhausted due to chronic or repeated antigen stimulation (Wei et al. 2013), thereby allowing infectious virions to incrementally establish a higher frequency of latently infected B cells.

In support of an incremental loss of EBV-specific T-cell immunosurveillance hypothesis in BL etiology is a study conducted by Moormann et al. (2009) in 2005–2006, which was the first to directly examine T-cell immunity to EBV and malaria in BL patients. This study compared IFN-γ responses to HLA Class I restricted EBV lytic and latent antigen peptides (Rickinson et al. 1992; Khanna and Burrows 2000) in addition to an overlapping pool of longer HLA Class II peptides to EBNA-1 (Heller et al. 2007) and a recombinant protein to malaria merozoite surface antigen 1 (MSP-1) (Singh et al. 2003). EBNA-1 is of particular interest since it is the only EBV antigen expressed in all BL tumor cells (Crawford 2001) and is only rarely detected by cytotoxic CD8+ T lymphocytes (Munz 2004). Children diagnosed with BL had a significant deficiency in IFN-γ responses to EBNA-1 compared to healthy age-matched controls and yet had robust responses to MSP-1 comparable to healthy malaria-exposed children (Moormann et al. 2009). Of note, the BL children had IgG_1 subclass to EBNA-1 similar to healthy EBV-seropositive children suggesting T-cell depletion in the BL patients rather than Th2 polarization. In addition, non-BL, malaria-exposed children lacking IFN-γ responses to EBNA-1 had higher median EBV loads compared to healthy children with EBNA1-specific T-cell immunity. Another study by Chattopadhyay et al. used HLA-A2 tetramers to phenotype CD8+ T cells specific to EBV lytic (BMFL1 and BRLF1) and latent (LMP1, LMP2, and EBNA3C) peptides and multidimensional analysis of CD45RO, CD27, CCR7, CD127, CD57, and PD-1 expression (Chattopadhyay et al. 2013). They found that CD8+ T cells against lytic antigens tended to display an exhausted phenotype lacking homeostatic potential and individuals residing in malaria holoendemic areas had more differentiated CD8+ T cells to EBV latent antigens with fewer central memory subsets compared to those living in regions with little to no malaria transmission. Malaria did not skew CMV-specific T-cell subsets nor affect the global CD8+ memory T-cell pool. These studies further solidified a malaria-associated detrimental impact on the generation and maintenance of EBV-specific T cells that may contribute to the etiology of BL.

5 Other Potential Co-factors in Endemic BL

The tumor promoter 12-O-tetradecanoylphorbol-13-acetate (TPA), a phorbol ester isolated from croton oil was found to efficiently induce the production of EBV from BL cell lines (zur Hausen et al. 1978) and enhance the ability of EBV to transform B cells (Mizuno et al. 1983; Yamamoto and zur Hausen 1979). As croton oil is extracted from a plant in the *Euphorbiaceae* family, extracts of other plants from the same plant family (*Croton tiglium*, *Euphorbia lathyris*, and *Euphorbia tirucalli*) were also found to induce EBV reactivation (Ito et al. 1981). Dose-dependent treatment of the Jijoye BL cell line with crude *E. tirucalli* latex (10-fold serial dilutions ranging from 10^{-3} to 10^{-7}) resulted in induction of lytic cycle gene expression (sixfold for BZLF1 to 19-fold for gp350) over background levels (MacNeil et al. 2003). Additionally, dual treatment of EBV+ cell lines with *n*-butyrate and extracts from *E. tirucalli* induced EBV antigen expression and treatment of cord blood lymphocytes with the extract following EBV infection enhanced the transforming ability of EBV over 10-fold. The active ingredient in these extracts was isolated by silica gel column chromatography and found to be the chemical 4-deoxyphorbol ester (Osato et al. 1987, 1990). Following infection of cord blood lymphocytes with EBV and treatment with purified 4-deoxyphorbol ester, this group noted a high frequency of chromosomal rearrangements, often involving chromosome 8, the chromosome commonly implicated in translocations in BL (Aya et al. 1991). A more recent study (Mannucci et al. 2012) treated EBV transformed lymphoblastoid cell lines with *E. tirucalli* extracts and found that within 5-day evidence of chromosomal abnormalities were present.

Armed with data indicating an interaction between extracts from plants of the *Euphorbiaceae* family and EBV, Osato et al. (1987) performed surveys from 1984 to 1986 in Kenya and Tanzania, looking at the geographic distribution of *Euphorbiaceae* with relation to the occurrence of BL. In a brief but compelling letter, this group reported that profusion of the plant *E. tirucalli* coincided with the endemicity of BL. According to their research, *E. tirucalli* was used daily in the Lake Victoria Basin of Kenya as a traditional medicine. In a separate study, 45 residents and 6 traditional healers in the Lake Victoria region of Kenya were interviewed about their uses for and exposure to *E. tirucalli* (MacNeil et al. 2003). *E. tirucalli* was an extremely common, domesticated plant with many traditional uses. Exposure to the plant among children could occur through a variety of ways including through the use of the latex as a topical medicine, an ingested medicine, a play item, and as glue. Epidemiologically, only one study to date has examined the association between BL and *E. tirucalli*. In this case–control study, conducted by van den Bosch et al. in Malawi (van den Bosch et al. 1993), the presence of *E. tirucalli* around households was investigated for 67 cases of BL and 228 matched controls. This group found there to be a significant association between the prevalence of *E. tirucalli* at homesteads and the occurrence of BL (OR = 7.96, p-value = 0.012). However, it should be noted that the plant was only identified at the homes of 5 cases and 2 controls, making it difficult to assess the validity of these results.

Other co-factors that could contribute to the increased risk of endemic BL have included micronutrient deficiency (Sumba et al. 2010) and arboviruses, specifically Chikungunya virus (van den Bosch and Lloyd 2000). Further studies are needed to determine whether these agents are causal in the etiology of endemic BL.

6 Future Directions

After 50 years since the discovery of EBV, its role in BL etiology and continued role in preventing apoptosis once B-cell oncogenesis has been initiated remain active areas of investigation. There appears to be a sequence of events whereby EBV infection occurs early in life, followed by repeated and often chronic, asymptomatic *P. falciparum* malaria infections that in turn modulates EBV persistence and erodes EBV-specific T-cell immunity. The peak incident age for the development of BL (5–9 years) indicates a cumulative effect of these two infectious agents; the chronology of which is depicted in Fig. 1. The possible involvement of other co-factors such as an arboviral infection or environmental exposure that would trigger malignant transformation after EBV and malaria has set the stage warrant further exploration. Future studies to prevent BL include the development of a prophylactic EBV vaccine to prevent infection (Balfour 2014; Cohen et al. 2013). Decreases in BL incidence may be coincidental with the implementation of a malaria vaccine to reduce the burden of malaria in young children and with other

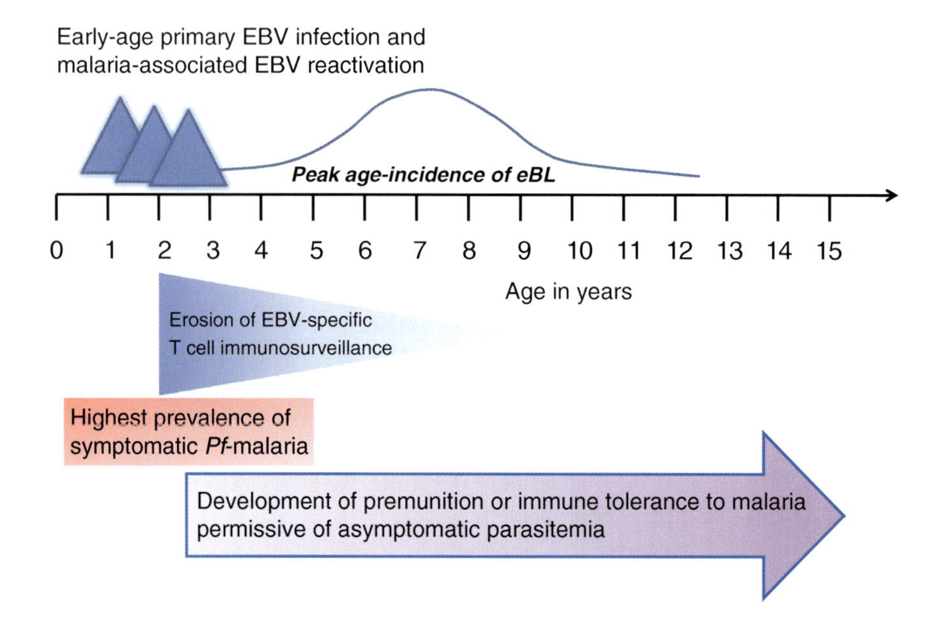

Fig. 1 Cumulative impact of Epstein–Barr virus and *Plasmodium falciparum* malaria co-infections in the etiology of endemic Burkitt's lymphoma. Proposed model indicating early-age EBV infection with repeated EBV reactivation illustrated by *blue triangle* mountain peaks of EBV viremia which precedes the development of eBL. *Blue line* represents distribution curve for eBL incidence by age, rising at 5 years of age and decreasing by 9 years of age with a range of 2–14 years of age. Average age of eBL clinical presentation is 7.5 years old. Malaria co-infections frequently occur before 5 years of age and contribute to repeated EBV reactivation. After cumulative exposure to malaria, children develop immune tolerance allowing malaria to become a chronic rather than acute infection which contributes to the detrimental impact of malaria on EBV-infected B cells and the erosion of EBV-specific T-cell immunosurveillance. Abbreviations: *EBV* Epstein–Barr virus; *Pf Plasmodium falciparum*; *eBL* endemic Burkitt's lymphoma

malaria control programs. There is also interest in a therapeutic EBV vaccine that could be used in combination with conventional chemotherapy to improve survival for children diagnosed with BL (Neparidze and Lacy 2014). These and other challenges await the next generation of scientists.

References

Asito AS, Moormann AM, Kiprotich C, Ng'ang'a ZW, Ploutz-Snyder R et al (2008) Alterations on peripheral B cell subsets following an acute uncomplicated clinical malaria infection in children. Malar J 7:238

Asito AS, Piriou E, Odada PS, Fiore N, Middeldorp JM et al (2010) Elevated anti-Zta IgG levels and EBV viral load are associated with site of tumor presentation in endemic Burkitt's lymphoma patients: a case control study. Infect Agent Cancer 5:13

Aya T, Kinoshita T, Imai S, Koizumi S, Mizuno F et al (1991) Chromosome translocation and c-MYC activation by Epstein-Barr virus and Euphorbia tirucalli in B lymphocytes. Lancet 337:1190

Balfour HH Jr (2014) Progress, prospects, and problems in Epstein-Barr virus vaccine development. Curr Opin Virol 6C:1–5

Balfour HH Jr, Sifakis F, Sliman JA, Knight JA, Schmeling DO et al (2013) Age-specific prevalence of Epstein-Barr virus infection among individuals aged 6–19 years in the United States and factors affecting its acquisition. J Infect Dis 208:1286–1293

Biggar RJ, Nkrumah FK (1979) Burkitt's lymphoma in Ghana: urban-rural distribution, time-space clustering and seasonality. Int J Cancer 23:330–336

Biggar RJ, Gardiner C, Lennette ET, Collins WE, Nkrumah FK et al (1981) Malaria, sex, and place of residence as factors in antibody response to Epstein-Barr virus in Ghana, West Africa. Lancet 2:115–118

Burkitt D (1958) A sarcoma involving the jaws in African children. Br J Surg 46:218–223

Burkitt D (1962a) Determining the climatic limitations of a children's cancer common in Africa. Br Med J 2:1019–1023

Burkitt D (1962b) A "tumour safari" in East and Central Africa. Br J Cancer 16:379–386

Burkitt D (1962c) A children's cancer dependent on climatic factors. Nature 194:232–234

Burkitt DP (1969) Etiology of Burkitt's lymphoma–an alternative hypothesis to a vectored virus. J Natl Cancer Inst 42:19–28

Burkitt D, O'Conor GT (1961) Malignant lymphoma in African children. I. A clinical syndrome. Cancer 14:258–269

Burkitt D, Wright D (1966) Geographical and tribal distribution of the African lymphoma in Uganda. Br Med J 1:569–573

Carpenter LM, Newton R, Casabonne D, Ziegler J, Mbulaiteye S et al (2008) Antibodies against malaria and Epstein-Barr virus in childhood Burkitt lymphoma: a case-control study in Uganda. Int J Cancer 122:1319–1323

Chattopadhyay PK, Chelimo K, Embury PB, Mulama DH, Sumba PO et al (2013) Holoendemic malaria exposure is associated with altered Epstein-Barr virus-specific CD8(+) T-cell differentiation. J Virol 87:1779–1788

Chene A, Donati D, Guerreiro-Cacais AO, Levitsky V, Chen Q et al (2007) A molecular link between malaria and Epstein-Barr virus reactivation. PLoS Pathog 3:e80

Cohen JI, Mocarski ES, Raab-Traub N, Corey L, Nabel GJ (2013) The need and challenges for development of an Epstein-Barr virus vaccine. Vaccine 31(Suppl 2):B194–B196

Crawford DH (2001) Biology and disease associations of Epstein-Barr virus. Philos Trans R Soc Lond B Biol Sci 356:461–473

Dalla-Favera R, Bregni M, Erikson J, Patterson D, Gallo RC et al (1982) Human c-myc onc gene is located on the region of chromosome 8 that is translocated in Burkitt lymphoma cells. Proc Natl Acad Sci USA 79:7824–7827

Dalldorf G, Linsell CA, Barnhart FE, Martyn R (1964) An epidemiologic approach to the lymphomas of African children and Burkitt's sacroma of the jaws. Perspect Biol Med 7:435–449

de-The G (1977) Is Burkitt's lymphoma related to perinatal infection by Epstein-Barr virus? Lancet 1:335–338

de-The G, Geser A, Day NE, Tukei PM, Williams EH et al (1978) Epidemiological evidence for causal relationship between Epstein-Barr virus and Burkitt's lymphoma from Ugandan prospective study. Nature 274:756–761

Donati D, Espmark E, Kironde F, Mbidde EK, Kamya M et al (2006) Clearance of circulating Epstein-Barr virus DNA in children with acute malaria after antimalaria treatment. J Infect Dis 193:971–977

Epstein MA, Achong BG, Barr YM (1964) Virus particles in cultured lymphoblasts from Burkitt's lymphoma. Lancet 1:702–703

Fachiroh J, Paramita DK, Hariwiyanto B, Harijadi A, Dahlia HL et al (2006) Single-assay combination of Epstein-Barr Virus (EBV) EBNA1- and viral capsid antigen-p18-derived synthetic peptides for measuring anti-EBV immunoglobulin G (IgG) and IgA antibody levels in sera from nasopharyngeal carcinoma patients: options for field screening. J Clin Microbiol 44:1459–1467

Fakunle YM, Greenwood BM (1976) A suppressor T-cell defect in tropical splenomegaly syndrome. Lancet 2:608–609

Geser A, Brubaker G (1985) A preliminary report of epidemiological studies of Burkitt's lymphoma, Epstein-Barr virus infection and malaria in North Mara, Tanzania. IARC Sci Publ:205–215

Geser A, Brubaker G, Draper CC (1989) Effect of a malaria suppression program on the incidence of African Burkitt's lymphoma. Am J Epidemiol 129:740–752

Greenwood BM, Bradley-Moore AM, Bryceson AD, Palit A (1972) Immunosuppression in children with malaria. Lancet 1:169–172

Gruhne B, Sompallae R, Marescotti D, Kamranvar SA, Gastaldello S et al (2009) The Epstein-Barr virus nuclear antigen-1 promotes genomic instability via induction of reactive oxygen species. Proc Natl Acad Sci U S A 106:2313–2318

Haddow AJ (1963) An improved map for the study of Burkitt's lymphoma syndrome in Africa. East Afr Med J 40:429–432

Haddow AJ (1964) Age incidence in Burkitt's lymphoma syndrome. East Afr Med J 41:1–6

Heath E, Begue-Pastor N, Chaganti S, Croom-Carter D, Shannon-Lowe C et al (2012) Epstein-Barr virus infection of naive B cells in vitro frequently selects clones with mutated immunoglobulin genotypes: implications for virus biology. PLoS Pathog 8:e1002697

Heller KN, Upshaw J, Seyoum B, Zebroski H, Munz C (2007) Distinct memory CD4+ T-cell subsets mediate immune recognition of Epstein Barr virus nuclear antigen 1 in healthy virus carriers. Blood 109:1138–1146

Henle G, Henle W, Clifford P, Diehl V, Kafuko GW et al (1969) Antibodies to Epstein-Barr virus in Burkitt's lymphoma and control groups. J Natl Cancer Inst 43:1147–1157

Ito Y, Kawanishi M, Harayama T, Takabayashi S (1981) Combined effect of the extracts from Croton tiglium, Euphorbia lathyris or Euphorbia tirucalli and n-butyrate on Epstein-Barr virus expression in human lymphoblastoid P3HR-1 and Raji cells. Cancer Lett 12:175–180

Kaneda A, Matsusaka K, Aburatani H, Fukayama M (2012) Epstein-Barr virus infection as an epigenetic driver of tumorigenesis. Cancer Res 72:3445–3450

Kelly GL, Stylianou J, Rasaiyaah J, Wei W, Thomas W et al (2013) Different patterns of Epstein-Barr virus latency in endemic Burkitt lymphoma (BL) lead to distinct variants within the BL-associated gene expression signature. J Virol 87:2882–2894

Khanna R, Burrows SR (2000) Role of cytotoxic T lymphocytes in Epstein-Barr virus-associated diseases. Annu Rev Microbiol 54:19–48

Labrecque LG, Xue SA, Kazembe P, Phillips J, Lampert I et al (1999) Expression of Epstein-Barr virus lytically related genes in African Burkitt's lymphoma: correlation with patient response to therapy. Int J Cancer 81:6–11

Lam KM, Syed N, Whittle H, Crawford DH (1991) Circulating Epstein-Barr virus-carrying B cells in acute malaria. Lancet 337:876–878

Lindahl T, Klein G, Reedman BM, Johansson B, Singh S (1974) Relationship between Epstein-Barr virus (EBV) DNA and the EBV-determined nuclear antigen (EBNA) in Burkitt lymphoma biopsies and other lymphoproliferative malignancies. Int J Cancer 13:764–772

MacNeil A, Sumba OP, Lutzke ML, Moormann A, Rochford R (2003) Activation of the Epstein-Barr virus lytic cycle by the latex of the plant Euphorbia tirucalli. Br J Cancer 88:1566–1569

Mannucci S, Luzzi A, Carugi A, Gozzetti A, Lazzi S et al (2012) EBV reactivation and chromosomal polysomies: euphorbia tirucalli as a possible cofactor in endemic Burkitt lymphoma. Adv Hematol 2012:149780

Mizuno F, Koizumi S, Osato T, Kokwaro JO, Ito Y (1983) Chinese and African Euphorbiaceae plant extracts: markedly enhancing effect on Epstein-Barr virus-induced transformation. Cancer Lett 19:199–205

Moormann AM, Chelimo K, Sumba OP, Lutzke ML, Ploutz-Snyder R et al (2005) Exposure to holoendemic malaria results in elevated Epstein-Barr virus loads in children. J Infect Dis 191:1233–1238

Moormann AM, Chelimo K, Sumba PO, Tisch DJ, Rochford R et al (2007) Exposure to holoendemic malaria results in suppression of Epstein-Barr virus-specific T cell immunosurveillance in Kenyan children. J Infect Dis 195:799–808

Moormann AM, Heller KN, Chelimo K, Embury P, Ploutz-Snyder R et al (2009) Children with endemic Burkitt lymphoma are deficient in EBNA1-specific IFN-gamma T cell responses. Int J Cancer 124:1721–1726
Morrow RH, Pike MC, Smith PG (1977) Further studies of space-time clustering of Burkitt's lymphoma in Uganda. Br J Cancer 35:668–673
Moss DJ, Burrows SR, Castelino DJ, Kane RG, Pope JH et al (1983) A comparison of Epstein-Barr virus-specific T-cell immunity in malaria-endemic and -nonendemic regions of Papua New Guinea. Int J Cancer 31:727–732
Munz C (2004) Epstein-barr virus nuclear antigen 1: from immunologically invisible to a promising T cell target. J Exp Med 199:1301–1304
Muramatsu M, Kinoshita K, Fagarasan S, Yamada S, Shinkai Y et al (2000) Class switch recombination and hypermutation require activation-induced cytidine deaminase (AID), a potential RNA editing enzyme. Cell 102:553–563
Mutalima N, Molyneux E, Jaffe H, Kamiza S, Borgstein E et al (2008) Associations between Burkitt lymphoma among children in Malawi and infection with HIV, EBV and malaria: results from a case-control study. PLoS ONE 3:e2505
Neparidze N, Lacy J (2014) Malignancies associated with epstein-barr virus: pathobiology, clinical features, and evolving treatments. Clin Adv Hematol Oncol 12:358–371
Neri A, Barriga F, Inghirami G, Knowles DM, Neequaye J et al (1991) Epstein-Barr virus infection precedes clonal expansion in Burkitt's and acquired immunodeficiency syndrome-associated lymphoma. Blood 77:1092–1095
Njie R, Bell AI, Jia H, Croom-Carter D, Chaganti S et al (2009) The effects of acute malaria on Epstein-Barr virus (EBV) load and EBV-specific T cell immunity in Gambian children. J Infect Dis 199:31–38
Orem J, Sandin S, Mbidde E, Mangen FW, Middeldorp J et al (2014) Epstein-Barr virus viral load and serology in childhood non-Hodgkin's lymphoma and chronic inflammatory conditions in Uganda: implications for disease risk and characteristics. J Med Virol 86:1796–1803
Osato T, Mizuno F, Imai S, Aya T, Koizumi S et al (1987) African Burkitt's lymphoma and an Epstein-Barr virus-enhancing plant Euphorbia tirucalli. Lancet 1:1257–1258
Osato T, Imai S, Kinoshita T, Aya T, Sugiura M et al (1990) Epstein-Barr virus, Burkitt's lymphoma, and an African tumor promoter. Adv Exp Med Biol 278:147–150
Parroche P, Lauw FN, Goutagny N, Latz E, Monks BG et al (2007) Malaria hemozoin is immunologically inert but radically enhances innate responses by presenting malaria DNA to Toll-like receptor 9. Proc Natl Acad Sci USA 104:1919–1924
Piriou E, Kimmel R, Chelimo K, Middeldorp JM, Odada PS et al (2009) Serological evidence for long-term Epstein-Barr virus reactivation in children living in a holoendemic malaria region of Kenya. J Med Virol 81:1088–1093
Piriou E, Asito AS, Sumba PO, Fiore N, Middeldorp JM et al (2012) Early age at time of primary Epstein-Barr virus infection results in poorly controlled viral infection in infants from Western Kenya: clues to the etiology of endemic Burkitt lymphoma. J Infect Dis 205:906–913
Portugal S, Doumtabe D, Traore B, Miller LH, Troye-Blomberg M et al (2012) B cell analysis of ethnic groups in Mali with differential susceptibility to malaria. Malar J 11:162
Potup P, Kumsiri R, Kano S, Kalambaheti T, Looareesuwan S et al (2009) Blood stage Plasmodium falciparum antigens induce immunoglobulin class switching in human enriched B cell culture. Southeast Asian J Trop Med Public Health 40:651–664
Rainey JJ, Omenah D, Sumba PO, Moormann AM, Rochford R et al (2007) Spatial clustering of endemic Burkitt's lymphoma in high-risk regions of Kenya. Int J Cancer 120:121–127
Ramiro AR, Jankovic M, Eisenreich T, Difilippantonio S, Chen-Kiang S et al (2004) AID is required for c-myc/IgH chromosome translocations in vivo. Cell 118:431–438
Ramiro AR, Jankovic M, Callen E, Difilippantonio S, Chen HT et al (2006) Role of genomic instability and p53 in AID-induced c-myc-Igh translocations. Nature 440:105–109

Rasti N, Falk KI, Donati D, Gyan BA, Goka BQ et al (2005) Circulating epstein-barr virus in children living in malaria-endemic areas. Scand J Immunol 61:461–465
Rickinson AB, Murray RJ, Brooks J, Griffin H, Moss DJ et al (1992) T cell recognition of Epstein-Barr virus associated lymphomas. Cancer Surv 13:53–80
Rickinson AB, Long HM, Palendira U, Munz C, Hislop AD (2014) Cellular immune controls over Epstein-Barr virus infection: new lessons from the clinic and the laboratory. Trends Immunol 35:159–169
Riley EM, Stewart VA (2013) Immune mechanisms in malaria: new insights in vaccine development. Nat Med 19:168–178
Robbiani DF, Bothmer A, Callen E, Reina-San-Martin B, Dorsett Y et al (2008) AID is required for the chromosomal breaks in c-myc that lead to c-myc/IgH translocations. Cell 135:1028–1038
Ruf IK, Rhyne PW, Yang H, Borza CM, Hutt-Fletcher LM et al (2001) EBV regulates c-MYC, apoptosis, and tumorigenicity in Burkitt's lymphoma. Curr Top Microbiol Immunol 258:153–160
Siemiatycki J, Brubaker G, Geser A (1980) Space-time clustering of Burkitt's lymphoma in East Africa: analysis of recent data and a new look at old data. Int J Cancer 25:197–203
Singh S, Kennedy MC, Long CA, Saul AJ, Miller LH et al (2003) Biochemical and immunological characterization of bacterially expressed and refolded Plasmodium falciparum 42-kilodalton C-terminal merozoite surface protein 1. Infect Immun 71:6766–6774
Snider CJ, Cole SR, Chelimo K, Sumba PO, Macdonald PD et al (2012) Recurrent Plasmodium falciparum malaria infections in Kenyan children diminish T-cell immunity to Epstein Barr virus lytic but not latent antigens. PLoS ONE 7:e31753
Sulzer AJ, Wilson M, Hall EC (1969) Indirect fluorescent-antibody tests for parasitic diseases. V. An evaluation of a thick-smear antigen in the IFA test for malaria antibodies. Am J Trop Med Hyg 18:199–205
Sumba PO, Kabiru EW, Namuyenga E, Fiore N, Otieno RO et al (2010) Microgeographic variations in Burkitt's lymphoma incidence correlate with differences in malnutrition, malaria and Epstein-Barr virus. Br J Cancer 103:1736–1741
Tao Q, Robertson KD, Manns A, Hildesheim A, Ambinder RF (1998) Epstein-Barr virus (EBV) in endemic Burkitt's lymphoma: molecular analysis of primary tumor tissue. Blood 91:1373–1381
Ten Seldam RE, Cooke R, Atkinson L (1966) Childhood Lymphoma in the territories of papua and new guinea. Cancer 19:437–446
Torgbor C, Awuah P, Deitsch K, Kalantari P, Duca KA et al (2014) A multifactorial role for P. falciparum malaria in endemic Burkitt's lymphoma pathogenesis. PLoS Pathog 10:e1004170
van den Bosch C, Lloyd G (2000) Chikungunya fever as a risk factor for endemic Burkitt's lymphoma in Malawi. Trans R Soc Trop Med Hyg 94:704–705
van den Bosch C, Griffin BE, Kazembe P, Dziweni C, Kadzamira L (1993) Are plant factors a missing link in the evolution of endemic Burkitt's lymphoma? Br J Cancer 68:1232–1235
Verra F, Simpore J, Warimwe GM, Tetteh KK, Howard T et al (2007) Haemoglobin C and S role in acquired immunity against Plasmodium falciparum malaria. PLoS ONE 2:e978
Wei F, Zhong S, Ma Z, Kong H, Medvec A et al (2013) Strength of PD-1 signaling differentially affects T-cell effector functions. Proc Natl Acad Sci USA 110:E2480–E2489
Weiss GE, Crompton PD, Li S, Walsh LA, Moir S et al (2009) Atypical memory B cells are greatly expanded in individuals living in a malaria-endemic area. J Immunol 183:2176–2182
Whittle HC, Brown J, Marsh K, Greenwood BM, Seidelin P et al (1984) T-cell control of Epstein-Barr virus-infected B cells is lost during P. falciparum malaria. Nature 312:449–450
Williamson WA, Greenwood BM (1978) Impairment of the immune response to vaccination after acute malaria. Lancet 1:1328–1329
Wilmore JR, Asito AS, Wei C, Piriou E, Sumba PO et al (2014) AID expression in peripheral blood of children living in a malaria holoendemic region is associated with changes in B cell subsets and Epstein-Barr virus. Int J Cancer

Wilson JB, Bell JL, Levine AJ (1996) Expression of Epstein-Barr virus nuclear antigen-1 induces B cell neoplasia in transgenic mice. EMBO J 15:3117–3126

Xue SA, Labrecque LG, Lu QL, Ong SK, Lampert IA et al (2002) Promiscuous expression of Epstein-Barr virus genes in Burkitt's lymphoma from the central African country Malawi. Int J Cancer 99:635–643

Yamamoto N, zur Hausen H (1979) Tumour promoter TPA enhances transformation of human leukocytes by Epstein-Barr virus. Nature 280:244–245

zur Hausen H, Schulte-Holthausen H, Klein G, Henle W, Henle G et al (1970) EBV DNA in biopsies of Burkitt tumours and anaplastic carcinomas of the nasopharynx. Nature 228:1056–1058

zur Hausen H, O'Neill FJ, Freese UK, Hecker E (1978) Persisting oncogenic herpesvirus induced by the tumour promotor TPA. Nature 272:373–375

Contribution of the Epstein-Barr Virus to the Pathogenesis of Hodgkin Lymphoma

Paul Murray and Andrew Bell

Abstract The morphology of the pathognomonic Hodgkin and Reed-Sternberg cells (HRS) of Hodgkin lymphoma was described over a century ago, yet it was only relatively recently that the B-cell origin of these cells was identified. In a proportion of cases, HRS cells harbour monoclonal forms of the B lymphotropic Epstein-Barr virus (EBV). This review summarises current knowledge of the pathogenesis of Hodgkin lymphoma with a particular emphasis on the contribution of EBV.

Contents

1 Introduction 288
2 Origin of HRS Cells 289
3 Suppression of the B-Cell Phenotype in HRS Cells 290
4 Deregulated Cellular Signalling in Classical HL 291
5 EBV and Classical HL 294
6 Contribution of EBV to the Pathogenesis of Classical HL 296
7 Loss of BCR Functions as a Potential Pathogenic Event in EBV-Positive HL 298
8 EBV and the HL Microenvironment 299
9 Conclusions 301
References 302

Abbreviations

BCR	B-cell receptor
B2m	Beta-2 microglobulin
CTL	Cytotoxic T lymphocyte

P. Murray (✉) · A. Bell
School of Cancer Sciences and Centre for Human Virology, College of Medical and Dental Sciences, University of Birmingham, Birmingham, Edgbaston B15 2TT, UK
e-mail: P.G.Murray@bham.ac.uk

A. Bell
e-mail: A.I.BELL@bham.ac.uk

C. Münz (ed.), *Epstein Barr Virus Volume 1*, Current Topics in Microbiology and Immunology 390, DOI 10.1007/978-3-319-22822-8_12

DDR1	Discoidin domain receptor 1
DLBCL	Diffuse large B-cell lymphoma
EBERS	EBV-encoded RNAs
EBNA	EBV nuclear antigen
EBF1	Early B-cell factor 1
EBV	Epstein-Barr virus
HAART	Highly active anti-retroviral therapy
HRS	Hodgkin and Reed-Sternberg
HL	Hodgkin lymphoma
HLA	Human leucocyte antigen
IM	Infectious mononucleosis
ITAM	Immunoreceptor tyrosine activation motif
JAK	Janus kinase
L&H	Lymphocytic and histiocytic
LMP	Latent membrane protein
NF-κB	Nuclear factor kappa B
NLP	Nodular lymphocyte predominant
PD-L1	Programmed cell death-ligand 1
PI3K	Phosphatidylinositol-3-kinase
REAL	Revised European American Lymphoma
RTK	Receptor tyrosine kinase
STAT	Signal transducer and activator of transcription
TGFβ	Transforming growth factor β
TNF	Tumour necrosis factor
WHO	World Health Organization

1 Introduction

Hodgkin lymphoma (HL) is one of the most common lymphomas in the Western world, with an annual incidence of approximately three new cases per 100,000. On the basis of morphological, immunophenotypic and clinical differences, the Revised European American Lymphoma (REAL)/World Health Organization (WHO) classification divides HL into two major types: classical HL and nodular lymphocyte predominant (NLP) HL (Swerdlow et al. 2008). In both cases, involved lymph nodes show a disrupted architecture in which the malignant cells represent a minority of the tumour mass, with the remaining cells comprising a cellular infiltrate of non-neoplastic cells including T lymphocytes and B lymphocytes. Cross talk between this tumour microenvironment and the malignant cells plays a crucial role in the growth, survival and immune escape of the tumour (Aldinucci et al. 2010). In NLPHL, the tumour cells are referred to as lymphocytic and histiocytic (L&H) cells, while classical HL is characterised by the presence of malignant Hodgkin/Reed-Sternberg (HRS) cells (Kuppers 2009). Classical HL is further subdivided into four morphological subtypes known as mixed cellularity, nodular sclerosis, lymphocyte

Table 1 Comparison of classical and nodular lymphocyte predominant Hodgkin lymphoma

REAL/WHO classification	Morphology and immunophenotype of HRS cells	EBV status	Immunoglobulin gene status
Classical HL Mixed cellularity Nodular sclerosis Lymphocyte rich Lymphocyte depletion	Typical HRS cells $CD15^{+}$ $CD20^{-}$ $CD30^{+}$ $CD45^{-}$	20–40 % positive	Lack of BCR expression Destructive or non-functional IgH rearrangements Loss of Ig-specific transcription factors
Nodular lymphocyte predominant HL	Atypical 'popcorn' cells $CD15^{-}$ $CD20^{+}$ $CD30^{-}$ $CD45^{+}$	Usually negative	Express BCR Functional Ig rearrangements Evidence of intra-clonal diversity indicating ongoing somatic hypermutation

rich and lymphocyte depletion (Table 1). While L&H cells retain the expression of typical B-cell antigens such as CD20 and CD19, HRS cells of classical HL have an unusual phenotype in which the B-cell gene expression programme is largely extinguished. Although recent studies suggest that a small proportion of NLPHL cases harbour the Epstein-Barr virus (EBV) (Wang et al. 2014; Huppmann et al. 2014), here we focus on classical HL in which the link with EBV is well established.

2 Origin of HRS Cells

The cellular origin of HL tumour cells remained elusive for many years with early studies suggesting that HRS cells might be derived from macrophages, dendritic cells or granulocytes. However, based on the detection of clonally rearranged immunoglobulin heavy and light chain genes, it is now clear that HRS cells of classical HL originate from mature B lymphocytes (Kuppers et al. 1994). Moreover, the immunoglobulin variable region sequences of HRS cells show evidence of somatic hypermutation, indicating that HRS cells are derived from post-germinal centre B cells (Kuppers et al. 1994; Kanzler et al. 1996; Vockerodt et al. 1998; Marafioti et al. 2000). Notably, in approximately one-quarter of cases, the rearranged immunoglobulin sequences carry 'crippling' mutations which render the surface immunoglobulin molecule or B-cell receptor (BCR) non-functional (Kanzler et al. 1996). Since a functional BCR is required for B-cell survival within the germinal centre, it is now widely believed that classical HL originates from pre-apoptotic germinal centre B cells that have been rescued from apoptosis by cellular transformation events (Fig. 1). However, in rare instances, classical HL can be of T-cell origin, as evidenced by the presence of T-cell receptor gene rearrangements (Muschen et al. 2000; Seitz et al. 2000).

The lack of expression of a functional BCR is one of the hallmarks of classical HL. While in some cases, this is a consequence of destructive immunoglobulin mutations, other mechanisms can also result in the loss of BCR expression.

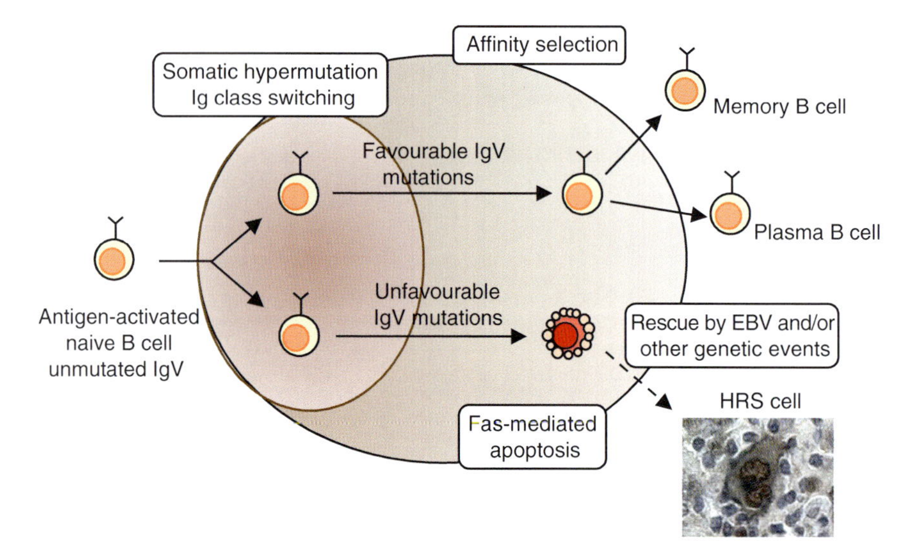

Fig. 1 HRS cells may originate from pre-apoptotic germinal centre B cells. When naive B cells are activated as a result of binding to their cognate antigen via the B-cell receptor (BCR), they migrate into B-cell follicles, proliferate rapidly and establish germinal centres. This B-cell proliferation is accompanied by somatic hypermutation of the immunoglobulin variable (IgV) region genes, thus generating BCR variants. Only B cells carrying a BCR with an increased affinity for antigen will survive and leave the germinal centre as either class-switched memory B cells or plasma cells. In contrast, B cells with unfavourable BCR mutations, such as reduced antigen affinity, premature stop codons or frameshifts, should be eliminated by Fas-mediated apoptosis. However, these pre-apoptotic germinal centre B cells may be rescued by a combination of EBV and cellular genetic alterations, generating a pool of cells which are thought to be the progenitors of HRS cells

For example, immunoglobulin gene expression can be downregulated by the epigenetic silencing of critical transcription factors such as BOB-1/POU2AF1, OCT2/POU2F2 and the Spi-1/PU.1 proto-oncogene (Re et al. 2001; Jundt et al. 2002a; Torlakovic et al. 2001; Stein et al. 2001; Hertel et al. 2002). Immunoglobulin transcription can also be suppressed by epigenetic modifications of the immunoglobulin promoter region (Doerr et al. 2005; Ushmorov et al. 2004), while in a minority of cases, mutations in the octamer region of the immunoglobulin promoter disrupt transcription factor binding (Theil et al. 2001). In addition, many downstream components of the BCR signalling pathway are either absent or expressed at low levels in HRS cells (Schwering et al. 2003).

3 Suppression of the B-Cell Phenotype in HRS Cells

HRS cells have a highly unusual phenotype characterised by a striking downregulation of global B-cell lineage gene expression and the aberrant co-expression of markers of other haematopoietic cell types including T cells, NK-cells and myeloid cells (Hertel et al. 2002; Ushmorov et al. 2004; Schwering et al. 2003;

Kuppers et al. 2003; Tiacci et al. 2012; Steidl et al. 2012). This loss of B-cell identity is linked to the disruption of a number of transcription factor networks involving PAX5, early B-cell factor 1 (EBF1) and TCF3/E2A that regulate B-cell development (Nutt and Kee 2007; Bohle et al. 2013). The downregulation of the B-cell commitment factor EBF1 is a critical event, since EBF1 cooperates with TCF3 and PAX5 to orchestrate the activation of numerous B-cell genes. Loss of EBF1 also contributes to the aberrant expression of the NK-cell-associated transcription factor, ID2, which in turn further represses the B-cell gene expression programme by binding and inhibiting TCF3 (Kuppers et al. 2003; Renne et al. 2006). ID2 overexpression may also result from amplification of the ID2 locus, which has been reported in approximately one-half of HL patients (Renne et al. 2006). In addition, HRS cells express the helix-loop-helix protein ABF1 which further inhibits TCF3 activity (Mathas et al. 2006). The importance of silencing TCF3 activity is underlined by a recent meta-analysis of genome-wide association studies that identified a TCF3 polymorphism which is protective against HL (Cozen et al. 2014). This polymorphism might increase the expression of TCF3, thereby preventing loss of the B-cell phenotype in HRS cell progenitors and reducing the risk of developing HL. In this respect, it is notable that the HL cell line, SUPHD1, contains the N551K substitution previously described to alter the DNA binding specificity of TCF3 in Burkitt lymphoma cells (Schmitz et al. 2012). This raises the possibility that certain mutations may redirect TCF3 to a different set of target genes in HRS cells.

Activation of the Notch pathway also leads to remodelling of the B-cell-specific transcription factor network and to the aberrant expression of T-cell genes in HRS cells (Jundt et al. 2002, 2008). Notch suppresses the B-cell programme by inducing the degradation of TCF3, while increasing expression of the TCF3 inhibitor ABF1 (Mathas et al. 2006; Jundt et al. 2008; Nie et al. 2003; Smith et al. 2005). Notch may also bind and negatively regulate the master B-cell regulator, PAX5, expression of which is usually retained in HL (Jundt et al. 2008). Several mechanisms contribute to Notch signalling in HRS cells. Notch activity is triggered by the Notch ligand, Jagged1, which is expressed both by HRS cells and by cells present in the tumour infiltrate (Jundt et al. 2008). HRS cells are also characterised by lack of expression of deltex1, the major Notch1 inhibitor, as well as high-level expression of the Notch co-activator, mastermind-like-2 (Jundt et al. 2008; Kochert et al. 2011). In addition, EBV-encoded LMP2A can induce Notch expression in human B-cell lines, suggesting that EBV infection may also contribute to increased Notch activity in HL (Portis et al. 2003; Portis and Longnecker 2004).

4 Deregulated Cellular Signalling in Classical HL

A critical step in the pathogenesis of HL is the ability of HRS progenitor cells to evade apoptosis that would be the normal fate of germinal centre B cells lacking BCR expression. In this regard, HRS cells show deregulation of multiple cell signalling pathways that contribute to HRS survival and proliferation (Fig. 2).

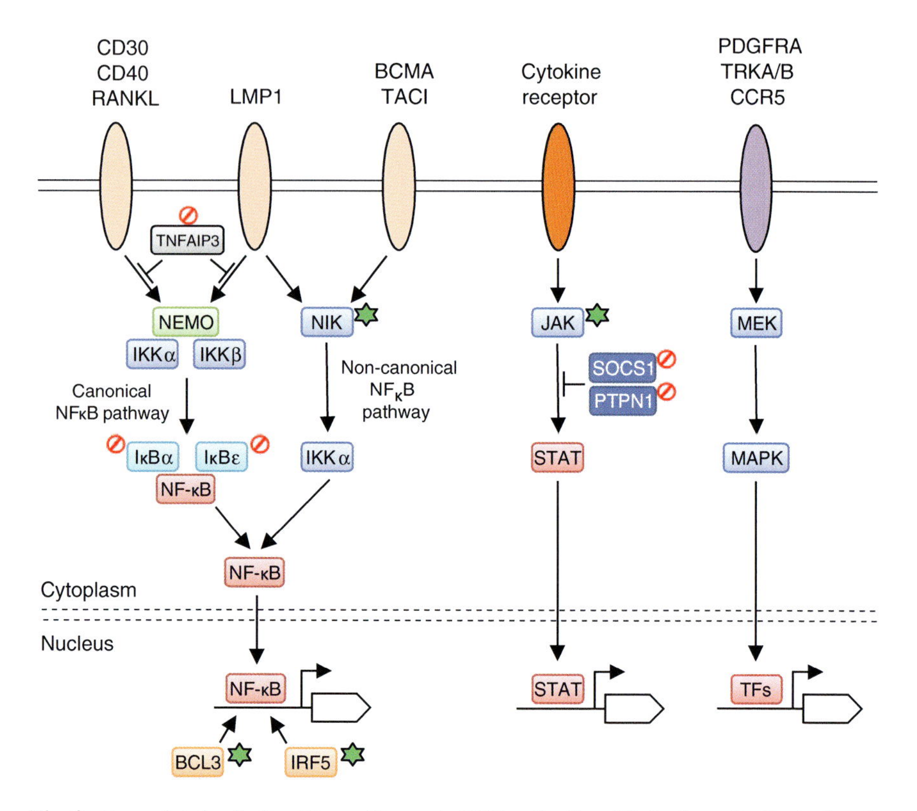

Fig. 2 Deregulated cell signalling pathways in HRS cells. Simplified schematic illustrating the major signalling pathways deregulated in HRS cells. HRS cells constitutively activate both the canonical and non-canonical NF-κB pathways. Multiple receptors including CD30, CD40 and RANKL constitutively activate the canonical pathway through the IKK complex (composed of IKKα, IKKβ and NEMO), leading to phosphorylation and degradation of the NF-κB inhibitors IκBα and IκBε. Consequently, active NF-κB complexes translocate to the nucleus where they activate target genes in cooperation with BCL3 and IRF5. In some cases, NF-κB activity can be induced by overexpression of the c-Rel subunit. Note this pathway is negatively regulated by TNFAIP3, which prevents activation of the IKK complex. In the non-canonical pathway, receptor stimulation leads to activation of NIK (MAPK3K14) and IKKα, enhancing the processing of NF-κB precursors which subsequently translocate to the nucleus. In EBV-positive cases of HL, LMP1 can activate both the canonical and non-canonical NF-κB pathways. Stimulation of cytokine receptors by various cytokines secreted by HRS cells and by other cells in the microenvironment results in activation of the JAK/STAT signalling pathway, which can be further augmented by mutation of the suppressor of cytokine signalling 1 (SOCS1) and the protein tyrosine phosphatase PTPN1/PTP1B. Various receptor tyrosine kinases, including PDGFRA, TRKA, TRKB and CCR5, are also aberrantly activated in HRS cells leading to transcription factor (TF) activation. Activating mutations which lead to increased cell signalling in HRS cells are indicated by *green stars*, while loss-of-function mutations are indicated by *red circles*

A key pathogenic feature of HRS cells is the constitutive activation of a family of transcription factors referred to as nuclear factor kappa B (NF-κB) (Bargou et al. 1996). NF-κB signalling contributes essential functions in HL, since inhibition of this pathway in HL cell lines leads to increased sensitivity to apoptosis after growth factor withdrawal and impaired tumorigenicity in severe combined immunodeficiency mice (Izban et al. 2001; Bargou et al. 1997). Several mechanisms contribute to the activation of both the canonical and non-canonical NF-κB pathways in HL. HRS cells abundantly express multiple TNF receptors, including CD30, CD40, TACI, BCMA and RANK which trigger NF-κB activation (Carbone et al. 1995a; Fiumara et al. 2001; Horie et al. 2002; Chiu et al. 2007). Importantly, the cellular infiltrate surrounding the HRS cells includes eosinophils, neutrophils and CD4 T cells which activate these receptors in a paracrine manner (Carbone et al. 1995; Pinto et al. 1997). A recent genomic analysis of transcription factor binding sites in HL cell lines also revealed that IRF5, aberrantly activated in HRS cells, cooperates with NF-κB to fully activate the HRS cell phenotype (Kreher et al. 2014). In addition, there may be cross talk between the Notch and NF-κB pathways which further induces NF-κB activity in HL (Schwarzer et al. 2012).

NF-κB activation also results from a wide variety of genetic lesions in HRS cells. The c-REL subunit of NF-κB is amplified in HL cell lines and in around one-half of all cases of classical HL (Barth et al. 2003; Joos et al. 2003; Martin-Subero et al. 2002). In around 20 % of HL cases, NF-κB activity is upregulated by mutation of the IκB inhibitor proteins, IκB alpha and IκB epsilon, which normally sequester NF-κB as an inactive complex in the cytoplasm (Cabannes et al. 1999; Emmerich et al. 1999, 2003; Jungnickel et al. 2000; Lake et al. 2009), while rare HL cases overexpress the NF-κB transcriptional co-activator BCL3 (Martin-Subero et al. 2006; Mathas et al. 2005). In addition, there are recurrent mutations that affect TNFAIP3/A20, a ubiquitin-modifying enzyme that negatively regulates NF-κB signalling (Schmitz et al. 2009; Reichel et al. 2015). The non-canonical NF-κB pathway also contributes to the survival of HL cell lines through NIK/MAP3K14-mediated activation of RelB (Ranuncolo et al. 2012), and chromosomal gains of NIK have been reported in around 30 % of primary HL cases (Otto et al. 2012).

JAK/STAT signalling, a central pathway involved in mediating the effects of cytokines, is also strongly implicated in the proliferation and survival of HRS cells. HRS cells abundantly produce multiple cytokines, including IL-7, IL-9 and IL-13 (Cattaruzza et al. 2009; Skinnider et al. 2001; Kapp et al. 1999), which induce JAK/STAT activation leading to elevated levels of phosphorylated STAT3, STAT4 and STAT6 (Hinz et al. 2002; Skinnider et al. 2002; Kube et al. 2001). In addition, HRS cells also respond in a paracrine manner to cytokines secreted by a number of different cell types in the tumour microenvironment, including IL3 produced by infiltrating T cells (Aldinucci et al. 2002). JAK/STAT signalling can be further dysregulated by genetic events including amplification of JAK2 and loss-of-function mutations of the negative regulators SOCS1 and PTPN1/PTPB1 (Joos et al. 2000; Weniger et al. 2006; Gunawardana et al. 2014). Numerous receptor tyrosine kinases (RTKs), including PDGFRA, TRKA, TRKB and TIE1, are also aberrantly activated in HRS cells, although co-expression of several RTKs appears to be largely restricted to EBV-negative cases of HL (Renne et al. 2005, 2007).

While the above findings provide important clues to the genetic events underpinning the pathogenesis of HL, genomic studies will undoubtedly increase our knowledge of the broader mutational landscape in HRS cells. In this regard, a recent exome analysis of seven HL cell lines identified 463 genes that were mutated in at least two of the cell lines (Liu et al. 2014a). While a number of these mutated genes had previously been described in the context of other B-cell lymphomas, the majority of mutations were unique, suggesting that the genetic events during the malignant transformation in HL are distinct from those in non-Hodgkin lymphomas. A second study, describing the exome analysis of flow-sorted primary HRS cells, recently identified novel alterations affecting genes involved in antigen presentation, transcription regulation and genome integrity, in addition to well-known HL-associated genes such as TNFAIP3 and SOCS1 (Reichel et al. 2015). Interestingly, this study also revealed frequent mutations in beta-2 microglobulin (B2M) and showed that the absence of B2M was associated with a lower stage of disease, younger age at diagnosis and better clinical outcome (Reichel et al. 2015). Beta-2 microglobulin forms part of the MHC class I complex, and loss of MHC class I expression is a common feature of HL (Oudejans et al. 1996a). Since previous reports have suggested a link between clinical outcome and the presence of certain cell types, including regulatory T cells and CD68^{+} macrophages, in the tumour infiltrate (Alvaro-Naranjo et al. 2005; Tzankov et al. 2008, 2010; Kamper et al. 2011), it is possible that the loss of MHC class I may confer a better clinical response due to differences in the interactions between HRS cells and immune cells in the tumour microenvironment.

5 EBV and Classical HL

The detection of raised antibody titers to EBV antigens present in HL patients both at the time of, and preceding, diagnosis provided the first evidence that EBV might be involved in the pathogenesis of HL (Levine et al. 1971; Mueller et al. 1989). The detection by in situ hybridisation of EBV DNA and expression of the abundant EBV-encoded RNAs (EBERs) within HRS cells provided strong evidence in support of an aetiological role for EBV in this disease (Weiss et al. 1989; Wu et al. 1990). The detection of clonal EBV DNA in 20–25 % of HL tissues suggests that EBV infection of the HRS tumour progenitor occurred prior to its clonal expansion (Anagnostopoulos et al. 1989).

The fraction of classical HL which harbours EBV genomes varies dramatically with age, gender, histological subtype, ethnicity and country of residence (Glaser et al. 1997). Thus, EBV rates in HL from North America and Europe vary between 20 and 50 %, but much higher rates are observed in underdeveloped countries (Chang et al. 1993; Weinreb et al. 1996). EBV-positive rates are also higher: in males compared with females; in Asians and Hispanics compared with whites or blacks (Glaser et al. 1997); and in the UK in South Asian children compared with non-South Asian children (Flavell et al. 2001). In developed countries, the proportion of EBV-positive cases is higher in older people and in children, especially in those under 10 years of age, whereas the lowest rates of EBV-positive disease are found in young adults (Glaser et al. 1997; Jarrett et al. 1991; Armstrong et al. 1998). These differences have led to the hypothesis that there are three forms of classical HL: paediatric HL (EBV-positive, mixed cellularity type), HL of young adults (EBV-negative, nodular sclerosis type) and HL of older adults (EBV-positive, mixed cellularity type) (Armstrong et al. 1998). EBV-positive classical HL in older adults has been attributed to an age-related decline in EBV-specific immunity (Jarrett et al. 1991; Armstrong et al. 1998). Senescence of EBV immunity is also suspected in a related tumour, known as EBV-positive diffuse large B-cell lymphoma (DLBCL) of the elderly (Oyama et al. 2003, 2007).

The incidence of classical HL is only modestly increased during human immunodeficiency virus (HIV) infection, and most cases are EBV-positive and of mixed cellularity type (Glaser et al. 2003; Uccini et al. 1990; Biggar et al. 2006). Classical HL occurs more commonly in HIV patients with intermediate levels of immune impairment (Biggar et al. 2006). This is in contrast to the peak in Burkitt lymphoma incidence which occurs early in HIV infection when circulating $CD4^+$ T cell numbers are normal or only slightly decreased, and the peak in the incidence of the 'immunoblastic' form of DLBCL which occurs when circulating $CD4^+$ T cells are very low and the patient is severely immunocompromised. Furthermore, the incidence of classical HL in HIV-positive patients has not fallen in the post-highly active anti-retroviral therapy (HAART) era; indeed, some studies suggest HL risk may be increased in the first few months following immune reconstitution on HAART (Kowalkowski et al. 2013; Gotti et al. 2013; Bohlius et al. 2011). These findings not only provide further evidence that defects in EBV-specific immunity are important for the development of EBV-positive HL, but also suggest tumour-promoting functions for $CD4^+$ T cells in the microenvironment.

First-degree relatives of patients with classical HL have a threefold to ninefold increased risk (Crump et al. 2012a, b), and monozygotic twins a 100-fold increase compared with dizygotic twins (Mack et al. 1995). Genetic association studies based on segregation and linkage analysis in families have identified susceptibility loci in the human leucocyte antigen (HLA) region (Kushekhar et al. 2014). Thus, an increased and decreased risk of EBV-positive HL is associated with HLA-A*01 and HLA-A*02 alleles, respectively (Diepstra et al. 2005; Niens et al. 2007; Hjalgrim et al. 2010). Both self-reported and laboratory-confirmed prior infectious mononucleosis (IM) is also associated with an increased risk of developing EBV-positive; an association not observed in EBV-negative HL (Hjalgrim et al. 2000, 2007). These data have led to an immunological model for the development of EBV-positive HL in which the levels of circulating EBV-infected lymphocytes regulated by cytotoxic T-cell responses are a critical determinant of disease risk (Farrell and Jarrett 2011).

6 Contribution of EBV to the Pathogenesis of Classical HL

EBV is a potentially oncogenic virus which can induce resting B cells into immortalised lymphoblastoid cell lines. This growth transformation process is driven by the cooperative action of a small number of viral gene products: six EBV nuclear antigens (EBNA1, -2, -3A, -3B, -3C and –LP), three latent membrane proteins (LMP1, LMP2A and LMP2B), two highly expressed non-polyadenylated RNAs (EBER1, EBER2) and around 40 miRNAs (Longnecker et al. 2013). In contrast, virus gene expression in EBV-infected HRS cells is limited to EBNA1, LMP1 and LMP2A/B, together with the non-coding EBERs and miRNAs (Deacon et al. 1993; Young et al. 1991; Niedobitek et al. 1997; Murray et al. 1992; Grasser et al. 1994). In the absence of pro-survival signals mediated by a functional BCR, these EBV latent gene products contribute to the survival of HRS progenitors through multiple routes, most of which converge on a limited number of cell signalling pathways. Importantly, LMP1 and LMP2 mimic the physiological functions of CD40 and BCR, respectively, which are normally required for the survival and proliferation of germinal centre B cells.

LMP1 is a member of the TNF receptor superfamily and functions as a constitutively active homologue of the cellular CD40 receptor (Lam and Sugden 2003). LMP1 expression is especially critical in the pathogenesis of HL since it activates multiple cell signalling pathway which could potentially rescue HRS progenitors from apoptosis (Gires et al. 1999; Huen et al. 1995; Kieser et al. 1997). These include the NF-κB, JAK/STAT and phospatidylinositol-3-kinase (PI3K)/AKT pathways which are known to be constitutively expressed in HRS cells (Bargou et al. 1997; Kube et al. 2001; Dutton et al. 2005). LMP1 induces the overexpression of FLICE-inhibitory protein (c-FLIP), a negative regulator of Fas signalling, thus providing further protection from apoptosis (Dutton et al. 2004). The importance of LMP1 in HL pathogenesis is further emphasised by the finding that LMP1 expression in primary germinal centre cells, the presumed progenitor of HL, induces many gene expression changes characteristic of HRS cells; these include downregulation of B-cell transcription factors and BCR signalling components required to maintain B-cell identity and upregulation of survival genes such as BCL2, BFL-1 and c-FLIP that protect B cells from apoptosis (Vockerodt et al. 2008; Henderson et al. 1991; D'Souza et al. 2004). Some of these transcriptional changes may be mediated through the induction of ID2 (Vockerodt et al. 2008), while others may be result from epigenetic mechanisms involving polycomb proteins, DNA methyltransferases and protein arginine methyltransferases (Anderton et al. 2011; Dutton et al. 2007; Leonard et al. 2011, 2012).

Unlike LMP1, LMP2A is not a classical transforming gene but engages a number of signalling pathways important for B-cell survival, including activation of RAS/PI3K/AKT (Dutton et al. 2005; Swart et al. 2000; Fukuda and Longnecker 2004). LMP2 exists as two isoforms, LMP2A and LMP2B, which share 8 common coding exons but have different 5′ exons. While the 5′ exon of LMP2B is non-coding, the unique N-terminus of LMP2A includes an ITAM motif, which resembles the signalling domain of the BCR (Longnecker and Kieff 1990). In transgenic mice experiments, LMP2A functions as a BCR surrogate, allowing B cells to develop in the absence of normal BCR signalling (Caldwell et al. 1998; Merchant et al. 2000). Gene expression profiling experiments have demonstrated that LMP2A interferes with the expression of numerous B-cell transcription factors, including EBF1 and E2A, and mimics many of the gene expression changes seen in HRS cells (Portis et al. 2003; Portis and Longnecker 2004; Vockerodt et al. 2013). LMP2A also activates the Notch pathway, and recent evidence suggests that these two factors may cooperate to further reinforce the loss of B-cell identity in HRS cells (Anderson and Longnecker 2008). Paradoxically, however, many of the adaptor molecules necessary for both BCR and LMP2A signalling are absent in HRS cells, and therefore, it remains unclear whether LMP2A acts as a BCR surrogate in HL. One possible explanation is that LMP2A provides essential survival signals to the HRS progenitor cells, which retain the downstream BCR signalling molecules, but subsequent transformation events followed by downregulation of B-cell-specific genes replace some of these critical LMP2A functions. However, given that EBV-positive cases of HL are consistently LMP2A positive, it is likely that LMP2A also has BCR-independent functions that are important for maintenance of the HRS phenotype.

Other viral genes expressed in HL include EBNA1, EBERs and the BART miRNAs. Little is known about the role of EBERs and viral miRNAs in the context of HL, although it is interesting to speculate that export of these non-coding RNAs from HRS cells via exosomes may affect gene expression in cells of the surrounding microenvironment (Pegtel et al. 2010). By contrast, EBNA1 potentially contributes several functions in HRS cells. EBNA1 is absolutely required for EBV infection as it acts as a viral replication factor and tethers the viral genome to the chromosomes of daughter cells during cell division (Smith and Sugden 2013). In addition to its genome maintenance function, EBNA1 is also a transcription factor that regulates both viral and cellular gene expression (Altmann et al. 2006; Frappier 2012). EBNA1 inhibits TGFβ signalling, in part through increasing the turnover of SMAD2, and promotes the growth and survival of HL cells by downregulating the TGFβ target gene PTPRK (Flavell et al. 2008). EBNA1 also enhances expression of the T-cell chemokine CCL20 which may dampen the immune response to HRS cells by recruiting regulatory T cells to the tumour microenvironment (Baumforth et al. 2008). In the context of another B-cell tumour, Burkitt lymphoma, EBNA1 has also been shown to contribute anti-apoptotic functions through destabilisation of p53 and induction of the anti-apoptotic gene, survivin (Holowaty et al. 2003; Lu et al. 2011).

While EBV only appears to have subtle effects on the overall transcriptional programme in the final HRS clone (Tiacci et al. 2012), it is clear that EBV can potentially make a major contribution to the pathogenesis of HL. Furthermore, there is increasing evidence that EBV-positive and EBV-negative cases of HL arise through different pathogenetic routes and that the role of EBV is to functionally substitute for cellular genetic changes. First, almost all cases of classical HL bearing destructive immunoglobulin gene mutations are EBV-positive (Brauninger et al. 2006), suggesting that progenitor cells carrying such mutations can only survive and undergo malignant transformation if infected with EBV. This is supported by the fact that EBV immortalises BCR-negative B cells in vitro, demonstrating that EBV has the potential to rescue pre-apoptotic germinal centre cells (Chaganti et al. 2005; Mancao et al. 2005; Bechtel et al. 2005). Second, TNFAIP3 mutations are more common in EBV-negative cases of the disease, suggesting that inactivation of TNFAIP3 and EBV infection may represent alternative pathways to deregulate NF-κB signalling during HL development (Schmitz et al. 2009). Third, the aberrant co-expression of several RTKs is largely restricted to EBV-negative HL cases, arguing that the functions of these RTKs can be replaced by LMP1 and LMP2 signalling (Renne et al. 2007).

7 Loss of BCR Functions as a Potential Pathogenic Event in EBV-Positive HL

As we have seen, EBV can provide the necessary anti-apoptotic functions required for the survival of BCR-negative HRS cell progenitors. However, it is not clear why the loss of a functional BCR should be important for the pathogenesis of EBV-associated classical HL. To attempt to answer this question, we need to consider another important aspect of EBV biology, namely regulation of the virus replication cycle. Productive EBV infection is associated with the temporal expression of immediate early, early and late viral genes which ultimately leads to the production of new virus particles. This switch from latency to virus replication can be triggered by two different mechanisms: plasma cell differentiation or activation of BCR signalling. Since productive infection leads eventually to cell death, it is assumed that suppression of EBV lytic replication is an important event in HL pathogenesis. To this end, both mechanisms appear to be disrupted in EBV-infected HRS cells.

Plasma cell differentiation is dependent on a series of transcription factors which coordinately silence genes associated with the germinal centre B-cell programme while activating genes required for terminal B-cell differentiation (Lin et al. 2003). One of these factors is BLIMP1, which has been reported to activate the EBV immediate early genes BZLF1 and BRLF1, thus providing a mechanistic link between plasma cell differentiation and EBV reactivation (Reusch et al. 2015). LMP1 suppresses plasma cell differentiation by abrogating BLIMP1α expression (Vrzalikova et al. 2011, 2012a) and inducing the expression of BLIMP1β, a BLIMP1 isoform which antagonises the activity of BLIMP1α (Vrzalikova et al. 2012b). Thus, it is likely that disruption of BLIMP1α function is an essential step in the pathogenesis of EBV-positive germinal centre-derived lymphomas, preventing not only the terminal differentiation of the tumour cells but also inhibiting virus replication.

EBV replication can also be induced following BCR activation, and therefore, loss of a functional BCR is likely to be an important event in HL development because it would prevent entry into the replicative cycle and the ensuing cell death. As described above, EBV-positive HRS cells frequently have non-functional immunoglobulin genes and lack expression of critical downstream molecules of the BCR signalling pathway. Importantly, while LMP2A can still induce lytic cycle entry in the absence of a functional BCR, it cannot do so when the essential downstream BCR components are missing (Vockerodt et al. 2013). Thus, the loss of BCR, as well as of BCR signalling components, could combine to prevent both BCR- and LMP2A-induced virus replications.

8 EBV and the HL Microenvironment

The HRS cells of classical HL are surrounded by a non-neoplastic cell infiltrate composed of T cells, B cells, macrophages, mast cells, eosinophils and fibroblasts which support the proliferation and survival of HRS cells (Aldinucci et al. 2010; Liu et al. 2014b). There is now increasing evidence that HRS cells actively shape this reactive microenvironment by secreting multiple chemokines and cytokines which aid in the recruitment of certain cell types. For example, Th2 and FoxP3^{+} regulatory T cells are attracted by CCL5 (RANTES), CCL17 (TARC), CCL20 and CCL22 (Skinnider and Mak 2002; Aldinucci et al. 2008; Fischer et al. 2003; van den Berg et al. 1999), eosinophils are attracted by IL5, CCL5 and CCL28 (Aldinucci et al. 2008; Hanamoto et al. 2004), and neutrophils are attracted by IL8 (Foss et al. 1996). In addition, activated fibroblasts in the tumour infiltrate produce CCL11 (eotaxin) and CCL5, which further contribute to the attraction of eosinophils, mast cells and regulatory T cells (Buri et al. 2001; Jundt et al. 1999). Multiple cytokines, including IL3, IL5, IL6, IL8, IL9 and IL13, are also abundantly expressed in HL tissues, either by the HRS cells themselves or by the cellular infiltrate (Aldinucci et al. 2010). In EBV-positive cases, EBV-encoded LMP1 may further alter this microenvironment (Fig. 3). Thus, LMP1 signalling through the p38/MAPK and PI3K pathways increases the expression of IL6, IL8 and IL10 (Vockerodt et al. 2008; Lambert and Martinez 2007; Eliopoulos et al. 1999) which can exert both autocrine and paracrine effects, while LMP1 also stimulates the production of immune modulators such as CCL5, CCL17 and CCL22 (Vockerodt et al. 2008; Nakayama et al. 2004).

In addition to supporting the growth and survival of the HRS cells (Liu et al. 2014b), the tumour infiltrate also plays an important role in suppressing the host immune response. In this regard, CCR4^{+} regulatory T cells, attracted by high levels of CCL17, CCL20 and CCL22, inhibit the activity of infiltrating effector CD4^{+} T cells (Ishida et al. 2006). T-cell effector functions are further blocked by engagement between the PD-1 receptor and the programmed cell death-ligand 1 (PD-L1) expressed on HRS cells, resulting in functional exhaustion of the T cells (Muenst et al. 2009; Yamamoto et al. 2008). Notably, amplification of the PD-L1

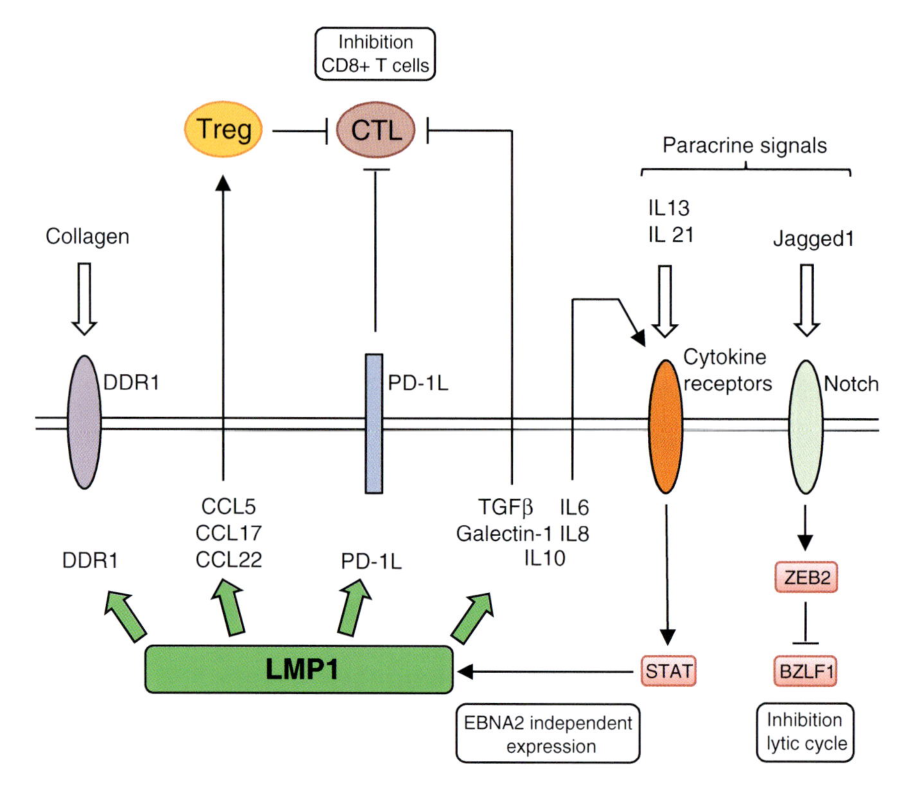

Fig. 3 Interactions between EBV and cellular environment in HL. LMP1 contributes to the highly suppressive immune environment of HL through several different mechanisms. LMP1-mediated NF-κB signalling increases the expression of numerous chemokines, including CCL5, CCL17 and CCL22, which dampen $CD8^+$ T cell activity by recruiting regulatory T cells (Treg). LMP1 also upregulates surface expression of PD-1L, which induces T cell anergy through interaction with the PD-1 receptor, and increases the secretion of immunosuppressive regulators such as IL10, galectin-1 and TGFβ. LMP1 signalling through the MAPK and PI3K pathways enhances the production of several cytokines including IL6, IL8 and IL10, leading to autocrine activation of JAK/STAT signalling in HRS cells. Importantly, IL10, IL13 and IL21 activate STAT6 which stimulates LMP1 transcription from an alternative promoter independent of EBNA2 expression. LMP1 also upregulates expression of the receptor tyrosine kinase and discoidin domain receptor 1 (DDR1) which in the presence of its ligand, collagen, may promote the survival of HRS cells through activation of NF-κB and the PI3K pathways. Notch signalling extinguishes the expression of many B-cell-specific genes and also prevents entry into lytic cycle by ZEB2-mediated repression of the viral immediate early gene BZLF1

and PD-L2 genes at 9p24 is common in nodular sclerosis HL (Green et al. 2010), while the PD-L1 gene can also be deregulated as a result of a reciprocal translocation involving CIITA in a proportion of classical HL cases (Steidl et al. 2011). In the context of EBV-positive HL, EBV upregulates the expression of numerous immunosuppressive factors, including IL10, galectin-1 and TGFβ, all of which can inhibit cytotoxic T-cell responses (Marshall et al. 2007; Juszczynski et al. 2007; Gandhi et al. 2007), while LMP1 has also been reported to induce

PD-L1 expression (Green et al. 2012). These immune evasion mechanisms may be particularly important in EBV-positive HL, which express virus antigens. The microenvironment of HL is an emerging therapeutic target as exemplified by a recent study in which the vast majority of patients with relapsed or refractory HL responded to PD-1 blockade therapy (Ansell et al. 2015).

Loss of HLA expression is another important mechanism that enables the tumour cells to avoid the host immune response. Immune evasion is particularly important in the context of EBV-positive HL, since virus-encoded LMP1 and LMP2A proteins are targets for $CD8^{+}$ cytotoxic T lymphocytes (CTLs) (Rickinson and Moss 1997). In vitro, HL cells can process and present epitopes from LMP1 and LMP2A in the context of multiple HLA class I alleles and are sensitive to lysis by EBV-specific CTLs (Khanna et al. 1998; Lee et al. 1993). Loss of HLA class I expression is frequently observed in classical HL, and a recent study showed that most cases of nodular sclerosis HL carry inactivating mutations in the B2M (Reichel et al. 2015). Notably, however, HLA loss is more common in EBV-negative cases of classical HL, while EBV-positive cases express normal or even elevated levels of HLA and contain more activated CTLs than EBV-negative cases (Oudejans et al. 1996b, 1997; Lee et al. 1998). This suggests that EBV-positive cases of HL must exploit additional mechanisms to avoid recognition by CTLs.

Recent evidence also suggests that the microenvironment of classical HL plays a role in modulating virus gene expression in HRS cells (Fig. 3). For example, LMP1 expression in HRS cells is driven from an alternative STAT-responsive promoter which is stimulated by IL-4, IL-10, IL-13 and IL-21 (Kis et al. 2006, 2010, 2011). Importantly, this provides a mechanism for LMP1 expression in the absence of EBNA2, which is a hallmark of the typical type II latency pattern seen in HRS cells. Recent data also suggest that Notch can inhibit EBV entry into the lytic cycle in a non-Hodgkin B-cell lymphoma cell line by upregulating the cellular transcription factor Zeb2, which subsequently represses the transcription of the lytic switch regulator BZLF1 (Rowe et al. 2014). There is also emerging evidence that the microenvironment can modulate EBV gene functions. Thus, LMP1 induces expression of a receptor tyrosine kinase known as discoidin domain receptor 1 (DDR1) (Cader et al. 2013), but this receptor is only active in the presence of its ligand, collagen, which is a major constituent of the HL microenvironment. Notably, ligation of DDR1 by collagen promotes the survival of lymphoma cells in vitro (Cader et al. 2013). These observations suggest that some of the oncogenic effects LMP1 may be dependent on cues from the tumour microenvironment.

9 Conclusions

The study of Hodgkin lymphoma has been a challenge, not least because of the low abundance of the malignant cells in the tumour; this is one of the reasons why the genetic investigation of this tumour has lagged behind that of many other solid cancers. Furthermore, although the association with EBV has been known for many years, we still do not fully understand how the virus contributes to the development of this tumour. Although there is already some evidence that the EBV-positive and EBV-negative forms are genetically distinct, future studies attempting to unravel these differences will have to take into account the complex epidemiology of Hodgkin lymphoma; EBV may have different aetiological roles in different patient groups, for example, at different ages. Other important questions remain. For example, we still do not fully understand why the loss of B-cell identity, including the loss of B-cell receptor signalling, is seemingly so important for Hodgkin lymphoma development, especially for those cases which are EBV-negative. Further studies are warranted to identify which are the critical driver mutations in HRS cells and investigate whether alterations in multiple signalling pathways can cooperate during lymphomagenesis. As we move into an era in which personalised medicine is becoming increasingly common, large-scale studies will be required to determine the clinical relevance of these different genetic alterations and how they impact upon survival and response to treatment. In addition, the successful application of approaches such as immunotherapy will require an improved understanding of how the microenvironment contributes to the immune evasion of HRS cells, especially in the context of EBV.

Acknowledgments We are grateful to Leukaemia Lymphoma Research and to Cancer Research UK for support.

References

Aldinucci D, Poletto D, Gloghini A, Nanni P, Degan M, Perin T et al (2002) Expression of functional interleukin-3 receptors on Hodgkin and Reed-Sternberg cells. Am J Pathol 160(2):585–596 Epub 2002/02/13

Aldinucci D, Lorenzon D, Cattaruzza L, Pinto A, Gloghini A, Carbone A et al (2008) Expression of CCR5 receptors on Reed-Sternberg cells and Hodgkin lymphoma cell lines: involvement of CCL5/Rantes in tumor cell growth and microenvironmental interactions. Int J Cancer 122(4):769–776 Epub 2007/10/16

Aldinucci D, Gloghini A, Pinto A, De Filippi R, Carbone A (2010) The classical Hodgkin's lymphoma microenvironment and its role in promoting tumour growth and immune escape. J Pathol 221(3):248–263 Epub 2010/06/09

Altmann M, Pich D, Ruiss R, Wang J, Sugden B, Hammerschmidt W (2006) Transcriptional activation by EBV nuclear antigen 1 is essential for the expression of EBV's transforming genes. Proc Natl Acad Sci USA 103(38):14188–14193 Epub 2006/09/13

Alvaro-Naranjo T, Lejeune M, Salvado-Usach MT, Bosch-Princep R, Reverter-Branchat G, Jaen-Martinez J et al (2005) Tumor-infiltrating cells as a prognostic factor in Hodgkin's lymphoma: a quantitative tissue microarray study in a large retrospective cohort of 267 patients. Leuk Lymphoma 46(11):1581–1591 Epub 2005/10/21

Anagnostopoulos I, Herbst H, Niedobitek G, Stein H (1989) Demonstration of monoclonal EBV genomes in Hodgkin's disease and Ki-1-positive anaplastic large cell lymphoma by combined Southern blot and in situ hybridization. Blood 74(2):810–816

Anderson LJ, Longnecker R (2008) An auto-regulatory loop for EBV LMP2A involves activation of Notch. Virology 371(2):257–266 Epub 2007/11/06

Anderton JA, Bose S, Vockerodt M, Vrzalikova K, Wei W, Kuo M et al (2011) The H3K27me3 demethylase, KDM6B, is induced by Epstein-Barr virus and over-expressed in Hodgkin's lymphoma. Oncogene 30(17):2037–2043 Epub 2011/01/19

Ansell SM, Lesokhin AM, Borrello I, Halwani A, Scott EC, Gutierrez M et al (2015) PD-1 blockade with nivolumab in relapsed or refractory Hodgkin's lymphoma. N Engl J Med 372:311–319 Epub 2014/12/09

Armstrong AA, Alexander FE, Cartwright R, Angus B, Krajewski AS, Wright DH et al (1998) Epstein-Barr virus and Hodgkin's disease: further evidence for the three disease hypothesis. Leukemia 12(8):1272–1276 Epub 1998/08/11

Bargou RC, Leng C, Krappmann D, Emmerich F, Mapara MY, Bommert K et al (1996) High-level nuclear NF-kappa B and Oct-2 is a common feature of cultured Hodgkin/Reed-Sternberg cells. Blood 87(10):4340–4347

Bargou RC, Emmerich F, Krappmann D, Bommert K, Mapara MY, Arnold W et al (1997) Constitutive nuclear factor-kappaB-RelA activation is required for proliferation and survival of Hodgkin's disease tumor cells. J Clin Invest 100(12):2961–2969

Barth TF, Martin-Subero JI, Joos S, Menz CK, Hasel C, Mechtersheimer G et al (2003) Gains of 2p involving the REL locus correlate with nuclear c-Rel protein accumulation in neoplastic cells of classical Hodgkin lymphoma. Blood 101(9):3681–3686 Epub 2003/01/04

Baumforth KR, Birgersdotter A, Reynolds GM, Wei W, Kapatai G, Flavell JR et al (2008) Expression of the Epstein-Barr virus-encoded Epstein-Barr virus nuclear antigen 1 in Hodgkin's lymphoma cells mediates Up-regulation of CCL20 and the migration of regulatory T cells. Am J Pathol 173(1):195–204 Epub 2008/05/27

Bechtel D, Kurth J, Unkel C, Kuppers R (2005) Transformation of BCR-deficient germinal-center B cells by EBV supports a major role of the virus in the pathogenesis of Hodgkin and posttransplantation lymphomas. Blood 106(13):4345–4350

Biggar RJ, Jaffe ES, Goedert JJ, Chaturvedi A, Pfeiffer R, Engels EA (2006) Hodgkin lymphoma and immunodeficiency in persons with HIV/AIDS. Blood 108(12):3786–3791 Epub 2006/08/19

Bohle V, Doring C, Hansmann ML, Kuppers R (2013) Role of early B-cell factor 1 (EBF1) in Hodgkin lymphoma. Leukemia 27(3):671–679 Epub 2012/11/24

Bohlius J, Schmidlin K, Boue F, Fatkenheuer G, May M, Caro-Murillo AM et al (2011) HIV-1-related Hodgkin lymphoma in the era of combination antiretroviral therapy: incidence and evolution of CD4(+) T-cell lymphocytes. Blood 117(23):6100–6108 Epub 2011/03/04

Brauninger A, Schmitz R, Bechtel D, Renne C, Hansmann ML, Kuppers R (2006) Molecular biology of Hodgkin's and Reed/Sternberg cells in Hodgkin's lymphoma. Int J Cancer 118(8):1853–1861

Buri C, Korner M, Scharli P, Cefai D, Uguccioni M, Mueller C et al (2001) CC chemokines and the receptors CCR3 and CCR5 are differentially expressed in the nonneoplastic leukocytic infiltrates of Hodgkin disease. Blood 97(6):1543–1548 Epub 2001/03/10

Cabannes E, Khan G, Aillet F, Jarrett RF, Hay RT (1999) Mutations in the IkBa gene in Hodgkin's disease suggest a tumour suppressor role for IkappaBalpha. Oncogene 18(20):3063–3070 Epub 1999/05/26

Cader FZ, Vockerodt M, Bose S, Nagy E, Brundler MA, Kearns P et al (2013) The EBV oncogene LMP1 protects lymphoma cells from cell death through the collagen-mediated activation of DDR1. Blood 122(26):4237–4245 Epub 2013/10/19

Caldwell RG, Wilson JB, Anderson SJ, Longnecker R (1998) Epstein-Barr virus LMP2A drives B cell development and survival in the absence of normal B cell receptor signals. Immunity 9(3):405–411 Epub 1998/10/13

Carbone A, Gloghini A, Gattei V, Aldinucci D, Degan M, De Paoli P et al (1995a) Expression of functional CD40 antigen on Reed-Sternberg cells and Hodgkin's disease cell lines. Blood 85(3):780–789 Epub 1995/02/01

Carbone A, Gloghini A, Gruss HJ, Pinto A (1995b) CD40 ligand is constitutively expressed in a subset of T cell lymphomas and on the microenvironmental reactive T cells of follicular lymphomas and Hodgkin's disease. Am J Pathol 147(4):912–922 Epub 1995/10/01

Cattaruzza L, Gloghini A, Olivo K, Di Francia R, Lorenzon D, De Filippi R et al (2009) Functional coexpression of Interleukin (IL)-7 and its receptor (IL-7R) on Hodgkin and Reed-Sternberg cells: involvement of IL-7 in tumor cell growth and microenvironmental interactions of Hodgkin's lymphoma. Int J Cancer 125(5):1092–1101 Epub 2009/04/25

Chaganti S, Bell AI, Pastor NB, Milner AE, Drayson M, Gordon J et al (2005) Epstein-Barr virus infection in vitro can rescue germinal center B cells with inactivated immunoglobulin genes. Blood 106(13):4249–4252

Chang KL, Albujar PF, Chen YY, Johnson RM, Weiss LM (1993) High prevalence of Epstein-Barr virus in the Reed-Sternberg cells of Hodgkin's disease occurring in Peru. Blood 81:496

Chiu A, Xu W, He B, Dillon SR, Gross JA, Sievers E et al (2007) Hodgkin lymphoma cells express TACI and BCMA receptors and generate survival and proliferation signals in response to BAFF and APRIL. Blood 109(2):729–739 Epub 2006/09/09

Cozen W, Timofeeva MN, Li D, Diepstra A, Hazelett D, Delahaye-Sourdeix M et al (2014) A meta-analysis of Hodgkin lymphoma reveals 19p13.3 TCF3 as a novel susceptibility locus. Nat Commun 5:3856 Epub 2014/06/13

Crump C, Sundquist K, Sieh W, Winkleby MA, Sundquist J (2012a) Perinatal and family risk factors for Hodgkin lymphoma in childhood through young adulthood. Am J Epidemiol 176(12):1147–1158 Epub 2012/11/23

Crump C, Sundquist K, Sieh W, Winkleby MA, Sundquist J (2012b) Perinatal and family risk factors for non-Hodgkin lymphoma in early life: a Swedish national cohort study. J Natl Cancer Inst 104(12):923–930 Epub 2012/05/25

Deacon EM, Pallesen G, Niedobitek G, Crocker J, Brooks L, Rickinson AB et al (1993) Epstein-Barr virus and Hodgkin's disease: transcriptional analysis of virus latency in the malignant cells. J Exp Med 177(2):339–349

Diepstra A, Niens M, Vellenga E, van Imhoff GW, Nolte IM, Schaapveld M et al (2005) Association with HLA class I in Epstein-Barr-virus-positive and with HLA class III in Epstein-Barr-virus-negative Hodgkin's lymphoma. Lancet 365(9478):2216–2224 Epub 2005/06/28

Doerr JR, Malone CS, Fike FM, Gordon MS, Soghomonian SV, Thomas RK et al (2005) Patterned CpG methylation of silenced B cell gene promoters in classical Hodgkin lymphoma-derived and primary effusion lymphoma cell lines. J Mol Biol 350(4):631–640 Epub 2005/06/22

D'Souza BN, Edelstein LC, Pegman PM, Smith SM, Loughran ST, Clarke A et al (2004) Nuclear factor kappa B-dependent activation of the antiapoptotic bfl-1 gene by the Epstein-Barr virus latent membrane protein 1 and activated CD40 receptor. J Virol 78(4):1800–1816 Epub 2004/01/30

Dutton A, O'Neil JD, Milner AE, Reynolds GM, Starczynski J, Crocker J et al (2004) Expression of the cellular FLICE-inhibitory protein (c-FLIP) protects Hodgkin's lymphoma cells from autonomous Fas-mediated death. Proc Natl Acad Sci USA 101(17):6611–6616 Epub 2004/04/21

Dutton A, Reynolds GM, Dawson CW, Young LS, Murray PG (2005) Constitutive activation of phosphatidyl-inositide 3 kinase contributes to the survival of Hodgkin's lymphoma cells through a mechanism involving Akt kinase and mTOR. J Pathol 205(4):498–506 Epub 2005/02/17

Dutton A, Woodman CB, Chukwuma MB, Last JI, Wei W, Vockerodt M et al (2007) Bmi-1 is induced by the Epstein-Barr virus oncogene LMP1 and regulates the expression of viral target genes in Hodgkin lymphoma cells. Blood 109(6):2597–2603 Epub 2006/12/07

Eliopoulos AG, Gallagher NJ, Blake SM, Dawson CW, Young LS (1999) Activation of the p38 mitogen-activated protein kinase pathway by Epstein-Barr virus-encoded latent membrane protein 1 coregulates interleukin-6 and interleukin-8 production. J Biol Chem 274(23):16085–16096 Epub 1999/05/29

Emmerich F, Meiser M, Hummel M, Demel G, Foss HD, Jundt F et al (1999) Overexpression of I kappa B alpha without inhibition of NF-kappaB activity and mutations in the I kappa B alpha gene in Reed-Sternberg cells. Blood 94(9):3129–3134

Emmerich F, Theurich S, Hummel M, Haeffker A, Vry MS, Dohner K et al (2003) Inactivating I kappa B epsilon mutations in Hodgkin/Reed-Sternberg cells. J Pathol 201(3):413–420 Epub 2003/11/05

Farrell K, Jarrett RF (2011) The molecular pathogenesis of Hodgkin lymphoma. Histopathology 58(1):15–25 Epub 2011/01/26

Fischer M, Juremalm M, Olsson N, Backlin C, Sundstrom C, Nilsson K et al (2003) Expression of CCL5/RANTES by Hodgkin and Reed-Sternberg cells and its possible role in the recruitment of mast cells into lymphomatous tissue. Int J Cancer 107(2):197–201 Epub 2003/09/02

Fiumara P, Snell V, Li Y, Mukhopadhyay A, Younes M, Gillenwater AM et al (2001) Functional expression of receptor activator of nuclear factor kappaB in Hodgkin disease cell lines. Blood 98(9):2784–2790 Epub 2001/10/25

Flavell KJ, Biddulph JP, Powell JE, Parkes SE, Redfern D, Weinreb M et al (2001) South Asian ethnicity and material deprivation increase the risk of Epstein-Barr virus infection in childhood Hodgkin's disease. Br J Cancer 85(3):350–356 Epub 2001/08/07

Flavell JR, Baumforth KR, Wood VH, Davies GL, Wei W, Reynolds GM et al (2008) Down-regulation of the TGF-beta target gene, PTPRK, by the Epstein-Barr virus encoded EBNA1 contributes to the growth and survival of Hodgkin lymphoma cells. Blood 111(1):292–301 Epub 2007/08/28

Foss HD, Herbst H, Gottstein S, Demel G, Araujo I, Stein H (1996) Interleukin-8 in Hodgkin's disease. Preferential expression by reactive cells and association with neutrophil density. Am J Pathol 148(4):1229–1236 Epub 1996/04/01

Frappier L (2012) Contributions of Epstein-Barr nuclear antigen 1 (EBNA1) to cell immortalization and survival. Viruses 4(9):1537–1547 Epub 2012/11/22

Fukuda M, Longnecker R (2004) Latent membrane protein 2A inhibits transforming growth factor-beta 1-induced apoptosis through the phosphatidylinositol 3-kinase/Akt pathway. J Virol 78(4):1697–1705 Epub 2004/01/30

Gandhi MK, Moll G, Smith C, Dua U, Lambley E, Ramuz O et al (2007) Galectin-1 mediated suppression of Epstein-Barr virus specific T-cell immunity in classic Hodgkin lymphoma. Blood 110(4):1326–1329 Epub 2007/04/18

Gires O, Kohlhuber F, Kilger E, Baumann M, Kieser A, Kaiser C et al (1999) Latent membrane protein 1 of Epstein-Barr virus interacts with JAK3 and activates STAT proteins. EMBO J 18(11):3064–3073 Epub 1999/06/05

Glaser SL, Lin RJ, Stewart SL, Ambinder RF, Jarrett RF, Brousset P et al (1997) Epstein-Barr virus-associated Hodgkin's disease: epidemiologic characteristics in international data. Int J Cancer 70(4):375–382 Epub 1997/02/07

Glaser SL, Clarke CA, Gulley ML, Craig FE, DiGiuseppe JA, Dorfman RF et al (2003) Population-based patterns of human immunodeficiency virus-related Hodgkin lymphoma in the Greater San Francisco Bay Area, 1988–1998. Cancer 98(2):300–309 Epub 2003/07/23

Gotti D, Danesi M, Calabresi A, Ferraresi A, Albini L, Donato F et al (2013) Clinical characteristics, incidence, and risk factors of HIV-related Hodgkin lymphoma in the era of combination antiretroviral therapy. Aids Patient Care STDS 27(5):259–265 Epub 2013/04/23

Grasser FA, Murray PG, Kremmer E, Klein K, Remberger K, Feiden W et al (1994) Monoclonal antibodies directed against the Epstein-Barr virus-encoded nuclear antigen 1 (EBNA1): immunohistologic detection of EBNA1 in the malignant cells of Hodgkin's disease. Blood 84(11):3792–3798

Green MR, Monti S, Rodig SJ, Juszczynski P, Currie T, O'Donnell E et al (2010) Integrative analysis reveals selective 9p24.1 amplification, increased PD-1 ligand expression, and further induction via JAK2 in nodular sclerosing Hodgkin lymphoma and primary mediastinal large B-cell lymphoma. Blood 116(17):3268–3277 Epub 2010/07/16

Green MR, Rodig S, Juszczynski P, Ouyang J, Sinha P, O'Donnell E et al (2012) Constitutive AP-1 activity and EBV infection induce PD-L1 in Hodgkin lymphomas and posttransplant lymphoproliferative disorders: implications for targeted therapy. Clin Cancer Res 18(6):1611–1618 Epub 2012/01/25

Gunawardana J, Chan FC, Telenius A, Woolcock B, Kridel R, Tan KL et al (2014) Recurrent somatic mutations of PTPN1 in primary mediastinal B cell lymphoma and Hodgkin lymphoma. Nat Genet 46(4):329–335 Epub 2014/02/18

Hanamoto H, Nakayama T, Miyazato H, Takegawa S, Hieshima K, Tatsumi Y et al (2004) Expression of CCL28 by Reed-Sternberg cells defines a major subtype of classical Hodgkin's disease with frequent infiltration of eosinophils and/or plasma cells. Am J Pathol 164(3):997–1006 Epub 2004/02/26

Henderson S, Rowe M, Gregory C, Croom-Carter D, Wang F, Longnecker R et al (1991) Induction of bcl-2 expression by Epstein-Barr virus latent membrane protein 1 protects infected B cells from programmed cell death. Cell 65(7):1107–1115 Epub 1991/06/28

Hertel CB, Zhou XG, Hamilton-Dutoit SJ, Junker S (2002) Loss of B cell identity correlates with loss of B cell-specific transcription factors in Hodgkin/Reed-Sternberg cells of classical Hodgkin lymphoma. Oncogene 21(32):4908–4920 Epub 2002/07/16

Hinz M, Lemke P, Anagnostopoulos I, Hacker C, Krappmann D, Mathas S et al (2002) Nuclear factor kappaB-dependent gene expression profiling of Hodgkin's disease tumor cells, pathogenetic significance, and link to constitutive signal transducer and activator of transcription 5a activity. J Exp Med 196(5):605–617 Epub 2002/09/05

Hjalgrim H, Askling J, Sorensen P, Madsen M, Rosdahl N, Storm HH et al (2000) Risk of Hodgkin's disease and other cancers after infectious mononucleosis. J Natl Cancer Inst 92:1522

Hjalgrim H, Smedby KE, Rostgaard K, Molin D, Hamilton-Dutoit S, Chang ET et al (2007) Infectious mononucleosis, childhood social environment, and risk of Hodgkin lymphoma. Cancer Res 67(5):2382–2388 Epub 2007/03/03

Hjalgrim H, Rostgaard K, Johnson PC, Lake A, Shield L, Little AM et al (2010) HLA-A alleles and infectious mononucleosis suggest a critical role for cytotoxic T-cell response in EBV-related Hodgkin lymphoma. Proc Natl Acad Sci USA 107(14):6400–6405 Epub 2010/03/24

Holowaty MN, Zeghouf M, Wu H, Tellam J, Athanasopoulos V, Greenblatt J et al (2003) Protein profiling with Epstein-Barr nuclear antigen-1 reveals an interaction with the herpesvirus-associated ubiquitin-specific protease HAUSP/USP7. J Biol Chem 278(32):29987–29994 Epub 2003/06/05

Horie R, Watanabe T, Morishita Y, Ito K, Ishida T, Kanegae Y et al (2002) Ligand-independent signaling by overexpressed CD30 drives NF-kappaB activation in Hodgkin-Reed-Sternberg cells. Oncogene 21(16):2493–2503 Epub 2002/04/24

Huen DS, Henderson SA, Croom-Carter D, Rowe M (1995) The Epstein-Barr virus latent membrane protein-1 (LMP1) mediates activation of NF-kappa B and cell surface phenotype via two effector regions in its carboxy-terminal cytoplasmic domain. Oncogene 10(3):549–560 Epub 1995/02/02

Huppmann AR, Nicolae A, Slack GW, Pittaluga S, Davies-Hill T, Ferry JA et al (2014) EBV may be expressed in the LP cells of nodular lymphocyte-predominant Hodgkin lymphoma (NLPHL) in both children and adults. Am J Surg Pathol 38(3):316–324 Epub 2014/02/15

Ishida T, Ishii T, Inagaki A, Yano H, Komatsu H, Iida S et al (2006) Specific recruitment of CC chemokine receptor 4-positive regulatory T cells in Hodgkin lymphoma fosters immune privilege. Cancer Res 66(11):5716–5722 Epub 2006/06/03

Izban KF, Ergin M, Huang Q, Qin JZ, Martinez RL, Schnitzer B et al (2001) Characterization of NF-kappaB expression in Hodgkin's disease: inhibition of constitutively expressed NF-kappaB results in spontaneous caspase-independent apoptosis in Hodgkin and Reed-Sternberg cells. Mod Pathol 14(4):297–310 Epub 2001/04/13

Jarrett RF, Gallagher A, Jones DB, Alexander FE, Krajewski AS, Kelsey A et al (1991) Detection of Epstein-Barr virus genomes in Hodgkin's disease: relation to age. J Clin Pathol 44(10):844–848 Epub 1991/10/01

Joos S, Kupper M, Ohl S, von Bonin F, Mechtersheimer G, Bentz M et al (2000) Genomic imbalances including amplification of the tyrosine kinase gene JAK2 in $CD30^+$ Hodgkin cells. Cancer Res 60(3):549–552 Epub 2000/02/17

Joos S, Granzow M, Holtgreve-Grez H, Siebert R, Harder L, Martin-Subero JI et al (2003) Hodgkin's lymphoma cell lines are characterized by frequent aberrations on chromosomes 2p and 9p including REL and JAK2. Int J Cancer 103(4):489–495 Epub 2002/12/13

Jundt F, Anagnostopoulos I, Bommert K, Emmerich F, Muller G, Foss HD et al (1999) Hodgkin/Reed-Sternberg cells induce fibroblasts to secrete eotaxin, a potent chemoattractant for T cells and eosinophils. Blood 94(6):2065–2071 Epub 1999/09/09

Jundt F, Kley K, Anagnostopoulos I, Schulze Probsting K, Greiner A, Mathas S et al (2002a) Loss of PU.1 expression is associated with defective immunoglobulin transcription in Hodgkin and Reed-Sternberg cells of classical Hodgkin disease. Blood 99(8):3060–3062 Epub 2002/04/04

Jundt F, Anagnostopoulos I, Forster R, Mathas S, Stein H, Dorken B (2002b) Activated Notch1 signaling promotes tumor cell proliferation and survival in Hodgkin and anaplastic large cell lymphoma. Blood 99(9):3398–3403 Epub 2002/04/20

Jundt F, Acikgoz O, Kwon SH, Schwarzer R, Anagnostopoulos I, Wiesner B et al (2008) Aberrant expression of Notch1 interferes with the B-lymphoid phenotype of neoplastic B cells in classical Hodgkin lymphoma. Leukemia 22(8):1587–1594 Epub 2008/05/02

Jungnickel B, Staratschek-Jox A, Brauninger A, Spieker T, Wolf J, Diehl V et al (2000) Clonal deleterious mutations in the IkappaBalpha gene in the malignant cells in Hodgkin's lymphoma. J Exp Med 191(2):395–402

Juszczynski P, Ouyang J, Monti S, Rodig SJ, Takeyama K, Abramson J et al (2007) The AP1-dependent secretion of galectin-1 by Reed Sternberg cells fosters immune privilege in classical Hodgkin lymphoma. Proc Natl Acad Sci USA 104(32):13134–13139 Epub 2007/08/03

Kamper P, Bendix K, Hamilton-Dutoit S, Honore B, Nyengaard JR, d'Amore F (2011) Tumor-infiltrating macrophages correlate with adverse prognosis and Epstein-Barr virus status in classical Hodgkin's lymphoma. Haematologica 96(2):269–276 Epub 2010/11/13

Kanzler H, Kuppers R, Hansmann ML, Rajewsky K (1996) Hodgkin and Reed-Sternberg cells in Hodgkin's disease represent the outgrowth of a dominant tumor clone derived from (crippled) germinal center B cells. J Exp Med 184(4):1495–1505

Kapp U, Yeh WC, Patterson B, Elia AJ, Kagi D, Ho A et al (1999) Interleukin 13 is secreted by and stimulates the growth of Hodgkin and Reed-Sternberg cells. J Exp Med 189(12):1939–1946 Epub 1999/06/22

Khanna R, Burrows SR, Nicholls J, Poulsen LM (1998) Identification of cytotoxic T cell epitopes within Epstein-Barr virus (EBV) oncogene latent membrane protein 1 (LMP1): evidence for HLA-A2 supertype-restricted immune recognition of EBV-infected cells by LMP1-specific cytotoxic T lymphocytes. Eur J Immunol 28:451

Kieser A, Kilger E, Gires O, Ueffing M, Kolch W, Hammerschmidt W (1997) Epstein-Barr virus latent membrane protein-1 triggers AP-1 activity via the c-Jun N-terminal kinase cascade. EMBO J 16(21):6478–6485 Epub 1997/11/14

Kis LL, Takahara M, Nagy N, Klein G, Klein E (2006) IL-10 can induce the expression of EBV-encoded latent membrane protein-1 (LMP-1) in the absence of EBNA-2 in B lymphocytes and in Burkitt lymphoma- and NK lymphoma-derived cell lines. Blood 107(7):2928–2935 Epub 2005/12/08

Kis LL, Salamon D, Persson EK, Nagy N, Scheeren FA, Spits H et al (2010) IL-21 imposes a type II EBV gene expression on type III and type I B cells by the repression of C- and activation of LMP-1-promoter. Proc Natl Acad Sci USA 107(2):872–877 Epub 2010/01/19

Kis LL, Gerasimcik N, Salamon D, Persson EK, Nagy N, Klein G et al (2011) STAT6 signaling pathway activated by the cytokines IL-4 and IL-13 induces expression of the Epstein-Barr virus-encoded protein LMP-1 in absence of EBNA-2: implications for the type II EBV latent gene expression in Hodgkin lymphoma. Blood 117(1):165–174 Epub 2010/09/30

Kochert K, Ullrich K, Kreher S, Aster JC, Kitagawa M, Johrens K et al (2011) High-level expression of Mastermind-like 2 contributes to aberrant activation of the NOTCH signaling pathway in human lymphomas. Oncogene 30(15):1831–1840 Epub 2010/12/02

Kowalkowski MA, Mims MP, Amiran ES, Lulla P, Chiao EY (2013) Effect of immune reconstitution on the incidence of HIV-related Hodgkin lymphoma. PLoS ONE 8(10):e77409. Epub 2013/10/08

Kreher S, Bouhlel MA, Cauchy P, Lamprecht B, Li S, Grau M et al (2014) Mapping of transcription factor motifs in active chromatin identifies IRF5 as key regulator in classical Hodgkin lymphoma. Proc Natl Acad Sci USA 111(42):E4513–E4522 Epub 2014/10/08

Kube D, Holtick U, Vockerodt M, Ahmadi T, Haier B, Behrmann I et al (2001) STAT3 is constitutively activated in Hodgkin cell lines. Blood 98(3):762–770 Epub 2001/07/27

Kuppers R (2009) The biology of Hodgkin's lymphoma. Nat Rev Cancer 9(1):15–27 Epub 2008/12/17

Kuppers R, Rajewsky K, Zhao M, Simons G, Laumann R, Fischer R et al (1994) Hodgkin disease: Hodgkin and Reed-Sternberg cells picked from histological sections show clonal immunoglobulin gene rearrangements and appear to be derived from B cells at various stages of development. Proc Natl Acad Sci USA 91(23):10962–10966

Kuppers R, Klein U, Schwering I, Distler V, Brauninger A, Cattoretti G et al (2003) Identification of Hodgkin and Reed-Sternberg cell-specific genes by gene expression profiling. J Clin Invest 111(4):529–537

Kushekhar K, van den Berg A, Nolte I, Hepkema B, Visser L, Diepstra A (2014) Genetic associations in classical hodgkin lymphoma: a systematic review and insights into susceptibility mechanisms. Cancer Epidemiol Biomarkers Prev 23(12):2737–2747 Epub 2014/09/11

Lake A, Shield LA, Cordano P, Chui DT, Osborne J, Crae S et al (2009) Mutations of NFKBIA, encoding IkappaB alpha, are a recurrent finding in classical Hodgkin lymphoma but are not a unifying feature of non-EBV-associated cases. Int J Cancer 125(6):1334–1342 Epub 2009/06/10

Lam N, Sugden B (2003) CD40 and its viral mimic, LMP1: similar means to different ends. Cell Signal 15(1):9–16 Epub 2002/10/29

Lambert SL, Martinez OM (2007) Latent membrane protein 1 of EBV activates phosphatidylinositol 3-kinase to induce production of IL-10. J Immunol 179(12):8225–8234 Epub 2007/12/07

Lee SP, Thomas WA, Murray RJ, Khanim F, Kaur S, Young LS et al (1993) HLA A2.1-restricted cytotoxic T cells recognizing a range of Epstein-Barr virus isolates through a defined epitope in latent membrane protein LMP2. J Virol 67(12):7428–7435

Lee SP, Constandinou CM, Thomas WA, Croom-Carter D, Blake NW, Murray PG et al (1998) Antigen presenting phenotype of Hodgkin Reed-Sternberg cells: analysis of the HLA class I processing pathway and the effects of interleukin-10 on epstein-barr virus-specific cytotoxic T-cell recognition. Blood 92(3):1020–1030 Epub 1998/07/29

Leonard S, Wei W, Anderton J, Vockerodt M, Rowe M, Murray PG et al (2011) Epigenetic and transcriptional changes which follow Epstein-Barr virus infection of germinal center B cells and their relevance to the pathogenesis of Hodgkin's lymphoma. J Virol 85(18):9568–9577 Epub 2011/07/15

Leonard S, Gordon N, Smith N, Rowe M, Murray PG, Woodman CB (2012) Arginine Methyltransferases are regulated by Epstein-Barr Virus in B Cells and are differentially expressed in Hodgkin's lymphoma. Pathogens 1(1):52–64 Epub 2012/01/01

Levine PH, Ablashi DV, Berard CW, Carbone PP, Waggoner DE, Malan L (1971) Elevated antibody titers to Epstein-Barr virus in Hodgkin's disease. Cancer 27(2):416–421 Epub 1971/02/01

Lin KI, Tunyaplin C, Calame K (2003) Transcriptional regulatory cascades controlling plasma cell differentiation. Immunol Rev 194:19–28 Epub 2003/07/09

Liu Y, Abdul Razak FR, Terpstra M, Chan FC, Saber A, Nijland M et al (2014a) The mutational landscape of Hodgkin lymphoma cell lines determined by whole-exome sequencing. Leukemia 28(11):2248–2251 Epub 2014/06/21

Liu Y, Sattarzadeh A, Diepstra A, Visser L, van den Berg A (2014b) The microenvironment in classical Hodgkin lymphoma: an actively shaped and essential tumor component. Semin Cancer Biol 24:15–22 Epub 2013/07/23

Longnecker R, Kieff E (1990) A second Epstein-Barr virus membrane protein (LMP2) is expressed in latent infection and colocalizes with LMP1. J Virol 64(5):2319–2326 Epub 1990/05/01
Longnecker D, Kieff E, Cohen J (2013) Epstein-Barr Virus. In: Knipe DM, Howley PM (eds) Field's virology, 6th edn. Lippincott William and Wilkins, p 1898–1959
Lu J, Murakami M, Verma SC, Cai Q, Haldar S, Kaul R et al (2011) Epstein-Barr Virus nuclear antigen 1 (EBNA1) confers resistance to apoptosis in EBV-positive B-lymphoma cells through up-regulation of survivin. Virology 410(1):64–75 Epub 2010/11/26
Mack TM, Cozen W, Shibata DK, Weiss LM, Nathwani BN, Hernandez AM et al (1995) Concordance for Hodgkin's disease in identical twins suggesting genetic susceptibility to the young-adult form of the disease. N Engl J Med 332(7):413–418 Epub 1995/02/16
Mancao C, Altmann M, Jungnickel B, Hammerschmidt W (2005) Rescue of "crippled" germinal center B cells from apoptosis by Epstein-Barr virus. Blood 106(13):4339–4344
Marafioti T, Hummel M, Foss HD, Laumen H, Korbjuhn P, Anagnostopoulos I et al (2000) Hodgkin and reed-sternberg cells represent an expansion of a single clone originating from a germinal center B-cell with functional immunoglobulin gene rearrangements but defective immunoglobulin transcription. Blood 95(4):1443–1450 Epub 2000/02/09
Marshall NA, Culligan DJ, Tighe J, Johnston PW, Barker RN, Vickers MA (2007) The relationships between Epstein-Barr virus latent membrane protein 1 and regulatory T cells in Hodgkin's lymphoma. Exp Hematol 35(4):596–604 Epub 2007/03/24
Martin-Subero JI, Gesk S, Harder L, Sonoki T, Tucker PW, Schlegelberger B et al (2002) Recurrent involvement of the REL and BCL11A loci in classical Hodgkin lymphoma. Blood 99(4):1474–1477 Epub 2002/02/07
Martin-Subero JI, Wlodarska I, Bastard C, Picquenot JM, Hoppner J, Giefing M et al (2006) Chromosomal rearrangements involving the BCL3 locus are recurrent in classical Hodgkin and peripheral T-cell lymphoma. Blood 108(1):401–402. author reply 2-3. Epub 2006/06/23
Mathas S, Johrens K, Joos S, Lietz A, Hummel F, Janz M et al (2005) Elevated NF-kappaB p50 complex formation and Bcl-3 expression in classical Hodgkin, anaplastic large-cell, and other peripheral T-cell lymphomas. Blood 106(13):4287–4293 Epub 2005/08/27
Mathas S, Janz M, Hummel F, Hummel M, Wollert-Wulf B, Lusatis S et al (2006) Intrinsic inhibition of transcription factor E2A by HLH proteins ABF-1 and Id2 mediates reprogramming of neoplastic B cells in Hodgkin lymphoma. Nat Immunol 7(2):207–215 Epub 2005/12/22
Merchant M, Caldwell RG, Longnecker R (2000) The LMP2A ITAM is essential for providing B cells with development and survival signals in vivo. J Virol 74(19):9115–9124
Mueller N, Evans A, Harris NL, Comstock GW, Jellum E, Magnus K et al (1989) Hodgkin's disease and Epstein-Barr virus. Altered antibody pattern before diagnosis. N Engl J Med 320(11):689–695 Epub 1989/03/16
Muenst S, Hoeller S, Dirnhofer S, Tzankov A (2009) Increased programmed death-1+ tumor-infiltrating lymphocytes in classical Hodgkin lymphoma substantiate reduced overall survival. Hum Pathol 40(12):1715–1722 Epub 2009/08/22
Murray PG, Young LS, Rowe M, Crocker J (1992) Immunohistochemical demonstration of the Epstein-Barr virus-encoded latent membrane protein in paraffin sections of Hodgkin's disease. J Pathol 166(1):1–5 Epub 1992/01/01
Muschen M, Rajewsky K, Brauninger A, Baur AS, Oudejans JJ, Roers A et al (2000) Rare occurrence of classical Hodgkin's disease as a T cell lymphoma. J Exp Med 191(2):387–394
Nakayama T, Hieshima K, Nagakubo D, Sato E, Nakayama M, Kawa K et al (2004) Selective induction of Th2-attracting chemokines CCL17 and CCL22 in human B cells by latent membrane protein 1 of Epstein-Barr virus. J Virol 78(4):1665–1674 Epub 2004/01/30
Nie L, Xu M, Vladimirova A, Sun XH (2003) Notch-induced E2A ubiquitination and degradation are controlled by MAP kinase activities. EMBO J 22(21):5780–5792 Epub 2003/11/01
Niedobitek G, Kremmer E, Herbst H, Whitehead L, Dawson CW, Niedobitek E et al (1997) Immunohistochemical detection of the Epstein-Barr virus-encoded latent membrane protein 2A in Hodgkin's disease and infectious mononucleosis. Blood 90(4):1664–1672

Niens M, Jarrett RF, Hepkema B, Nolte IM, Diepstra A, Platteel M et al (2007) HLA-A*02 is associated with a reduced risk and HLA-A*01 with an increased risk of developing EBV+ Hodgkin lymphoma. Blood 110(9):3310–3315 Epub 2007/07/17

Nutt SL, Kee BL (2007) The transcriptional regulation of B cell lineage commitment. Immunity 26(6):715–725 Epub 2007/06/22

Otto C, Giefing M, Massow A, Vater I, Gesk S, Schlesner M et al (2012) Genetic lesions of the TRAF3 and MAP3K14 genes in classical Hodgkin lymphoma. Br J Haematol 157(6):702–708 Epub 2012/04/04

Oudejans JJ, Jiwa NM, Kummer JA, Horstman A, Vos W, Baak JP et al (1996a) Analysis of major histocompatibility complex class I expression on Reed-Sternberg cells in relation to the cytotoxic T-cell response in Epstein-Barr virus-positive and -negative Hodgkin's disease. Blood 87(9):3844–3851 Epub 1996/05/01

Oudejans JJ, Jiwa NM, Kummer JA, Horstman A, Vos W, Baak JPA et al (1996b) Analysis of MHC class I expression on Reed-Sternberg cells in relation to the cytotoxic T-cell response in Epstein-Barr virus positive and negative Hodgkin's disease. Blood 87:3844

Oudejans JJ, Jiwa NM, Kummer JA, Ossenkoppele GJ, van Heerde P, Baars JW et al (1997) Activated cytotoxic T cells as prognostic marker in Hodgkin's disease. Blood 89(4):1376–1382 Epub 1997/02/15

Oyama T, Ichimura K, Suzuki R, Suzumiya J, Ohshima K, Yatabe Y et al (2003) Senile EBV+ B-cell lymphoproliferative disorders: a clinicopathologic study of 22 patients. Am J Surg Pathol 27(1):16–26 Epub 2002/12/28

Oyama T, Yamamoto K, Asano N, Oshiro A, Suzuki R, Kagami Y et al (2007) Age-related EBV-associated B-cell lymphoproliferative disorders constitute a distinct clinicopathologic group: a study of 96 patients. Clin Cancer Res 13(17):5124–5132 Epub 2007/09/06

Pegtel DM, Cosmopoulos K, Thorley-Lawson DA, van Eijndhoven MA, Hopmans ES, Lindenberg JL et al (2010) Functional delivery of viral miRNAs via exosomes. Proc Natl Acad Sci USA 107(14):6328–6333 Epub 2010/03/23

Pinto A, Aldinucci D, Gloghini A, Zagonel V, Degan M, Perin V et al (1997) The role of eosinophils in the pathobiology of Hodgkin's disease. Ann Oncol 8(Suppl 2):89–96 Epub 1997/01/01

Portis T, Longnecker R (2004) Epstein-Barr virus (EBV) LMP2A alters normal transcriptional regulation following B-cell receptor activation. Virology 318(2):524–533 Epub 2004/02/20

Portis T, Dyck P, Longnecker R (2003) Epstein-Barr Virus (EBV) LMP2A induces alterations in gene transcription similar to those observed in Reed-Sternberg cells of Hodgkin lymphoma. Blood 102:4166

Ranuncolo SM, Pittaluga S, Evbuomwan MO, Jaffe ES, Lewis BA (2012) Hodgkin lymphoma requires stabilized NIK and constitutive RelB expression for survival. Blood 120(18):3756–3763 Epub 2012/09/13

Re D, Muschen M, Ahmadi T, Wickenhauser C, Staratschek-Jox A, Holtick U et al (2001) Oct-2 and Bob-1 deficiency in Hodgkin and Reed Sternberg cells. Cancer Res 61(5):2080–2084 Epub 2001/03/31

Reichel J, Chadburn A, Rubinstein PG, Giulino-Roth L, Tam W, Liu Y et al (2015) Flow-sorting and exome sequencing reveals the oncogenome of primary Hodgkin and Reed-Sternberg cells. Blood 125:1061–1072 Epub 2014/12/10

Renne C, Willenbrock K, Kuppers R, Hansmann ML, Brauninger A (2005) Autocrine- and paracrine-activated receptor tyrosine kinases in classic Hodgkin lymphoma. Blood 105(10):4051–4059 Epub 2005/01/29

Renne C, Martin-Subero JI, Eickernjager M, Hansmann ML, Kuppers R, Siebert R et al (2006) Aberrant expression of ID2, a suppressor of B-cell-specific gene expression, in Hodgkin's lymphoma. Am J Pathol 169(2):655–664

Renne C, Hinsch N, Willenbrock K, Fuchs M, Klapper W, Engert A et al (2007) The aberrant coexpression of several receptor tyrosine kinases is largely restricted to EBV-negative cases of classical Hodgkin's lymphoma. Int J Cancer 120(11):2504–2509 Epub 2007/03/03

Reusch JA, Nawandar DM, Wright KL, Kenney SC, Mertz JE (2015) Cellular differentiation regulator BLIMP1 induces Epstein-Barr virus lytic reactivation in epithelial and B cells by activating transcription from both the R and Z promoters. J Virol 89:731–1743 Epub 2014/11/21

Rickinson AB, Moss DJ (1997) Human cytotoxic T lymphocyte responses to Epstein-Barr virus infection. Annu Rev Immunol 15:405–431 Epub 1997/01/01

Rowe M, Raithatha S, Shannon-Lowe C (2014) Counteracting effects of cellular Notch and Epstein-Barr virus EBNA2: implications for stromal effects on virus-host interactions. J Virol 88(20):12065–12076 Epub 2014/08/15

Schmitz R, Hansmann ML, Bohle V, Martin-Subero JI, Hartmann S, Mechtersheimer G et al (2009) TNFAIP3 (A20) is a tumor suppressor gene in Hodgkin lymphoma and primary mediastinal B cell lymphoma. J Exp Med 206(5):981–989 Epub 2009/04/22

Schmitz R, Young RM, Ceribelli M, Jhavar S, Xiao W, Zhang M et al (2012) Burkitt lymphoma pathogenesis and therapeutic targets from structural and functional genomics. Nature 490(7418):116–120 Epub 2012/08/14

Schwarzer R, Dorken B, Jundt F (2012) Notch is an essential upstream regulator of NF-kappaB and is relevant for survival of Hodgkin and Reed-Sternberg cells. Leukemia 26(4):806–813 Epub 2011/09/29

Schwering I, Brauninger A, Klein U, Jungnickel B, Tinguely M, Diehl V et al (2003) Loss of the B-lineage-specific gene expression program in Hodgkin and Reed-Sternberg cells of Hodgkin lymphoma. Blood 101(4):1505–1512

Seitz V, Hummel M, Marafioti T, Anagnostopoulos I, Assaf C, Stein H (2000) Detection of clonal T-cell receptor gamma-chain gene rearrangements in Reed-Sternberg cells of classic Hodgkin disease. Blood 95(10):3020–3024

Skinnider BF, Mak TW (2002) The role of cytokines in classical Hodgkin lymphoma. Blood 99(12):4283–4297 Epub 2002/05/31

Skinnider BF, Elia AJ, Gascoyne RD, Trumper LH, von Bonin F, Kapp U et al (2001) Interleukin 13 and interleukin 13 receptor are frequently expressed by Hodgkin and Reed-Sternberg cells of Hodgkin lymphoma. Blood 97(1):250–255 Epub 2001/01/03

Skinnider BF, Elia AJ, Gascoyne RD, Patterson B, Trumper L, Kapp U et al (2002) Signal transducer and activator of transcription 6 is frequently activated in Hodgkin and Reed-Sternberg cells of Hodgkin lymphoma. Blood 99(2):618–626 Epub 2002/01/10

Smith DW, Sugden B (2013) Potential cellular functions of Epstein-Barr Nuclear Antigen 1 (EBNA1) of Epstein-Barr Virus. Viruses 5(1):226–240. Epub 2013/01/18

Smith EM, Akerblad P, Kadesch T, Axelson H, Sigvardsson M (2005) Inhibition of EBF function by active Notch signaling reveals a novel regulatory pathway in early B-cell development. Blood 106(6):1995–2001 Epub 2005/05/28

Steidl C, Shah SP, Woolcock BW, Rui L, Kawahara M, Farinha P et al (2011) MHC class II transactivator CIITA is a recurrent gene fusion partner in lymphoid cancers. Nature 471(7338):377–381 Epub 2011/03/04

Steidl C, Diepstra A, Lee T, Chan FC, Farinha P, Tan K et al (2012) Gene expression profiling of microdissected Hodgkin Reed-Sternberg cells correlates with treatment outcome in classical Hodgkin lymphoma. Blood 120(17):3530–3540 Epub 2012/09/08

Stein H, Marafioti T, Foss HD, Laumen H, Hummel M, Anagnostopoulos I et al (2001) Down-regulation of BOB.1/OBF.1 and Oct2 in classical Hodgkin disease but not in lymphocyte predominant Hodgkin disease correlates with immunoglobulin transcription. Blood 97(2):496–501 Epub 2001/01/12

Swart R, Ruf IK, Sample J, Longnecker R (2000) Latent membrane protein 2A-mediated effects on the phosphatidylinositol 3-Kinase/Akt pathway. J Virol 74(22):10838–10845 Epub 2000/10/24

Swerdlow SH, Campo E, Harris NL, Jaffe E, Pileri S, Stein H et al (2008) WHO classification of tumours of haematopoietic and lymphoid tissues, 4th edn. IARC Press, Lyon

Theil J, Laumen H, Marafioti T, Hummel M, Lenz G, Wirth T et al (2001) Defective octamer-dependent transcription is responsible for silenced immunoglobulin transcription in Reed-Sternberg cells. Blood 97(10):3191–3196 Epub 2001/05/09

Tiacci E, Doring C, Brune V, van Noesel CJ, Klapper W, Mechtersheimer G et al (2012) Analyzing primary Hodgkin and Reed-Sternberg cells to capture the molecular and cellular pathogenesis of classical Hodgkin lymphoma. Blood 120(23):4609–4620 Epub 2012/09/08

Torlakovic E, Tierens A, Dang HD, Delabie J (2001) The transcription factor PU.1, necessary for B-cell development is expressed in lymphocyte predominance, but not classical Hodgkin's disease. Am J Pathol 159(5):1807–1814 Epub 2001/11/07

Tzankov A, Meier C, Hirschmann P, Went P, Pileri SA, Dirnhofer S (2008) Correlation of high numbers of intratumoral FOXP3$^+$ regulatory T cells with improved survival in germinal center-like diffuse large B-cell lymphoma, follicular lymphoma and classical Hodgkin's lymphoma. Haematologica 93(2):193–200 Epub 2008/01/29

Tzankov A, Matter MS, Dirnhofer S (2010) Refined prognostic role of CD68-positive tumor macrophages in the context of the cellular micromilieu of classical Hodgkin lymphoma. Pathobiology 77(6):301–308 Epub 2011/01/27

Uccini S, Monardo F, Stoppacciaro A, Gradilone A, Agliano AM, Faggioni A et al (1990) High frequency of Epstein-Barr virus genome detection in Hodgkin's disease of HIV-positive patients. Int J Cancer 46(4):581–585 Epub 1990/10/15

Ushmorov A, Ritz O, Hummel M, Leithauser F, Moller P, Stein H et al (2004) Epigenetic silencing of the immunoglobulin heavy-chain gene in classical Hodgkin lymphoma-derived cell lines contributes to the loss of immunoglobulin expression. Blood 104(10):3326–3334 Epub 2004/07/31

van den Berg A, Visser L, Poppema S (1999) High expression of the CC chemokine TARC in Reed-Sternberg cells. A possible explanation for the characteristic T-cell infiltratein Hodgkin's lymphoma. Am J Pathol 154(6):1685–1691 Epub 1999/06/11

Vockerodt M, Soares M, Kanzler H, Kuppers R, Kube D, Hansmann ML et al (1998) Detection of clonal Hodgkin and Reed-Sternberg cells with identical somatically mutated and rearranged VH genes in different biopsies in relapsed Hodgkin's disease. Blood 92(8):2899–2907 Epub 1998/10/09

Vockerodt M, Morgan SL, Kuo M, Wei W, Chukwuma MB, Arrand JR et al (2008) The Epstein-Barr virus oncoprotein, latent membrane protein 1, reprograms germinal centre B cells towards a Hodgkin's Reed-Sternberg-like phenotype. J Pathol 216(1):83–92 Epub 2008/06/21

Vockerodt M, Wei W, Nagy E, Prouzova Z, Schrader A, Kube D et al (2013) Suppression of the LMP2A target gene, EGR-1, protects Hodgkin's lymphoma cells from entry to the EBV lytic cycle. J Pathol 230(4):399–409 Epub 2013/04/18

Vrzalikova K, Vockerodt M, Leonard S, Bell A, Wei W, Schrader A et al (2011) Down-regulation of BLIMP1alpha by the EBV oncogene, LMP-1, disrupts the plasma cell differentiation program and prevents viral replication in B cells: implications for the pathogenesis of EBV-associated B-cell lymphomas. Blood 117(22):5907–5917 Epub 2011/03/18

Vrzalikova K, Woodman CB, Murray PG (2012a) BLIMP1alpha, the master regulator of plasma cell differentiation is a tumor supressor gene in B cell lymphomas. Biomed Papers (Medical Faculty of the University Palacky, Olomouc, Czechoslovakia) 156(1):1–6 Epub 2012/05/15

Vrzalikova K, Leonard S, Fan Y, Bell A, Vockerodt M, Flodr P et al (2012b) Hypomethylation and over-expression of the beta isoform of BLIMP1 is Induced by Epstein-Barr virus infection of B Cells; potential implications for the pathogenesis of EBV-associated lymphomas. Pathogens 1(2):83–101 Epub 2012/01/01

Wang S, Medeiros LJ, Xu-Monette ZY, Zhang S, O'Malley DP, Orazi A et al (2014) Epstein-Barr virus-positive nodular lymphocyte predominant Hodgkin lymphoma. Ann Diagn Pathol 18(4):203–209 Epub 2014/05/24

Weinreb M, Day PJR, Niggli F, Green EK, Nyongo AO, Othieno-Abinya NA et al (1996) The consistent association between Epstein-Barr virus and Hodgkin's disease in children in Kenya. Blood 87:3828

Weiss LM, Movahed LA, Warnke RA, Sklar J (1989) Detection of Epstein-Barr viral genomes in Reed-Sternberg cells of Hodgkin's disease. N Engl J Med 320(8):502–506

Weniger MA, Melzner I, Menz CK, Wegener S, Bucur AJ, Dorsch K et al (2006) Mutations of the tumor suppressor gene SOCS-1 in classical Hodgkin lymphoma are frequent and associated with nuclear phospho-STAT5 accumulation. Oncogene 25(18):2679–2684 Epub 2006/03/15

Wu TC, Mann RB, Charache P, Hayward SD, Staal S, Lambe BC et al (1990) Detection of EBV gene expression in Reed-Sternberg cells of Hodgkin's disease. Int J Cancer 46(5):801–804 Epub 1990/11/15

Yamamoto R, Nishikori M, Kitawaki T, Sakai T, Hishizawa M, Tashima M et al (2008) PD-1-PD-1 ligand interaction contributes to immunosuppressive microenvironment of Hodgkin lymphoma. Blood 111(6):3220–3224 Epub 2008/01/22

Young LS, Deacon EM, Rowe M, Crocker J, Herbst H, Niedobitek G et al (1991) Epstein-Barr virus latent genes in tumour cells of Hodgkin's disease. Lancet 337(8757):1617

The Role of EBV in the Pathogenesis of Diffuse Large B Cell Lymphoma

Jane A. Healy and Sandeep S. Dave

Abstract Epstein-Barr virus (EBV) infection is a common feature of B cell lymphoproliferative disorders (LPDs), including diffuse large B cell lymphoma. Approximately 10 % of DLBCLs are EBV-positive, with the highest incidence in immunocompromised and elderly patients. Here, we review the clinical, genetic, and pathologic characteristics of DLBCL and discuss the molecular role of EBV in lymphoma tumorigenesis. Using EBV-positive DLBCL of the elderly as a model, we describe the key features of EBV-positive DLBCL. Studies of EBV-positive DLBCL of the elderly demonstrate that EBV-positive DLBCL has a distinct biology, related to both viral and host factors. The pathogenic mechanisms noted in EBV-positive DLBCL of the elderly, including enhanced NFκB activity, are likely to be a generalizable feature of EBV-positive DLBCL. Therefore, we review how this information might be used to target the EBV or its host response for the development of novel treatment strategies.

Contents

1 Introduction 316
2 Classification Challenges of DLBCL 316
2.1 Disease Etiology 321
2.2 EBV and Lymphoma 322
3 Clinical and Pathologic Aspects of EBV-Positive DLBCL 325
3.1 EBV-Positive DLBCL as a Disease Model 325
3.2 Historical Perspective 326

J.A. Healy
Department of Hematology/Oncology, Duke University, DUMC Box 3841, Durham, NC 27710, USA
e-mail: jane.healy@duke.edu

S.S. Dave (✉)
Center for Genomic and Computational Biology, Duke University, CIEMAS, 2177C, 101 Science Drive, Box 3382, Durham, NC 27708, USA
e-mail: sandeep.dave@duke.edu

C. Münz (ed.), *Epstein Barr Virus Volume 1*, Current Topics in Microbiology and Immunology 390, DOI 10.1007/978-3-319-22822-8_13

3.3 Epidemiology 326
3.4 Clinical Characteristics 327
3.5 Histopathology 328
3.6 Genetics 328
3.7 Biology 329
4 Novel Treatment Approaches for EBV-Positive DLBCL 330
4.1 Antiviral Therapy 330
4.2 EBV-Targeted Adoptive Immunotherapy 330
4.3 Biology of Targeted Therapies 331
5 Conclusions 331
References 332

1 Introduction

Epstein-Barr virus (EBV) has been linked to a wide number of human cancers. Among these neoplasms, B cell lymphoproliferative disorders (LPDs) are the most frequent and strongly associated. Diffuse large B cell lymphoma is the most common lymphoid malignancy in adults, accounting for nearly a third of non-Hodgkin's lymphoma cases (NHL) globally (Fisher and Fisher 2004; Menon et al. 2012; Jemal et al. 2011).

DLBCL is characterized by rapidly proliferating cells expressing B cell-associated antigens CD19, CD20, CD22, and CD79a (Martelli et al. 2013; Swerdlow et al. 2008). Despite these common features, DLBCL is a heterogeneous disease from the standpoints of biology and clinical outcome (Dave 2010). Prognosis reflects this heterogeneity, with long-term survival rates ranging between 30 and 90 %, dependent on clinical stage and disease subtype (Ziepert et al. 2010; Corti et al. 2010).

Roughly 10 % of DLBCLs are EBV-positive, with a significantly higher incidence in setting of immunocompromised and elderly patients (Heslop 2005). There are additional biological differences that distinguish EBV-positive DLBCLs from other DLBCLs, particularly the activation of the NF-kB and JAK/STAT signaling pathways (Montes-Moreno et al. 2012; Kato et al. 2014).

This chapter will review the challenges of classifying DLBCL, particularly with regard to EBV-positive lymphoid tumors. We will further review the clinical, genetic, and biological studies that have led to our current understanding of EBV-positive DLBCL. Finally, implications for these findings on treatment strategies and patient care will be considered.

2 Classification Challenges of DLBCL

The general organizing principles of DLBCL as it understood today were initially presented in the REAL classification of 1994, and were subsequently incorporated into the WHO classification (Menon et al. 2012) which is now the widely accepted standard. While these evolving classifications have provided a clearer categorization of these tumors, the diagnosis does not capture all the observed clinicopathologic heterogeneity that is commonly observed in the disease.

Significant effort has been dedicated to the subclassification of DLBCL into more clearly delineated disease entities (Swerdlow et al. 2008; Balague Ponz et al. 2008). Table 1 depicts the subtypes of DLBCL defined in the 2008 WHO classification. Details of each category, including cellular immunophenotype, cytogenetics, gene expression patterns, and EBV association, are provided. The complex subdivisions in this scheme highlight diverse characteristics such as anatomic location (DLBCL leg type), patient age (EBV-positive DLBCL of the elderly), and presence of viral coinfections (HHV8+ DLBCL arising in the setting of multicentric Castleman's disease) (Swerdlow et al. 2008).

In addition to immunophenotyping and cytogenetics, microarray technology has provided new insight into molecular patterns of DLBCL through the examination of global gene expression alterations occurring in a given tumor. Gene expression profiling (GEP) of DLBCL tumors demonstrates that these tumors can be divided into two major subtypes based on their gene signatures (Alizadeh et al. 2000; Rosenwald et al. 2002; Lenz et al. 2008a). These two entities, germinal center B cell (GCB) and activated B cell (ABC), show distinct clinical behavior and treatment outcomes. The five-year overall survival of ABC and GCB DLBCL is 30 and 59 %, respectively (Lenz et al. 2008a). The complexity of processing and interpreting microarray data has restricted its global applicability and has thus far prevented GEP classification from being incorporating into the WHO classification (Balague Ponz et al. 2008). However, immunohistochemistry provides a useful surrogate to GEP (Hans et al. 2004; Choi et al. 2009). GEP surrogates are a reproducible, less expensive alternative to microarrays and have been widely adopted by pathologists. A combination of five immunohistochemical markers, GCTE1, CD10, BCL-6, IRF4, and FOXP1, predicted clinical outcome and achieved 90 % concordance with GEP (Choi et al. 2009).

GEP data suggest that most DLBCLs originate either during a B cell's transit through the germinal center reaction (GCB type) or in post-germinal center B cells (ABC type). Next-generation sequencing studies of DLBCL, including whole exome sequencing, genome copy number analysis, and RNA sequencing show that, in addition to having unique gene expression signatures, GCB and ABC tumors have different patterns of genetic mutation that likely reflect their distinct biology (Zhang et al. 2013; Morin et al. 2013; Pasqualucci et al. 2011). Distinct features of GCB and ABC DLBCL will be reviewed next.

Germinal centers (GCs) are transient structures that form in secondary lymphoid tissue in response to antigenic stimulation. During their passage through the GC, B cells undergo rapid proliferation, somatic hypermutation (SHM) of the variable chains of their immunoglobulin genes, class-switch recombination (CSR), and affinity selection (Victora and Nussenzweig 2012). The GC reaction is the process by which mature B cells are generated, and is thus absolutely critical to adaptive immunity. It is also a common site of lymphomagenesis. B cell transformation can occur during the GC reaction from the acquisition of mutations that

Table 1 DLBCL subtypes from the WHO 2008 classification of lymphomas

Subtype	Defining features	Immunophenotype	Cytogenetics	GEP subtype	% EBV tumor positivity	EBV latency pattern	References
DLBCL, NOS	– Most common NHL subtype – Diffuse immunoblasts effacing normal tissue architecture – No defining age/disease location/virus	– (+)CD19/20, CD22, CD79a, PAX5 (pan B cell markers)	– 3q27 rearrangement (30 %)– t(14:18) (20 %)– MYC rearrangement (10 %)	ABC or GCB	10 %	I/II	(Martelli et al. 2013; Swerdlow et al. 2008; Stein et al. 1984)
T cell/histiocyte-rich large B cell lymphoma	– Scattered immunoblasts (<10 % cells) with predominant T cell/histiocyte background – Splenomegaly BM involvement – Can progress to DLBCL NOS	– Neoplastic cells express pan B cell markers – (−) CD30/15 – T cells (+) CD3/4, TIA-1, (−) Granzyme B	– High burden of chromosomal alterations – 3q27 rearrangement (50 %) – X gain – 17p−	Intermed. between GCB and Reed–Sternberg cells	20 %	I/II	(Lim et al. 2002; Franke et al. 2002; Pittaluga and Jaffe 2010)
DLBCL associated with chronic inflammation	– Arising at sites of chronic inflammation (e.g., metallic prostheses, pyothorax)– “Local immune deficiency” due to chronic inflammation	– Pan B cell markers	– See DLBCL NOS	ABC	100 %	III	(Loong et al. 2010; Narimatsu et al. 2007; Nakatsuka et al. 2002)
EBV-positive DLBCL of the elderly	– Age >50y, immunocompetent, no previous lymphoid malignancy	– Pan B cell markers – $CD30^{+}$ (50 %)	– See DLBCL NOS	ABC	100 %	II/III	(Oyama et al. 2003, 2007; Ok et al. 2013)

(continued)

Table 1 (continued)

Subtype	Defining features	Immunophenotype	Cytogenetics	GEP subtype	% EBV tumor positivity	EBV latency pattern	References
Primary mediastinal large B cell lymphoma	– Commonly young adults – Derived from medullary thymic B cells	– (+) Pan B cell markers – (+) CD30 – (−) Surface Ig	– 9p24 gain common	Thymic pattern, not ABC or GCB	0 %	N/A	(Addis and Isaacson 1986; Cazals-Hatem et al. 1996)
Primary cutaneous DLBCL, leg type	– Aggressive cutaneous lymphoma – Poor prognosis	– (+) Pan B cell markers – (+) BCL2, IRF4 – (−) CD10	– See DLBCL NOS	ABC	Unknown	Unknown	(Martelli et al. 2013; Swerdlow et al. 2008; Nakatsuka et al. 2002)
Plasmablastic lymphoma	– HIV associated – Typically arises in oral cavity	– (−) CD20, PAX5 – (+) CD79a, CD138	– MYC rearrangement common (50 %)	ABC	50–70 %	II/III	(Castillo et al. 2008; Valera et al. 2010; Capello et al. 2010)
Intravascular large B cell lymphoma	– Large B cells in blood vessel lumens – Symptoms due to microvascular infarcts – Rapidly fatal	– Pan B cell markers – (−) CD29/54	– DLBCL NOS	ABC	0 %	N/A	(Murase et al. 2007; Ponzoni et al. 2007)
Large B cell lymphoma arising in HHV8-associated multicentric Castleman's disease	– Coincident HHV8 and (±) and HIV infections – Arises in severely immunocompromised	– Pan B cell markers – IRF4/MUM1	– DLBCL NOS	ABC	Rare	III	(Corti et al. 2010; Carbone and Gloghini 2005)

either promote sustained proliferation or impair apoptosis. These mutations may be acquired during repeated cycle cellular replication or may result from off target effects of activation-induced cytidine deaminase (AID), the enzyme that initiates the processes of SHM and CSR in GC B cells (Orthwein and Di Noia 2012).

Gene mutations that occur more frequently in GCB compared to ABC-type DLBCL include C-MYC, EZH2, and GNA13 (Zhang et al. 2013). BCL2 translocations have also been identified in roughly one-quarter of GCB tumors (Schuetz et al. 2012). BCL2 activation protects cells from programmed cell death (Kroemer 1997). Chromosomal rearrangements involving C-MYC, or gain of function mutations, promote unregulated cellular proliferation in affected cells (Ott et al. 2013). Gain of function mutations in EZH2, a histone methyltransferase, promote lymphoma by silencing cell cycle regulation genes and tumor suppressor genes (Béguelin et al. 2013). Loss of function mutations in GNA13, a G protein involved in cell–cell adhesion, enhance AKT signaling and cellular motility and are strongly associated with GCB DLBCL (Morin et al. 2013; Muppidi et al. 2014).

In contrast to GCB DLBCL, ABC tumors have gene expression profiles similar to those seen in activated B cells (Lenz et al. 2008a). The hallmark of ABC biology is believed to be chronically active B cell receptor signaling, leading to upregulated NFκB activity (Davis et al. 2001; Lenz et al. 2008b). Gene expression profiling in primary tumor samples, as well as ABC lymphoma-derived cell lines, demonstrates enhanced expression of NFκB target genes (Bea et al. 2005; Rosenwald and Staudt 2003). Augmented NFκB activity is driven by activating mutations in signaling proteins downstream of the B cell receptor and/or Toll-like receptors. Common genetic defects in the ABC subtype of DLBCL include gain of function mutations in CD79B (B cell receptor), MyD88, MALT1, and Card11, all of which promote canonical NFκB activity (Zhang et al. 2013; Morin et al. 2013). A20, a negative NFκB regulator, is subject to inactivating mutations (Kato et al. 2009). These events promote oncogenesis by enhancing cell proliferation and suppressing apoptotic signals. Constitutive NFκB activity may also explain the post-germinal center phenotype of ABC DLBCL. NFκB promotes enhanced expression of IRF4 (interferon regulatory factor-4), a transcription factor that drives the plasmablastic differentiation (Staudt 2010). Another important aspect of ABC biology is IL-6 and IL-10 generation. These cytokines exert autocrine effects on the tumor cells, resulting in the activation of STAT3 (Ding et al. 2008). A subset of DLBCL tumors have high STAT3 target gene expression and nuclear localization of phosphorylated STAT3. These tumors also demonstrate higher expression of NFκB target genes (Lam et al. 2008).

At a molecular level, constitutive NFκB activity is related to activating mutations in proteins upstream of NFκB. In B cells, CARD11, MALT1, and BCL10 form a signaling complex regulated by activation of the B cell receptor (Schulze-Luehrmann and Ghosh 2006). Upon antigen stimulation, cytoplasmic CARD11 is phosphorylated by PKCβ (Tan and Parker 2003). CARD11 is then recruited to the plasma membrane, where it serves as a molecular scaffold for the assembly of MALT1 and BCL10. This recruits the NFκB IKK complex, eventually leading to the activation of IKKβ. IKKβ is a kinase that phosphorylates the tonic NFκB inhibitor IκBα. The inhibitor then dissociates with NFκB subunits p50/p65, leaving them free to dimerize, translocate to the nucleus, and activate target genes (Staudt 2010). This process is critical for clonally expanding populations of antibody-producing B cells in response to an antigenic challenge. ABC tumors co-opt BCR signaling with activating mutations in CARD11 or CD79A/B (Lenz et al. 2008c; Davis et al. 2010). Unbound to antigenic stimulation, affected B cells are now capable of sustained NFκB activity. This leads to the enhanced proliferation and evasion of cell death, both hallmarks of cancer.

2.1 Disease Etiology

DLBCL usually occurs as a de novo malignancy, or less frequently through the "transformation" of an indolent B cell neoplasm, such as follicular lymphoma or chronic lymphocytic leukemia (CLL). In latter instance, acquisition of additional mutations results in transformation to a more aggressive neoplasm, often referred to as Richter's transformation (Giles et al. 1998).

A causative agent is not identifiable for most cases of DLBCL, though there are a number of known risk factors associated with its development, including mutagens, toxins, immune dysfunction, and infections (Martelli et al. 2013). Chemical exposures including alkylating chemotherapeutics, industrial chemicals, pesticides, and fertilizers have been shown to increase a person's risk of developing DLBCL (Fisher and Fisher 2004). This is due to the ability of these agents to mutagenize DNA. The high proliferation rate of hematopoietic cells renders them particularly vulnerable to these toxins.

Problems of immune dysregulation are common in patients with DLBCL (Smedby et al. 2006; Miranda et al. 2013). These include genetic and acquired causes of immunodeficiency, chronic inflammation, and autoimmune disease. The cause of this association is likely multifactorial, but T cell suppression resulting in impaired antitumor immunity is felt to play an important role (Yang et al. 2006). Immune suppression also permits the reactivation of lymphoma-associated viral pathogens, particularly EBV (Rickinson 2014).

Viral infections that increase the risk of DLBCL development include EBV, human immunodeficiency virus (HIV), hepatitis C (HCV), and human herpesvirus-8(HHV8) (Rickinson 2014). Viral infections can increase the risk of lymphoma through diverse mechanisms. They can alter T cell suppressor function, such as in the case of HIV, which diminishes the body's immune-mediated antitumor surveillance (Carbone and Gloghini 2005). Viruses can also promote lymphoma by driving B cell hyperstimulation, as is the case with EBV infection (discussed in detail below), HCV, and HHV8 (Rickinson 2014).

Environmental exposures, immune dysfunction, and infections can operate synergistically to promote mutagenesis, suppress T cell function, and activate B cell stimulation. The cumulative effects of these processes increase the likelihood of lymphoma development.

2.2 EBV and Lymphoma

Over 90 % of the world's population is infected with EBV, though this infection is asymptomatic in the vast majority of individuals (Niederman et al. 1970). However, the oncogenic potential of EBV is undeniable, as it has been linked to a broad range of tumor subtypes, most of which are of B cell origin (De Martel et al. 2012). The addition of EBV to primary B cells grown in culture leads to growth transformation and the generation of immortalized lymphoblastoid B cell lines (Nilsson et al. 1971). B cell neoplasms linked to EBV infection include DLBCL, post-transplant lymphoproliferative disease (PTLD), Burkitt lymphoma (BL), AIDS-related lymphoma (Primary CNS Lymphoma), plasmablastic lymphoma, and primary effusion lymphoma. These tumors are characterized by malignant B cells that frequently express EBV transcripts and proteins indicating EBV infection.

In PTLD, rates of EBV infection range from 70 to 100 % and are related to the length and degree of immunosuppression (Juvonen et al. 2003; Taylor et al. 2005). In the HIV-positive population, 80 % of DLBCL is EBV-positive (Park et al. 2007) and 100 % of primary CNS lymphomas are EBV-positive (Swerdlow et al. 2008; Gloghini et al. 2013). Primary effusion and plasmablastic lymphomas, which mostly occur in HIV-positive patients, are also positive for EBV in most cases (Hsi et al. 2011; Verma et al. 2005). Immunosuppression is a common pathogenic cofactor in these B cell neoplasms, and the relevance of this association requires a discussion of EBV virology.

EBV is a gamma-1-type herpesvirus that first discovered fifty years ago in Burkitt lymphoma tumors from pediatric patients in equatorial Africa (Epstein et al. 1964). EBV is similar to other herpesviridae in its capacity to persist in a latent state in infected cells. It is distinctive from other herpesvirus genera in its restriction to primate hosts, its tropism for establishing latency in B lymphocytes, and its ability to promote oncogenic transformation of B lymphocytes through its latent gene expression repertoire (Rickinson 2014; Nilsson et al. 1971).

EBV infection in humans occurs in three distinct stages: lytic phase, latent phase, and reactivation. First, EBV enters a host by infecting the polarized respiratory epithelial cells of the nasopharynx. This initial infection is followed by the entry of EBV into the surrounding mucosal lymphoid cells through transcytosis, leading to infection of B cells. This initial lytic phase results in the cell death, sloughing of the respiratory epithelial cells and release of high titers of virus, a process that can also occur during viral reactivation (Lemon et al. 1977). During this phase, EBV binds to CD21 receptors on naive B lymphocytes (Carel et al. 1990). Once EBV infects a B lymphocyte, the expression of viral genes initially promotes growth transformation in the infected cell. Transformed B cells are then believed to transit though the germinal center reaction, differentiating into long-lived quiescent memory B cells (Thorley-Lawson and Gross 2004; Roughan and Thorley-Lawson 2009). At this stage, the viral genome persists in an episomal state, expressing only a limited number of viral antigens (Young and Rickinson 2004). This is the latent phase of infection.

The latent infection is often punctuated by brief periods of viral reactivation caused by perturbations in the EBV-infected memory B cell pool. EBV-positive memory B cells are thought to function similarly to other memory B cells. Thus, if a given cell encounters cognate antigen, it will awake from its resting state and undergo plasma cell differentiation. Plasma cell differentiation is a trigger of viral reactivation (Laichalk and Thorley-Lawson 2005). Reactivation is believed to occur as a result of the actions of XBP-1, a B cell transcription factor critical for plasma cell differentiation. XBP-1 is capable of activating the viral BZLF promoter, which controls the expression of viral lytic genes (Sun and Thorley-Lawson 2007). During this process, infected cells reinitiate the expression of viral antigens on their cell surface. In immunocompetent individuals, viral reactivation results in brisk humoral and cell-mediated immune responses (Jones and Straus 1987; Rickinson and Moss 1997). While antibodies to viral membrane proteins decrease viral shedding and infectivity, CD4 and $CD8^+$ T cells are primarily responsible for suppressing lytic and latent EBV infections (Jones and Straus 1987). The immune system ensures that EBV-positive B cells that reactivate are promptly eliminated, thus re-establishing the steady state of latent infection. This is the usual cycle of EBV infection present in a normal, healthy individual.

Immunosuppressive states upset the virus–host balance by weakening the body's principle defense against EBV reactivation: cell-mediated immunity. In the absence of the $CD8^+$ T cell response, EBV-positive B cells are able to proliferate and express viral antigens. Hence, patients suffering from immune-deficient states associated with T cell dysfunction are particularly vulnerable to EBV reactivation. These include HIV infection and post-transplant immunodeficiency, where medication-induced T cell suppression is necessary to prevent graft rejection. Aging can also result in T cell dysfunction due to reduced numbers of $CD4/8^+$ T cells and reduction in naïve T cell receptor diversity (Miller 1996). Immune senescence is defined as age-related immune alterations that result in increased infections, autoimmunity, cancer, and reduced response to prophylactic vaccines. Consistent with these observations, the incidence of polyclonal EBV-positive lymphoproliferative disease increases with age, as does the risk of DLBCL[133]. Figure 1 summarizes the effect of EBV on the pathogenesis of lymphoma formation.

The oncogenic effect of EBV on B cells occurs through the action of a number of viral microRNAs and the protein LMP-1 (Rickinson 2014). EBV produces 44 viral microRNAs, which are believed to regulate viral and cellular mRNA during the latent phase of infection (Lopes et al. 2013). These noncoding RNAs promote cell growth and immune evasion and prevent the transcription of proapoptotic signaling molecules. LMP-1 is a well-studied viral oncogene that is believed to be the prime actor in EBV-mediated B cell transformation (Kaye et al. 1993).

LMP-1 is expressed during the latent and lytic phases of infection and is present in tumors corresponding to higher degrees of immune compromise (latency pattern II and III). LMP-1 is an integral membrane protein that behaves as a functional mimic of CD40, a costimulatory receptor required for B cell activation (Uchida et al. 1999; Eliopoulos et al. 1997; Huen et al. 1995). Under physiologic conditions, helper T cells recognize antigens presented by B cells. During this

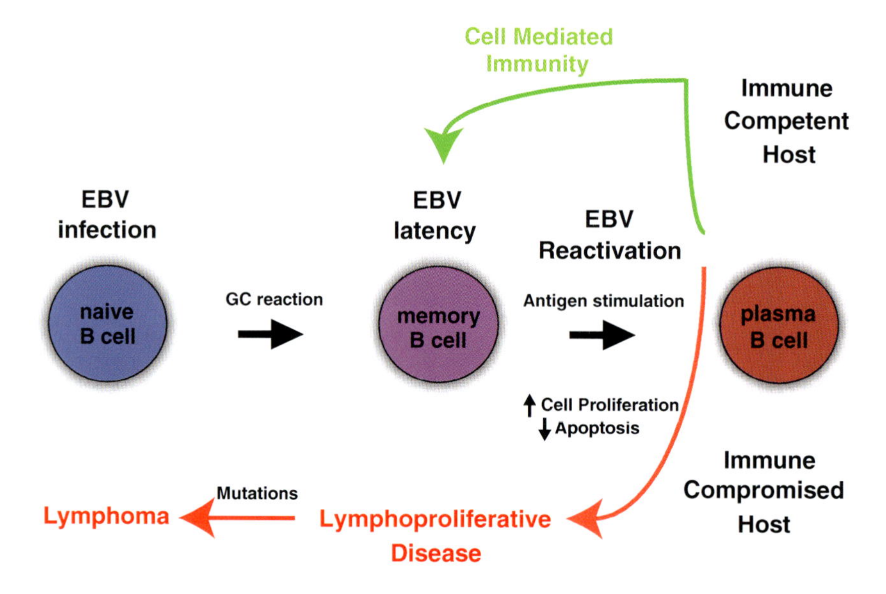

Fig. 1 The relationship of EBV to the pathogenesis of lymphoma. Naive B lymphocytes are infected with EBV and subsequently undergo transformation to memory cells via transit through the GC reaction. EBV establishes a latent infection in memory B cells. Upon memory B cell reactivation, EBV enters the lytic stage of infection, where it expresses viral antigens on the surface of the infected cell. In immune competent hosts (*green*), this activates cell-mediated immunity and the lytic phase cells are targeted for destruction. By contrast, immune compromise is associated with deficiency in cell-mediated immunity. In these individuals, EBV sets up a program of proliferation and increased cell survival which promotes polyclonal lymphocyte expansions or lymphoproliferative disease (*red*). Ongoing rounds of cellular division lead to the acquisition of additional mutations. If a sufficient number of oncogenic driver and tumor suppressor mutations are achieved, then lymphoma occurs

interaction, the extracellular portion of CD40 binds its ligand on the T cell membrane. This activates CD40, inducing conformational shape changes that promote receptor oligomerization and nucleation of TRAF signaling proteins to its cytoplasmic domain (Tsubata et al. 1993). This leads to activation of canonical NFκB pathway signaling, cell cycle entry, and protection against apoptosis. Under normal conditions, these processes lead to adaptive immunity, resulting in the rapid generation of high affinity antibodies against foreign antigens. The viral LMP-1 protein co-opts this process by behaving as a constitutively active CD40 (Uchida et al. 1999). In doing so, LMP1 uncouples B cell activation from antigen selection and activates AICDA (activation-induced cytidine deaminase), and leads to polyclonal lymphocytosis and the acquisition of additional mutations that increase the likelihood of transformation into overt lymphoma.

Dependent on the degree of host immune compromise, transformed B cells may express all, or just a portion of the EBV latency genes (Tierney et al. 1994). Latency I pattern corresponds to the expression profile present in a typical active

Table 2 EBV latency patterns and associated malignancies

Latency pattern	Associated EBV proteins	Tumors
I	LMP2A[a] EBNA1[a]	Burkitt lymphoma DBCL NOS T cell-rich DLBCL
II	*#I proteins* + LMP1 LMP2B	Classic Hodgkin's lymphoma Angioimmunoblastic T cell lymphoma NK/T cell lymphoma, N Nasopharyngeal carcinoma Gastric carcinoma
III	*#I/II proteins* + LP[a] EBNA2 EBNA3A EBNA3B EBNA3C	Primary EBV infection Post-transplant lymphoproliferative disease AIDS-related lymphomas (plasmablastic DLBCL, primary CNS lymphoma, primary effusion lymphoma) EBV + DLBCL of the elderly

[a]*EBNA*, EBV virus nuclear antigen. EBNAs promote EBV genome maintenance and regulate gene expression. *LMP*, latent membrane protein. LMPs interfere with signaling pathways from various receptors in the B cell membrane to induce cellular proliferation and inhibit programmed cell death. *LP*, leader protein. LPs co-opt hormone receptors in the B cell nucleus to promote growth transformation (Igarashi et al. 2003)

infection, latency II pattern is marked by the presence of a subset of viral antigens, and latency III pattern results in the expression of the entire EBV repertoire (Young and Murray 2003). There may also be a role for viral genetic variation in the efficiency of B cell transformation; however, this remains to be fully defined through a careful study of cases and controls. Latency patterns may be significant as they may reflect distinct aspects of tumor biology. Table 2 depicts the proteins associated with each latency pattern and shows the malignancies possessing each pattern. The EBV latency patterns associated with different DLBCL subsets are indicated in Table 1.

3 Clinical and Pathologic Aspects of EBV-Positive DLBCL

3.1 EBV-Positive DLBCL as a Disease Model

Much of our current understanding of the clinical impact of EBV in DLBCL comes from studies of EBV-positive DLBCL of the elderly. This DLBCL classification appeared first in the WHO classification of tumors in 2008 as a provisional entity (Swerdlow et al. 2008). EBV-positive DLBCL is defined as an EBV-positive monoclonal large B cell lymphoproliferation occurring in an immunocompetent patient greater than 50 years of age with no history of prior lymphoma. The age cutoff emphasizes the tendency of these tumors to arise in individuals of advanced age (Balague Ponz et al. 2008; Cho et al. 2008; Cohen et al. 2014), though

EBV-positive DLBCL has been reported in younger immunocompetent patients, albeit rarely (Cohen et al. 2014; Ao et al. 2014; Beltran et al. 2011a). It is presently unclear whether elderly patients can be truly considered immune competent, and thus distinct from other EBV-positive DLBCL subtypes, or whether the EBV reactivation in these patients is due to age-related T cell dysfunction (Miller 1996). Regardless, compared to other EBV-positive DLBCL subtypes, EBV-positive DLBCL of the elderly is not associated with concomitant immunosuppression, or use of transplant rejection medications, HIV or HHV8 coinfection, secondary malignancies, or chronic inflammatory disease. For this reason, EBV-positive DLBCL of the elderly appears to be a good model for the study of how EBV affects DLBCL disease course and treatment response and it is instructive to review this disease entity in some detail.

3.2 Historical Perspective

EBV-positive DLBCL was initially described by Oyama et al. in 2003 as "senile EBV-positive lymphoproliferative disorder" (Oyama et al. 2003). In that case series of 22 Japanese patients, the authors described a spectrum of EBV-positive lymphoproliferative disease (LPD) ranging from polyclonal B cell lymphocytosis to DLBCL. Compared to EBV-negative LPD, patients with EBV-positive tumors had a higher rate of extranodal involvement, a more aggressive clinical course, frequent refractory disease or early relapse, and worse overall survival. Since this initial observation, groups in other countries have confirmed the existence of DLBCL patients with EBV-positive tumors and without known immunodeficiency (Gibson and Hsi 2009; Beltran et al. 2011; Hoeller et al. 2010; Hofscheier et al. 2011; Uner et al. 2011; Al-Humood et al. 2014). Consistent with the initial report, these EBV-positive immune competent individuals are almost exclusively elderly.

3.3 Epidemiology

Though there have been a number of studies assessing the prevalence rate of EBV positivity in DLBCL tumors from elderly, immune competent patients, the geographic prevalence of the disease appears variable. Groups from the USA and European countries have reported incidences <5 % (Gibson and Hsi 2009; Hoeller et al. 2010; Hofscheier et al. 2011) and some Asian and Latin American countries report rates as high as 10–15 % of DLBCLs (Beltran et al. 2011b; Hofscheier et al. 2011), implicating potential genetic polymorphisms, coinfections, or environmental factors in these geographic differences. However, some studies have demonstrated significant variability within the same geographic region (Wada et al. 2011), suggesting that there might be an additional confounding variable in the lack of standardized criteria for determining EBV positivity in clinical cases.

Laboratories use differing cutoffs for the percentage of antigen expressing cells necessary for a tumor to be deemed EBV-positive (20 % vs. 50 %) (Wada et al. 2011). Furthermore, sample processing is not uniform. For example, there are various methods for separating EBV-positive tumor cells from contaminating background cells (Ok et al. 2013) and these methods are not always applied consistently. Finally, there are different means of detecting EBV infection (EBERin situ hybridization or LMP-1 immunohistochemistry) that may affect the assay sensitivity. Universal standardization of EBV testing is needed before definitive conclusions about geographic variation of this disease can be made.

3.4 Clinical Characteristics

The median age for EBV-positive DLBCL is 71, with the greatest proportion (20–25 %) of cases occurring in patients greater than age 90 (Ok et al. 2013; Castillo et al. 2011). Initial descriptions of EBV-positive DLBCL of the elderly stressed that patients presented in later stage of the disease, measured by IPI and Ann Arbor stage, and possessed a high degree of extranodal involvement (Oyama et al. 2003, 2007). Extranodal extension to GI tract, lung, liver, skin, soft tissue and bone were described. However, later studies in North American patients showed that, similar to EBV-negative DLBCL, both nodal and extranodal disease sites are common (Gibson and Hsi 2009; Hoeller et al. 2010; Hofscheier et al. 2011). No distinguishing clinical features have been reliably associated with EBV positivity, except a trend toward higher Ann Arbor stage at presentation (Gibson and Hsi 2009).

Studies in Asia and Europe demonstrate that patients with EBV-positive DLBCL of the elderly respond poorly to treatment and have worse overall survival compared to those with EBV-negative tumors (Park et al. 2007; Hofscheier et al. 2011; Oyama et al. 2007; Chang et al. 2014). Age, by itself, is a risk for poorer outcomes in DLBCL. However, EBV-positive DLBCL is also associated with an ABC GEP, which is known to have a worse outcome than GCB tumors (Kato et al. 2014; Ok et al. 2014). Montes-Moreno et al. (2012) explored whether the difference in survival was merely due to a higher prevalence of ABC phenotype. They compared elderly patients with either EBV-positive DLBCL or EBV-negative DLBCL stratified by ABC versus GCB GEP and found that EBV positivity conferred a worse outcome than ABC subtype alone, suggesting that EBV is an independent risk factor for poor outcome.

One caveat to the survival studies is that most patients were treated prior to the time when rituximab, a monoclonal antibody directed against CD20, became a standard addition to chemotherapy regimens targeting B cell lymphoma. A study performed on DLBCL patients treated with R-CHOP demonstrated no survival difference between EBV-positive and EBV-negative patients, suggesting that rituximab alone may overcome the survival difference noted in previous reports (Montes-Moreno et al. 2012). The possibility of rituximab having activity in EBV-positive DLBCL is not unexpected, since rituximab monotherapy effectively eliminates the majority of mature B cells and is a highly effective treatment for EBV-positive PTLD (Taylor et al. 2005). More studies are needed to clarify the impact of rituximab on the clinical outcomes of this lymphoma.

3.5 Histopathology

EBV-positive DLBCLs typically demonstrate an effacement of nodal and extranodal tissue architecture by large, rapidly proliferating immunoblasts with interdispersed areas of geographic necrosis (Al-Humood et al. 2014; Oyama et al. 2007; Dojcinov et al. 2011). The cellular makeup of the tumor is variable, with both polymorphic and monomorphic subtypes described. Polymorphic tumors display numerous reactive cells, including histiocytes, plasma cells, and normal lymphocytes intermingled with malignant large cells. The monomorphic subtype is characterized by sheets of uniform-appearing large cells with minimal reactive component. Both entities may contain Reed–Sternberg cells (Oyama et al. 2007), which are commonly present in Hodgkin's lymphoma, another B cell tumor associated with EBV. Despite differing appearances, histological subtypes do not have prognostic significance (Oyama et al. 2007).

The immunophenotype of EBV-positive DLBCL is that of an aggressive B cell tumor of post-germinal center origin. Malignant cells are typically positive for B cell markers CD19, CD20, CD79a, and PAX-5. Ki67, a marker of proliferation, is usually present in greater than 50 % of tumor cells (Swerdlow et al. 2008; Montes-Moreno et al. 2012; Al-Humood et al. 2014). Using immunohistochemical markers, post-germinal center (ABC-associated) proteins IRF4, MUM1, are typically positive. GC markers CD10 and BCL6 are usually negative (Montes-Moreno et al. 2012). In comparison with Reed–Sternberg cells found in Hodgkin's lymphoma, CD15 immunostaining of the neoplastic cells in EBV-positive DLBCL is negative, though most specimens (50–89 %) are CD30 positive (Montes-Moreno et al. 2012).

EBV positivity is measured either by fluorescent in situ hybridization of the EBERRNA (Chang et al. 1992) or by immunohistochemical detection of the LMP-1 protein (Gulley 2001). The expression of EBV latency genes in EBV-positive DLBCL reveals EBNA-1 and LMP-1 expression in >90 % of tumors, and 28 % positive for EBNA2, which is consistent with a viral latency II or III pattern, similar to that seen in PTLD (Oyama et al. 2003; Hofscheier et al. 2011; Oyama et al. 2007; Nguyen-Van et al. 2011; Shimoyama et al. 2008).

3.6 Genetics

Cytogenetic and FISH studies in tumors from EBV-positive DLBCL of the elderly have revealed no characteristic abnormalities (Al-Humood et al. 2014). Chromosomal translocations involving the heavy chain locus occur in approximately 15 % of samples (Montes-Moreno et al. 2012). Cytogenetic alterations have been reported, including copy number gains of the C-MYC, BCL2, and BCL6 loci (Dojcinov et al. 2011). Al-Humood et al. (2014) reported that the mean total number of chromosomal alternations per case was less than that seen for EBV-negative disease. This suggests that the EBV itself, rather than acquired mutations, may be driving the pathogenesis of infected tumor cells.

B cell clonality in EBV-positive DLBCL is common, as measured by PCR of VH-JH rearrangements in the Ig locus. Most patients demonstrate light chain restriction by Kappa and Lambda immunostaining (Al-Humood et al. 2014). The EBV genomes of these specimens also demonstrate clonality, as evidenced by FISH probes design to detect the EBV terminal repeat regions (Oyama et al. 2007).

There is also a high incidence of T cell clonality in EBV-positive DLBCL patients, with 24 % of cases demonstrating monoclonality in at least one TCR-gamma gene (Oyama et al. 2007). The significance of this finding is unclear, however, since clonal T cells have also been demonstrated in many healthy elderly individuals. Some reports show a prevalence of T cell clones greater than 80 % in asymptomatic patients over the age of 75 years (Hadrup et al. 2006). These clones are believed to reflect the reduction of T cell diversity inherent to age-related immune senescence.

3.7 Biology

EBV-positive DLBCL is associated with ABC phenotype, which is characterized by upregulated NFκB signaling (Staudt 2010). In 2012, Montes Moreno et al. assessed the state of NFκB activation in EBV-positive DLBCL tumors by Western blot analysis and subcellular localization of classical NFκB subunits p105/p50 and alternative pathway subunits p100/p52 (Montes-Moreno et al. 2012). They found nuclear localization of these factors in 79 and 74 % of tumors, respectively. Over half of tumors demonstrated nuclear expression of both canonical and alternative NFκB pathways, significantly greater NFκB activity than that seen in ABC lymphoma alone. Furthermore, Kato et al. (2014) found that infecting human ABC DLBCL-derived cell lines with EBV enhanced NFκB activity measured by electrophoretic mobility shift assay.

There is only one report of gene expression profiling for EBV-positive DLBCL thus far. Kato et al. (2014) studied a total of 61 patients meeting criteria for the diagnosis of EBV-positive DLBCL of the elderly and compared these tumors to 36 EBV-negative DLBCL specimens. The authors found that immune and inflammatory gene pathways are highly expressed in EBV-positive DLBCL of the elderly, including NFκB, JAK/STAT, NOD receptor, and Toll-like receptor signaling pathways. Expression analysis of the transcriptional targets of NFκB and STAT3 signaling revealed that these pathways are overactive in EBV-positive tumors. The authors went on to evaluate the effect of EBV status on the subset of ABC tumors. They found that NFκB and STAT3 expression gene sets were enriched in EBV-positive tumors. Finally, the authors evaluated other EBV-positive lymphoma subtypes to determine whether EBV positivity promoted NFκB and JAK/STAT signaling in other tumor types. NFκB target gene enrichment was the characteristic of HIV-associated DLBCL, EBV-positive Hodgkin's lymphoma, and NK cell lymphoma, whereas the STAT3 signature was only associated with B cell lymphoma subtypes. These data suggest that NFκB and STAT3 activity may be defining features of EBV pathogenesis as it relates to lymphoma.

4 Novel Treatment Approaches for EBV-Positive DLBCL

Recent insights into the biology of EBV-positive DLBCL of the elderly have revealed the distinct biology of B cell lymphomas that arise in the setting of EBV infection. Further, there are many reports suggesting that patients with EBV-positive lymphoma have worse prognosis than their EBV-negative counterparts. For these reasons, new treatments are needed to address the unique pathogenesis of this disorder. Possible therapeutic approaches include the following: antiviral strategies, EBV-targeted adoptive immunotherapy, and/or agents that target the NFκB or STAT3 signaling pathways.

4.1 Antiviral Therapy

Antiviral drugs offer clear potential for the treatment of EBV-positive lymphoma. There are other lymphoma subtypes that have previously demonstrated response to iradication of an associated microorganism. Gastric MALT (mucosa-associated lymphoma) is an extranodal marginal zone lymphoma that is highly associated with *Helicobacter pylori* infection and can be effectively treated with antibiotics in 70 % of patients (Bayerdörffer et al. 1995). Owing to its potential for latent infection in resting B cells, EBV is a less straightforward treatment target than *H. pylori*. Treatment of EBV with antiviral medications would first require activation of the virus into the lytic phase of infection. Known EBV lytic phase inducers include DNA methylase transferase inhibitors, HDAC inhibitors, and chemotherapeutics (Feng et al. 2004). Recently, HDAC inhibitors panobinostat and vorinostat have demonstrated potent induction of EBV lytic genes in cell lines (Ghosh et al. 2012), as well as activity in EBV-associated lymphoma (Piekarz et al. 2011; Younes 2009). Induction therapy with an EBV lytic phase inducing agent, followed by EBV antiviral therapy, would provide a potential solution to the latency issue.

4.2 EBV-Targeted Adoptive Immunotherapy

EBV-targeted adoptive immunotherapy is a strategy in which T cells isolated from a patient's peripheral blood are expanded in vitro and activated by exposure to EBV-specific antigens. These cells are then re-introduced into the patient, where they colonize tissues and attack lymphoma cells expressing EBV antigens (Gattinoni et al. 2006). Adoptive transfer of EBV-specific $CD8^+$ T cells in solid organ transplant recipients has been undertaken successfully (Sherritt et al. 2003). Adoptive immunotherapy used in combination with DLBCL chemotherapy regimens may result in improved response compared to chemotherapy alone for patients with EBV-postive DLBCL.

4.3 Biology of Targeted Therapies

The NFκB and JAK/STAT pathways are attractive therapeutic targets in EBV-positive DLBCL. If given in combination with traditional DLBCL regimens, targeted agents may mitigate the survival differences seen between EBV-positive and EBV-negative tumors. Therapeutic strategies that directly inhibit NFκB signaling have been fraught with difficulties. NFκB is critical to physiologic processes in many cells. Deficiency in genes IKKβ and p65 provokes massive hepatocyte apoptosis during development (Strnad and Burke 2007). Adult hepatocytes are less perturbed by reductions in these proteins, but still show high sensitivity to toxin and cytokine-related injury. It is still possible that these inhibitors could be useful in tumors that are highly reliant on NFκB activity and provided that the appropriate concentration of inhibitor can be achieved. Bortezomib is a proteasome inhibitor that is capable of inhibiting both canonical and noncanonical NFκB signaling (Staudt 2010). Bortezomib is widely used in the treatment of multiple myeloma as is well tolerated both alone and in combination with other agents. Bortezomib suppresses NFκB activation by degrading IαBα, an inhibitor of NFκB nuclear translocation. Bortezomib is cytotoxic to human EBV-infected lymphoblastoid cell lines (Zou et al. 2007). Drugs that target JAK/STAT signaling would also be of potential therapeutic benefit to EBV-positive DLBCL. At present, there is much interest in the development of potent, selective STAT3 inhibitors for the treatment of lymphoid malignancies.

5 Conclusions

In this chapter, the clinical, genetic, and pathologic characteristics of DLBCL were presented, followed by an explanation of the role of EBV in DLBCL tumorigenesis. Using EBV-positive DLBCL of the elderly as a model, we describe the key clinical and pathologic characteristics of EBV-positive DLBCL. We also discussed the recent insights into EBV-positive lymphoma biology and potential treatment strategies.

Studies of EBV-positive DLBCL of the elderly have provided key insight into the pathogenic role that EBV plays in DLBCL. These data demonstrate that EBV-positive DLBCL has a distinct tumor biology, which is related to the tenuous relationship that the EBV virus establishes with its B cell host. The pathogenic mechanisms noted in EBV-positive DLBCL of the elderly, including enhanced NFκB activity, are likely to play a role in all forms of EBV-positive DLBCL.

More work is needed to determine whether EBV-positive DLBCLs occurring in distinct contexts of immune dysfunction are biologically different tumors. The current WHO classification scheme includes subgroups for plasmablastic DLBCL, DLBCL associated with chronic inflammation, and EBV-positive DLBCL of the elderly, all of which are EBV-positive DLBCL tumors. It would be interesting and informative to compare gene expression profiles from these subtypes to see whether they are similar. If EBV is contributing to the tumor pathogenesis in each case, the other subtypes may also demonstrate marked NFκB and JAK/STAT activation. This work would confirm that EBV-positive DLBCL has a unique biology and provide new clues to treating this disease by methods that disrupt the life cycle of EBV viral infection or the signaling pathways induced in these tumors.

References

Addis B, Isaacson P (1986) Large cell lymphoma of the mediastinum: a B-cell tumour of probable thymic origin. Histopathology 10:379–390

Al-Humood S, Alqallaf A, Al-Shemmari S, Al-Faris L, Al-Ayadhy B (2014) Genetic and immunohistochemical characterization of Epstein-Barr virus-associated diffuse large B-cell lymphoma. Acta Haematol 131:1–10

Alizadeh AA et al (2000) Distinct types of diffuse large B-cell lymphoma identified by gene expression profiling. Nature 403:503–511

Ao Q, Wang Y, Xu S, Tian Y, Huang W (2014) A case of EBV positive diffuse large B-cell lymphoma of the adolescent. Int J Clin Exp Med 7:307–311

Balague Ponz O et al (2009) Commentary on the WHO classification of tumors of lymphoid tissues (2008): aggressive B-cell lymphomas. J Hematop 2:83–87

Bayerdörffer E et al (1995) Regression of primary gastric lymphoma of mucosa-associated lymphoid tissue type after cure of Helicobacter pylori infection. The Lancet 345:1591–1594

Bea S et al (2005) Diffuse large B-cell lymphoma subgroups have distinct genetic profiles that influence tumor biology and improve gene-expression-based survival prediction. Blood 106:3183–3190

Béguelin W et al (2013) EZH2 is required for germinal center formation and somatic EZH2 mutations promote lymphoid transformation. Cancer Cell 23:677–692

Beltran BE et al (2011a) EBV-positive diffuse large B-cell lymphoma in young immunocompetent individuals. Clin Lymphoma Myeloma Leuk 11:512–516

Beltran BE et al (2011b) EBV-positive diffuse large B-cell lymphoma of the elderly: a case series from Peru. Am J Hematol 86:663–667

Capello D et al (2010) Genome wide DNA-profiling of HIV-related B-cell lymphomas. Br J Haematol 148:245–255

Carbone A, Gloghini A (2005) AIDS-related lymphomas: from pathogenesis to pathology. Br J Haematol 130:662–670

Carel J, Myones B, Frazier B, Holers VM (1990) Structural requirements for C3d, g/Epstein-Barr virus receptor (CR2/CD21) ligand binding, internalization, and viral infection. J Biol Chem 265:12293–12299

Castillo J, Pantanowitz L, Dezube BJ (2008) HIV-associated plasmablastic lymphoma: Lessons learned from 112 published cases. Am J Hematol 83:804–809

Castillo JJ et al (2011) Epstein-barr virus-positive diffuse large B-cell lymphoma of the elderly: what we know so far. Oncologist 16:87–96
Cazals-Hatem D et al (1996) Primary mediastinal large B-cell lymphoma: A clinicopathologic study of 141 cases compared with 916 nonmediastinal large B-cell lymphomas, a GELA ("Groupe d'Etude des Lymphomes de l'Adulte") study. Am J Surg Pathol 20:877–888
Chang KL, Chen Y-Y, Shibata D, Weiss LM (1992) Description of an in situ hybridization methodology for detection of Epstein-Barr virus RNA in paraffin-embedded tissues, with a survey of normal and neoplastic tissues. Diagn Mol Pathol 1:246–255
Chang ST et al (2014) Epstein-Barr virus is rarely associated with diffuse large B cell lymphoma in Taiwan and carries a trend for a shorter median survival time. J Clin Pathol 67:326–332
Cho EY et al (2008) The spectrum of Epstein-Barr virus-associated lymphoproliferative disease in Korea: incidence of disease entities by age groups. J Korean Med Sci 23:185–192
Choi WW et al (2009) A new immunostain algorithm classifies diffuse large B-cell lymphoma into molecular subtypes with high accuracy. Clin Cancer Res Off J Am Assoc Cancer Res 15:5494–5502
Cohen M et al (2014) Epstein-Barr virus-positive diffuse large B-cell lymphoma association is not only restricted to elderly patients. Int J Cancer
Corti M et al (2010) AIDS related lymphomas: histopathological subtypes and association with Epstein Barr virus and human herpes virus type-8. Medicina 70:151–158
Dave SS (2010) Host factors for risk and survival in lymphoma. In: ASH Education Program Book, pp. 255–258
Davis RE, Brown KD, Siebenlist U, Staudt LM (2001) Constitutive nuclear factor κB activity is required for survival of activated B cell-like diffuse large B cell lymphoma cells. J Exp Med 194:1861–1874
Davis RE et al (2010) Chronic active B-cell-receptor signalling in diffuse large B-cell lymphoma. Nature 463:88–92
De Martel C et al (2012) Global burden of cancers attributable to infections in 2008: a review and synthetic analysis. Lancet Oncol 13:607–615
Ding BB et al (2008) Constitutively activated STAT3 promotes cell proliferation and survival in the activated B-cell subtype of diffuse large B-cell lymphomas. Blood 111:1515–1523
Dojcinov SD et al (2011) Age-related EBV-associated lymphoproliferative disorders in the Western population: a spectrum of reactive lymphoid hyperplasia and lymphoma. Blood 117:4726–4735
Eliopoulos AG et al (1997) Epstein-Barr virus-encoded LMP1 and CD40 mediate IL-6 production in epithelial cells via an NF-κB pathway involving TNF receptor-associated factors. Oncogene 14:2899–2916
Epstein MA, Achong BG, Barr YM (1964) Virus particles in cultured lymphoblasts from Burkitt's lymphoma. The Lancet 283:702–703
Feng W-H, Hong G, Delecluse H-J, Kenney SC (2004) Lytic induction therapy for Epstein-Barr virus-positive B-cell lymphomas. J Virol 78:1893–1902
Fisher SG, Fisher RI (2004) The epidemiology of non-Hodgkin's lymphoma. Oncogene 23:6524–6534
Franke S et al (2002) Comparative genomic hybridization pattern distinguishes T-cell/histiocyte-rich B-cell lymphoma from nodular lymphocyte predominance Hodgkin's lymphoma. Am J Pathol 161:1861–1867
Gattinoni L, Powell DJ, Rosenberg SA, Restifo NP (2006) Adoptive immunotherapy for cancer: building on success. Nat Rev Immunol 6:383–393
Ghosh SK, Perrine SP, Williams RM, Faller DV (2012) Histone deacetylase inhibitors are potent inducers of gene expression in latent EBV and sensitize lymphoma cells to nucleoside antiviral agents. Blood 119:1008–1017
Gibson SE, Hsi ED (2009) Epstein-Barr virus-positive B-cell lymphoma of the elderly at a United States tertiary medical center: an uncommon aggressive lymphoma with a nongerminal center B-cell phenotype. Hum Pathol 40:653–661

Giles FJ, O'Brien SM, Keating MJ (1998) Chronic lymphocytic leukemia in (Richter's) transformation. Semin Oncol 25:117–125
Gloghini A, Dolcetti R, Carbone A (2013) Lymphomas occurring specifically in HIV-infected patients: from pathogenesis to pathology. Semin Cancer Biol 23:457–467
Gulley ML (2001) Molecular diagnosis of Epstein-Barr virus-related diseases. J Mol Diagn 3:1–10
Hadrup SR et al (2006) Longitudinal studies of clonally expanded CD8 T cells reveal a repertoire shrinkage predicting mortality and an increased number of dysfunctional cytomegalovirus-specific T cells in the very elderly. J Immunol 176:2645–2653
Hans CP et al (2004) Confirmation of the molecular classification of diffuse large B-cell lymphoma by immunohistochemistry using a tissue microarray. Blood 103:275–282
Heslop HE (2005) Biology and treatment of Epstein-Barr virus-associated non-Hodgkin lymphomas. In: Hematology/The Education Program of the American Society of Hematology. American Society of Hematology. Education Program, pp 260–266
Hoeller S, Tzankov A, Pileri SA, Went P, Dirnhofer S (2010) Epstein-Barr virus–positive diffuse large B-cell lymphoma in elderly patients is rare in Western populations. Hum Pathol 41:352–357
Hofscheier A et al (2011) Geographic variation in the prevalence of Epstein-Barr virus-positive diffuse large B-cell lymphoma of the elderly: a comparative analysis of a Mexican and a German population. Mod Pathol 24:1046–1054
Hsi ED, Lorsbach RB, Fend F, Dogan A (2011) Plasmablastic lymphoma and related disorders. Am J Clin Pathol 136:183–194
Huen D, Henderson S, Croom-Carter D, Rowe M (1995) The Epstein-Barr virus latent membrane protein-1 (LMP1) mediates activation of NF-κB and cell surface phenotype via two effector regions in its carboxy-terminal cytoplasmic domain. Oncogene 10:549–560
Igarashi M, Kawaguchi Y, Hirai K, Mizuno F (2003) Physical interaction of Epstein-Barr virus (EBV) nuclear antigen leader protein (EBNA-LP) with human oestrogen-related receptor 1 (hERR1): hERR1 interacts with a conserved domain of EBNA-LP that is critical for EBV-induced B-cell immortalization. J General Virol 84:319–327
Jemal A et al (2011) Global cancer statistics. CA Cancer J Clin 61:69–90
Jones JF, Straus SE (1987) Chronic epstein-barr virus infection. Annu Rev Med 38:195–209
Juvonen E et al (2003) High incidence of PTLD after non-T-cell-depleted allogeneic haematopoietic stem cell transplantation as a consequence of intensive immunosuppressive treatment. Bone Marrow Transplant 32:97–102
Kato M et al (2009) Frequent inactivation of A20 in B-cell lymphomas. Nature 459:712–716
Kato H et al (2014) Gene expression profiling of Epstein-Barr virus-positive diffuse large B-cell lymphoma of the elderly reveals alterations of characteristic oncogenetic pathways. Cancer Sci 105:537–544
Kaye KM, Izumi KM, Kieff E (1993) Epstein-Barr virus latent membrane protein 1 is essential for B-lymphocyte growth transformation. Proc Natl Acad Sci 90:9150–9154
Kroemer G (1997) The proto-oncogene Bcl-2 and its role in regulating apoptosis. Nat Med 3:614–620
Laichalk LL, Thorley-Lawson DA (2005) Terminal differentiation into plasma cells initiates the replicative cycle of Epstein-Barr virus in vivo. J Virol 79:1296–1307
Lam LT et al (2008) Cooperative signaling through the signal transducer and activator of transcription 3 and nuclear factor-κB pathways in subtypes of diffuse large B-cell lymphoma. Blood 111:3701–3713
Lenz G et al (2008a) Stromal gene signatures in large-B-cell lymphomas. N Engl J Med 359:2313–2323
Lenz G et al (2008b) Molecular subtypes of diffuse large B-cell lymphoma arise by distinct genetic pathways. Proc Natl Acad Sci USA 105:13520–13525
Lenz G et al (2008c) Oncogenic CARD11 mutations in human diffuse large B cell lymphoma. Science 319:1676–1679
Lemon SM, Hutt LM, Shaw JE, Li, J-LH, Pagano JS (1977) Replication of EBV in epithelial cells during infectious mononucleosis

Lim MS et al (2002) T-cell/histiocyte-rich large B-cell lymphoma: a heterogeneous entity with derivation from germinal center B cells. Am J Surg Pathol 26:1458–1466
Loong F et al (2010) Diffuse large B-cell lymphoma associated with chronic inflammation as an incidental finding and new clinical scenarios. Mod Pathol 23:493–501
Lopes LF et al (2013) Epstein-Barr virus (EBV) microRNAs: involvement in cancer pathogenesis and immunopathology. Int Rev Immunol 32:271–281
Martelli M et al (2013) Diffuse large B-cell lymphoma. Critical Rev Oncol Hematol 87:146–171
Menon MP, Pittaluga S, Jaffe ES (2012) The histological and biological spectrum of diffuse large B-cell lymphoma in the World Health Organization classification. Cancer J (Sudbury, Mass.) 18:411–420
Miller RA (1996) The aging immune system: primer and prospectus. Science 273:70–74
Miranda RN, Khoury JD, Medeiros LJ (2013) Lymphoproliferative disorders associated with primary immune disorders. Atlas Lymph Node Pathol 383–397
Montes-Moreno S et al (2012) EBV-positive diffuse large B-cell lymphoma of the elderly is an aggressive post-germinal center B-cell neoplasm characterized by prominent nuclear factor-kB activation. Mod Pathol 25:968–982
Morin RD et al (2013) Mutational and structural analysis of diffuse large B-cell lymphoma using whole-genome sequencing. Blood 122:1256–1265
Muppidi JR et al (2014) Loss of signalling via Galpha13 in germinal centre B-cell-derived lymphoma. Nature
Murase T et al (2007) Intravascular large B-cell lymphoma (IVLBCL): a clinicopathologic study of 96 cases with special reference to the immunophenotypic heterogeneity of CD5. Blood 109:478–485
Nakatsuka S-I et al (2002) Pyothorax-associated lymphoma: a review of 106 cases. J Clin Oncol 20:4255–4260
Narimatsu H et al (2007) Clinicopathological features of pyothorax-associated lymphoma; a retrospective survey involving 98 patients. Ann Oncol 18:122–128
Nguyen-Van D et al (2011) Epstein-Barr virus-positive diffuse large B-cell lymphoma of the elderly expresses EBNA3A with conserved CD8 T-cell epitopes. Am J Blood Res 1:146–159
Niederman J, Evans A, Subrahmanyan L, McCollum R (1970) Prevalence, incidence and persistence of EB virus antibody in young adults. N Engl J Med 282:361–365
Nilsson K, Klein G, Henle W, Henle G (1971) The establishment of lymphoblastoid lines from adult and fetal human lymphoid tissue and its dependence on EBV. Int J cancer 8:443–450
Ok CY, Papathomas TG, Medeiros LJ, Young KH (2013) EBV-positive diffuse large B-cell lymphoma of the elderly. Blood 122:328–340
Ok CY et al (2014) Prevalence and clinical implications of epstein-barr virus infection in de novo diffuse large B-cell lymphoma in Western countries. Clin Cancer Res Off J Am Assoc Cancer Res 20:2338–2349
Orthwein A, Di Noia JM (2012) Activation induced deaminase: how much and where? Semin Immunol 24:246–254
Ott G, Rosenwald A, Campo E (2013) Understanding MYC-driven aggressive B-cell lymphomas: pathogenesis and classification. Blood 122:3884–3891
Oyama T et al (2003) Senile EBV+ B-cell lymphoproliferative disorders: a clinicopathologic study of 22 patients. Am J Surg Pathol 27:16–26
Oyama T et al (2007) Age-related EBV-associated B-cell lymphoproliferative disorders constitute a distinct clinicopathologic group: a study of 96 patients. Clin Cancer Res Off J Am Assoc Cancer Res 13:5124–5132
Park S et al (2007) The impact of Epstein-Barr virus status on clinical outcome in diffuse large B-cell lymphoma. Blood 110:972–978
Pasqualucci L et al (2011) Analysis of the coding genome of diffuse large B-cell lymphoma. Nat Genet 43:830–837
Piekarz RL et al (2011) Phase 2 trial of romidepsin in patients with peripheral T-cell lymphoma. Blood 117:5827–5834

Pittaluga S, Jaffe ES (2010) T-cell/histiocyte-rich large B-cell lymphoma. Haematologica 95:352–356
Ponzoni M et al (2007) Definition, diagnosis, and management of intravascular large B-cell lymphoma: proposals and perspectives from an international consensus meeting. J Clin Oncol 25:3168–3173
Rickinson AB (2014) Co-infections, inflammation and oncogenesis: future directions for EBV research. Semin Cancer Biol 26:99–115
Rickinson AB, Moss DJ (1997) Human cytotoxic T lymphocyte responses to Epstein-Barr virus infection. Annu Rev Immunol 15:405–431
Rosenwald A, Staudt LM (2003) Gene expression profiling of diffuse large B-cell lymphoma. Leuk Lymphoma 44(Suppl 3):S41–S47
Rosenwald A et al (2002) The use of molecular profiling to predict survival after chemotherapy for diffuse large-B-cell lymphoma. N Engl J Med 346:1937–1947
Roughan JE, Thorley-Lawson DA (2009) The intersection of Epstein-Barr virus with the germinal center. J Virol 83:3968–3976
Schuetz JM et al (2012) BCL2 mutations in diffuse large B-cell lymphoma. Leukemia 26:1383–1390
Schulze-Luehrmann J, Ghosh S (2006) Antigen-receptor signaling to nuclear factor κB. Immunity 25:701–715
Sherritt MA et al (2003) Reconstitution of the latent T-lymphocyte response to Epstein-Barr virus is coincident with long-term recovery from posttransplant lymphoma after adoptive immunotherapy. Transplantation 75:1556–1560
Shimoyama Y et al (2008) Age-related Epstein-Barr virus-associated B-cell lymphoproliferative disorders: special references to lymphomas surrounding this newly recognized clinicopathologic disease. Cancer Sci 99:1085–1091
Smedby KE et al (2006) Autoimmune and chronic inflammatory disorders and risk of non-Hodgkin lymphoma by subtype. J Natl Cancer Inst 98:51–60
Staudt LM (2010) Oncogenic activation of NF-κB. Cold Spring Harb Perspect Biol 2:a000109
Stein H, Lennert K, Feller AC (1984) Immunohistological analysis of human lymphoma: correlation of histological and immunological categories. Adv Cancer Res 42:67–147
Strnad J, Burke JR (2007) IκB kinase inhibitors for treating autoimmune and inflammatory disorders: potential and challenges. Trends Pharmacol Sci 28:142–148
Sun CC, Thorley-Lawson DA (2007) Plasma cell-specific transcription factor XBP-1 s binds to and transactivates the Epstein-Barr virus BZLF1 promoter. J Virol 81:13566–13577
Swerdlow SH, Campo E, Harris NL et al (2008) WHO classification of tumours of haematopoetic and lymphoid tissues, 4th edn. IARC Press, Lyon
Tan S, Parker P (2003) Emerging and diverse roles of protein kinase C in immune cell signalling. Biochem J 376:545–552
Taylor AL, Marcus R, Bradley JA (2005) Post-transplant lymphoproliferative disorders (PTLD) after solid organ transplantation. Crit Rev Oncol Hematol 56:155–167
Thorley-Lawson DA, Gross A (2004) Persistence of the Epstein-Barr virus and the origins of associated lymphomas. N Engl J Med 350:1328–1337
Tierney R, Steven N, Young L, Rickinson A (1994) Epstein-Barr virus latency in blood mononuclear cells: analysis of viral gene transcription during primary infection and in the carrier state. J Virol 68:7374–7385
Tsubata T, Wu J, Honjo T (1993) B-cell apoptosis induced by antigen receptor crosslinking is blocked by a T-cell signal through CD40
Uchida J et al (1999) Mimicry of CD40 signals by Epstein-Barr virus LMP1 in B lymphocyte responses. Science 286:300–303
Uner A et al (2011) The presence of Epstein-Barr virus (EBV) in diffuse large B-cell lymphomas (DLBCLs) in Turkey: special emphasis on 'EBV-positive DLBCL of the elderly'. APMIS : acta pathologica, microbiologica, et immunologica Scandinavica 119:309–316

Valera A et al (2010) IG/MYC rearrangements are the main cytogenetic alteration in plasmablastic lymphomas. Am J Surg Pathol 34:1686

Verma S et al (2005) Epstein-Barr virus- and human herpesvirus 8-associated primary cutaneous plasmablastic lymphoma in the setting of renal transplantation. J Cutan Pathol 32:35–39

Victora GD, Nussenzweig MC (2012) Germinal centers. Annu Rev Immunol 30:429–457

Wada N et al (2011) Epstein-barr virus in diffuse large B-Cell lymphoma in immunocompetent patients in Japan is as low as in Western Countries. J Med Virol 83:317–321

Yang Z-Z, Novak AJ, Stenson MJ, Witzig TE, Ansell SM (2006) Intratumoral $CD4^+$ $CD25^+$ regulatory T-cell-mediated suppression of infiltrating $CD4^+$ T cells in B-cell non-Hodgkin lymphoma. Blood 107:3639–3646

Younes A (2009) Novel treatment strategies for patients with relapsed classical Hodgkin lymphoma. ASH Education Program Book 2009:507–519

Young LS, Murray PG (2003) Epstein-Barr virus and oncogenesis: from latent genes to tumours. Oncogene 22:5108–5121

Young LS, Rickinson AB (2004) Epstein-Barr virus: 40 years on. Nat Rev Cancer 4:757–768

Zhang J et al (2013) Genetic heterogeneity of diffuse large B-cell lymphoma. Proc Natl Acad Sci USA 110:1398–1403

Ziepert M et al (2010) Standard International prognostic index remains a valid predictor of outcome for patients with aggressive $CD20^+$ B-cell lymphoma in the rituximab era. J Clini Oncol Off J Am Soc Clin Oncol 28:2373–2380

Zou P, Kawada J, Pesnicak L, Cohen JI (2007) Bortezomib induces apoptosis of Epstein-Barr virus (EBV)-transformed B cells and prolongs survival of mice inoculated with EBV-transformed B cells. J Virol 81:10029–10036

Nasopharyngeal Carcinoma: An Evolving Role for the Epstein–Barr Virus

Nancy Raab-Traub

Abstract The Epstein–Barr herpesvirus (EBV) is an important human pathogen that is closely linked to several major malignancies including the major epithelial tumor, undifferentiated nasopharyngeal carcinoma (NPC). This important tumor occurs with elevated incidence in specific areas, particularly in southern China but also in Mediterranean Africa and some regions of the Middle East. Regardless of tumor prevalence, undifferentiated NPC is consistently associated with EBV. The consistent detection of EBV in all cases of NPC, the maintenance of the viral genome in every cell, and the continued expression of viral gene products suggest that EBV is a necessary factor for the malignant growth in vivo. However, the molecular characterization of the infection and identification of critical events have been hampered by the difficulty in developing in vitro models of NPC. Epithelial cell infection is difficult in vitro and in contrast to B-cell infection does not result in immortalization and transformation. Cell lines established from NPC usually do not retain the genome, and the successful establishment of tumor xenografts is difficult. However, critical genetic changes that contribute to the onset and progression of NPC and key molecular properties of the viral genes expressed in NPC have been identified. In some cases, viral expression becomes increasingly restricted during tumor progression and tumor cells may express only the viral nuclear antigen EBNA1 and viral noncoding RNAs. As NPC develops in the immunocompetent, the continued progression of deregulated growth likely reflects the combination of expression of viral oncogenes in some cells and viral noncoding RNAs that likely function synergistically with changes in cellular RNA and miRNA expression.

N. Raab-Traub (✉)
Department of Microbiology, Lineberger Comprehensive Cancer Center, CB#7295, University of North Carolina, Chapel Hill, NC 27599-7295, USA
e-mail: nrt@med.unc.edu

C. Münz (ed.), *Epstein Barr Virus Volume 1*, Current Topics in Microbiology and Immunology 390, DOI 10.1007/978-3-319-22822-8_14

Contents

1 Introduction ... 341
2 Aspects of EBV and Cancer Biology Discovered in NPC ... 342
3 Classes of EBV-Associated Malignancies ... 345
4 Latent Membrane Protein 1 ... 347
5 Latent Membrane Protein 2 ... 350
6 EBNA1 ... 352
7 EBV BART Noncoding RNAS ... 353
8 Conclusions ... 354
8.1 EBV and Carcinoma: A Contest of Hide and Seek ... 354
References ... 355

Abbreviations

EBV	Epstein–Barr Virus
NPC	Nasopharyngeal carcinoma
mRNA	Messenger ribonucleic acid
BL	Burkitt's lymphoma
IM	Infectious mononucleosis
VCA	Viral capsid antigen
EAd	Early antigen diffuse
IgA	Immunoglobulin subtype A
AIDS	Acquired immunodeficiency syndrome
EBNA	EBV nuclear antigen
LP	Leader protein
LMP	Latent membrane protein
CTL	Cytotoxic lymphocyte
TR	Terminal repeat
CIS	Carcinoma in situ
LOH	Loss of heterozygosity
FHIT	Fragile histidine triad protein
RASSF1A	Ras association domain family 1 isoform A
Rb	Retinoblastoma gene
Cdk4	Cyclin-dependent kinase 4
BART	BamHI A rightward transcript
EBER	Epstein–Barr encoded RNA
TNFR	Tumor necrosis factor receptor
TRAF	TNFR-associated factors
NFkB	Nuclear factor kappa B locus
CTAR	Carboxy terminal activation region
IRF7	Interferon response factor 7
NIK	NFkB inducing kinase
EGFR	Epidermal growth factor receptor
PI3kinase	Phosphoinositol 3 kinase

Akt	Protein kinase B—v-akt murine thymoma viral oncogene homolog 1
ERK	Extracellular signal-regulated kinases
ICAM	Intercellular adhesion molecule 1
ID	Inhibitor of DNA binding

1 Introduction

The Epstein–Barr herpesvirus (EBV) is an important human pathogen that is closely linked to several major malignancies (Raab-Traub 2007). After the discovery of EBV in Burkitt's lymphoma (BL) and the determination that primary infection could cause infectious mononucleosis (IM), an association of EBV with nasopharyngeal carcinoma (NPC) was revealed in early studies of EBV serology (Henle et al. 1968, 1970). These studies showed that patients with NPC had exceptionally high titers to the EBV antigens, viral capsid antigen (VCA) , and early antigen diffuse (EAd) (Henle et al. 1970). The viral genome was subsequently detected in the malignant epithelial cells by in situ hybridization and renaturation kinetics (Wolf et al. 1975; Pagano et al. 1975). The finding of EBV in NPC revealed that EBV, in addition to B lymphocytes, also could infect epithelial cells. This finding was supported by later studies identifying EBV and viral RNAs in sloughed epithelial cells in throat washings (Sixbey et al. 1984; Lemon et al. 1977). An additional important aspect of these findings was that IgA titers were specifically indicative of NPC, suggesting that these antigens were produced within the tumor and inducing a mucosal antibody response (Henle and Henle 1976). Detailed subsequent studies using EBV serology have shown that IgA antibodies to multiple antigens associated with EBV replication are specific markers for the development, prognosis, and reoccurrence of NPC (Dardari et al. 2000). IgA titers to these antigens begin to rise 1–2 years prior to the development of NPC (Zeng et al. 1985; Levine et al. 1981). More recent studies also show that detection of viral DNA in the sera of NPC patients is also informative (Shao et al. 2004).

Lifelong infection with EBV begins with infection and replication in oral epithelial and lymphoid cells followed by persistence in memory B lymphocytes (Babcock et al. 1998). EBV-associated cancers develop in both epithelial and lymphoid cells and include NPC, gastric cancer, Hodgkin's lymphoma, BL, and lymphomas that develop in the immunocompromised, including both AIDS and post-transplant lymphomas (Raab-Traub 2007). EBV readily infects B lymphocytes in vivo and in vitro and usually persists in the infected cell as an extrachromosomal episome (Gussander and Adams 1984; Nonoyama and Pagano 1972). The infected cells that have acquired the ability to continue to proliferate in vitro are growth immortalized (Pope et al. 1973). The majority of cells do not produce virus and are considered latently infected. However, the cells express multiple viral proteins including the EBV nuclear antigens (EBNA) 1, 2, 3A, 3B, and 3C, and leader protein (LP) and the latent membrane proteins (LMP) 1, 2A, and 2B

(Kieff and Rickinson 2001). The EBNA3 proteins are major targets of cytotoxic lymphocytes (Murray et al. 1992). The ability to expand and proliferate EBV-specific cytotoxic T cells (CTL) has led to immunotherapy that is highly effective in the prevention and treatment of immunodeficiency-associated lymphomas (Rooney et al. 1998).

In the absence of immunodeficiency, EBV-associated malignancies often have distinct geographical distributions. BL was quickly shown to have elevated incidence in areas of high malarial infection, while NPC is highly prevalent in southern China, but also occurs with elevated incidence among Alaskan Inuits, and in Mediterranean Africa (Burkitt 1971; Chang and Adami 2006). In southeast Asia, NPC develops with a peak incidence of 20–50 cases per 100,000 and constitutes approximately 18 % of all cancers in that area. The World Health Organization distinguishes 3 subtypes of NPC based on their degree of differentiation (Shanmugaratnam 1978). EBV is found in all three subtypes but is consistently associated worldwide with the most common subtype (subtype 3), characterized as an undifferentiated carcinoma (Raab-Traub et al. 1987).

The consistent detection of EBV in all cases of NPC, the maintenance of the viral genome in every cell, and the continued expression of viral gene products suggest that EBV is a necessary factor for the malignant growth in vivo. Surprisingly, the viral genome is frequently lost in epithelial cell lines established from EBV (Dittmer et al. 2008; Lin et al. 1990). In BL cell lines, the loss of EBV results in cell death (Vereide and Sugden 2009; Kennedy et al. 2003). This chapter will summarize our understanding of the potential contribution of EBV to the development and growth of malignant epithelial cells.

2 Aspects of EBV and Cancer Biology Discovered in NPC

Studies of NPC have been highly informative in the characterization of the biologic and molecular properties of EBV and the contribution of genetic changes during cancer progression. Multiple studies have shown that despite major differences in NPC occurrence, all cases of undifferentiated NPC, including those that develop with low incidence in nonendemic areas, contain EBV (Pathmanathan et al. 1995a). Interestingly, multiple studies have identified EBV in more differentiated forms of NPC, especially in areas of high incidence (Pathmanathan et al. 1995a; Teng et al. 1996). Early studies using renaturation kinetics determined that the viral genome was essentially the same as that found in BL and during IM and that the viral genome was maintained as an extrachromosomal episome within the nucleus (Pagano et al. 1975; Nonoyama and Pagano 1972). However, unlike EBV infection of B cells, epithelial cells are extremely difficult to infect and the viral genome is frequently not maintained in stable cell lines established from the tumors (Dittmer et al. 2008; Cheung et al. 1999; Lin et al. 1990). Thus, much of the early molecular biology of NPC utilized biopsy material and rare examples of NPC that could grow as xenografts in nude mice (Raab-Traub et al. 1983; Busson et al. 1988).

Early restriction enzyme mapping and cloning of the viral genome facilitated these studies (Raab-Traub et al. 1987; Dambaugh et al. 1980). Identification of the terminal fragments of the viral genome revealed that virion DNA had heterogeneous terminal fragments with varying number of terminal repeats (TRs) (Dambaugh et al. 1980). Additionally, the fused terminal fragments of the episomes could be distinguished by hybridization to Southern blots with probes from either end of the linear genome (Raab-Traub and Flynn 1986). This led to the surprising finding that in NPC and BL, the episomal DNA was homogeneous with a single restriction fragment representing the heterogeneous terminal restriction fragments that are fused in the episome. The identification of clonal viral DNA suggested by extension that the cells were also clonal and that the tumor had arisen from a single EBV-infected cell (Raab-Traub and Flynn 1986). Previously, only the clonality of lymphomas could be determined based on immunoglobulin gene (Ig) rearrangements (Arnold et al. 1983). However, the clonality of the viral episome enabled the identification of clonality in carcinomas. Analyses of Ig rearrangement and EBV termini confirmed the relationship between viral and cellular clonality (Katz et al. 1989; Walling et al. 2004). Multiple subsequent studies have identified clonal EBV in most of malignancies associated with EBV including Hodgkin's lymphoma, salivary gland carcinoma, and gastric cancer, and clonal termini have become one of the criteria indicating a etiologic role for EBV in these cancers (Raab-Traub et al. 1991; Weiss et al. 1988). This first identification of the terminal fragments also identified faint ladder arrays of fragments, indicating that virus replication did occur in a subset of cells and was likely the source of the link between elevated antibody titers to replicative antigens and development or reoccurrence of NPC (Raab-Traub and Flynn 1986). This was confirmed in later studies that identified the Z immediate early protein and EA (Martel-Renoir et al. 1995; Luka et al. 1988).

In contrast to many malignancies, particularly cervical cancer, identification of early dysplastic or premalignant lesions was rare in NPC. In an initial survey, only 11 examples of atypical hyperplasia or carcinoma in situ (CIS) were identified in screening over 5000 biopsies. Clonal EBV was detected in the few examples of CIS with detection of the EBERs and latent membrane protein 1 (LMP1) in all cells (Pathmanathan et al. 1995b). In subsequent studies, examples of dysplasia and CIS were identified that contained EBV in a subset of the cells and intriguing findings revealed genetic changes, such as loss of p16 expression, that preceded cancer development (Chan et al. 2000, 2002). These findings have led to a model of carcinogen-induced genetic changes that initiate dysplastic growth. These environmental exposures also potentially activate EBV replication resulting in production of IgA antibodies to replicative antigens. A recent study using epithelial raft cultures has shown that EBV replication is activated as cells differentiate; however, evidence of latent infection was not detected within the basal cells in this in vitro model (Temple et al. 2014). In the genesis of NPC, the potential increase in viral replication induced by environmental factors likely results in increased epithelial infection. The dysplastic growth induced by potential genetic changes may enable the establishment of a latent epithelial cell infection. However, after infection with EBV and expression of EBV transforming genes, the cells rapidly progress to aggressive malignancy. Thus, it is a matter of discussion whether transforming EBV infection should be considered a late event in the development of NPC or perhaps the precipitating event leading to rapid progression to invasive malignancy.

The endemic pattern of incidence of NPC led to many thorough epidemiologic studies that identified potential contributing environmental and genetic contributions. Indeed, very early insightful studies by Dr. John Ho revealed a strong correlation between consumption of preserved food at an early age and subsequent development of NPC, 30–50 years later (Ho et al. 1978). One major factor thought to contribute to the high incidence of NPC among Hong Kong fisherman was the consumption and weaning of children to foods containing preserved, salted fish. Perhaps, this is the first environmental insult that begins the malignant process. Dr. Ho's lifelong campaign to stop this custom is now likely coming to fruition with dropping incidence of NPC in the affected populations (Li et al. 2014). Indeed, preserved foods containing carcinogens and compounds that lead to EBV reactivation and replication are also consumed in other areas with elevated incidence, including Alaskan and Greenland Inuits, and Mediterranean countries such as Tunisia and Algeria (Jeannel et al. 1990).

Dr. Dolly Huang continued these studies and initiated the first genetic studies of NPC to identify common genetic changes. Although very difficult to perform on solid tumors, cytogenetic studies identified consistent changes on multiple chromosomes including 1, 3p, 5p, 9p, 11q, and 12q (Huang et al. 1989). The development of comparative genomic hybridization and identification of loss of heterozygosity (LOH) further pinpointed important changes. LOH was identified with very high frequency at many of the same sites identified by cytogenetics, including 3p, 9p, and 14q in almost all samples (Wong et al. 2003). Many of these regions were subsequently shown to contain important tumor suppressor genes. One major factor is either homozygous loss of the p16 gene at 9p or p16 promoter methylation (Lo et al. 1996). These genetic and epigenetic changes result in the loss of p16 protein in the majority of primary NPC samples and in early EBV negative dysplasia. A recent study has shown that a region adjacent to p16 that encodes a cellular miRNA, miR31, was also deleted in NPC and early lesions (Cheung et al. 2014). These genetic changes may work in combination with viral proteins and miRNAs to alter cell growth regulation. Potential tumor suppressor genes that contribute to NPC development include FHIT and RASSF1A at the 3p locus and additional genes at multiple other loci (Chow et al. 2004; Ko et al. 2002). Interestingly, mutation of the canonical tumor suppressor genes, p53 and Rb, is rarely detected in Chinese NPC (Effert et al. 1992; Sun et al. 1993). This finding suggested that viral proteins may reduce the selection for mutation of these tumor suppressors. It was shown that LMP1 could specifically block p53-mediated apoptosis (Fries et al. 1996; Okan et al. 1995). Additionally, cyclin D1 overexpression has been detected in premalignant lesions, and its expression or expression of p16 resistant cdk4 facilitates EBV infection (Tsang et al. 2012). Identification of chromosomal amplification at specific regions also suggested that PI3kinase and p63 may contribute to NPC development (Chiang et al. 2009). It is intriguing that EBV genes, LMP1 and LMP2, expressed in NPC also affect these pathways (Dawson et al. 2003; Scholle et al. 2000; Fotheringham et al. 2010).

The use of cDNA cloning enabled the identification of viral transcription in tumor samples. Surprisingly, the pattern of EBV expression in NPC was quite distinct from that determined in transformed lymphocytes (Raab-Traub et al. 1983). The DNA sequences subsequently shown to encode the viral transforming proteins EBNA2, 3A, 3B, and 3C were not transcribed, and transcription was restricted to the sequences that encode EBNA1, LMP1, and LMP2, and abundant transcription from the BamHI A restriction fragment (BART RNA) was not transcribed in transformed lymphocytes (Kieff and Rickinson 2001).

This novel pattern of expression led to the identification of distinct states of viral latency in EBV-associated tumors which could also be recapitulated in some cell lines (Rowe et al. 1992). The most restricted form of latency, Type I, was characteristic of BL with expression limited to EBNA1 and the EBV-encoded RNAs (EBERs). Type II latency was considered characteristic of NPC and included expression of the latent membrane proteins, LMP1 and LMP2, and BART transcription. In Type III latency, which is characteristic of transformed B lymphocytes, all of the EBNAs, LMP1, LMP2, and the EBERs are expressed (Kieff and Rickinson 2001). This pattern of expression is also found in post-transplant lymphoma (Young et al. 1989). However, continued study has revealed that additional patterns of expression can occur and that a malignancy may have some cells in different patterns of latency (Niedobitek et al. 1995).

Further dissection of the mechanisms underlying these differences in expression led to the discovery of different promoters used for latent genes in different forms of latency. An important difference was the identification of a previously unidentified promoter, Qp, that initiated transcription of an RNA that only encoded EBNA1 (Sample et al. 1991). This is in contrast to the RNAs that are initiated from B-cell specific promoters, Cp and Wp, that encode EBNA2, 3A, 3B, 3C, and leader protein (Zetterberg et al. 1999; Kieff and Rickinson 2001). Additionally, Northern blotting and RNA mapping revealed that the LMP1 RNA initiated from a promoter within the TR (Sadler and Raab-Traub 1995a). It has subsequently been suggested that the number of TRs may affect transcription of both LMP1 and LMP2 (Repic et al. 2010; Moody et al. 2003). These differences in viral regulation reflect the clearly distinct interactions of EBV with epithelial or lymphoid cells.

3 Classes of EBV-Associated Malignancies

These first studies in NPC led to the understanding that the EBV-associated malignancies are distinguished by distinct patterns of expression. Interestingly, it was quickly determined that the lymphomas that develop after immune suppression post-organ or bone marrow transplantation express the same genes expressed in growth transformed lymphocytes including the EBNA2 and EBNA3 genes (Young et al. 1989). Characterization of the EBV immune response had revealed that the EBNA2 and 3 proteins were the major targets for recognition by cytotoxic lymphocytes (Murray et al. 1992). Thus, in the absence of recognition by cytotoxic lymphocytes, EBV-infected lymphocytes are able to switch to the transforming growth mode of expression resulting in lymphoma. In some cases, multiple clonal tumors are identified in different organs (Walling et al. 2004). This detailed understanding of the EBV immune response has led to the first clearly effective viral-specific immunotherapy or prophylaxis where EBV-specific CTLs are expanded ex vivo by exposure to EBV transformed lymphocytes (Heslop et al. 1999). These CTLs can be infused back into the patient and cure or prevent tumor development. This confirms the etiologic role for EBV in these lymphomas that develop during immunosuppression with the lymphomas dependent on EBV transforming genes.

In contrast to NPC, gastric cancer, Hodgkin's lymphoma, and many EBV-associated cancers that develop in the immunocompetent, the major CTL targets, the EBNAs 2 and 3 genes, are not expressed. Expression is restricted to EBNA1 which is essential for maintenance of the viral genome and also contributes to the regulation of expression of both viral and cellular genes (Yates et al. 1984). Additionally, the viral membrane proteins, LMP1 and LMP2, are expressed although the consistency and strength of their expression has clear variation, and the proteins are not necessarily detected in all cells or all samples. This may indicate that during cancer progression, genetic changes may occur that can substitute for the functions of the viral proteins. These proteins are also much less immunogenic than the EBNA proteins which might contribute to their successful expression in immunocompetent individuals (Hislop et al. 2007). Multiple studies have distinguished consistent markers of distinct EBV genomes with many studies that have analyzed sequence variation within LMP1 (Edwards et al. 1999; Miller et al. 1994; Lung et al. 1991). One study identified at least 6 consistent variants with a predominance of sequence changes within the epitopes predicted to be presented by their HLA types (Edwards et al. 2004). The successful enrichment of CTLs directed against these proteins has been achieved, and some efficacy has been shown in clinical trials (Haigh et al. 2008). Thus, the potentially reduced recognition of cells expressing a LMP1 variant within a person of a specific HLA type could be one contributing factor to the ability to evade immune recognition.

Additionally, both NPC and gastric cancer have very high levels of expression of the EBV BART RNAs (Gilligan et al. 1990b; Hitt et al. 1989). These RNAs have since been shown to be the primary RNA template for the production of viral miRNAs (Cai et al. 2006; Pfeffer et al. 2004). Interestingly, the BART RNAs are not expressed or expressed at very low levels in transformed lymphocytes (Cai et al. 2006; Gilligan et al. 1990b). Thus, the elevated expression of these RNAs in the epithelial tumors that lack many of the EBV transforming proteins suggests that the effects of BART miRNAs contribute to transformation in these restricted forms of infection. This leads to the intriguing possibility that in the presence of the highly tuned immune response to EBV in the immunocompetent individual, deregulated growth continues due to the evolving transforming powers of nonimmunogenic, noncoding RNAs.

4 Latent Membrane Protein 1

Latent membrane protein 1 has long been considered the EBV oncogene as it was the first viral protein shown to have transforming properties in vitro (Wang et al. 1985). Genetic studies have shown that LMP1 is also essential for EBV transformation of B lymphocytes (Kaye et al. 1999). Importantly, transgenic mice that express LMP1 in B lymphocytes develop lymphoma, while mice that express LMP1 using epithelial specific promoters develop hyperplasia and carcinoma (Kulwichit et al. 1998; Curran et al. 2001). The LMP1 mRNA was shown early on to be an abundant transcript expressed in NPC and is frequently transcribed from a distinct promoter from that utilized in lymphocytes (Sadler and Raab-Traub 1995a). The promoter is within the TR which results in a larger sized 3.5 kb mRNA. This larger mRNA is also detected in the permissive EBV infection, hairy leukoplakia, suggesting that its utilization is more frequent in epithelial cells (Gilligan et al. 1990a). Although the mRNA is abundant, the detection of the viral protein is less consistent. However, the aggressive growth and invasive properties of NPC, which is a highly metastatic malignancy, have been linked to both EBV LMP1 and 2.

LMP1 has the ability to alter many cellular growth properties and cellular regulatory networks. The first molecular clues to its mechanisms of action were the identification of interaction of LMP1, in yeast two hybrid assays, with the tumor necrosis factor receptor (TNFR) associated factors (TRAFs) (Mosialos et al. 1995). This finding led to the understanding that LMP1 functioned as a constitutively active member of the TNFR family. Two domains of LMP1 were identified that could activate NFκB and interacted with different TRAF proteins (Huen et al. 1995). The carboxy terminal activation region 1 (CTAR1) is thought to interact with TRAF1, 2, 3, and 5 with different avidities, while CTAR2 may interact with RIP, TRADD, and BS69 and also affects the binding of IRF7 to LMP1 (Fig. 1) (Mosialos et al. 1995; Izumi and Kieff 1997; Song et al. 2008; Miller et al. 1998). Different biological and molecular properties are linked to either CTAR1 or CTAR2. CTAR2 has the strongest activation of the canonical NFκB pathways, and inhibition of NFκB leads to irrevocable cell death (Cahir-McFarland et al. 2004). Additionally, CTAR2 is required for activation of the jun kinase pathway (JNK) and the p38 MAPK (Fig. 1) (Eliopoulos et al. 1999). Activation of CTAR2 is also required for the binding of IRF7 to an adjacent region of LMP1 (Song et al. 2008). Interestingly, EBV containing LMP1 deleted for CTAR2 can still transform B lymphocytes (Kaye et al. 1999). Additionally, LMP1-mediated rodent and human fibroblast transformation requires LMP1-CTAR1 but not CTAR2 (Mainou et al. 2005). Initial studies showed that LMP1-CTAR1 activated multiple forms of NFκB including canonical p50/p65 heterodimers, abundant homodimers of p50, and increased processing of p100 to p52 (Paine et al. 1995; Eliopoulos et al. 2003). The formation of p52/relB heteroimers is now termed "noncanonical NFκB activation" and is dependent on LMP1-CTAR1 and activation of the NFκB inducing kinase (NIK). LMP1-CTAR1 also uniquely activates the PI3kinase/Akt pathway and ERK

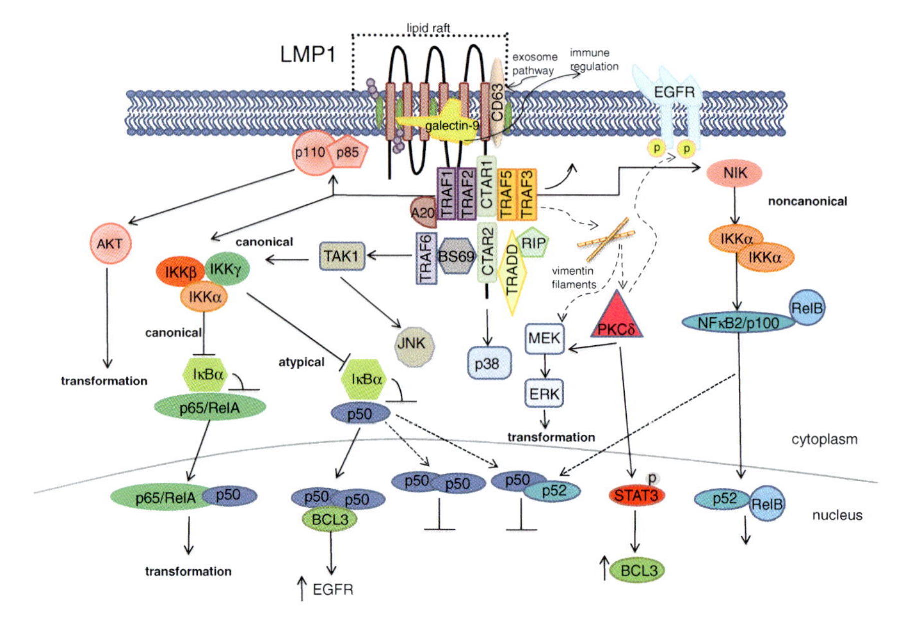

Fig. 1 Latent membrane protein 1: structure and molecular interactions. The proteins that have been identified to interact with LMP1 and the resultant effects of cellular signaling pathways including distinct forms of NFκB and activation of kinases are indicated

activation, both of which are required for rodent fibroblast transformation (Mainou et al. 2007; Dawson et al. 2003). Surprisingly, although activation of NFκB is critically required for continued B-lymphocyte proliferation, it is not required for the altered growth properties of rodent or human fibroblasts (Mainou et al. 2005).

One of the first unique properties identified for CTAR1 was the ability to induce expression of the epidermal growth factor receptor (EGFR) (Miller et al. 1997). This induction is thought to reflect the ability of LMP1-CTAR1 to induce a unique form of NFκB consisting of p50 homodimers in a complex with bcl3 (Thornburg et al. 2003). Additionally, LMP1 increases bcl3 expression through activation of STAT3 (Fig. 1) (Kung and Raab-Traub 2008). Activation of STAT3 and ERK are linked to PKCδ activation (Fig. 1) (Kung et al. 2011). Additional genes that are specifically induced by LMP1-CTAR1 include TRAF1, ICAM, and the ID proteins (Devergne et al. 1998; Everly et al. 2004). A recent study has shown that in NPC, the primary forms of NFκB consist of p50/p50/bcl3 and p50/relB heterodimers, both of which are NFκB forms uniquely induced by LMP1-CTAR1 (Chung et al. 2013). Intriguingly, this study also showed that most NPC had missense mutations in the TRAF3, TRAF2, NFκB p105, or A20, suggesting that NFκB is constitutively activated in NPC through effects of LMP1 or perhaps genetic mutation. It is possible that in the early stages of cancer such as carcinoma in situ, LMP1 expression is required for growth while as the tumor evolves, genetic changes occur that reduce the requirement for the viral transforming proteins.

The constitutive activity of LMP1, which is independent of any ligand binding, reflects the aggregation of LMP1 within membranes due to its six hydrophobic transmembrane domains. LMP1 has been shown to localize to lipid-rich membrane domains, termed lipid rafts (Ardila-Osorio et al. 2005). LMP1 is associated with TRAF3 in lipid rafts and also modifies raft contents. Subsequent studies have shown that LMP1 induces raft localization of the p85 subunit of PI3kinase and the cellular intermediate filament, vimentin, which was one of the first proteins shown to interact with LMP1 (Meckes et al. 2013b). The integrity of LMP1 raft structures and polymerization of vimentin filaments are required for rodent fibroblast transformation as disruption of rafts with cholesterol or inhibition of vimentin polymerization blocked activation of Akt, ERK, and growth transformation (Meckes et al. 2013b). It is likely that the association of cellular proteins with LMP1 is dynamic and perhaps distinct in different subcellular components. Most of the TRAF proteins that associate with LMP1 are ubiquitin ligases, and the A20 protein which also binds LMP1 is both a ubiquitin ligase and deubiquitinase (Fries et al. 1999). The interaction of LMP1 with the ubiquitin pathway could likely modulate the levels and location of various critical signaling molecules.

The LMP1 effects on lipid rafts are also reflected in exosomes produced by LMP1-expressing cells (Meckes et al. 2010). Exosomes are microvesicles thought to be specifically produced through budding into multivesicular bodies. Initial studies showed that LMP1 was excreted from infected B cells in exosomes and that exosomes produced by EBV-infected cells impaired T-cell function (Flanagan et al. 2003; Keryer-Bibens et al. 2006). Using mass spectrometry of purified lipid rafts, an interaction of LMP1 with galectin 9 was identified (Fig. 1) (Keryer-Bibens et al. 2006). It was shown that LMP1 induced galectin 9 expression resulting in the secretion of exosomes containing LMP1 and galectin 9. Subsequent studies showed that exosomes produced by LMP1 expressing cells could transfer LMP1 and activated signaling molecules into recipient cells leading to Akt and ERK activation (Meckes et al. 2010). Recent studies have shown that LMP1 induces exosomal transfer of HIF1α and activation of HIF1α-regulated transcription (Aga et al. 2014). Further characterization of the effects of LMP1 on exosome content analyzed purified exosomes from EBV and KSHV, and EBV/KSHV dually infected B cell lines using 2D difference gel electrophoresis and spectral counting (Meckes et al. 2013a). Multiple significant changes were identified with 360 of 870 identified proteins specific to the viral exosomes. Bioinformatic analysis suggested that the viral exosomes likely modulate cell death and survival, ribosomal function, and mTOR signaling. Viral distinct effects on exosomes suggested that the KSHV exosomes would alter cellular metabolism, while the EBV exosomes would activate cell signaling through effects on integrins, actin, and NFκB. LMP1 was identified as a significant principal component that affected exosome content (Meckes et al. 2013a). These findings suggest that virally induced changes in exosome content likely modulate the tumor environment. This would be critical for inhibiting immune function but also could enable a rare subset of cells expressing LMP1 to affect cell growth regulation in neighboring cells. In many cases of NPC, LMP1 expression is found in a small subset of cells; however exosomal transfer of LMP1 itself or its activated signaling molecules could contribute to the growth of neighboring cells. Additionally, it is likely that the subset of cells within NPC that become permissively infected also modulate the tumor environment and possibly contribute to infected cell growth. Several studies have revealed that replicative infection enhances tumor induction in humanized mice which may reflect effects on cytokine production or potential exosome transfer of important growth inducing factors (Ma et al. 2011; Jones et al. 2007).

5 Latent Membrane Protein 2

LMP2 is also considered an EBV transforming protein although it is not strictly required for B-lymphocyte transformation (Longnecker and Miller 1996; Longnecker et al. 1993). LMP2 is transcribed across the terminal repeats of the episome with two forms that contain the first coding exon which contains the critical signaling motifs (LMP2A) or do not contain this exon (LMP2B) (Laux et al. 1989). LMP2A is consistently detected in NPC at the RNA and protein level and is also expressed in many EBV-associated lymphomas with high levels of expression in HD (Busson et al. 1992; Morrison et al. 2004). Recent studies show synergy between LMP2A and activated c-myc and suggest that LMP2 may also contribute to BL (Bieging et al. 2010).

LMP2A contains 12 hydrophobic transmembrane domains with multiple tyrosines that likely form signaling motifs within the intracellular amino terminus of LMP2A (Longnecker and Kieff 1990). The protein contains a putative src kinase binding motif, YEEA, and an immunoreceptor tyrosine-based activation motif (ITAM) that binds the syk kinase (Longnecker et al. 1991). Additionally, LMP2A contains two PY domains that interact with WW-containing ubiquitin ligases such as Itch (Ikeda et al. 2003). In B cells, LMP2A is thought to induce survival by mimicking B-cell receptor signaling. In both B lymphocytes and epithelial cells, LMP2A activates the Akt pathway (Swart et al. 2000; Scholle et al. 2000). In epithelial cells, Akt activation leads to translocation of B-catenin to the nucleus (Morrison et al. 2003). Expression of LMP2A can induce growth transformation in some epithelial cell lines, perhaps in association with ras activation, resulting in anchorage independence and tumorigenicity (Scholle et al. 2000; Fukuda and Longnecker 2007). LMP2A also inhibits epithelial cell differentiation possibly reflecting the induction of ΔNp63 by LMP2A (Fotheringham et al. 2010; Scholle et al. 2000). Additionally, LMP2 induces cell migration, and this induction of migration is dependent on effects on integrin expression and localization and activation of src, syk, and focal adhesion kinase (Fotheringham et al. 2012; Lu et al. 2006).

In transgenic mice where LMP2 is expressed in B cells under the control of the Ig promoter, LMP2 can provide prosurvival signals enabling survival of B cells lacking a functional B-cell receptor (Swanson-Mungerson et al. 2006). However, transgenic mice expressing LMP2 in epithelial cells lacked a discernable phenotype

(Longan and Longnecker 2000). In classical skin painting experiments in mice that distinguish tumor initiation, promotion, and progression, LMP1 functioned as a weak tumor promoter and increased the formation of papillomas (Curran et al. 2001; Shair et al. 2012). In contrast, expression of LMP2 alone did not affect papilloma or carcinoma formation. Interestingly, the formation of squamous cell carcinoma was significantly increased in the LMP1 and LMP2 doubly transgenic animals with elevated levels of activated ERK and STAT3 in all tumors (Shair et al. 2012). This was the first identification of synergy between LMP1 and LMP2. These studies all indicate that the effects of LMP2 are manifest in specific situations.

An interesting system that has been used to characterize the properties of multiple viral and cellular oncogenes is based on the nontumorigenic mammary epithelial cell line, MCF10A (Debnath and Brugge 2005). When grown in Matrigel, the normal cells differentiate and form hollow, spherical acinar structures that maintain normal glandular features. The effects of multiple cellular oncogenes and requirements for specific signaling pathways involved in the sequential cell proliferation, apoptosis, and potential malignant progression have been identified in this system. Deregulation of the cell cycle through overexpression of cyclin D1 or inactivation of the retinoblastoma tumor suppressor gene by the human papilloma virus (HPV) E7 protein resulted in continued proliferation of apical cells resulting in enlarged acini; however, formation of the hollow lumen, a process dependent upon apoptosis, was not impaired. In contrast, activation of Akt repressed the formation of acini lumen, while expression of the ErbB2 oncogene inhibited cell death within the acinar lumen and also induced continued proliferation leading to the formation of irregularly shaped spheroids (Debnath et al. 2003). Expression of LMP2A was also evaluated in this system to determine the relationship of the effects of LMP2A to previously identified properties of other oncogenes (Fig. 2). Surprisingly, LMP2A had profound effects on MCF10A growth and differentiation and induced the formation of filled lobular structures, similar to those induced by the potent oncogene, ErbB2 (Fotheringham and Raab-Traub 2013). The block in differentiation and lumen formation and enhanced proliferation all required LMP2-mediated activation of src and Akt. The ITAM, src binding, and PY motifs were also required to block the anoikis-mediated lumen formation and the increased proliferation (Fig. 2). Surprisingly, expression of LMP2A with mutation of the PY motif alone resulted in the formation of large, hollow acini with continued proliferation of the apical cells, a phenotype reminiscent of the E7 oncogene that inhibits the retinoblastoma tumor suppressor gene. These findings reveal that the activation of multiple pathways by LMP2 can induce a highly transformed, proliferative phenotype with loss of growth control and inhibition of the induction of cell death during differentiation. Similarly to LMP1, the ability to bind ubiquitin ligases is an essential component in the LMP2A-induced uncontrolled proliferation and protection from cell death. As both LMP1 and LMP2 have profound effects on cellular signaling networks and can affect differentiation, migration, anchorage independence, and tumorigenicity, it is likely that the combined effects of LMP1 and LMP2 contribute to the pathogenesis of undifferentiated and highly metastatic NPC.

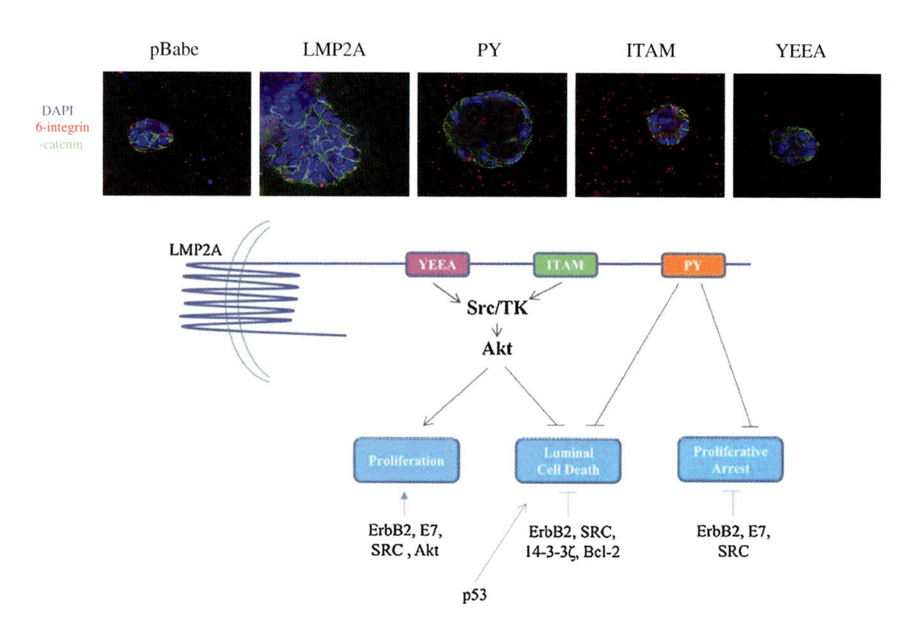

Fig. 2 Latent membrane protein 2: structure and activation of signaling pathways. *Top section* The effects of LMP2 and LMP2 signaling mutants on acini formation in MCF10a cells are shown. The vector control (pBabe), wild type (LMP2A), mutation in the ubiquitin ligase binding domains (PY), mutation of the ITAM domain, and mutation of the src family kinase binding domain (YEEA) are compared. Nuclei are identified by DAPI staining (*blue*), and cell membranes and junctions are shown by staining for α6-integrin (*red*) and β-catenin (*green*). *Bottom section* Schematic representation of the requirement for the YEEA, ITAM, and PY mutants and specific signaling pathways on distinct processes in acini formation. Effects of known oncogenes in this system are indicated. Adapted from Fotheringham and Raab-Traub (2013)

6 EBNA1

EBNA1 is clearly a critical viral protein which is essential for maintenance of the EBV genome and for the controlled segregation with the dividing cells (Yates et al. 1984; Sivachandran et al. 2011). EBNA was the first viral protein detected in NPC and is expressed within all cells (Huang et al. 1974). In addition to its essential role in maintaining the viral genome, EBNA1 also has signaling activity and transforming effects. EBNA interacts with USP7, a protein that prevents degradation of p53 (Saridakis et al. 2005). EBNA1 has also been shown to induce disruption of promyelocytic nuclear bodies (PML-NB) in NPC cells (Sivachandran et al. 2010). PML bodies contribute to DNA repair, and the loss induced by EBNA1 may contribute to increase genetic instability in NPC cells (Sivachandran et al. 2008). These same properties have been identified in EBV-infected AGS gastric carcinoma cells (Sivachandran et al. 2012). EBNA1 may also regulate expression of other EBV proteins through its enhancer function (Altmann et al. 2006). It is likely that the molecular properties of EBNA1 contribute to altered growth regulation perhaps synergistically with viral oncogenes, miRNAs, and cellular genetic changes.

7 EBV BART Noncoding RNAS

The BamHI A rightward transcripts (BARTs) were originally identified as the most abundant transcripts in studies of EBV transcription in NPC biopsy samples (Hitt et al. 1989; Gilligan et al. 1990b). These RNAs were shown to be a set of alternatively spliced transcripts (Sadler and Raab-Traub 1995b). The BART RNAs can be detected by PCR in all forms of EBV latency, although expression is considerably more abundant in Type II latency, and particularly in the infected epithelial cells in NPC and gastric carcinomas (Brooks et al. 1993; Gilligan et al. 1990b; Cai et al. 2006). The variable splicing of the RNA forms multiple open reading frames (Sadler and Raab-Traub 1995b). The potential protein products have intriguing properties; however, the proteins have not been identified in infected cells and the transcripts have been shown to remain in the nucleus (Kusano and Raab-Traub 2001; Thornburg et al. 2004; Smith et al. 2000). Importantly, the BART transcripts have been shown to be the template for the precursors for as many as 44 microRNAs (miRNAs) (Cai et al. 2006). MicroRNAs are ~22 nucleotide RNAs that are processed and function much like siRNAs in inhibiting protein translation through base pair interactions with target mRNAs. Interestingly, the EBV B95-8 laboratory strain that readily transforms primary B lymphocytes is deleted for most of the BART miRNAs which indicates that they are not required for B-cell transformation (Raab-Traub et al. 1980; Pfeffer et al. 2004). This is in agreement with the minimal BART expression in B-cell lines (Cai et al. 2006). In contrast, a cluster of three miRNAs that are produced from the primary EBNA2 transcript and are not expressed in NPC have been shown to contribute to B-cell transformation (Feederle et al. 2011). As the function of miRNAs is dependent upon their relative abundance to a given target, multiple studies have attempted to determine patterns of EBV miRNA expression that distinguish the diseases associated with EBV and even the type of B-cell infection (Qiu et al. 2011; Chen et al. 2010; Amoroso et al. 2011). The relative abundance of the BART miRNAs determined by different methods varies greatly across studies (Marquitz and Raab-Traub 2012). This variation appears to be due to more than just differences in the samples used, as the relative abundance of miRNAs present in the C666-1 cell line is very different depending on the method of PCR detection or direct sequencing.

Considering the limited viral protein expression in NPC and gastric carcinoma and the abundance of the BART miRNAs, it is likely that these miRNAs contribute to the development of EBV-associated carcinomas. Several genes involved in apoptosis are potential targets of various BART miRNAs, including PUMA, Bim, and TOMM22 (Marquitz et al. 2011; Choy et al. 2008; Dolken et al. 2010). These targets have been identified in different cells in different studies but point to a common targeting of mitochondrial proteins. It is possible that the targets differ between cell lines and varies from one clonal growth to another.

The AGS gastric cell line is a rare epithelial cell line that can be consistently infected with EBV. It has been shown that EBV infection of this cell line results in altered growth properties (Marquitz et al. 2012; Kassis et al. 2002). Although the EBV negative AGS cell line normally grows very poorly in soft agar, cells latently by EBV become anchorage independent, a hallmark of transformation. The expression pattern of these cells is basically Type I latency, with very limited protein expression but high levels of the BART miRNAs. Analysis of changes in cellular expression using microarray analysis revealed dramatic changes after infection with EBV. The majority of changes reflected decreased expression of genes after EBV infection. Additionally, the downregulated genes were significantly enriched for potential BART miRNA targets, suggesting that the BART miRNAs are major contributors to the dramatic expression changes (Marquitz et al. 2012). Ingenuity Pathway Analysis identified genes involved with migration, cellular movement, invasion, growth, and proliferation as highly and significantly enriched in the affected gene set. The genes affected by the BARTs may be distinct in different infections and also be different in the presence or absence of viral protein expression. The effect of EBV infection on cellular miRNA expression was also determined using RNA-Seq. This study revealed that EBV had significant effects on cellular miRNA expression. In EBV-infected AGS cells, 15 % of all miRNAs were BART miRNAs. In the C15 NPC tumor, 57 % were BART miRNAs, and in the C666 cell line established from an NPC xenograft, the BART miRNAs represented 40 % of all miRNAs. These high levels of BART miRNAs detected in the NPC tumors may indicate that tumorigenicity selects for high levels of BART miRNA expression (Marquitz et al. 2014). Interestingly, multiple tumor suppressor miRNAs were consistently downregulated. Additionally, other studies have identified tumor suppressor genes such as WIF1 and APC as miRNA targets (Wong et al. 2012). A recent study provided evidence for selection for high levels of BART miRNA expression in EBV-positive cells that form tumors in immunodeficient mice (Qiu et al. 2015).

However, many likely targets of the BART miRNAs and their biologic effects have yet to be determined. As our analyses of tumors become more accurately detailed, unique features of tumors are increasingly identified. It is known that miRNAs have many potential targets; thus, the actual effects on growth during infection are likely a combination of the abundance of viral proteins, the cellular target mRNAs, the viral miRNA(s), and cellular miRNAs. Indeed, a recent study indicated that the cellular miR31 which is close to the p16 locus is frequently deleted or not expressed in NPC and in premalignant epithelial tissue (Cheung et al. 2014). It is possible that in different cell lines and individual tumors, different targets may be responsible for malignant growth. However, it is likely that the same cellular pathways are affected either directly by viral proteins or indirectly through viral noncoding RNAs.

8 Conclusions

8.1 EBV and Carcinoma: A Contest of Hide and Seek

The many years of study of EBV and NPC have revealed many unique properties of EBV and its proteins. The epidemiology and the studies of viral infection reveal a required contribution for genetic changes likely induced by carcinogenic insult. These may modulate the ability of EBV to infect these cells and also influence the type of infection from replication to latent and transforming. In the early stages of tumor development, the viral proteins in combination with the genetic changes likely are essential for deregulated cell growth. The viral proteins expressed in this restricted form of latency are poorly immunogenic. Additionally, variants of these proteins are found within NPC that have sequence changes that would affect recognition by cytotoxic lymphocytes.

As the tumor continues to evolve, the requirement for viral expression may be reduced such that viral oncogenes may be expressed in a subset of cells and transferred through exosomes to induce growth. Alternatively, the occurrence of additional genetic changes may supplant the need for viral protein expression through effects on the same critical cellular signaling pathways.

Additionally, it appears that the viral noncoding RNAs in combination with the EBNA1 may be sufficient to induce deregulated growth by targeting distinct cell functions through a combination of changes in cellular expression and cellular and viral miRNAs. However, the continued retention of the viral genome in vivo despite this ongoing evolution points to a continued requirement for EBV despite its evolving role in contributing to oncogenic growth. The ability to transform cells without expression of immunogenic viral proteins would be a major and significant mechanism through which oncogenic viruses induce cancer. Importantly, the identification of specific cellular functions and pathways that may be affected through multiple mechanisms could lead to specific therapies that target the activated pathways.

Acknowledgments These studies represent the work of many graduate students and postdoctoral fellows from my laboratory who are indicated in the references. I would also like to thank Rachel Edwards and Anuja Mathur for assistance with figure preparation and Dr. Aron Marquitz for helpful comments. These studies have been supported by the NCI through grants RO1 CA32979, RO1138811, and PO1 19014.

References

Aga M, Bentz GL, Raffa S, Torrisi MR, Kondo S, Wakisaka N, Yoshizaki T, Pagano JS, Shackelford J (2014) Exosomal HIF1alpha supports invasive potential of nasopharyngeal carcinoma-associated LMP1-positive exosomes. Oncogene 33:4613–4622

Altmann M, Pich D, Ruiss R, Wang J, Sugden B, Hammerschmidt W (2006) Transcriptional activation by EBV nuclear antigen 1 is essential for the expression of EBV's transforming genes. Proc Natl Acad Sci USA 103:14188–14193

Amoroso R, Fitzsimmons L, Thomas WA, Kelly GL, Rowe M, Bell AI (2011) Quantitative studies of Epstein-Barr virus-encoded microRNAs provide novel insights into their regulation. J Virol 85:996–1010

Ardila-Osorio H, Pioche-Durieu C, Puvion-Dutilleul F, Clausse B, Wiels J, Miller W, Raab-Traub N, Busson P (2005) TRAF interactions with raft-like buoyant complexes, better than TRAF rates of degradation, differentiate signaling by CD40 and EBV latent membrane protein 1. Int J Cancer 113:267–275

Arnold A, Cossman J, Bakhshi A, Jaffe ES, Waldmann TA, Korsmeyer SJ (1983) Immunoglobulin-gene rearrangements as unique clonal markers in human lymphoid neoplasms. N Engl J Med 309:1593–1599
Babcock GJ, Decker LL, Volk M, Thorley-Lawson DA (1998) EBV persistence in memory B cells in vivo. Immunity 9:395–404
Bieging KT, Swanson-Mungerson M, Amick AC, Longnecker R (2010) Epstein-Barr virus in Burkitt's lymphoma: a role for latent membrane protein 2A. Cell Cycle 9:901–908
Brooks LA, Lear AL, Young LS, Rickinson AB (1993) Transcripts from the Epstein-Barr virus BamHI A fragment are detectable in all three forms of virus latency. J Virol 67:3182–3190
Burkitt DP (1971) Epidemiology of Burkitt's lymphoma. Proc R Soc Med 64:909–910
Busson P, Ganem G, Flores P, Mugneret F, Clausse B, Caillou B, Braham K, Wakasugi H, Lipinski M, Tursz T (1988) Establishment and characterization of three transplantable EBV-containing nasopharyngeal carcinomas. Int J Cancer 42:599–606
Busson P, McCoy R, Sadler R, Gilligan K, Tursz T, Raab-Traub N (1992) Consistent transcription of the Epstein-Barr virus LMP2 gene in nasopharyngeal carcinoma. J Virol 66:3257–3262
Cahir-Mcfarland ED, Carter K, Rosenwald A, Giltnane JM, Henrickson SE, Staudt LM, Kieff E (2004) Role of NF-kappa B in cell survival and transcription of latent membrane protein 1-expressing or Epstein-Barr virus latency III-infected cells. J Virol 78:4108–4119
Cai X, Schafer A, Lu S, Bilello JP, Desrosiers RC, Edwards R, Raab-Traub N, Cullen BR (2006) Epstein-Barr virus microRNAs are evolutionarily conserved and differentially expressed. PLoS Pathog 2:e23
Chan AS, To KF, Lo KW, Mak KF, Pak W, Chiu B, Tse GM, Ding M, Li X, Lee JC, Huang DP (2000) High frequency of chromosome 3p deletion in histologically normal nasopharyngeal epithelia from southern Chinese. Cancer Res 60:5365–5370
Chan AS, To KF, Lo KW, Ding M, Li X, Johnson P, Huang DP (2002) Frequent chromosome 9p losses in histologically normal nasopharyngeal epithelia from southern Chinese. Int J Cancer 102:300–303
Chang ET, Adami HO (2006) The enigmatic epidemiology of nasopharyngeal carcinoma. Cancer Epidemiol Biomarkers Prev 15:1765–1777
Chen SJ, Chen GH, Chen YH, Liu CY, Chang KP, Chang YS, Chen HC (2010) Characterization of Epstein-Barr virus miRNAome in nasopharyngeal carcinoma by deep sequencing. PLoS One 5(9):e12745
Cheung ST, Huang DP, Hui AB, Lo KW, Ko CW, Tsang YS, Wong N, Whitney BM, Lee JC (1999) Nasopharyngeal carcinoma cell line (C666-1) consistently harbouring Epstein-Barr virus. Int J Cancer 83:121–126
Cheung CC, Chung GT, Lun SW, To KF, Choy KW, Lau KM, Siu SP, Guan XY, Ngan RK, Yip TT, Busson P, Tsao SW, Lo KW (2014) miR-31 is consistently inactivated in EBV-associated nasopharyngeal carcinoma and contributes to its tumorigenesis. Mol Cancer 13:184
Chiang CT, Chu WK, Chow SE, Chen JK (2009) Overexpression of delta Np63 in a human nasopharyngeal carcinoma cell line downregulates CKIs and enhances cell proliferation. J Cell Physiol 219:117–122
Chow LS, Lo KW, Kwong J, To KF, Tsang KS, Lam CW, Dammann R, Huang DP (2004) RASSF1A is a target tumor suppressor from 3p21.3 in nasopharyngeal carcinoma. Int J Cancer 109:839–847
Choy EY, Siu KL, Kok KH, Lung RW, Tsang CM, To KF, Kwong DL, Tsao SW, Jin DY (2008) An Epstein-Barr virus-encoded microRNA targets PUMA to promote host cell survival. J Exp Med 205:2551–2560
Chung GT, Lou WP, Chow C, To KF, Choy KW, Leung AW, Tong CY, Yuen JW, Ko CW, Yip TT, Busson P, Lo KW (2013) Constitutive activation of distinct NF-kappaB signals in EBV-associated nasopharyngeal carcinoma. J Pathol 231:311–322
Curran JA, Laverty FS, Campbell D, Macdiarmid J, Wilson JB (2001) Epstein-Barr virus encoded latent membrane protein-1 induces epithelial cell proliferation and sensitizes transgenic mice to chemical carcinogenesis. Cancer Res 61:6730–6738

Dambaugh T, Beisel C, Hummel M, King W, Fennewald S, Cheung A, Heller M, Raab-Traub N, Kieff E (1980) Epstein-Barr virus (B95-8) DNA VII: molecular cloning and detailed mapping. Proc Natl Acad Sci USA 77:2999–3003

Dardari R, Khyatti M, Benider A, Jouhadi H, Kahlain A, Cochet C, Mansouri A, el Gueddari B, Benslimane A, Joab I (2000) Antibodies to the Epstein-Barr virus transactivator protein (ZEBRA) as a valuable biomarker in young patients with nasopharyngeal carcinoma. Int J Cancer 86:71–75

Dawson CW, Tramountanis G, Eliopoulos AG, Young LS (2003) Epstein-Barr virus latent membrane protein 1 (LMP1) activates the phosphatidylinositol 3-kinase/Akt pathway to promote cell survival and induce actin filament remodeling. J Biol Chem 278:3694–3704

Debnath J, Brugge JS (2005) Modelling glandular epithelial cancers in three-dimensional cultures. Nat Rev Cancer 5:675–688

Debnath J, Walker SJ, Brugge JS (2003) Akt activation disrupts mammary acinar architecture and enhances proliferation in an mTOR-dependent manner. J Cell Biol 163:315–326

Devergne O, Cahir Mcfarland ED, Mosialos G, Izumi KM, Ware CF, Kieff E (1998) Role of the TRAF binding site and NF-kappaB activation in Epstein-Barr virus latent membrane protein 1-induced cell gene expression. J Virol 72:7900–7908

Dittmer DP, Hilscher CJ, Gulley ML, Yang EV, Chen M, Glaser R (2008) Multiple pathways for Epstein-Barr virus episome loss from nasopharyngeal carcinoma. Int J Cancer 123:2105–2112

Dolken L, Malterer G, Erhard F, Kothe S, Friedel CC, Suffert G, Marcinowski L, Motsch N, Barth S, Beitzinger M, Lieber D, Bailer SM, Hoffmann R, Ruzsics Z, Kremmer E, Pfeffer S, Zimmer R, Koszinowski UH, Grasser F, Meister G, Haas J (2010) Systematic analysis of viral and cellular microRNA targets in cells latently infected with human gamma-herpesviruses by RISC immunoprecipitation assay. Cell Host Microbe 7:324–334

Edwards RH, Seillier-Moiseiwitsch F, Raab-Traub N (1999) Signature amino acid changes in latent membrane protein 1 distinguish Epstein-Barr virus strains. Virology 261:79–95

Edwards RH, Sitki-Green D, Moore DT, Raab-Traub N (2004) Potential selection of LMP1 variants in nasopharyngeal carcinoma. J Virol 78:868–881

Effert P, McCoy R, Abdel-Hamid M, Flynn K, Zhang Q, Busson P, Tursz T, Liu E, Raab-Traub N (1992) Alterations of the p53 gene in nasopharyngeal carcinoma. J Virol 66:3768–3775

Eliopoulos AG, Blake SM, Floettmann JE, Rowe M, Young LS (1999) Epstein-Barr virus-encoded latent membrane protein 1 activates the JNK pathway through its extreme C terminus via a mechanism involving TRADD and TRAF2. J Virol 73:1023–1035

Eliopoulos AG, Caamano JH, Flavell J, Reynolds GM, Murray PG, Poyet JL, Young LS (2003) Epstein-Barr virus-encoded latent infection membrane protein 1 regulates the processing of p100 NF-kappaB2 to p52 via an IKKgamma/NEMO-independent signalling pathway. Oncogene 22:7557–7569

Everly DN Jr, Mainou BA, Raab-Traub N (2004) Induction of Id1 and Id3 by latent membrane protein 1 of Epstein-Barr virus and regulation of p27/Kip and cyclin-dependent kinase 2 in rodent fibroblast transformation. J Virol 78:13470–13478

Feederle R, Linnstaedt SD, Bannert H, Lips H, Bencun M, Cullen BR, Delecluse HJ (2011) A viral microRNA cluster strongly potentiates the transforming properties of a human herpesvirus. PLoS Pathog 7:e1001294

Flanagan J, Middeldorp J, Sculley T (2003) Localization of the Epstein-Barr virus protein LMP 1 to exosomes. J Gen Virol 84:1871–1879

Fotheringham JA, Raab-Traub N (2013) Epstein-Barr virus latent membrane protein 2 effects on epithelial acinus development reveal distinct requirements for the PY and YEEA motifs. J Virol 87:13803–13815

Fotheringham JA, Mazzucca S, Raab-Traub N (2010) Epstein-Barr virus latent membrane protein-2A-induced DeltaNp63alpha expression is associated with impaired epithelial-cell differentiation. Oncogene 29:4287–4296

Fotheringham JA, Coalson NE, Raab-Traub N (2012) Epstein-Barr virus latent membrane protein-2A induces ITAM/Syk- and Akt-dependent epithelial migration through alphav-integrin membrane translocation. J Virol 86:10308–10320

Fries KL, Miller WE, Raab-Traub N (1996) Epstein-Barr virus latent membrane protein 1 blocks p53-mediated apoptosis through the induction of the A20 gene. J Virol 70:8653–8659
Fries KL, Miller WE, Raab-Traub N (1999) The A20 protein interacts with the Epstein-Barr virus latent membrane protein 1 (LMP1) and alters the LMP1/TRAF1/TRADD complex. Virology 264:159–166
Fukuda M, Longnecker R (2007) Epstein-Barr virus latent membrane protein 2A mediates transformation through constitutive activation of the Ras/PI3-K/Akt Pathway. J Virol 81:9299–9306
Gilligan K, Rajadurai P, Resnick L, Raab-Traub N (1990a) Epstein-Barr virus small nuclear RNAs are not expressed in permissively infected cells in AIDS-associated leukoplakia. Proc Natl Acad Sci USA 87:8790–8794
Gilligan K, Sato H, Rajadurai P, Busson P, Young L, Rickinson A, Tursz T, Raab-Traub N (1990b) Novel transcription from the Epstein-Barr virus terminal EcoRI fragment, DIJhet, in a nasopharyngeal carcinoma. J Virol 64:4948–4956
Gussander E, Adams A (1984) Electron microscopic evidence for replication of circular Epstein-Barr virus genomes in latently infected Raji cells. J Virol 52:549–556
Haigh TA, Lin X, Jia H, Hui EP, Chan AT, Rickinson AB, Taylor GS (2008) EBV latent membrane proteins (LMPs) 1 and 2 as immunotherapeutic targets: LMP-specific CD4+ cytotoxic T cell recognition of EBV-transformed B cell lines. J Immunol 180:1643–1654
Henle G, Henle W (1976) Epstein-Barr virus-specific IgA antibodies as an outstanding feature of nasopharyngeal carcinoma. Int J Cancer 17:1–7
Henle G, Henle W, Diehl V (1968) Relation of Burkitt's tumor-associated herpes-type virus to infectious mononucleosis. Proc Natl Acad Sci USA 59:94–101
Henle W, Henle G, Ho HC, Burtin P, Cachin Y, Clifford P, de Schryver A, De-The G, Diehl V, Klein G (1970) Antibodies to Epstein-Barr virus in nasopharyngeal carcinoma, other head and neck neoplasms, and control groups. J Natl Cancer Inst 44:225–231
Heslop HE, Perez M, Benaim E, Rochester R, Brenner MK, Rooney CM (1999) Transfer of EBV-specific CTL to prevent EBV lymphoma post bone marrow transplant. J Clin Apher 14:154–156
Hislop AD, Taylor GS, Sauce D, Rickinson AB (2007) Cellular responses to viral infection in humans: lessons from Epstein-Barr virus. Annu Rev Immunol 25:587–617
Hitt MM, Allday MJ, Hara T, Karran L, Jones MD, Busson P, Tursz T, Ernberg I, Griffin BE (1989) EBV gene expression in an NPC-related tumour. EMBO J 8:2639–2651
Ho JH, Huang DP, Fong YY (1978) Salted fish and nasopharyngeal carcinoma in southern Chinese. Lancet 2:626
Huang DP, Ho JH, Henle W, Henle G (1974) Demonstration of Epstein-Barr virus-associated nuclear antigen in nasopharyngeal carcinoma cells from fresh biopsies. Int J Cancer 14:580–588
Huang DP, Ho JH, Chan WK, Lau WH, Lui M (1989) Cytogenetics of undifferentiated nasopharyngeal carcinoma xenografts from southern Chinese. Int J Cancer 43:936–939
Huen DS, Henderson SA, Croom-Carter D, Rowe M (1995) The Epstein-Barr virus latent membrane protein-1 (LMP1) mediates activation of NF-kappa B and cell surface phenotype via two effector regions in its carboxy-terminal cytoplasmic domain. Oncogene 10:549–560
Ikeda A, Caldwell RG, Longnecker R, Ikeda M (2003) Itchy, a Nedd4 ubiquitin ligase, downregulates latent membrane protein 2A activity in B-cell signaling. J Virol 77:5529–5534
Izumi KM, Kieff ED (1997) The Epstein-Barr virus oncogene product latent membrane protein 1 engages the tumor necrosis factor receptor-associated death domain protein to mediate B lymphocyte growth transformation and activate NF-kappaB. Proc Natl Acad Sci USA 94:12592–12597
Jeannel D, Hubert A, de Vathaire F, Ellouz R, Camoun M, Ben Salem M, Sancho-Garnier H, De-The G (1990) Diet, living conditions and nasopharyngeal carcinoma in Tunisia–a case-control study. Int J Cancer 46:421–425
Jones RJ, Seaman WT, Feng WH, Barlow E, Dickerson S, Delecluse HJ, Kenney SC (2007) Roles of lytic viral infection and IL-6 in early versus late passage lymphoblastoid cell lines and EBV-associated lymphoproliferative disease. Int J Cancer 121:1274–1281

Kassis J, Maeda A, Teramoto N, Takada K, Wu C, Klein G, Wells A (2002) EBV-expressing AGS gastric carcinoma cell sublines present increased motility and invasiveness. Int J Cancer 99:644–651

Katz BZ, Raab-Traub N, Miller G (1989) Latent and replicating forms of Epstein-Barr virus DNA in lymphomas and lymphoproliferative diseases. J Infect Dis 160:589–598

Kaye KM, Izumi KM, Li H, Johannsen E, Davidson D, Longnecker R, Kieff E (1999) An Epstein-Barr virus that expresses only the first 231 LMP1 amino acids efficiently initiates primary B-lymphocyte growth transformation. J Virol 73:10525–10530

Kennedy G, Komano J, Sugden B (2003) Epstein-Barr virus provides a survival factor to Burkitt's lymphomas. Proc Natl Acad Sci USA 100:14269–14274

Keryer-Bibens C, Pioche-Durieu C, Villemant C, Souquere S, Nishi N, Hirashima M, Middeldorp J, Busson P (2006) Exosomes released by EBV-infected nasopharyngeal carcinoma cells convey the viral latent membrane protein 1 and the immunomodulatory protein galectin 9. BMC Cancer 6:283

Kieff E, Rickinson AB (2001) Epstein-Barr virus and its replication. In: Knipe DM, Howley PM (eds) Field's virology, 4th edn. Lippincott Williams & Wilkins Publishers, Philadelphia

Ko JY, Lee TC, Hsiao CF, Lin GL, Yen SH, Chen KY, Hsiung CA, Chen PJ, Hsu MM, Jou YS (2002) Definition of three minimal deleted regions by comprehensive allelotyping and mutational screening of FHIT, p16(INK4A), and p19(ARF) genes in nasopharyngeal carcinoma. Cancer 94:1987–1996

Kulwichit W, Edwards RH, Davenport EM, Baskar JF, Godfrey V, Raab-Traub N (1998) Expression of the Epstein-Barr virus latent membrane protein 1 induces B cell lymphoma in transgenic mice. Proc Natl Acad Sci USA 95:11963–11968

Kung CP, Raab-Traub N (2008) Epstein-Barr virus latent membrane protein 1 induces expression of the epidermal growth factor receptor through effects on Bcl-3 and STAT3. J Virol 82:5486–5493

Kung CP, Meckes DG Jr, Raab-Traub N (2011) Epstein-Barr virus LMP1 activates EGFR, STAT3, and ERK through effects on PKCdelta. J Virol 85:4399–4408

Kusano S, Raab-Traub N (2001) An Epstein-Barr virus protein interacts with Notch. J Virol 75:384–395

Laux G, Economou A, Farrell PJ (1989) The terminal protein gene 2 of Epstein-Barr virus is transcribed from a bidirectional latent promoter region. J Gen Virol 70(Pt 11):3079–3084

Lemon SM, Hutt LM, Shaw JE, Li JL, Pagano JS (1977) Replication of EBV in epithelial cells during infectious mononucleosis. Nature 268:268–270

Levine PH, Pearson GR, Armstrong M, Bengali Z, Berenberg J, Easton J, Goepfert H, Henle G, Henle W, Heffner D, Huang A, Hyams VJ, Lanier A, Neel HB, Pilch B, Pointek N, Taylor W, Terebelo H, Weiland L (1981) The reliability of IgA antibody to Epstein-Barr virus (EBV) capsid antigen as a test for the diagnosis of nasopharyngeal carcinoma (NPC). Cancer Detect Prev 4:307–312

Li K, Lin GZ, Shen JC, Zhou Q (2014) Time trends of nasopharyngeal carcinoma in urban Guangzhou over a 12-year period (2000–2011): declines in both incidence and mortality. Asian Pac J Cancer Prev 15:9899–9903

Lin CT, Wong CI, Chan WY, Tzung KW, Ho JK, Hsu MM, Chuang SM (1990) Establishment and characterization of two nasopharyngeal carcinoma cell lines. Lab Invest 62:713–724

Lo KW, Cheung ST, Leung SF, van Hasselt A, Tsang YS, Mak KF, Chung YF, Woo JK, Lee JC, Huang DP (1996) Hypermethylation of the p16 gene in nasopharyngeal carcinoma. Cancer Res 56:2721–2725

Longan L, Longnecker R (2000) Epstein-Barr virus latent membrane protein 2A has no growth-altering effects when expressed in differentiating epithelia. J Gen Virol 81:2245–2252

Longnecker R, Kieff E (1990) A second Epstein-Barr virus membrane protein (LMP2) is expressed in latent infection and colocalizes with LMP1. J Virol 64:2319–2326

Longnecker R, Miller CL (1996) Regulation of Epstein-Barr virus latency by latent membrane protein 2. Trends Microbiol 4:38–42

Longnecker R, Druker B, Roberts TM, Kieff E (1991) An Epstein-Barr virus protein associated with cell growth transformation interacts with a tyrosine kinase. J Virol 65:3681–3692

Longnecker R, Miller CL, Miao XQ, Tomkinson B, Kieff E (1993) The last seven transmembrane and carboxy-terminal cytoplasmic domains of Epstein-Barr virus latent membrane protein 2 (LMP2) are dispensable for lymphocyte infection and growth transformation in vitro. J Virol 67:2006–2013

Lu J, Lin WH, Chen SY, Longnecker R, Tsai SC, Chen CL, Tsai CH (2006) Syk tyrosine kinase mediates Epstein-Barr virus latent membrane protein 2A-induced cell migration in epithelial cells. J Biol Chem 281:8806–8814

Luka J, Deeb ZE, Hartmann DP, Jenson B, Pearson GR (1988) Detection of antigens associated with Epstein-Barr virus replication in extracts from biopsy specimens of nasopharyngeal carcinomas. J Natl Cancer Inst 80:1164–1167

Lung ML, Lam WP, Sham J, Choy D, Yong-Sheng Z, Guo HY, Ng MH (1991) Detection and prevalence of the "f" variant of Epstein-Barr virus in southern China. Virology 185:67–71

Ma SD, Hegde S, Young KH, Sullivan R, Rajesh D, Zhou Y, Jankowska-Gan E, Burlingham WJ, Sun X, Gulley ML, Tang W, Gumperz JE, Kenney SC (2011) A new model of Epstein-Barr virus infection reveals an important role for early lytic viral protein expression in the development of lymphomas. J Virol 85:165–177

Mainou BA, Everly DN Jr, Raab-Traub N (2005) Epstein-Barr virus latent membrane protein 1 CTAR1 mediates rodent and human fibroblast transformation through activation of PI3K. Oncogene 24:6917–6924

Mainou BA, Everly DN Jr, Raab-Traub N (2007) Unique signaling properties of CTAR1 in LMP1-mediated transformation. J Virol 81:9680–9692

Marquitz AR, Raab-Traub N (2012) The role of miRNAs and EBV BARTs in NPC. Semin Cancer Biol 22:166–172

Marquitz AR, Mathur A, Nam CS, Raab-Traub N (2011) The Epstein-Barr Virus BART microRNAs target the pro-apoptotic protein Bim. Virology 412:392–400

Marquitz AR, Mathur A, Shair KH, Raab-Traub N (2012) Infection of Epstein-Barr virus in a gastric carcinoma cell line induces anchorage independence and global changes in gene expression. Proc Natl Acad Sci USA 109:9593–9598

Marquitz AR, Mathur A, Chugh PE, Dittmer DP, Raab-Traub N (2014) Expression profile of microRNAs in Epstein-Barr virus-infected AGS gastric carcinoma cells. J Virol 88:1389–1393

Martel-Renoir D, Grunewald V, Touitou R, Schwaab G, Joab I (1995) Qualitative analysis of the expression of Epstein-Barr virus lytic genes in nasopharyngeal carcinoma biopsies. J Gen Virol 76(Pt 6):1401–1408

Meckes DG Jr, Shair KH, Marquitz AR, Kung CP, Edwards RH, Raab-Traub N (2010) Human tumor virus utilizes exosomes for intercellular communication. Proc Natl Acad Sci USA 107:20370–20375

Meckes DG Jr, Gunawardena HP, Dekroon RM, Heaton PR, Edwards RH, Ozgur S, Griffith JD, Damania B, Raab-Traub N (2013a) Modulation of B-cell exosome proteins by gamma herpesvirus infection. Proc Natl Acad Sci USA 110:E2925–E2933

Meckes DG Jr, Menaker NF, Raab-Traub N (2013b) Epstein-Barr virus LMP1 modulates lipid raft microdomains and the vimentin cytoskeleton for signal transduction and transformation. J Virol 87:1301–1311

Miller WE, Edwards RH, Walling DM, Raab-Traub N (1994) Sequence variation in the Epstein-Barr virus latent membrane protein 1. J Gen Virol 75(Pt 10):2729–2740

Miller WE, Mosialos G, Kieff E, Raab-Traub N (1997) Epstein-Barr virus LMP1 induction of the epidermal growth factor receptor is mediated through a TRAF signaling pathway distinct from NF-kappaB activation. J Virol 71:586–594

Miller WE, Cheshire JL, Raab-Traub N (1998) Interaction of tumor necrosis factor receptor-associated factor signaling proteins with the latent membrane protein 1 PXQXT motif is essential for induction of epidermal growth factor receptor expression. Mol Cell Biol 18:2835–2844

Moody CA, Scott RS, Su T, Sixbey JW (2003) Length of Epstein-Barr virus termini as a determinant of epithelial cell clonal emergence. J Virol 77:8555–8561
Morrison JA, Klingelhutz AJ, Raab-Traub N (2003) Epstein-Barr virus latent membrane protein 2A activates beta-catenin signaling in epithelial cells. J Virol 77:12276–12284
Morrison JA, Gulley ML, Pathmanathan R, Raab-Traub N (2004) Differential signaling pathways are activated in the Epstein-Barr virus-associated malignancies nasopharyngeal carcinoma and Hodgkin lymphoma. Cancer Res 64:5251–5260
Mosialos G, Birkenbach M, Yalamanchili R, Vanarsdale T, Ware C, Kieff E (1995) The Epstein-Barr virus transforming protein LMP1 engages signaling proteins for the tumor necrosis factor receptor family. Cell 80:389–399
Murray RJ, Kurilla MG, Brooks JM, Thomas WA, Rowe M, Kieff E, Rickinson AB (1992) Identification of target antigens for the human cytotoxic T cell response to Epstein-Barr virus (EBV): implications for the immune control of EBV-positive malignancies. J Exp Med 176:157–168
Niedobitek G, Agathanggelou A, Rowe M, Jones EL, Jones DB, Turyaguma P, Oryema J, Wright DH, Young LS (1995) Heterogeneous expression of Epstein-Barr virus latent proteins in endemic Burkitt's lymphoma. Blood 86:659–665
Nonoyama M, Pagano JS (1972) Separation of Epstein-Barr virus DNA from large chromosomal DNA in non-virus-producing cells. Nat New Biol 238:169–171
Okan I, Wang Y, Chen F, Hu LF, Imreh S, Klein G, Wiman KG (1995) The EBV-encoded LMP1 protein inhibits p53-triggered apoptosis but not growth arrest. Oncogene 11:1027–1031
Pagano JS, Huang CH, Klein G, De-The G, Shanmugaratnam K, Yang CS (1975) Homology of Epstein-Barr virus DNA in nasopharyngeal carcinomas from Kenya, Taiwan, Singapore and Tunisia. IARC Sci Publ 179–190
Paine E, Scheinman RI, Baldwin AS Jr, Raab-Traub N (1995) Expression of LMP1 in epithelial cells leads to the activation of a select subset of NF-kappa B/Rel family proteins. J Virol 69:4572–4576
Pathmanathan R, Prasad U, Chandrika G, Sadler R, Flynn K, Raab-Traub N (1995a) Undifferentiated, nonkeratinizing, and squamous cell carcinoma of the nasopharynx. Variants of Epstein-Barr virus-infected neoplasia. Am J Pathol 146:1355–1367
Pathmanathan R, Prasad U, Sadler R, Flynn K, Raab-Traub N (1995b) Clonal proliferations of cells infected with Epstein-Barr virus in preinvasive lesions related to nasopharyngeal carcinoma. N Engl J Med 333:693–698
Pfeffer S, Zavolan M, Grasser FA, Chien M, Russo JJ, Ju J, John B, Enright AJ, Marks D, Sander C, Tuschl T (2004) Identification of virus-encoded microRNAs. Science 304:734–736
Pope JH, Scott W, Moss DJ (1973) Human lymphoid cell transformation by Epstein-Barr virus. Nat New Biol 246:140–141
Qiu J, Cosmopoulos K, Pegtel M, Hopmans E, Murray P, Middeldorp J, Shapiro M, Thorley-Lawson DA (2011) A novel persistence associated EBV miRNA expression profile is disrupted in neoplasia. PLoS Pathog 7:e1002193
Qiu J, Smith P, Leahy L, Thorley-Lawson DA (2015) The Epstein-Barr Virus encoded BART miRNAs potentiate tumor growth in vivo. PLoS Pathog 11:e1004561
Raab-Traub N (2007) EBV-induced oncogenesis. In: Arvin A, Campadelli-Fiume G, Mocarski E, Moore PS, Roizman B, Whitley R, Yamanishi K (eds) Human herpesviruses: biology, therapy, and immunoprophylaxis. Cambridge University Press, Cambridge
Raab-Traub N, Flynn K (1986) The structure of the termini of the Epstein-Barr virus as a marker of clonal cellular proliferation. Cell 47:883–889
Raab-Traub N, Dambaugh T, Kieff E (1980) DNA of Epstein-Barr virus VIII: B95-8, the previous prototype, is an unusual deletion derivative. Cell 22:257–267
Raab-Traub N, Hood R, Yang CS, Henry B 2nd, PAGANO JS (1983) Epstein-Barr virus transcription in nasopharyngeal carcinoma. J Virol 48:580–590
Raab-Traub N, Flynn K, Pearson G, Huang A, Levine P, Lanier A, Pagano J (1987) The differentiated form of nasopharyngeal carcinoma contains Epstein-Barr virus DNA. Int J Cancer 39:25–29

Raab-Traub N, Rajadurai P, Flynn K, Lanier AP (1991) Epstein-Barr virus infection in carcinoma of the salivary gland. J Virol 65:7032–7036
Repic AM, Shi M, Scott RS, Sixbey JW (2010) Augmented latent membrane protein 1 expression from Epstein-Barr virus episomes with minimal terminal repeats. J Virol 84:2236–2244
Rooney CM, Heslop HE, Brenner MK (1998) EBV specific CTL: a model for immune therapy. Vox Sang 74(Suppl 2):497–498
Rowe M, Lear AL, Croom-Carter D, Davies AH, Rickinson AB (1992) Three pathways of Epstein-Barr virus gene activation from EBNA1-positive latency in B lymphocytes. J Virol 66:122–131
Sadler RH, Raab-Traub N (1995a) The Epstein-Barr virus 3.5-kilobase latent membrane protein 1 mRNA initiates from a TATA-Less promoter within the first terminal repeat. J Virol 69:4577–4581
Sadler RH, Raab-Traub N (1995b) Structural analyses of the Epstein-Barr virus BamHI A transcripts. J Virol 69:1132–1141
Sample J, Brooks L, Sample C, Young L, Rowe M, Gregory C, Rickinson A, Kieff E (1991) Restricted Epstein-Barr virus protein expression in Burkitt lymphoma is due to a different Epstein-Barr nuclear antigen 1 transcriptional initiation site. Proc Natl Acad Sci USA 88:6343–6347
Saridakis V, Sheng Y, Sarkari F, Holowaty MN, Shire K, Nguyen T, Zhang RG, Liao J, Lee W, Edwards AM, Arrowsmith CH, Frappier L (2005) Structure of the p53 binding domain of HAUSP/USP7 bound to Epstein-Barr nuclear antigen 1 implications for EBV-mediated immortalization. Mol Cell 18:25–36
Scholle F, Bendt KM, Raab-Traub N (2000) Epstein-Barr virus LMP2A transforms epithelial cells, inhibits cell differentiation, and activates Akt. J Virol 74:10681–10689
Shair KH, Bendt KM, Edwards RH, Nielsen JN, Moore DT, Raab-Traub N (2012) Epstein-Barr virus-encoded latent membrane protein 1 (LMP1) and LMP2A function cooperatively to promote carcinoma development in a mouse carcinogenesis model. J Virol 86:5352–5365
Shanmugaratnam K (1978) Histological typing of nasopharyngeal carcinoma. IARC Sci Publ 3–12
Shao JY, Li YH, Gao HY, Wu QL, Cui NJ, Zhang L, Cheng G, Hu LF, Ernberg I, Zeng YX (2004) Comparison of plasma Epstein-Barr virus (EBV) DNA levels and serum EBV immunoglobulin A/virus capsid antigen antibody titers in patients with nasopharyngeal carcinoma. Cancer 100:1162–1170
Sivachandran N, Sarkari F, Frappier L (2008) Epstein-Barr nuclear antigen 1 contributes to nasopharyngeal carcinoma through disruption of PML nuclear bodies. PLoS Pathog 4:e1000170
Sivachandran N, Cao JY, Frappier L (2010) Epstein-Barr virus nuclear antigen 1 Hijacks the host kinase CK2 to disrupt PML nuclear bodies. J Virol 84:11113–11123
Sivachandran N, Thawe NN, Frappier L (2011) Epstein-Barr virus nuclear antigen 1 replication and segregation functions in nasopharyngeal carcinoma cell lines. J Virol 85:10425–10430
Sivachandran N, Dawson CW, Young LS, Liu FF, Middeldorp J, Frappier L (2012) Contributions of the Epstein-Barr virus EBNA1 protein to gastric carcinoma. J Virol 86:60–68
Sixbey JW, Nedrud JG, Raab-Traub N, Hanes RA, Pagano JS (1984) Epstein-Barr virus replication in oropharyngeal epithelial cells. N Engl J Med 310:1225–1230
Smith PR, de Jesus O, Turner D, Hollyoake M, Karstegl CE, Griffin BE, Karran L, Wang Y, Hayward SD, Farrell PJ (2000) Structure and coding content of CST (BART) family RNAs of Epstein-Barr virus. J Virol 74:3082–3092
Song YJ, Izumi KM, Shinners NP, Gewurz BE, Kieff E (2008) IRF7 activation by Epstein-Barr virus latent membrane protein 1 requires localization at activation sites and TRAF6, but not TRAF2 or TRAF3. Proc Natl Acad Sci USA 105:18448–18453
Sun Y, Hegamyer G, Colburn NH (1993) Nasopharyngeal carcinoma shows no detectable retinoblastoma susceptibility gene alterations. Oncogene 8:791–795
Swanson-Mungerson M, Bultema R, Longnecker R (2006) Epstein-Barr virus LMP2A enhances B-cell responses in vivo and in vitro. J Virol 80:6764–6770

Swart R, Ruf IK, Sample J, Longnecker R (2000) Latent membrane protein 2A-mediated effects on the phosphatidylinositol 3-Kinase/Akt pathway. J Virol 74:10838–10845

Temple RM, Zhu J, Budgeon L, Christensen ND, Meyers C, Sample CE (2014) Efficient replication of Epstein-Barr virus in stratified epithelium in vitro. Proc Natl Acad Sci USA 111:16544–16549

Teng ZP, Ooka T, Huang DP, Zeng Y (1996) Detection of Epstein-Barr virus DNA in well and poorly differentiated nasopharyngeal carcinoma cell lines. Virus Genes 13:53–60

Thornburg NJ, Pathmanathan R, Raab-Traub N (2003) Activation of nuclear factor-kappaB p50 homodimer/Bcl-3 complexes in nasopharyngeal carcinoma. Cancer Res 63:8293–8301

Thornburg NJ, Kusano S, Raab-Traub N (2004) Identification of Epstein-Barr virus RK-BARF0-interacting proteins and characterization of expression pattern. J Virol 78:12848–12856

Tsang CM, Yip YL, Lo KW, Deng W, To KF, Hau PM, Lau VM, Takada K, Lui VW, Lung ML, Chen H, Zeng M, Middeldorp JM, Cheung AL, Tsao SW (2012) Cyclin D1 overexpression supports stable EBV infection in nasopharyngeal epithelial cells. Proc Natl Acad Sci USA 109:E3473–E3482

Vereide D, Sugden B (2009) Proof for EBV's sustaining role in Burkitt's lymphomas. Semin Cancer Biol 19:389–393

Walling DM, Andritsos LA, Etienne W, Payne DA, Aronson JF, Flaitz CM, Nichols CM (2004) Molecular markers of clonality and identity in Epstein-Barr virus-associated B-cell lymphoproliferative disease. J Med Virol 74:94–101

Wang D, Liebowitz D, Kieff E (1985) An EBV membrane protein expressed in immortalized lymphocytes transforms established rodent cells. Cell 43:831–840

Weiss LM, Warnke RA, Sklar J (1988) Clonal antigen receptor gene rearrangements and Epstein-Barr viral DNA in tissues of Hodgkin's disease. Hematol Oncol 6:233–238

Wolf H, Werner J, Zur Hausen H (1975) EBV DNA in nonlymphoid cells of nasopharyngeal carcinomas and in a malignant lymphoma obtained after inoculation of EBV into cottontop marmosets. Cold Spring Harb Symp Quant Biol 39(Pt 2):791–796

Wong N, Hui AB, Fan B, Lo KW, Pang E, Leung SF, Huang DP, Johnson PJ (2003) Molecular cytogenetic characterization of nasopharyngeal carcinoma cell lines and xenografts by comparative genomic hybridization and spectral karyotyping. Cancer Genet Cytogenet 140:124–132

Wong AM, Kong KL, Tsang JW, Kwong DL, Guan XY (2012) Profiling of Epstein-Barr virus-encoded microRNAs in nasopharyngeal carcinoma reveals potential biomarkers and oncomirs. Cancer 118:698–710

Yates J, Warren N, Reisman D, Sugden B (1984) A cis-acting element from the Epstein-Barr viral genome that permits stable replication of recombinant plasmids in latently infected cells. Proc Natl Acad Sci USA 81:3806–3810

Young L, Alfieri C, Hennessy K, Evans H, O'Hara C, Anderson KC, Ritz J, Shapiro RS, Rickinson A, Kieff E et al (1989) Expression of Epstein-Barr virus transformation-associated genes in tissues of patients with EBV lymphoproliferative disease. N Engl J Med 321:1080–1085

Zeng Y, Zhang LG, Wu YC, Huang YS, Huang NQ, Li JY, Wang YB, Jiang MK, Fang Z, Meng NN (1985) Prospective studies on nasopharyngeal carcinoma in Epstein-Barr virus IgA/VCA antibody-positive persons in Wuzhou City, China. Int J Cancer 36:545–547

Zetterberg H, Stenglein M, Jansson A, Ricksten A, Rymo L (1999) Relative levels of EBNA1 gene transcripts from the C/W, F and Q promoters in Epstein-Barr virus-transformed lymphoid cells in latent and lytic stages of infection. J Gen Virol 80(Pt 2):457–466

EBV and Autoimmunity

Alberto Ascherio and Kassandra L. Munger

Abstract Although a role of EBV in autoimmunity is biologically plausible and evidence of altered immune responses to EBV is abundant in several autoimmune diseases, inference on causality requires the determination that disease risk is higher in individuals infected with EBV than in those uninfected and that in the latter it increases following EBV infection. This determination has so far been possible only for multiple sclerosis (MS) and, to some extent, for systemic lupus erythematosus (SLE), whereas evidence is either lacking or not supportive for other autoimmune conditions. In this chapter, we present the main epidemiological findings that justify the conclusion that EBV is a component cause of MS and SLE and possible mechanisms underlying these effects.

Contents

1 Introduction 366
2 Multiple Sclerosis 368
  2.1 Definition and Epidemiology 368
  2.2 Epidemiological Evidence that EBV Is a Cause of MS 369
  2.3 Potential Mechanisms Relating EBV to MS 372
  2.4 Implications for MS Prevention and Treatment 374
  2.5 Summary on EBV and MS 375

A. Ascherio (✉)
Channing Division of Network Medicine, Department of Medicine,
Brigham and Women's Hospital, Boston, MA, USA
e-mail: aascheri@hsph.harvard.edu

A. Ascherio
Harvard Medical School, Boston, MA, USA

A. Ascherio · K.L. Munger
Department of Epidemiology and Nutrition, Harvard T.H. Chan School of Public Health,
Boston, MA, USA

C. Münz (ed.), *Epstein Barr Virus Volume 1*, Current Topics in Microbiology
and Immunology 390, DOI 10.1007/978-3-319-22822-8_15

3 Systemic Lupus Erythematosus ... 376
3.1 Definition and Epidemiology ... 376
3.2 Epidemiological Evidence that EBV Is a Cause of SLE ... 376
3.3 Potential Mechanisms Relating EBV to SLE ... 378
3.4 Implications for SLE Prevention and Treatment ... 379
References ... 379

Abbreviations

CMV	Cytomegalovirus
CNS	Central nervous system
CSF	Cerebrospinal fluid
DoDSR	Department of defense serum repository
EBERs	Epstein–Barr virus-encoded small RNAs
EBNA	Epstein–Barr virus nuclear antigen
EBV	Epstein–Barr virus
HLA	Human leukocyte antigen
IM	Infectious mononucleosis
MS	Multiple sclerosis
OR	Odds ratio
SLE	Systemic lupus erythematosus
VCA	Epstein–Barr virus viral capsid antigen

1 Introduction

Autoimmune diseases encompass a broad range of conditions that collectively affect about 5 % of the population (Davidson and Diamond 2001). Recognition of self-antigens is normally required to maintain the broad repertoire of adaptive immunity (Goldrath and Bevan 1999), but disease occurs when the strength and persistence of autoimmune responses disrupt normal physiological functions or cause tissue damage. Autoimmune disease can be induced experimentally by injecting self-antigens with strong adjuvants, but human autoimmune diseases occur spontaneously, as a result of interactions between genetic susceptibility, related mostly to variations in human leukocyte antigen (HLA) class I and II genes, and poorly understood environmental factors. Associations between infectious agents and autoimmune disease suggest that infections play an important role in autoimmunity, but there are only a few well-documented conditions in which a specific infection can be identified as the cause. For example, rheumatic heart disease following infection with group A streptococci is caused by molecular mimicry between streptococcal and cardiac antigens such that the antibodies the immune system develops against the streptococcal infection also recognize and attack cardiac self-antigens (Marijon et al. 2012).

Epstein–Barr virus (EBV), by causing a persistent latent infection with periodic reactivations, immortalizing B lymphocytes, and eliciting a strong T-cell response, seems a uniquely plausible cause of autoimmunity (Pender 2003), and numerous claims of causality have been made based on altered humoral and cellular immune responses to EBV in several autoimmune diseases. These alterations, however, could as well be a consequence rather than a cause of the disease. A demonstration of causality, broadly defined as a state of nature in which a reduction in the frequency or severity of infection would be followed by a reduction in the frequency of the autoimmune disease (MacMahon and Trichopolous 1996), would require proof that prevention of EBV infection is followed by a reduction in disease incidence. Lacking a suitable vaccine for such an experiment, the strongest evidence of causality rests on the results of an experiment of nature: Are those individuals who are not infected with EBV protected from the disease, and do they lose protection after EBV infection? This protection does not need to be complete, because EBV could be the underlying cause in only a subset of cases. The relation between the hepatitis B virus and liver cancer provides a useful analogy: Liver cancer occurs in non-infected individuals, but infection causes a 50- to 100-fold increase in risk (Beasley et al. 1981). On the other hand, if the frequency of liver cancer was the same among infected and non-infected individuals in the same population, we would conclude that causality is unlikely.

More generally, we would say that EBV is a cause of a disease X if the future frequency of X in a healthy and virus naïve population would be higher if the members of this population became infected with EBV than if they remained uninfected. This counterfactual occurrence is unobservable, but given two groups of EBV-negative individuals with similar characteristics, we would expect a higher disease incidence in those who become infected with EBV than in those who do not. If so, and lacking any credible alternative explanation for the difference in frequency of X, we could infer that EBV is a cause of X, even if we do not fully understand the underlying mechanisms.

Most individuals worldwide are EBV infected, making determination of the autoimmune disease frequency in those who are EBV-negative challenging. To overcome this difficulty, two complementary strategies have been pursued:

(i) meta-analyses combining the data from several case–control studies and
(ii) case-control studies of pediatric populations in which prevalence of EBV positivity is still low at ~75 % or less.

Although EBV has been suspected as an etiological factor in multiple autoimmune diseases, evidence of a higher disease frequency in EBV infected as compared to non-infected individuals has been found only for multiple sclerosis (MS) and systemic lupus erythematosus (SLE). There is little to no epidemiological evidence in favor or against a role of EBV in most autoimmune diseases, whereas for a few others, including juvenile rheumatoid arthritis and myasthenia gravis, evidence does not support causality. The dysregulation of immune responses to EBV observed in these conditions is thus likely to be secondary to autoimmune and inflammatory reactions (Costenbader and Karlson 2006; Ferraccioli and Tolusso 2007). A summary of selected data is shown in Table 1.

The results in Table 1 support, but do not demonstrate, a causal link between EBV infection and MS or SLE. One of the main limitations of these data is their cross-sectional nature, because of which the possibility of reverse causation (i.e., MS or SLE increasing susceptibility to EBV infection) cannot be ruled out.

Table 1 Association between EBV infection and risk of autoimmune diseases in selected case–control studies

Age group	Disease	Cases *N* (% negative)	Controls *N* (% negative)	Odds ratio for EBV negative (*p* value)
Adults	Multiple sclerosis[a]	1779 (0.5)	2499 (6.4)	0.06 ($p < 10^{-8}$)
Pediatric onset	SLE (James et al. 1997) SLE (Harley and James 1999)	117 (0.9) 26 (0)	153 (30.0) 26 (30.0)	0.02 ($p < 10^{-10}$) 0.04 ($p < 10^{-3}$)
	Multiple sclerosis[b]	281 (8.9)	350 (35.7)	0.18 ($p < 10^{-6}$)
	Myasthenia gravis[c] (Klavinskis et al. 1985)	na (41)	na	na
	Myositis (James et al. 1997)	36 (28.0)	153 (30.0)	0.91 (ns)
	Juvenile rheumatoid arthritis (James et al. 1997)	38 (28.0)	153 (30.0)	0.91 (ns)

[a]Meta-analysis (Ascherio and Munger 2007a, b)
[b]Combined data from three investigations (Alotaibi et al. 2004; Pohl et al. 2006; Banwell et al. 2007a), approximate *p* value estimated by authors of this review
[c]Cases with onset <20 years; number of cases not provided, and no matched controls available, but the 41 % negativity is similar to expected among UK adolescents

Longitudinal studies demonstrating that EBV infection indeed precedes the first clinical manifestation of disease have been conducted only for MS and SLE. Thus, in the remainder of this chapter, we will focus on these conditions, specifically discussing the epidemiological evidence supporting EBV causality and some of the possible mechanisms that seem to converge with the epidemiological findings.

2 Multiple Sclerosis

2.1 Definition and Epidemiology

MS is a chronic and disabling inflammatory and neurodegenerative disease of the central nervous system that affects 350,000 people in the USA and approximately 2 million worldwide (Pugliatti et al. 2002). In over 80 % of cases, the disease starts with a relapsing–remitting course. Relapses are caused by discrete demyelinating lesions pathologically consistent with an immune-mediated process. Macrophages, B cells, T cells, and immune complexes with evidence of complement activation have been recognized in biopsy or autopsy material, with evidence of some heterogeneity across individuals (Lucchinetti et al. 2000). There is also evidence in peripheral blood cells of altered cellular immunity against myelin antigens (Lovett-Racke et al. 1998; Markovic-Plese et al. 2001) and of impaired function of regulatory T cells (Viglietta et al. 2004), which support the autoimmune etiology. Symptoms of MS are variable depending on the localization of the lesions within the brain or spinal cord. Although most lesions resolve completely or nearly completely within weeks, and relapses decrease over time and eventually cease completely, an underlying slow loss of brain tissue ultimately leads to progressive disability.

MS reaches a peak incidence between 20 and 40 years of age and affects women more often than men, and it is common in Europe, USA, Canada, Australia, and New Zealand. There is evidence that incidence is increasing among women in both Europe and North America and in both sexes in some countries, but MS remains rare in Africa and most of Asia. Although genetic factors, particularly the prevalence of the MS-risk allele HLA-DR15, contribute to this geographical distribution, the results of studies among migrants provide compelling evidence for a role of environmental factors (Gale and Martyn 1995). The most consistent risk factors are vitamin D insufficiency (Ascherio et al. 2010), cigarette smoking (Ascherio and Munger 2007b), and, as described below, infectious mononucleosis (IM) and EBV infection.

2.2 Epidemiological Evidence that EBV Is a Cause of MS

MS and IM share many epidemiological similarities—both diseases are uncommon in the tropics, as well as in Japan, China, and among Eskimos, display a latitude gradient in incidence within temperate zones in North America, Europe, and Oceania, and occur at somewhat younger age in women than in men (Warner and Carp 1988; Ascherio and Munger 2007a, b). Further, MS risk is about 2.3-fold higher among individuals with clinical history of IM, as compared with those without such history, as demonstrated in a meta-analysis of case-control studies (Thacker et al. 2006) and confirmed in a longitudinal investigation based on the Danish national data (Nielsen et al. 2007). One interpretation of this finding is that IM and MS share a common cause: a high level of hygiene in childhood that predisposes to IM, by delaying the age of primary EBV infection, and MS, by reducing exposure to helminthes and other immune-modulating infections. This hypothesis, known as the hygiene hypothesis, could explain why MS is rare in the tropics and subtropics where sanitation tends to be low. If hygiene was the common cause of IM and MS, however, individuals who escape EBV infection (lack of EBV infection being a marker of high hygiene) should have a high MS risk (Fig. 1). In striking contrast with this prediction, there is compelling evidence that individuals who are not infected with EBV rarely, if ever, develop MS. This evidence includes the following:

(i) Consistent results of numerous case–control studies demonstrating extremely low odds of MS in individuals with negative EBV serology. In a meta-analysis including 13 studies, the odds ratio for MS in EBV-negative individuals was 0.06 (95 % confidence interval: 0.03–0.13; p value $<10^{-8}$) (Ascherio and Munger 2007a). This low MS risk in EBV-negative individuals and the increased risk in individuals with a history of IM imply a dramatic increase in MS risk following IM, which cannot be attributed to hygiene alone (Fig. 2).

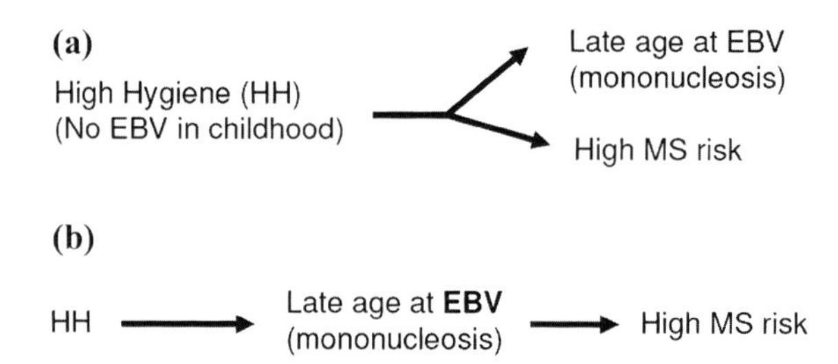

Fig. 1 The hygiene hypothesis of MS causation. **a** The formulation of the hygiene hypothesis that states MS and IM arise from the common cause of "high hygiene" in childhood and are associated due to this common cause. **b** The "EBV variant" of the hygiene hypothesis, which states that high hygiene in childhood increases the likelihood of a late age at infection with EBV (IM), which then leads to an increased risk of MS. Current epidemiological data support this latter formulation of the hypothesis. *Source* Ascherio and Munger, J Neuroimmune Pharmacol 2010

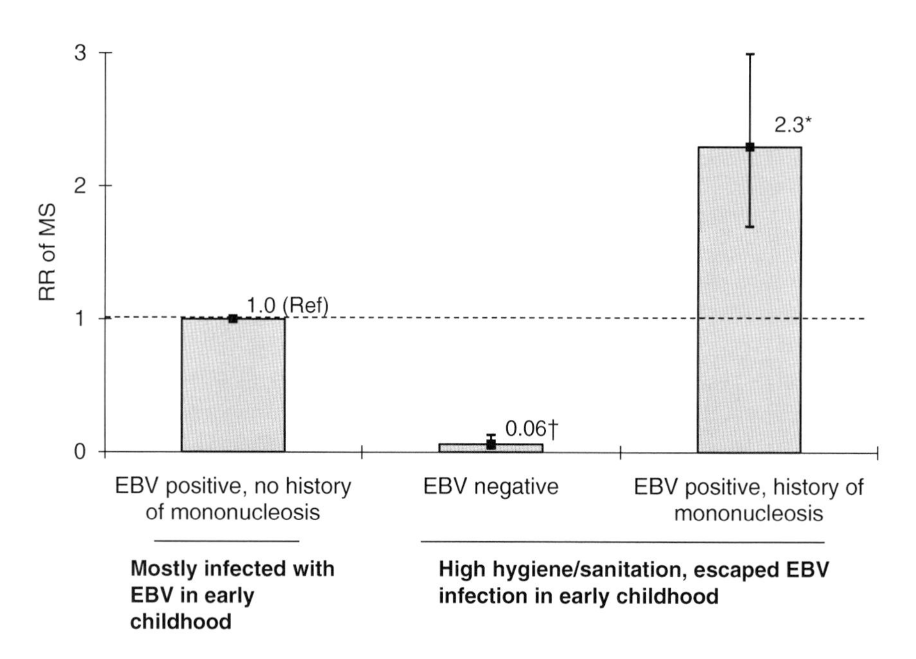

Fig. 2 Relative risk (RR) of developing MS according to EBV infection and history of mononucleosis. Bars represent the 95 % confidence intervals of the RR estimates. $^{\dagger,*}p < 10^{-8}$. Data from Thacker et al. (2006), Ascherio and Munger (2007a). Reprinted from Thieme Medical Publishers, Inc., Ascherio and Munger, Semin Neurol 2008:28(1);17–28

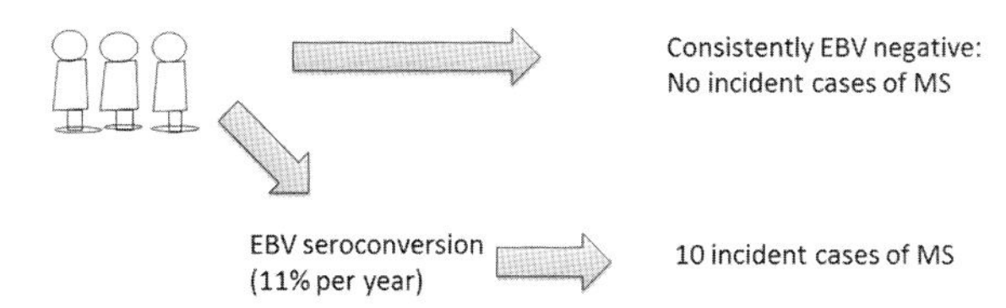

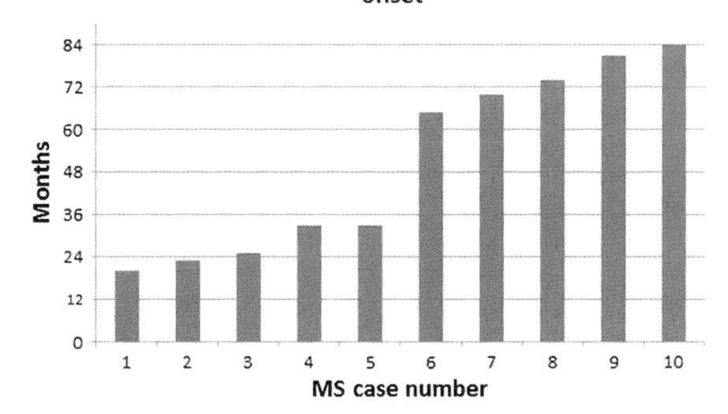

Fig. 3 **a** No incident cases of MS were observed among individuals who were EBV seronegative at recruitment and did not seroconvert during follow-up. In contrast, ten incident MS cases were confirmed after EBV seroconversion. **b** Time between first EBV positive serum and MS onset. The shortest interval between seroconversion and onset of first MS symptoms was >21 months. Data from Levin et al. (2010)

(ii) Investigations in pediatric onset MS. The odds ratio comparing children with negative EBV serology to children with evidence of past EBV infection was 0.11 ($p < 0.001$) in a study in Canada (Alotaibi et al. 2004), 0.04 ($p < 0.001$) in a study in Germany (Pohl et al. 2006), and 0.36 ($p = 0.02$) in a multisite international study (Banwell et al. 2007b).

(iii) The demonstration in a longitudinal study based on the Department of Defense Serum Repository with samples from over seven million young adults that individuals who are EBV negative do not appear to develop MS until after they seroconvert for EBV and that the onset of MS symptoms occurs only several months after EBV seroconversion (Fig. 3) (Levin et al. 2010).

Because the extremely low incidence of MS in EBV-negative individuals is critical to determine causality, it is important to explore alternative explanations, which include:

(i) *MS causes an activation of the immune system and increased antibody titers against multiple antigens. The higher titer of anti-EBV antibodies among cases reduces the number of false negative serology, creating a spurious difference in prevalence of EBV infection between MS cases and controls.* This argument has some foundation, because it has been documented that some individuals infected with EBV are indeed serologically negative (Savoldo et al. 2002). However, longitudinal studies of adults with negative EBV serology have demonstrated that these individuals seroconvert at high rates (10–11 % per year) and have high IM incidence when exposed to EBV (Balfour et al. 2013), which is consistent with their experiencing a primary infection. Similarly, EBV-negative adolescents are at high risk of IM and thus unlikely to have false negative serology (Balfour et al. 2013).

Finally, this difference in prevalence of infection is unique to EBV and is not seen for other viruses, including cytomegalovirus (CMV) and other herpes viruses (Ascherio and Munger 2007a).

(ii) *Higher rate of EBV positivity among MS cases is due to false positives.* This explanation is unlikely because serological results are based on a combination of highly specific tests. Further, there is no evidence of IM occurring in individuals with MS.

(iii) *Individuals who are EBV negative are genetically resistant to both EBV infection and MS.* This possibility is ruled out by the results of a longitudinal study that demonstrated that EBV-negative young adults are susceptible to both EBV infection and MS (Levin et al. 2010). Pediatric MS data also provide compelling evidence against this explanation, because almost all EBV-negative children will become EBV-positive adults, which proves their genetic susceptibility. Some HLA-DR alleles have been linked to risk of IM (Ramagopalan et al. 2011) or anti-Epstein–Barr virus nuclear antigen (EBNA)1 IgG titers (Waubant et al. 2013; Rubicz et al. 2013), but none of these explain the associations of these factors with MS. Rather, MS-risk alleles appear to have additive or multiplicative effects with history of IM (Disanto et al. 2013; Simon et al. 2014) and anti-EBNA1 titers (De Jager et al. 2008; Sundqvist et al. 2012; Sundstrom et al. 2009).

(iv) *Increased susceptibility to EBV infection is a feature of early, preclinical MS—i.e., MS causes EBV infection.* This theoretical possibility is difficult to exclude empirically, but it is rather implausible. There is no evidence that MS increases susceptibility to infection in general or to EBV specifically. Also, there is no evidence of EBV infection after the onset of MS. Most importantly, it has been demonstrated in a longitudinal study of EBV-negative young adults that MS occurs only after EBV infection, with the onset of the first MS symptoms occurring at least 21 months after seroconversion (Fig. 3) (Levin et al. 2010).

Artifacts and alternative explanations excluded, the exceedingly low MS incidence in EBV-negative individuals provides compelling evidence that EBV plays a causal role in most cases of MS.

2.3 Potential Mechanisms Relating EBV to MS

The molecular mechanisms linking EBV infection to MS have not been elucidated though numerous hypotheses have been proposed. Some of these hypotheses, however, do not explain several observations emerging from epidemiological and clinical studies, which include:

(i) MS has not been reported as a complication of immunosuppression in post-transplant lymphoproliferative disease or after prolonged treatment with immunosuppressive drugs. In fact, immunosuppression is used to treat MS that is resistant to first-line treatment.

(ii) There is a lag of at least several months between EBV infection and onset of the first MS symptoms, as demonstrated in a longitudinal study (Levin et al. 2010), and, indirectly, by lack of reported MS cases in children or adults with evidence of recent EBV infection (Yea et al. 2013).
(iii) EBV viral load in serum or peripheral blood cells is only modestly increased in MS (Wagner et al. 2004; Buljevac et al. 2005; Lindsey et al. 2009) (although children with MS shed EBV in saliva more frequently than healthy controls) (Yea et al. 2013).
(iv) Monoclonal antibodies (natalizumab) that block α4 integrin and thus prevent T-cell migration into the central nervous system (CNS) provide a very effective treatment against MS relapses (Polman et al. 2006). Notably, interruption of treatment is associated with a rebound of disease activity (Fox et al. 2014).
(v) Monoclonal anti-CD20 antibodies that deplete B cells (rituximab) are extremely effective in reducing MS relapses (Hauser et al. 2008). Rituximab does not directly deplete plasma cells, and therefore, its effectiveness within a few weeks and before any reduction in immunoglobulin titers suggests that the role of B cells in MS goes beyond antibody production (von Budingen et al. 2011). This role may include antigen-presenting activity and release of cytokines or cytotoxic factors (Lisak et al. 2012).
(vi) Serum antibody titers against the EBV nuclear antigens (EBNA complex, EBNA-1, and EBNA-2) in healthy young adults are strongly related to risk of developing MS (Ascherio et al. 2001). This association has been confirmed in numerous independent longitudinal studies (Sundstrom et al. 2004; Levin et al. 2005; DeLorenze et al. 2006), and it is not explained by the MS-risk haplotype HLA-DR1501 (De Jager et al. 2008). In the largest investigation, MS risk was up to 30-fold higher in individuals with the highest titers of anti-EBNA complex antibodies as compared to those with the lowest (Munger et al. 2011).

The above observations suggest that uncontrolled lytic viral replication or overproliferation of EBV-infected B cells is not the primary cause of MS, but rather the pathological process is driven by the immune response to EBV, including T cells, B cells, and antibodies. The rebound of inflammatory activity following interruption of natalizumab, which keeps T cells, including EBV-specific T cells, out of the CNS, may result from increased EBV replication in the CNS during treatment. The several months interval between EBV infection and MS onset is consistent with an important specific role of anti-EBNA1 antibodies and anti-EBNA1 CD4+ T cells, which are usually absent during acute infection and increase over a period of months after recovery (Long et al. 2013). Notably, CD4+ T cells specific for EBNA1 peptides, which are an important part of immune control of EBV in healthy individuals, have been found to be increased in frequency and to have enhanced proliferation capacity, interferon-γ production, and broadened specificity in individuals with MS than in HLA-DR and demographically matched healthy controls (Lunemann et al. 2006). Further, these cells have been found to have a T-helper 1 central memory or effector memory phenotype and to recognize myelin antigens more frequently than other antigens not associated with MS (Lunemann et al. 2008). There is thus a convergence of epidemiological, clinical, and immunological evidence that anti-EBNA1 antibodies and CD4+ positive EBNA1-specific T cells cross-reacting with myelin antigens contribute to the pathological process in MS. CD4+ T cells recognizing other EBV antigens and myelin epitopes have also been reported in blood (Lang et al. 2002) and the cerebrospinal fluid (Holmoy and Vartdal 2004).

These findings do not exclude an important role for EBV-specific CD8+ T cells, although comparisons of their frequency and function in blood and cerebrospinal fluid (CSF) from individuals with MS and matched healthy controls have given somewhat mixed results, which could be in part attributable to methodological differences (Hollsberg et al. 2003; Gronen et al. 2006; Jilek et al. 2008, 2012; Jaquiery et al. 2010; Angelini et al. 2013; Pender et al. 2014a; Lossius et al. 2014). B cells must also play an important role, either through their antigen-presentation activity or other mechanisms. In particular, EBV-infected B cells have a survival advantage and even when encoding cross-reacting antibodies they could pass the checkpoints that eliminate most autoreactive cells (Pender 2003). Other mechanisms that could relate EBV infection to MS include activation of superantigens such as HERV-K18 (Tai et al. 2008) or induction of autoimmune responses against alpha–beta crystallin, an important antigen in CNS myelin (van Sechel et al. 1999).

According to a report in 2007, large numbers of EBV-infected B cells were found postmortem in meningeal follicles and MS lesions in the brain of the majority of patients with relapsing–remitting or secondary progressive MS (Serafini et al. 2007). The more active lesions also showed evidence of lytic infection and cytotoxic responses to EBV-infected B cells, suggesting that these cells drive MS pathology. Several discordant reports (Willis et al. 2009; Peferoen et al. 2010; Sargsyan et al. 2010) raised questions on these findings (Ascherio and Bar-Or 2010), but the presence of latently infected EBV-positive cells in perivascular infiltrate of all active MS lesions has been recently confirmed in a new rigorous investigation (Tzartos et al. 2012). By itself, the presence of EBV-positive cells in MS lesions does not prove that these cells have a causal role, because these cells could be attracted to areas of inflammation by locally produced cytokines, but when considered in the context of the epidemiological evidence, it provides a plausible scenario of MS causation. Infected B cells, in addition to activating EBV-specific T cells, can also promote inflammation by releasing Epstein–Barr virus-encoded small RNAs (EBERs) that bind to the Toll-like receptor 3 resulting in activation of innate immunity and interferon-α production (Iwakiri et al. 2009; Tzartos et al. 2012).

2.4 Implications for MS Prevention and Treatment

If, as we propose, EBV infection is a component cause of most cases of MS, prevention of EBV infection would be expected to substantially reduce MS incidence. Although complete prevention of EBV infection is not in the foreseeable future, a vaccine under development was found to be effective in preventing IM in young adults exposed to EBV (Moutschen et al. 2007; Sokal et al. 2007). Conceivably, by reducing the intensity of the immune response to EBV such a vaccine could reduce MS risk, but this may depend on the specific effects of the vaccine on the relevant immune responses. Theoretically, the incidence of both IM and MS could also be reduced by facilitating EBV infection in early childhood, but in the case of MS, this intervention would only be partially effective. Whether EBV continues to play a pathological role after MS onset or whether it contributes to initiate a self-perpetuating autoimmune response remains uncertain. The presence of EBV-infected cells in MS lesions suggests that EBV is an important factor in determining relapses and MS progression, but evidence relating serological signs of EBV reactivation with MS relapses or disease activity (Wandinger et al. 2000; Buljevac et al. 2005; Lindsey et al. 2009) remains sparse and unconvincing. This, however, does not exclude the possibility that EBV reactivation within the CNS drives MS pathology. Conflicting results have also been reported on the relation between EBNA-1 IgG antibodies and MS outcomes (Farrell et al. 2009; Lunemann and Ascherio 2009). The results of three placebo-controlled trials of the antiviral alacyclovir or its precursor valacyclovir suggested overall a nonsignificant benefit in the treated patients, but these studies were too small to be conclusive, and most importantly, alacyclovir does not decrease the number of latently infected B cells which are likely to drive the immune response and thus MS pathology (Bech et al. 2002; Friedman et al. 2005; Lycke et al. 1996). Treatment of progressive MS with in vitro expanded autologous EBV-specific CD8+ T cells has been proposed and attempted in a single patient with progressive MS (Pender et al. 2014b). The patient did well, but clearly more data are needed to determine whether this is a valuable therapeutic approach.

2.5 Summary on EBV and MS

The overall epidemiological evidence and the demonstration in a longitudinal study that EBV-negative individuals do not get MS unless they first acquire EBV provides unquestionable evidence that EBV is an important component cause of MS, but it is neither necessary, as proven by rare cases of pediatric MS in EBV-negative children, nor sufficient. Rather, in most cases, MS is a rare complication of EBV infection in genetically predisposed individuals. Other factors are likely to influence MS risk. Some are known, such as vitamin D insufficiency or cigarette smoking, whereas others remain to be discovered. However, neither EBV infection nor other known risk factors provide a convincing explanation for several features of MS epidemiology. Among the observations that remained unexplained are the change in risk among migrants (Gale and Martyn 1995), a probable MS outbreak in the Faroe islands (Kurtzke and Heltberg 2001), and the increasing MS frequency among African-Americans in the USA (Wallin et al. 2012) and among women in several countries in North America (Orton et al. 2006) and Europe

(Koch-Henriksen and Sorensen 2010). One possibility is that there are genetic variations in EBV itself that modulate its propensity to cause MS. A few reports have addressed this hypothesis (Munch et al. 1998; Lindsey et al. 2008; Mechelli et al. 2011) as well as interactions with genetic or environmental factors, which could include other infectious agents.

3 Systemic Lupus Erythematosus

3.1 Definition and Epidemiology

SLE is a multisystem autoimmune disease with varied presentations, progression, and symptoms experienced by patients. Some common manifestations include skin rash (malar and/or discoid), photosensitivity, and arthritis, and many patients suffer with general pain and fatigue (Tsokos 2011). Despite the heterogeneity of the disease, the majority of SLE patients have detectable autoantibodies, the most common being antinuclear antibodies in nearly 100 %, but others include anti-Sm, anti-Ro (~40 % combined), and anti-dsDNA (Arbuckle et al. 2003). Immune complexes between autoantibodies and their respective antigens have a key role in tissue injury (Tsokos 2011), but multiple mechanisms contribute to pathogenesis, including abnormalities in B-cell and T-cell signaling and transcription, IL-2 production, and a deficiency in cytotoxic T-cell activity, which predispose to an increased risk of infection (Tsokos 2011). The incidence of SLE is about ninefold higher in women than in men and higher among African-Americans and Asians than Caucasians. The etiology of SLE is unknown, but many potential lifestyle and infectious agents, including EBV, have been studied (Simard and Costenbader 2007). As for MS, genetic susceptibility plays an important role, as demonstrated by higher concordance rates in monozygotic twins (24–60 %) than in dizygotic twins (2–5 %) (Simard and Costenbader 2007); numerous genes have been found to be associated with risk, but, as for MS, the strongest associations are with polymorphisms within the HLA-II region (Armstrong et al. 2014).

3.2 Epidemiological Evidence that EBV Is a Cause of SLE

While, as discussed above, history of IM is a strong, consistent risk factor for MS, there has been no similar association found with SLE (Strom et al. 1994; Cooper et al. 2002; Ulff-Moller et al. 2010). Thus, neither hygiene nor age at infection with EBV seems to be important factors in SLE etiology. There is evidence, however, that individuals who are not infected with EBV have a low risk of SLE. This evidence includes the following:

(i) In a meta-analysis of 16 case–control studies, there was a modest but significant association between EBV infection as assessed by positive serology for anti-EBV viral capsid antigen (VCA) IgG and SLE (odds ratio [OR] for EBV negative = 0.48; $p = 0.007$) (Hanlon et al. 2014). This result, however, should be interpreted cautiously, because most of the studies were small and of questionable quality and there was significant heterogeneity between studies.

(ii) The results of two case–control studies of EBV and pediatric SLE: The first study included children (age 4–19 years) with SLE in Oklahoma City ($n = 59$ SLE, $n = 95$ controls) and San Diego ($n = 58$, $n = 58$ controls) (James et al. 1997). Few details on control selection were given other than that the Oklahoma City controls were similar to cases on a variety of demographic factors and the San Diego controls were siblings of the cases. Seropositivity to anti-VCA IgG was determined; 99 % of SLE cases versus 70 % of controls were positive for VCA IgG (OR for EBV negative vs. EBV positive = 0.02; $p < 10^{-10}$). The second study in a smaller group of children ($n = 26$ cases, $n = 26$ age-, race-, and sex-matched controls) confirmed these findings (OR = 0.04, $p = 0.00024$) (Harley and James 1999). Odds ratios of this magnitude, which suggest a 25- to 50-fold increase in risk of SLE following EBV infection, rarely occur as a result of confounding or other sources of bias. Further, as with the pediatric MS studies, the lower prevalence of EBV infection in children as compared to adults makes these investigations particularly robust.

(iii) Evidence from a longitudinal study based on the Department of Defense Serum Repository (DoDSR) that EBV infection tends to precede not only the clinical onset of SLE, but also the appearance of autoantibodies. In this study, which included 130 incident cases of SLE with multiple blood samples collected before onset of clinical symptoms, positivity for anti-VCA antibodies preceded or was simultaneously detected with the circulating autoantibodies that support the diagnosis of SLE (McClain et al. 2005). In four cases, it was possible to observe a sequence of positivity with the cases becoming positive for anti-VCA, then anti-EBNA1, and then anti-Ro (McClain et al. 2005). While this is a small series of cases, these observations, coupled with the experimental data (see below), support a causal role for EBV in SLE development.

Because SLE is a systemic disease that affects virtually all the components of the immune system and increases susceptibility to infection, it is difficult to exclude the possibility that the higher prevalence of EBV infection among cases is a consequence of SLE rather than its cause and that the increased viral titers and EBV-specific immune responses in SLE merely reflect the sensitivity of the virus to perturbation of the immune system (Gross et al. 2005). The conclusion that EBV infection indeed precedes the first serological evidence of SLE is based mainly on the four cases from the longitudinal study described above, and it is thus less compelling than the comparable evidence for MS. On the other hand, as discussed below, there is clear mechanistic evidence supporting a link between EBV and SLE.

3.3 Potential Mechanisms Relating EBV to SLE

EBV was first proposed as a possible causal agent in SLE in the early 1970s when a small study among 100 SLE patients (34 of whom were age-, sex-, and race-matched to 34 tuberculosis patients) found that 62 % had high titers to EBV and among the matched subset, six times as many SLE as TB patients had elevated titers to EBV (Evans et al. 1971). However, results of similar studies over the following years were mixed, and the hypothesis largely fell out of favor. About 20 years later, a series of studies began to elucidate the peptide sequences of some of the targeted SLE antigens that the autoantibodies recognized, and whether there was cross-reaction with infectious agents including EBV.

One of the first studies to utilize this approach focused on the anti-Sm B/B′ antibodies that develop in some SLE patients (James et al. 1994). Investigators systematically deconstructed the Sm B/B′ protein into overlapping octapeptides and discovered that the peptide sequence PPPGMRPP was the most strongly antigenic, as shown by reactivity with SLE patients' anti-Sm B/B′ antibodies. The EBNA-1 protein contains a similar peptide sequence, PPPGRRP, also recognized by the anti-Sm B/B′ antibodies from SLE patients, suggesting that molecular mimicry could be a mechanism of SLE development in some patients (James et al. 1994). James et al. went on to show that this antigenic Sm B/B′ octapeptide can induce autoimmunity and epitope spreading in rabbits, leading to an SLE-like disease, and that rabbits immunized with the EBNA-1 peptide PPPGRRP develop an early auto-antigen profile followed by epitope spreading similar to that seen in SLE patients positive for anti-Sm B/B′ antibodies, further supporting a link between EBV and SLE (James et al. 1997; Poole et al. 2008). A high sequence similarity has also been identified between a glycine–arginine repeat in the C-terminal region of SmD protein and an EBNA-1 peptide. Similarly to results with Sm B/B′, it has been demonstrated that anti-Sm D antibodies bind this glycine–arginine-rich sequence of EBNA-1 and mice immunized with the EBNA-1 peptide develop autoimmunity (Sabbatini et al. 1993).

In the longitudinal study utilizing the DoDSR mentioned above, the development of autoimmune antibodies in 130 SLE cases with serial pre-diagnostic samples was examined (Arbuckle et al. 2003). One of the earliest appearing antibodies was to 60 kDa Ro. In a subsequent study building on this, and also including cases from the Oklahoma Clinical Immunology Serum Repository, 29 SLE cases who were anti-Ro negative and then became anti-Ro positive were identified. In nine of these cases, the earliest sample positive for anti-Ro recognized only one epitope (amino acids 169–180) where in the majority of the remaining cases, the first positive anti-Ro sample already recognized multiple epitopes including $Ro_{169-180}$ (McClain et al. 2005). The anti-$Ro_{169-180}$ antibodies, but not antibodies to the other Ro epitopes, cross reacted with EBNA-1. Interestingly, there is no overlap between the $Ro_{169-180}$ and the EBNA-1 peptide sequence recognized by the anti-Ro$_{169-180}$ antibody; however, immunization of rabbits with the EBNA-1 peptide also developed antibodies to Ro and symptoms consistent with SLE (McClain et al. 2005).

With experimental evidence suggesting molecular mimicry between antibodies against EBNA-1 and at least three targeted antigens in SLE (Sm B/B′, Sm D, and Ro), there is a strong case for biological plausibility for EBV as a causal agent in at least some cases of SLE. Interestingly, however, anti-VCA and anti-early antigen antibody titers tend to be more prominently elevated in SLE patients than anti-EBNA1 titers (Hanlon et al. 2014). This observation is consistent with the hypothesis that there is some degree of defective immune control of EBV in SLE, as further suggested by a ~40-fold increase in EBV viral load in peripheral blood cells not explained by immunosuppressive treatment (Kang et al. 2004), an increase in the frequency of EBV-specific CD4+ T cells producing gamma interferon (Kang et al. 2004), and evidence of defective cytotoxic T-cell activity against EBV (Kang et al. 2004; Larsen et al. 2011).

3.4 Implications for SLE Prevention and Treatment

Unlike in MS, late age at EBV infection and IM do not appear to be risk factors for SLE; thus, early exposure to EBV is unlikely to be beneficial. The potential effects of vaccines that do not provide sterilizing immunity are also uncertain, as the outcome is likely to depend on the specific effect of the vaccine on the generation of cross-reacting antibodies. On the other hand, elimination of the infection or reduction of the viral load by reducing the antigenic stimulation could reduce titers of autoantibodies and possibly have a clinical benefit.

References

Alotaibi S, Kennedy J, Tellier R, Stephens D, Banwell B (2004) Epstein-Barr virus in pediatric multiple sclerosis. JAMA 291(15):1875–1879

Angelini DF, Serafini B, Piras E, Severa M, Coccia EM, Rosicarelli B, Ruggieri S, Gasperini C, Buttari F, Centonze D, Mechelli R, Salvetti M, Borsellino G, Aloisi F, Battistini L (2013) Increased CD8+ T cell response to Epstein-Barr virus lytic antigens in the active phase of multiple sclerosis. PLoS Pathog 9(4):e1003220

Arbuckle MR, McClain MT, Rubertone MV, Scofield RH, Dennis GJ, James JA, Harley JB (2003) Development of autoantibodies before the clinical onset of systemic lupus erythematosus. N Engl J Med 349(16):1526–1533

Armstrong DL, Zidovetzki R, Alarcon-Riquelme ME, Tsao BP, Criswell LA, Kimberly RP, Harley JB, Sivils KL, Vyse TJ, Gaffney PM, Langefeld CD, Jacob CO (2014) GWAS identifies novel SLE susceptibility genes and explains the association of the HLA region. Genes Immun 15(6):347–354

Ascherio A, Bar-Or A (2010) EBV and brain matter(s)? Neurology 74:1092–1095

Ascherio A, Munger KL (2007a) Environmental risk factors for multiple sclerosis. Part I: the role of infection. Ann Neurol 61(4):288–299

Ascherio A, Munger KL (2007b) Environmental risk factors for multiple sclerosis. Part II: noninfectious factors. Ann Neurol 61(6):504–513
Ascherio A, Munger KL, Lennette ET, Spiegelman D, Hernán MA, Olek MJ, Hankinson SE, Hunter DJ (2001) Epstein-Barr virus antibodies and risk of multiple sclerosis: a prospective study. JAMA 286(24):3083–3088
Ascherio A, Munger KL, Simon KC (2010) Vitamin D and multiple sclerosis. Lancet Neurol 9(6):599–612
Balfour HH Jr, Odumade OA, Schmeling DO, Mullan BD, Ed JA, Knight JA, Vezina HE, Thomas W, Hogquist KA (2013) Behavioral, virologic, and immunologic factors associated with acquisition and severity of primary Epstein-Barr virus infection in university students. J Infect Dis 207(1):80–88
Banwell B, Ghezzi A, Bar-Or A, Mikaeloff Y, Tardieu M (2007a) Multiple sclerosis in children: clinical diagnosis, therapeutic strategies, and future directions. Lancet Neurol 6(10):887–902
Banwell B, Krupp L, Kennedy J, Tellier R, Tenembaum S, Ness J, Belman A, Boiko A, Bykova O, Waubant E, Mah JK, Stoian C, Kremenchutzky M, Bardini MR, Ruggieri M, Rensel M, Hahn J, Weinstock-Guttman B, Yeh EA, Farrell K, Freedman M, Iivanainen M, Sevon M, Bhan V, Dilenge ME, Stephens D, Bar-Or A (2007b) Clinical features and viral serologies in children with multiple sclerosis: a multinational observational study. Lancet Neurol 6(9):773–781
Beasley RP, Hwang LY, Lin CC, Chien CS (1981) Hepatocellular carcinoma and hepatitis B virus. A prospective study of 22,707 men in Taiwan. Lancet 2(8256):1129–1133
Bech E, Lycke J, Gadeberg P, Hansen HJ, Malmestrom C, Andersen O, Christensen T, Ekholm S, Haahr S, Hollsberg P, Bergstrom T, Svennerholm B, Jakobsen J (2002) A randomized, double-blind, placebo-controlled MRI study of anti-herpes virus therapy in MS. Neurology 58(1):31–36
Buljevac D, van Doornum GJ, Flach HZ, Groen J, Osterhaus AD, Hop W, van Doorn PA, van der Meche FG, Hintzen RQ (2005) Epstein-Barr virus and disease activity in multiple sclerosis. J Neurol Neurosurg Psychiatry 76(10):1377–1381
Cooper GS, Dooley MA, Treadwell EL, St Clair EW, Gilkeson GS (2002) Risk factors for development of systemic lupus erythematosus: allergies, infections, and family history. J Clin Epidemiol 55(10):982–989
Costenbader KH, Karlson EW (2006) Epstein-Barr virus and rheumatoid arthritis: is there a link? Arthritis Res Ther 8(1):204
Davidson A, Diamond B (2001) Autoimmune diseases. N Engl J Med 345(5):340–350
De Jager PL, Simon KC, Munger KL, Rioux JD, Hafler DA, Ascherio A (2008) Integrating risk factors: HLA-DRB1*1501 and Epstein-Barr virus in multiple sclerosis. Neurology 70(13 Pt 2):1113–1118
DeLorenze GN, Munger KL, Lennette E, Orentreich N, Vogelman J, Ascherio A (2006) Epstein-Barr virus and multiple sclerosis: evidence of association from a prospective study with long-term follow-up. Arch Neurol 63(6):839–844
Disanto G, Hall C, Lucas R, Ponsonby AL, Berlanga-Taylor AJ, Giovannoni G, Ramagopalan SV, The Ausimmune Investigator Group (2013) Assessing interactions between HLA-DRB1*15 and infectious mononucleosis on the risk of multiple sclerosis. Multiple Sclerosis J 19(10):1355–1358
Evans AS, Rothfield NF, Niederman JC (1971) Raised antibody titres to E.B. virus in systemic lupus erythematosus. Lancet 1(7691):167–168
Farrell RA, Antony D, Wall GR, Clark DA, Fisniku L, Swanton J, Khaleeli Z, Schmierer K, Miller DH, Giovannoni G (2009) Humoral immune response to EBV in multiple sclerosis is associated with disease activity on MRI. Neurology 73(1):32–38
Ferraccioli G, Tolusso B (2007) Infections, B cell receptor activation and autoimmunity: different check-point impairments lead to autoimmunity, clonal B cell expansion and fibrosis in different immunological settings. Autoimmun Rev 7(2):109–113

Fox RJ, Cree BA, De Seze J, Gold R, Hartung HP, Jeffery D, Kappos L, Kaufman M, Montalban X, Weinstock-Guttman B, Anderson B, Natarajan A, Ticho B, Duda P (2014) MS disease activity in RESTORE: a randomized 24-week natalizumab treatment interruption study. Neurology 82(17):1491–1498

Friedman JE, Zabriskie JB, Plank C, Ablashi D, Whitman J, Shahan B, Edgell R, Shieh M, Rapalino O, Zimmerman R, Sheng D (2005) A randomized clinical trial of valacyclovir in multiple sclerosis. Multiple Sclerosis J 11(3):286–295

Gale CR, Martyn CN (1995) Migrant studies in multiple sclerosis. Prog Neurobiol 47:425–448

Goldrath AW, Bevan MJ (1999) Selecting and maintaining a diverse T-cell repertoire. Nature 402(6759):255–262

Gronen F, Ruprecht K, Weissbrich B, Klinker E, Kroner A, Hofstetter HH, Rieckmann P (2006) Frequency analysis of HLA-B7-restricted Epstein-Barr virus-specific cytotoxic T lymphocytes in patients with multiple sclerosis and healthy controls. J Neuroimmunol 180(1–2):185–192

Gross AJ, Hochberg D, Rand WM, Thorley-Lawson DA (2005) EBV and systemic lupus erythematosus: a new perspective. J Immunol 174(11):6599–6607

Hanlon P, Avenell A, Aucott L, Vickers MA (2014) Systematic review and meta-analysis of the sero-epidemiological association between Epstein-Barr virus and systemic lupus erythematosus. Arthritis Res Ther 16(1):R3

Harley JB, James JA (1999) Epstein-Barr virus infection may be an environmental risk factor for systemic lupus erythematosus in children and teenagers. Arthritis Rheum 42(8):1782–1783

Hauser SL, Waubant E, Arnold DL, Vollmer T, Antel J, Fox RJ, Bar-Or A, Panzara M, Sarkar N, Agarwal S, Langer-Gould A, Smith CH (2008) B-cell depletion with rituximab in relapsing-remitting multiple sclerosis. N Engl J Med 358(7):676–688

Hollsberg P, Hansen HJ, Haahr S (2003) Altered CD8+ T cell responses to selected Epstein-Barr virus immunodominant epitopes in patients with multiple sclerosis. Clin Exp Immunol 132:137–143

Holmoy T, Vartdal F (2004) Cerebrospinal fluid T cells from multiple sclerosis patients recognize autologous Epstein-Barr virus-transformed B cells. J Neurovirol 10(1):52–56

Iwakiri D, Zhou L, Samanta M, Matsumoto M, Ebihara T, Seya T, Imai S, Fujieda M, Kawa K, Takada K (2009) Epstein-Barr virus (EBV)-encoded small RNA is released from EBV-infected cells and activates signaling from toll-like receptor 3. J Exp Med 206(10):2091–2099

James JA, Mamula MJ, Harley JB (1994) Sequential autoantigenic determinants of the small nuclear ribonucleoprotein Sm D shared by human lupus autoantibodies and MRL lpr/lpr antibodies. Clin Exp Immunol 98(3):419–426

James JA, Kaufman KM, Farris AD, Taylor-Albert E, Lehman TJA, Harley JB (1997) An increased prevalence of Epstein-Barr virus infection in young patients suggests a possible etiology for systemic lupus erythematosus. J Clin Invest 100(12):3019–3026

Jaquiery E, Jilek S, Schluep M, Meylan P, Lysandropoulos A, Pantaleo G, Du Pasquier RA (2010) Intrathecal immune responses to EBV in early MS. Eur J Immunol 40(3):878–887

Jilek S, Schluep M, Meylan P, Vingerhoets F, Guignard L, Monney A, Kleeberg J, Le Goff G, Pantaleo G, Du Pasquier RA (2008) Strong EBV-specific CD8+ T-cell response in patients with early multiple sclerosis. Brain 131(Pt 7):1712–1721

Jilek S, Schluep M, Harari A, Canales M, Lysandropoulos A, Zekeridou A, Pantaleo G, Du Pasquier RA (2012) HLA-B7-restricted EBV-specific CD8+ T cells are dysregulated in multiple sclerosis. J Immunol 88(9):4671–4680

Kang I, Quan T, Nolasco H, Park SH, Hong MS, Crouch J, Pamer EG, Howe JG, Craft J (2004) Defective control of latent Epstein-Barr virus infection in systemic lupus erythematosus. J Immunol 172(2):1287–1294

Klavinskis LS, Willcox N, Oxford JS, Newsom-Davis J (1985) Antivirus antibodies in myasthenia gravis. Neurology 35(9):1381–1384

Koch-Henriksen N, Sorensen PS (2010) The changing demographic pattern of multiple sclerosis epidemiology. Lancet Neurol 9(5):520–532
Kurtzke JF, Heltberg A (2001) Multiple sclerosis in the Faroe Islands: an epitome. J Clin Epidemiol 54(1):1–22
Lang HL, Jacobsen H, Ikemizu S, Andersson C, Harlos K, Madsen L, Hjorth P, Sondergaard L, Svejgaard A, Wucherpfennig K, Stuart DI, Bell JI, Jones EY, Fugger L (2002) A functional and structural basis for TCR cross-reactivity in multiple sclerosis. Nat Immunol 3(10):940–943
Larsen M, Sauce D, Deback C, Arnaud L, Mathian A, Miyara M, Boutolleau D, Parizot C, Dorgham K, Papagno L, Appay V, Amoura Z, Gorochov G (2011) Exhausted cytotoxic control of Epstein-Barr virus in human lupus. PLoS Pathog 7(10):e1002328
Levin LI, Munger KL, Rubertone MV, Peck CA, Lennette ET, Spiegelman D, Ascherio A (2005) Temporal relationship between elevation of Epstein Barr virus antibody titers and initial onset of neurological symptoms in multiple sclerosis. JAMA 293(20):2496–2500
Levin LI, Munger KL, O'Reilly EJ, Falk KI, Ascherio A (2010) Primary infection with the Epstein-Barr virus and risk of multiple sclerosis. Ann Neurol 67(6):824–830
Lindsey JW, Patel S, Zou J (2008) Epstein-Barr virus genotypes in multiple sclerosis. Acta Neurol Scand 117(2):141–144
Lindsey J, Hatfield L, Crawford M, Patel S (2009) Quantitative PCR for Epstein-Barr virus DNA and RNA in multiple sclerosis. Multiple Sclerosis J 15(2):153–158
Lisak RP, Benjamins JA, Nedelkoska L, Barger JL, Ragheb S, Fan B, Ouamara N, Johnson TA, Rajasekharan S, Bar-Or A (2012) Secretory products of multiple sclerosis B cells are cytotoxic to oligodendroglia in vitro. J Neuroimmunol 246(1–2):85–95
Long HM, Chagoury OL, Leese AM, Ryan GB, James E, Morton LT, Abbott RJ, Sabbah S, Kwok W, Rickinson AB (2013) MHC II tetramers visualize human CD4+ T cell responses to Epstein-Barr virus infection and demonstrate atypical kinetics of the nuclear antigen EBNA1 response. J Exp Med 210(5):933–949
Lossius A, Johansen JN, Vartdal F, Robins H, Jurate Saltyte B, Holmoy T, Olweus J (2014) High-throughput sequencing of TCR repertoires in multiple sclerosis reveals intrathecal enrichment of EBV-reactive CD8(+) T cells. Eur J Immunol 44(11):3439–3452
Lovett-Racke AE, Trotter JL, Lauber J, Perrin PJ, June CH, Racke MK (1998) Decreased dependence of myelin basic protein-reactive T cells on CD28-mediated costimulation in multiple sclerosis patients. A marker of activated/memory T cells. J Clin Invest 101(4):725–730
Lucchinetti C, Bruck W, Parisi J, Scheithauer B, Rodriguez M, Lassmann H (2000) Heterogeneity of multiple sclerosis lesions: implications for the pathogenesis of demyelination. Ann Neurol 47(6):707–717
Lunemann JD, Ascherio A (2009) Immune responses to EBNA1. Biomarkers in MS? Neurology 73(1):13–14
Lunemann JD, Edwards N, Muraro PA, Hayashi S, Cohen JI, Munz C, Martin R (2006) Increased frequency and broadened specificity of latent EBV nuclear antigen-1-specific T cells in multiple sclerosis. Brain 129(Pt 6):1493–1506
Lunemann JD, Jelcic I, Roberts S, Lutterotti A, Tackenberg B, Martin R, Munz C (2008) EBNA1-specific T cells from patients with multiple sclerosis cross react with myelin antigens and co-produce IFN-gamma and IL-2. J Exp Med 205(8):1763–1773
Lycke J, Svennerholm B, Hjelmquist E, Frisen L, Badr G, Andersson M, Vahlne A, Andersen O (1996) Acyclovir treatment of relapsing-remitting multiple sclerosis. A randomized, placebo-controlled, double-blind study. J Neurol 243(3):214–224
MacMahon B, Trichopolous D (1996) Epidemiology: principles and methods, 2nd edn. Little, Brown, & Co, Boston
Marijon E, Mirabel M, Celermajer DS, Jouven X (2012) Rheumatic heart disease. Lancet 379(9819):953–964

Markovic-Plese S, Cortese I, Wandinger KP, McFarland HF, Martin R (2001) CD4 + CD28− costimulation-independent T cells in multiple sclerosis. J Clin Invest 108(8):1185–1194
McClain MT, Heinlen LD, Dennis GJ, Roebuck J, Harley JB, James JA (2005) Early events in lupus humoral autoimmunity suggest initiation through molecular mimicry. Nat Med 11(1):85–89
Mechelli R, Anderson J, Vittori D, Coarelli G, Annibali V, Cannoni S, Aloisi F, Salvetti M, James JA, Ristori G (2011) Epstein-Barr virus nuclear antigen-1 B-cell epitopes in multiple sclerosis twins. Multiple Sclerosis J 17(11):1290–1294
Moutschen M, Leonard P, Sokal EM, Smets F, Haumont M, Mazzu P, Bollen A, Denamur F, Peeters P, Dubin G, Denis M (2007) Phase I/II studies to evaluate safety and immunogenicity of a recombinant gp350 Epstein-Barr virus vaccine in healthy adults. Vaccine 25(24):4697–4705
Munch M, Hvas J, Christensen T, Møller-Larsen A, Haahr S (1998) A single subtype of Epstein-Barr virus in members of multiple sclerosis clusters. Acta Neurol Scand 98:395–399
Munger KL, Levin LI, O'Reilly EJ, Falk KI, Ascherio A (2011) Anti-Epstein-Barr virus antibodies as serological markers of multiple sclerosis: a prospective study among United States military personnel. Multiple Sclerosis J 17(10):1185–1193
Nielsen TR, Rostgaard K, Nielsen NM, Koch-Henriksen N, Haahr S, Sorensen PS, Hjalgrim H (2007) Multiple sclerosis after infectious mononucleosis. Arch Neurol 64(1):72–75
Orton SM, Herrera BM, Yee IM, Valdar W, Ramagopalan SV, Sadovnick AD, Ebers GC (2006) Sex ratio of multiple sclerosis in Canada: a longitudinal study. Lancet Neurol 5(11):932–936
Peferoen LA, Lamers F, Lodder LN, Gerritsen WH, Huitinga I, Melief J, Giovannoni G, Meier U, Hintzen RQ, Verjans GM, van Nierop GP, Vos W, Peferoen-Baert RM, Middeldorp JM, van der Valk P, Amor S (2010) Epstein Barr virus is not a characteristic feature in the central nervous system in established multiple sclerosis. Brain 133(Pt 5):e137
Pender MP (2003) Infection of autoreactive B lymphocytes with EBV, causing chronic autoimmune diseases. Trends Immunol 24(11):584–588
Pender MP, Csurhes PA, Pfluger CM, Burrows SR (2014a) Deficiency of CD8+ effector memory T cells is an early and persistent feature of multiple sclerosis. Mult Scler J 20(14):1825–1832
Pender MP, Csurhes PA, Smith C, Beagley L, Hooper KD, Raj M, Coulthard A, Burrows SR, Khanna R (2014b) Epstein-Barr virus-specific adoptive immunotherapy for progressive multiple sclerosis. Multiple Sclerosis J 20(11):1541– 1544
Pohl D, Krone B, Rostasy K, Kahler E, Brunner E, Lehnert M, Wagner HJ, Gartner J, Hanefeld F (2006) High seroprevalence of Epstein-Barr virus in children with multiple sclerosis. Neurology 67(11):2063–2065
Polman CH, O'Connor PW, Havrdova E, Hutchinson M, Kappos L, Miller DH, Phillips JT, Lublin FD, Giovannoni G, Wajgt A, Toal M, Lynn F, Panzara MA, Sandrock AW (2006) A randomized, placebo-controlled trial of natalizumab for relapsing multiple sclerosis. N Engl J Med 354(9):899–910
Poole BD, Gross T, Maier S, Harley JB, James JA (2008) Lupus-like autoantibody development in rabbits and mice after immunization with EBNA-1 fragments. J Autoimmun 31(4):362–371
Pugliatti M, Sotgiu S, Rosati G (2002) The worldwide prevalence of multiple sclerosis. Clin Neurol Neurosurg 104(3):182–191
Ramagopalan SV, Meier UC, Conacher M, Ebers GC, Giovannoni G, Crawford DH, McAulay KA (2011) Role of the HLA system in the association between multiple sclerosis and infectious mononucleosis. Arch Neurol 68(4):469–472
Rubicz R, Yolken R, Drigalenko E, Carless MA, Dyer TD, Bauman L, Melton PE, Kent JW Jr, Harley JB, Curran JE, Johnson MP, Cole SA, Almasy L, Moses EK, Dhurandhar NV, Kraig E, Blangero J, Leach CT, Goring HH (2013) A genome-wide integrative genomic study localizes genetic factors influencing antibodies against Epstein-Barr virus nuclear antigen 1 (EBNA-1). PLoS Genet 9(1):e1003147

Sabbatini A, Bombardieri S, Migliorini P (1993) Autoantibodies from patients with systemic lupus erythematosus bind a shared sequence of SmD and Epstein-Barr virus-encoded nuclear antigen EBNA I. Eur J Immunol 23(5):1146–1152
Sargsyan SA, Shearer AJ, Ritchie AM, Burgoon MP, Anderson S, Hemmer B, Stadelmann C, Gattenloehner S, Owens GP, Gilden D, Bennett JL (2010) Absence of Epstein-Barr virus in the brain and CSF of patients with multiple sclerosis. Neurology 74(14):1127–1135
Savoldo B, Cubbage ML, Durett AG, Goss J, Huls MH, Liu Z, Teresita L, Gee AP, Ling PD, Brenner MK, Heslop HE, Rooney CM (2002) Generation of EBV-specific CD4+ cytotoxic T cells from virus naive individuals. J Immunol 168(2):909–918
Serafini B, Rosicarelli B, Franciotta D, Magliozzi R, Reynolds R, Cinque P, Andreoni L, Trivedi P, Salvetti M, Faggioni A, Aloisi F (2007) Dysregulated Epstein-Barr virus infection in the multiple sclerosis brain. J Exp Med 204(12):2899–2912
Simard JF, Costenbader KH (2007) What can epidemiology tell us about systemic lupus erythematosus? Int J Clin Pract 61(7):1170–1180
Simon K, Schmidt H, Loud S, Ascherio A (2014) Risk factors for multiple sclerosis, neuromyelitis optica and transverse myelitis. Multiple Sclerosis J 21(6):703– 709
Sokal EM, Hoppenbrouwers K, Vandermeulen C, Moutschen M, Leonard P, Moreels A, Haumont M, Bollen A, Smets F, Denis M (2007) Recombinant gp350 vaccine for infectious mononucleosis: a phase 2, randomized, double-blind, placebo-controlled trial to evaluate the safety, immunogenicity, and efficacy of an Epstein-Barr virus vaccine in healthy young adults. J Infect Dis 196(12):1749–1753
Strom BL, Reidenberg MM, West S, Snyder ES, Freundlich B, Stolley PD (1994) Shingles, allergies, family medical history, oral contraceptives, and other potential risk factors for systemic lupus erythematosus. Am J Epidemiol 140(7):632–642
Sundqvist E, Sundstrom P, Linden M, Hedstrom AK, Aloisi F, Hillert J, Kockum I, Alfredsson L, Olsson T (2012) Epstein-Barr virus and multiple sclerosis: interaction with HLA. Genes Immun 13(1):14–20
Sundstrom P, Juto P, Wadell G, Hallmans G, Svenningsson A, Nystrom L, Dillner J, Forsgren L (2004) An altered immune response to Epstein-Barr virus in multiple sclerosis: a prospective study. Neurology 62(12):2277–2282
Sundstrom P, Nystrom M, Ruuth K, Lundgren E (2009) Antibodies to specific EBNA-1 domains and HLA DRB11501 interact as risk factors for multiple sclerosis. J Neuroimmunol 215(1–2):102–107
Tai A, O'Reilly E, Alroy K, Simon K, Munger K, Huber B, Ascherio A (2008) Human endogenous retrovirus-K18 Env as a risk factor in multiple sclerosis. Multiple Sclerosis J 14(9):1175–1180
Thacker EL, Mirzaei F, Ascherio A (2006) Infectious mononucleosis and risk for multiple sclerosis: a meta-analysis. Ann Neurol 59(3):499–503
Tsokos GC (2011) Systemic lupus erythematosus. N Engl J Med 365(22):2110–2121
Tzartos JS, Khan G, Vossenkamper A, Cruz-Sadaba M, Lonardi S, Sefia E, Meager A, Elia A, Middeldorp JM, Clemens M, Farrell PJ, Giovannoni G, Meier UC (2012) Association of innate immune activation with latent Epstein-Barr virus in active MS lesions. Neurology 78(1):15–23
Ulff-Moller CJ, Nielsen NM, Rostgaard K, Hjalgrim H, Frisch M (2010) Epstein-Barr virus-associated infectious mononucleosis and risk of systemic lupus erythematosus. Rheumatology (Oxford) 49(9):1706–1712
van Sechel AC, Bajramovic JJ, van Stipdonk MJ, Persoon-Deen C, Geutskens SB, van Noort JM (1999) EBV-induced expression and HLA-DR-restricted presentation by human B cells of alpha B-crystallin, a candidate autoantigen in multiple sclerosis. J Immunol 162(1):129–135
Viglietta V, Baecher-Allan C, Weiner HL, Hafler DA (2004) Loss of functional suppression by CD4+ CD25+ regulatory T cells in patients with multiple sclerosis. J Exp Med 199(7):971–979

von Budingen HC, Bar-Or A, Zamvil SS (2011) B cells in multiple sclerosis: connecting the dots. Curr Opin Immunol 23(6):713–721

Wagner H-J, Munger KL, Ascherio A (2004) Plasma viral load of Epstein-Barr virus and risk of multiple sclerosis. Eur J Neurol 11:833–834

Wandinger K, Jabs W, Siekhaus A, Bubel S, Trillenberg P, Wagner H, Wessel K, Kirchner H, Hennig H (2000) Association between clinical disease activity and Epstein-Barr virus reactivation in MS. Neurology 55(2):178–184

Warner HB, Carp RI (1988) Multiple sclerosis etiology—an Epstein-Barrr virus hypothesis. Med Hypotheses 25:93–97

Waubant E, Mowry EM, Krupp L, Chitnis T, Yeh EA, Kuntz N, Ness J, Belman A, Milazzo M, Gorman M, Weinstock-Guttman B, Rodriguez M, James JA (2013) Antibody response to common viruses and human leukocyte antigen-DRB1 in pediatric multiple sclerosis. Multiple Sclerosis J 19(7):891–895

Willis SN, Stadelmann C, Rodig SJ, Caron T, Gattenloehner S, Mallozzi SS, Roughan JE, Almendinger SE, Blewett MM, Bruck W, Hafler DA, O'Connor KC (2009) Epstein-Barr virus infection is not a characteristic feature of multiple sclerosis brain. Brain 132(Pt 12):3318–3328

Wallin MT, Culpepper WJ, Coffman P, Pulaski S, Maloni H, Mahan CM, Haselkorn JK, Kurtzke JF and for the Veterans Affairs Multiple Sclerosis Centres of Excellence Epidemiology Group (2012) The Gulf War era multiple sclerosis cohort: age and incidence rates by race, sex and service. Brain 135(Pt 6):1778–1785

Yea C, Tellier R, Chong P, Westmacott G, Marrie RA, Bar-Or A, Banwell B, On behalf of the Canadian Pediatric Demyelinating Disease Network (2013) Epstein-Barr virus in oral shedding of children with multiple sclerosis. Neurology 81(16):1392–1399

Index

A
α5β1, 177
A20, 293, 320, 348, 349
Achong, Bert, 7, 11, 13, 271
Activation-induced deaminase (AID), 275, 320
Acyclovir, 230, 375
Adenovirus (Ad), 232
AIDS, 21, 52, 54, 58, 121, 153, 185, 271, 325, 341
Antibody
  anti-capsid antibody (VCA), 24–27, 213, 214, 220, 225, 226, 231, 271–273, 341, 377
  anti-CD20 antibody (Rituxan, rituximab), 251, 373
  anti-early antigen (EA) antibody, 226
  anti-EBNA1 antibody, 372–374, 377, 379
  anti-glycoprotein (gp340) antibody, 26
  anti-membrane antigen (MA) antibody, 25
  anti-Ro antibody, 376–378
  anti-Sm antibody, 255
  heterophile antibody, 216, 225
  IgA, 27, 165, 257, 341, 343
  IgM, 25, 165, 220, 221, 225, 226, 243, 257, 271, 272
ATM, 76, 87, 88, 260
ATR, 86, 87

B
Bacterial artificial chromosome (BAC), 232
BARF1
  rhBARF1, 55, 232
Barr, Yvonne, 7, 11, 271
BART, 51, 54, 58, 78, 79, 272, 297, 345, 346, 353, 354
BATF, 77
B cell
  B cell factor 1 (EBF1), 291, 297
  B cell receptor (BCR), 165, 168–170, 173, 176, 186, 194, 195, 289–291, 296–298, 321
Bcl2, 51, 55, 296, 320, 329
Bcl6, 328, 329
BcRF1, 107, 111, 128
BGLF4, 88
BHRF1, 46, 51, 54, 55, 107
BILF1, 223
Bim, 188, 353
BLIMP1, 175, 298
BMRF1, 107, 113, 128
BMRF2, 128
BNLF2a, 50, 55, 107
BNRF1, 74–76, 112
BPLF1, 76, 128
BRLF1, 55, 79, 107, 110, 128, 278, 298
Burkitt, Denis, 5, 268
Burkitt's lymphoma (BL)
  endemic, 27
  sporadic, 20, 271
BZLF1, 51, 55, 83, 84, 86, 88, 104–107, 109–111, 113, 122, 128, 129, 132, 142, 175, 176, 221, 224, 273, 278, 298, 301

C
Caspase recruitment domain-containing protein 11 (CARD11), 246, 260
CBF1, 77
CCR5, 292

C. Münz (ed.), *Epstein Barr Virus Volume 1*, Current Topics in Microbiology and Immunology 390, DOI 10.1007/978-3-319-22822-8

CCR7, 166, 278
Chronic active EBV (CAEBV), 216, 217, 255, 256
Clifford, Peter, 18
Cluster of differentiation (CD)
  CD19, 161, 174, 254, 289, 316, 328
  CD27, 161, 171, 172, 190, 191, 252, 258, 278
  CD30, 292, 293, 328
  CD40, 165, 167, 168, 243, 292, 293, 296, 323, 324
  CD79a, 316, 321, 328
  CD81, 161
C-myc, 20, 51, 167, 187, 271, 272, 275, 350
Complement receptor 1 (CD35), 142
Complement receptor 2 (CD21), 74, 142, 161, 322
Coronin, 254
CtBP, 83
CTCF, 77–80, 83, 84, 109
CXCR4, 165, 166
Cytidine 5'-triphosphate synthase 1 (CTPS1), 258

D
Dendritic cell (DC)
  conventional (cDC), 228
  plasmacytoid (pDC), 232
De The, Guy, 36, 271, 272
Diffuse large B cell lymphoma (DLBCL)
  ABC type, 317, 320
  germinal center B cell (GCB) type, 317, 320
Discoidin domain receptor 1 (DDR1), 300, 301
DNA damage response (DDR), 74, 76, 86, 87, 105
DNA methylation, 78, 83, 85, 88, 105, 108–110, 112, 113
DNA polymerase (BALF5), 110
DNAse (BGLF5), 80

E
EBER, 54, 79–81, 104, 109, 163, 187, 188, 294, 296, 297, 327, 328, 343, 345, 374
EBP2, 82
Episome, 21, 73, 75, 77, 78, 81, 88–90, 157, 164, 184, 341–343, 350
Epstein, Anthony, 271
Epstein Barr nuclear antigen (EBNA)
  EBNA1
    gly–ala repeat domain, 53
    gly–arg-rich region, 224
  EBNA2
    associated cellular proteins, 20
    cellular target gene, 56
    C-TAD, 57
    N-TAD, 57
    protein, 18, 20
    responsive promoter elements, 301
  EBNA3
    EBNA3A, 77, 135, 141
    EBNA3B, 126, 135, 136, 141, 143
    EBNA3C, 141
  EBNA-LP
    expression, 76, 107
    IR1 repeats, 55
    protein, 76
Electron microscopy, 4, 7, 11, 12, 14, 24, 271
Extracellular signal-regulated kinases (ERK), 52, 347–349, 351
Exosome, 106, 297, 349, 350, 355
EZH2, 85, 89, 108, 320

F
Fcγ receptor 3A (CD16a), 259

G
Galectin-1, 300
Galectin-9, 349
Ganciclovir, 230
GATA Binding Protein 2 (GATA2), 259
Genome
  Cp, 54, 77, 78, 80, 86, 105, 107, 110, 345
  EBV type 1, 57, 133
  EBV type 2, 133
  Fp, 109
  dyad symmetry element (DS), 80
    family of repeats (FR), 80
  Qp, 54, 77, 78, 80, 86, 109, 167, 273, 345
  Wp, 46, 55, 76, 107, 163, 273, 345
Germinal center model (GCM)
  default program, 159, 166–168, 185, 193
  growth program, 159, 160, 167, 185, 188, 189
  latency program, 104, 172, 182, 187, 191, 272
Glycoprotein
  gp42, 161
  gp110, 128, 142

gp340/gp350, 26, 50, 55, 128, 142, 143, 161, 229, 231, 232, 278
gL, 50

H
HDAC, 83–85, 87–89, 330
Hemophagocytic lymphohistiocytosis (HLH), 217, 222, 244, 254
Henle, Gertrude, 13, 271
Henle, Werner, 13, 18, 36, 271
Herpes virus, 11, 12, 105, 106, 108, 111, 153–155, 166, 171, 177, 232, 322, 372
Heterologous immunity, 182
Heterophile antibody (HA), 216, 225
Hodgkin's lymphoma (HD or HL)
classical
lymphocyte depleted, 288
lymphocyte rich, 288
mixed cellularity, 295
nodular sclerosis, 288, 295, 300, 301, 288, 289, 293–295, 297, 299–301
Hodgkin and Reed Sternberg cell (HRS), 186, 288–291, 293, 294, 296–299
nodular lymphocyte predominant (NLP), 288
pediatric, 223, 269, 276, 322
Human leucocyte antigen (HLA)
HLA-A*01, 295
HLA-A*02 (HLA-A2), 295
HLA-A11, 54, 136, 141
HLA-DRB1*1501, 372
Human immunodeficiency virus (HIV), 20, 22, 52, 121, 135, 153, 154, 295, 321, 322, 326, 330
Humanized mice (huMice)
NSG, 222, 230, 231

I
IFI16, 75, 76
IKK, 292, 320
IL-2-inducible T cell kinase (ITK), 252
IL-6, 320
IL-10, 167, 250, 301, 320
Immunosurveillance, 172, 174, 185, 267, 276, 277
Infectious mononucleosis (IM, AIM)
Lymphocytosis, 23, 30, 181, 222, 257, 260, 324, 326
INKT, 243, 250–254, 258
Interferon
type I (alpha and beta), 220, 232, 374
type II (gamma), 223, 373, 379
IRF
IRF4, 77, 317, 320, 328
IRF5, 292, 293
IRF7, 77, 347

J
Janus kinase (JAK), 292, 293, 296, 316, 329, 331, 332
JNK, 112, 347

K
Kaposi Sarcoma associated herpes virus (KSHV, HHV8), 75, 79, 82, 85, 88, 89, 108, 349
Klein, Eva, 30, 36
Klein, George, 25

L
Latency
latency 0, 158
latency I, 158
latency II, 158
latency III, 158
Latent membrane protein (LMP)
LMP1
AP1, 53, 74, 82–84
carboxy terminal activation region 1 (CTAR1), 347, 348
CTAR2, 52, 347
LMP2
ITAM, 168, 296, 350–352
proline-rich motif (PY motif), 351
LPS-responsive beige-like anchor (LRBA) protein, 254
Lymphoblastoid cell line (LCL), 46–48
Lymphocryptovirus (LCV)
rhesus LCV (rhLCV), 232
Lymphoma in the immunosuppressed (IL), 185
Lyn, 53
Lytic EBV infection
early (E) gene, 79, 107
immediate early (IE) gene, 79, 83, 174, 178, 181, 298, 300
late (L) gene, 85, 86, 111

M
MagT1, 253
Major capsid protein (p160, BcLF1), 131
Malaria
Plasmodium falciparum, 270

MAPK, 299, 300, 347
MHC class II, 161
Micro RNA (miRNA), 51, 54, 58, 78, 79, 128, 163, 296, 297, 344, 346, 352–355
Middlesex Hospital, 4, 5
Minichromosome maintenance complex component 4 (MCM4), 259
Moss, Denis, 30, 36
Multiple sclerosis (MS), 183, 214, 367
Munc
  Munc13-4, 255
  Munc18-2, 255, 256

N
Nasopharyngel carcinoma (NPC)
  EBV latency pattern, 325
  geography, 19, 122
Natural killer (NK) cells, 222
NF-Kb
  canonical pathway, 292
  non-canonical pathway, 292
NF-kB inducing kinase (NIK), 347
NK cell lymphoma, 330
NKG2D, 253
NKT cells, 250–252
Non-Hodgkin's lymphoma (NHL), 316
Notch, 77, 168, 291, 293, 297, 300, 301

O
Oral hairy leukoplakia, 177

P
P16^{INK4a}, 344
P53, 21, 87, 88, 297, 344, 352
PAX5, 76, 80, 291
PCNA, 76, 85, 113
PD-L1, 299–301
Perforin, 254, 255
PI3K, 128, 256, 257, 296, 299, 300
Plasma cell, 159, 162, 165, 169, 174–176, 183, 194, 250, 253, 290, 298, 323, 328, 373
PML nuclear bodies (PML-NB), 82
Pope, John, 18, 32
Posttransplant lymphoproliferative disease (PTLD), 121, 133, 135, 144, 185, 322, 328
Primary CNS lymphoma, 322
Primary effusion lymphoma, 322
Primary immunodeficiency
  X-linked agammaglobulinemia (XLA), 189
  X-linked lymphoproliferative disease (XLP), 188, 195, 218, 222, 244, 250, 251
  X-linked immunodeficiency with magnesium defect, EBV infection, and neoplasia (XMEN), 222, 253
Proteasome, 331
PU.1, 57, 78, 80, 290
PUMA, 353

R
Ras, 296, 344
RBP-Jk (CBF1), 57, 77, 80
Regression assay, 33, 276
Rheumatoid arthritis (RA), 183, 367
Rickinson, Alan, 32, 35, 142

S
Saliva, 48, 52, 58, 156, 157, 159, 160, 174–176, 189, 220, 373
SAP, 188, 218, 244, 250, 251
Serine/threonine kinase 4 (STK4), 57, 87, 258, 260
Severe combined immunodeficiency (SCID), 230, 260, 293
Sjögren's syndrome, 183
Small non-coding RNA (scRNA), 163
SP1, 83, 84, 87
Src, 250, 350–352
STAT, 77, 86, 292, 293, 296, 300, 301, 316, 329, 331, 332
Strains
  Akata, 48, 85, 89, 142
  B95-8, 46, 47, 50, 51, 53, 125, 128, 136, 141, 142, 144, 163, 231, 353
  CAO, 52
  M81, 46, 48, 142, 143
  Raji, 47, 48, 51, 84, 85, 89
SUMO, 75
Survivin, 297
Syk, 350
Systemic lupus erythematosus (SLE), 367

T
T cells
  CD4+ T cell, 295, 299

CD8+ T cell, 23, 181, 223, 224, 232, 277, 278, 300, 323, 331, 375
cytotoxic T lymphocytes (CTL), 30, 33, 34, 36, 37, 50, 141, 166, 174, 178, 181, 182, 186, 190, 192, 195, 253, 255, 301, 342, 346
memory T Cell, 23, 30, 33, 37, 258, 278
T cell lymphoma, 121, 244
Template activating factor Iβ (TAF-Iβ), 82
TNF-receptor-associated death domain protein (TRADD), 347
Toll-like receptor (TLR)
TLR9, 86, 275
TOMM22, 353
Tonsil, 155–157, 159, 161, 166, 174–177, 182, 189–191, 193, 195, 220, 223, 224, 231, 275
Transforming growth factor β (TGFβ), 176, 297, 300
Transporter associated with antigen processing (TAP), 50
Tumor necrosis factor receptor (TNFR) associated factors (TRAFs), 347

U
Ubiquitin specific protease 7 (USP7), 82, 83, 352

V
Valacyclovir, 230, 375
Virus like particle (VLP), 106

W
Waldeyer's ring, 155, 157, 159, 173–176, 182, 185, 192
Wiskott–Aldrich syndrome (WAS), 260

X
XBP1, 84
X-linked inhibitor of apoptosis (XIAP), 222, 251

Z
ZEB1/2, 83
Zur Hausen, Harald, 13, 19, 24

Printed by Printforce, the Netherlands